ENZYME-MEDIATED IMMUNOASSAY

ENZYME-MEDIATED IMMUNOASSAY

Edited by
T. T. Ngo and H. M. Lenhoff
University of California, Irvine
Irvine, California

Plenum Press · New York and London

Library of Congress Cataloging in Publication Data

Main entry under title:

Enzyme-mediated immunoassay.

Includes bibliographies and index.
1. Immunoenzyme technique. I. Ngo, T. T. (That Tjien), 1944– . II. Lenhoff, Howard M. [DNLM: 1. Immunoenzyme Technics. QW 525 E615]
QP519.9.I44E585 1985 616.07′56 85-16988
ISBN 0-306-42085-6

A Division of Plenum Publishing Corporation
233 Spring Street, New York, N.Y. 10013

Printed in the United States of America

CONTENTS

SEPERATION-REQUIRED (HETEROGENOUS) ENZYME IMMUNOASSAY

INTRODUCTION

T. T. Ngo and H. M. Lenhoff

Department of Developmental and Cell Biology
University of California, Irvine, CA 92717

In 1959, Yalow and Berson used insulin labeled with radioactive iodine to develop a quantitative immunological method for determining the amount of insulin in human plasma. Their method depends upon a competition between insulin labeled with radioactive iodine (I^{131}) and unlabeled insulin from plasma for a fixed and limited number of specific binding sites on the antibody to insulin. The amount of the labeled insulin bound to the antibody is inversely proportional to the amount of insulin in the plasma sample. Their method, which is so elegantly simple in concept, is made possible by the ability to detect with ease extremely low levels of radioactivity, and by the exquisite specificity of an antibody capable of specifically binding the analyte. Such a combination of sensitivity and specificity is the basis of this versatile analytical tool called radioimmunoassay (RIA).

Twelve years later, Engvall and Perlmann (1971) and Van Weemen and Schuurs (1971) independently introduced the use of enzymes as another category of sensitive and even more versatile labels for use in immunoassays. Engvall and Perlmann (1971) coined the term ELISA, which stands for Enzyme Linked Immunosorbent Assay.

In both RIA and ELISA, the labeled and unlabeled antigens bound to the antibody must first be separated from the labeled and unlabeled antigens that are not bound before the activity of the label can be measured in either fraction. These methods are collectively called separation-required immunoassays. Because that separation inevitably involves the use of heterogeneous phases in the assay mixture, the term "heterogeneous" immunoassay is interchangeably used for separation-required immunoassay.

In 1972, Rubenstein *et al*. developed a new immunoassay using an enzyme as the label which did not require the separation of the antigen bound to antibody from the unbound fraction before measuring the enzyme activity. The method depends on a change in the specific activity of the enzyme when the antibody binds to antigen labeled with enzyme. The enzyme activity in the unfractionated assay solution is proportional to amount of antigen labeled with enzyme that is not bound by the antibody. Because the assay does not require any heterogeneous phase to separate the bound and free antigens, it is called "homogeneous" immunoassay. We feel that the term "homogeneous" does not accurately describe the salient features of the assay and recommend that the more appropriate and more descriptive term "separation-free" to be used instead. Separation-free is a more accurate term to use, because some separation-free enzyme mediated immunoassays are not homogeneous and involve heterogeneous phases (e.g., see Chapters by Gibbons and Litman).

Since the first uses of enzymes as the label in immunoassays in 1971, there have appeared over 10,000 publications dealing with various aspects and aplications of enzyme-mediated immunoassays. The major aim of this volume is to collect state of the art reviews on this subject from scientists in the North American Continent. Reviews by European (Avrameas *et al*., 1983; Malvano, 1980) and Japanese (Ishikawa, *et al*., 1981) scientists have already been assembled.

REFERENCES

Avrameas, S., P. Druet, R. Masseyeff, and G. Feldmann, Eds. (1983). *Immunoenzymatic Techniques*. Elsevier Science Publisher, Amsterdam.

Engvall, E. and P. Perlmann (1971). Enzyme-linked immunosorbent assay (ELISA). Quantitative assay of immunoglobulin G. *Immunochem*. *8*: 871-874.

Ishikawa, E., T. Kawai and K. Miyai, Eds. (1981). *Enzyme Immunoassay*. Igaku-Shoin, Tokyo.

Malvano, R., Ed. (1980). *Immunoenzymatic Assay Techniques*, Martinus Nijhoff Publisher, The Hague.

Van Weemen, B. K. and A. H. W. M. Schuurs (1971). Immunoassay using antigen-enzyme conjugate. FEBS Letters. *15*: 232-236.

Yalow, R. S. and S. A. Berson (1959). Assay of plasma insulin in human subjects by immunological methods. *Nature* 184: 1648-1649.

ENZYME MEDIATED IMMUNOASSAY: AN OVERVIEW

T. T. Ngo

Department of Developmental and Cell Biology
University of California, Irvine, CA 92717

INTRODUCTION

Since its introduction by Yalow and Berson (1959), radioimmunoassay (RIA) has played important roles in facilitating and accelerating basic discoveries in biomedical sciences and has become a standard tool in laboratory medicine. Millions of clinical tests are now routinely performed by using RIA. The rapid and wide acceptance of RIA as a routine clinical method can be attributed to (1) the general applicability of the method, i.e., any compound can be analyzed by a RIA so long as an antibody specific to that compound is available; (2) the selectivity and specificity of the method; (3) the sensitivity of a radioactive isotope as the label in RIA and (4) the minimal inference and ease of performing the test.

Notwithstanding the advantages of RIA, there are, however, several drawbacks in using RIA. They include (a) the relatively short half-life of gamma-ray emitting isotopes, (b) the health hazards involved in preparing the isotopic labels and in handling and performing the test, (c) radiation induced structural damage on the labeled molecules, (d) the need of separating antibody bound labeled and unlabeled antigens from unbound ones and (e) therefore the difficulty in automating RIA.

In attempts to replace the use of radioactive materials as labels and to develop separation-free immunoassays, a number of non-isotopic labels have been used. These include bacteriopha ges (Haimovich and Sela, 1969), free radicals (Leute et al, 1972), fluorescent groups (Soini and Hemmila, 1979),

Chemiluminescent and bioluminescent groups (Simpson et al., 1979; Kricka and Carter, 1982; Serio and Pazzagli, 1982), synthetic particles (Rembaum and Dreyer, 1980), red blood cells (Adler and Liu, 1971), electron dense materials (Singer and Schick, 1961 and Leuvering et al., 1980), liposome (Chan et al., 1978; Litchfield et al., 1984), metals (Cais et al., 1977), enzyme substrates (Burd et al., 1977; Wong et al., 1979; Ngo et al., 1979; Ngo et al., 1981; Ngo and Wong, 1985), prosthetic group (Ngo, 1985), enzyme modulators (Ngo and Lenhoff, 1980a; Ngo, 1983; Ngo, 1985) and enzymes (for a review of this subject, please read Schuurs and Van Weemen, 1977; Borrebaech and Mattiasson, 1979; Ngo and Lenhoff, 1981 and Ngo and Lenhoff, 1982).

ENZYME MEDIATED LABELS

Enzyme mediated labels (EML) are defined as compounds which can be quantitatively measured via enzymatic processes. They include enzyme substrates, enzyme cofactors, enzyme prosthetic groups, enzyme modulators (i.e., inhibitors or activators), enzyme fragments, apoenzymes and enzymes.

Advantages of EML

Advantages in using EML for immunoassay are: (1) long-term stability of most EML's, their half-lives at 4°C are generally longer than six months; (2) their concentrations can be determined rapidly by using instruments commonly available in most analytical laboratories; (3) no radiation hazards; (4) possibilities of developing rapid, separation-free (homogeneous) immunoassays which can be readily automated; (5) the development of qualitative visual tests for mass-screening and (6) amplification of detection signal by some EML's when used in separation-required (heterogeneous) mode.

Limitations of EML

Limitations of using EML in immunoassay include: (1) limited sensitivity of some separation-free systems; (2) carcinogenicity and instability of some enzyme substrates; (3) interference by some endogeneous EML's and other substances found in biological fluids and environmental samples; (4) long assay time for some EML immunoassays and (5) procedures for measuring enzyme activity can be more complicated and demanding than the counting of radioactivity of an isotope.

Sensitivity and versatility of enzyme label

Among EML's, enzymes appear to be the most sensitive and versatile labels because they are protein molecules endowed with an extraordinary efficient catalytic power. The presence of a minute amount of enzymes can be detected and quantified by measuring products of the reaction catalyzed by the enzyme. Most of the commonly used enzyme labels are capable of converting 10^6 molecules of substrate into products within one minute by one enzyme molecule at ambient temperature and pressure. The catalytic efficiency of an enzyme depends strongly on its three dimensional structure (conformation). The three dimensional structure of an enzyme or protein is maintained through numerous non-covalent interactions, such as hydrophopic and hydrogen bondings, ionic interactions and covalent linkages, such as disulfide bonds. The three-dimensional structure of an enzyme allows the juxtaposition of certain amino acid residues at spatially most strategic positions for bringing about catalysis. Non-covalent interactions, being weak chemical bonds are easily broken or altered by thermal energy or other non-covalent interactions such as by binding of ions, chaotropic agents, detergents, lipids, etc. It is known that the binding of a molecule (such as an allosteric effector) by an enzyme at a site far remote from its active site (i.e., catalytic site) can bring about conformational changes that alter the spatial position of active site amino acid residues. Alterations in the non-covalent interactions of an enzyme which lead to a new and different conformation may significantly alter the catalytic efficiency of the enzyme. This conformational flexibility of an enzyme is one of the drawbacks in using enzyme as a label. However, it can also be used to advantage for developing a separtion-free enzyme immunoassay which is based on an antibody induced changes in the conformation of a ligand conjugated enzyme. Another advantage of using enzyme as a label is the numerous functional groups in an enzyme molecule such as the amino, sulfhydryl, carboxyl, carboxamide and tyrosyl groups that are available for covalent linking to ligand molecules.

Separation-free (homogeneous) versus separation-required (heterogeneous) enzyme mediated immunoassays (EMI)

From an operational standpoint, enzyme mediated immunoassays (EMI) can be divided into two categories: (1) separation-free (homogeneous) or (2) separation-required (heterogeneous) systems. In the separation-free system, the enzyme activity of the assay solution is measured without a prior physical separation of the antibody-bound labeled ligands from the free, unbound ones. Such a procedure is made possible because the activity of the antibody-bound labeled ligands is significantly different from the ones unbound by the antibody. In the separation-required assay, however, a procedure is required for separating the labeled ligand into antibody bound and free, unbound franctions. The enzyme activity of these fractions is then measured. A physical separation of the bound and unbound fractions is required, because the assay depends on the partitioning of the labeled conjugates into antibody bound and unbound fractions.

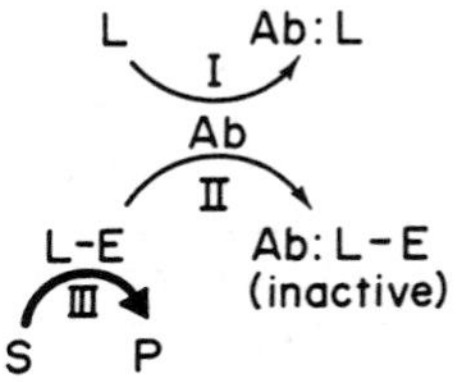

Fig. 1. Principle of separation-free EMI based on enzyme labeled ligand.

SEPARATION-FREE (HOMOGENEOUS) ENZYME MEDIATED IMMUNOASSAYS

Separation-free EMI using enzyme as the label

Rubenstein et al. (1972) described a separation-free EMI for morphine detection using lysozyme as the enzyme label. They coined the term "homogeneous" enzyme immunoassay to indicate that no physical separtion of the antibody bound enzyme labeled ligand from the unbound form was necessary. The principle of the assay is schematically shown in Fig. 1. The covalent enzyme labeled ligands (E-L) compete with ligands (L) from the sample for a limited concentration of antibody (Ab) to that ligand to form the complex, E-L:Ab (reaction II). The resultant E-L:Ab complex exhibits very little enzyme activity because of either steric hindrance (Rubenstein et al., 1972) or allosteric inhibition (Rowley et al., 1975) caused by the bound antibody. In the absence of L, reaction I would not take place and E-L and Ab would form the enzymatically inactive E-L:Ab complex (reaction II). In the presence of L, however, there would be a competition for the antibody leaving more E-L uncomplexed and free to catalyze the conversion of substrates to products (reaction III). Thus the enzyme activity would be directly proportional to the amount of free L in the sample.

Separation-free EMI based on enzyme labeled ligand has been developed for a number of therapeutically important drugs and for hormones. The drugs include phenytoin, phenobarbital primidone, ethosuximide, carbamazepine, digoxin, gentamicin, valproic acid, and methotrexate. The assay is rapid. It takes less than 1 minute to perform one test. It is also very sensitive; substances at picomole levels can be detected with ease (Jaklitsch, 1985).

The precision and accuracy of separation-free EMI based on enzyme-ligand conjugates are comparable with other immunological (e.g., RIA) and non-immunological methods (e.g., GC, HPLC, and TLC) (Jaklitsch, 1985).

Gibbons et al. (1980) reported a separation-free EMI for proteins employing β-galactosidase labeled protein antigens. When the antibody to the antigen bound to the enzyme labeled antigen, it did not alter the catalytic action of the enzyme toward its small molecular weight substrate. The activity of the antibody bound enzyme labled antigen toward a macromolecular substrate, however, was inhibited up to 95 percent because of the steric exclusion of the substrate from the active site of the enzyme. As the concentration of antigen increased, more antibodies are tied up, leaving more antigen-enzyme conjugates free to hydrolyze the macromolecular substrate. The method developed by Gibbons et al. (1980) is very sensitive. With the prototypical system, they were able to assay human IgG down to 25 ng/ml with ease. A fluorometric macromolecular substrate for β-galactosidase has recently been developed. Using umbelliferyl-β-galactoside substituted dextran as the substrate, nanogram quantity of β-galactosidase can be quantified in less than 1 minute (Gibbon, 1985).

<u>Separation-free EMI using enzyme modulator as the label</u>

Enzyme modulator mediated immunoassay (EMMIA) is based on the ability of a ligand labeled enzyme modulator (M-L) to modify the activity of an indicator enzyme (E). The M-L competed with free ligand (L), the analyte, for a limited amount of antibody to the ligand (Ab). The antibody bound M-L is unable to modulate the activity of the indicator enzyme. In the absence of analyte (L), reaction I (Figure 2) would not occur and the M-L and Ab to the analyte (L) would combine through reaction II, making M-L unavailable to modulate the enzyme activity. As the concentration of analyte increases, however, it would compete successfully for binding sites on Ab (reaction I) leaving more modulator free to complex with the indicator enzyme (reaction

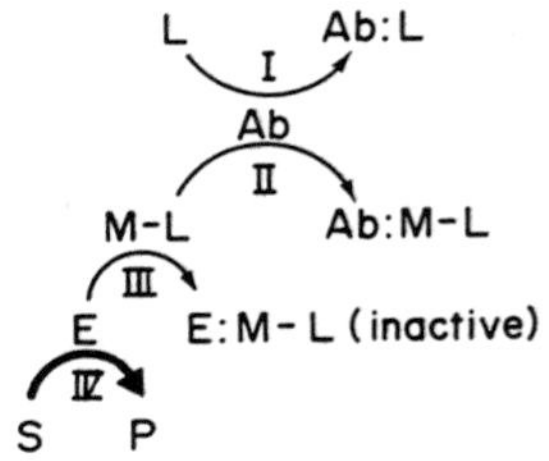

Fig. 2. Principle of separation-free EMI using enzyme modulator as the label.

III), thereby modulating its activity (Ngo and Lenhoff, 19801 Ngo, 1983; Ngo, 1985). Depending on the properties of the modulator, the enzyme activity will be increased by a positive modulator or it will be decreased by a negative modulator. Thus, for EMMIA with a positive modulator, the enzyme activity will be directly proportional to the concentration of the analyte. For EMMIA developed with a negative modulator, the activity will be inversely proportional to the concentration of the analyte.

Based on the principle of EMMIA, a practical assay for human serum thyroxine has been developed. The assay used an acetylcholinesterase as the indicator enzyme and a thyroxine labeled cholinesterase inhibitor as the enzyme modulator (Finley et al., 1980). Seventy-five samples were analyzed in one hour. The results were clear cut and showed excellent precision and accuracy. A theophylline assay based on the same principle has also been developed (Blecka et al., 1983).

EMMIA based on fragments of ribonuclease, such as S-peptide labeled ligand and S-protein, has been developed for thyroxine assay (Gonnelli et al., 1981 and Ngo, 1985).

Separation-free EMI using enzyme prosthetic group as the label

In this assay, an enzyme prosthetic group is covalently linked to a ligand and this conjugate is able to reconstitute an active holoenzyme from its apoenzyme. Flavin adenine dinucleotide (FAD) is a prosthetic group for glucose oxidase. A recombination of FAD and apoglucose oxidase, neither of which is active by itself, yield an active glucose oxidase. When FAD is used to covalently label a ligand (L) the resulting stable FAD-L conjugate serves two functinal roles: (a) FAD-L, as a ligand analog, competing with the analyte (L) for antibodies to L (reaction II, Fig. 3); or (b) as a modified prosthetic group that can bind to apoglucose oxidase (AG) through a high-affinity binding to form (reaction III) an enzymatically active hologlucose oxidase (HG) (Ngo and Lenhoff, 1980). On the other hand, once the FAD-L form FAD-L:Ab as in reaction II, the FAD moiety can no longer combine with the apoenzyme due to the steric hindrance imposed on FAD-L by the complexation with the antibody. In the absence of L, the analyte, reaction I would not occur; thus the antibody would combine the FAD-L instead (reaction II) making the FAD of the complex incapable of combining with AG (reaction III). Conversely, as the concentration of L increases, it would compete more successfully for Ab (reaction I) leaving more FAD-L free to combine with AG (reaction III), thereby increasing the amount of hologlucose oxidase to catalyze reaction IV (Morris et al., 1980).

Based on the principle shown in Fig. 3, a theophylline assay with a sensitivity of 2 μg/ml was developed. A novel application of FAD as label for macromolecule assay was first demonstrated by Ngo. Aminohexyl-FAD was covalently linked, via a bisimidate cross-linker, to human IgG to form a stable FAD-IgG conjugate. This conjugate can combine either with the antibody to IgG (Fig. 3, reaction II) or with apoglucose oxidase (Fig. 3, reaction III). In the absence of IgG, most of the FAD-IgG combined with the antibody to form FAD-IgG:Ab, rendering the FAD moiety of FAD-IgG incapable of serving as a prosthetic group for apoglucose oxidase; hence no holoenzyme was formed and no enzyme activity was observed. As the concentration of IgG increased, more FAD-IgG's were uncomplexed from the FAD-Ig:Ab and became free to act as the prosthetic group of an apoglucose oxidase. Consequently, the enzyme activity formed is directly proportional to the concentration of free IgG present in the sample.

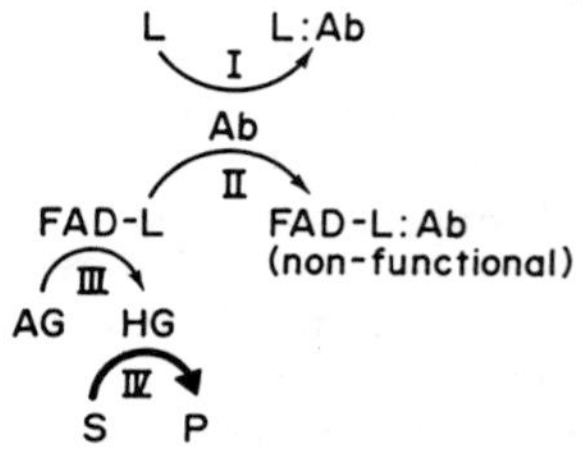

Fig. 3. Principle of separation-free EMI using enzyme prosthetic group as the label.

Nakane (1980) has recently described the development of an interesting separation-free enzyme immunoassay based on the above principle. The technique involves the reconstitution of an apoperoxidase with a heme labeled ligand that served as a prosthetic group for apoperoxidase.

Separation-free EMI using fluorogenic enzyme substrate as the label

Separation-free enzyme immunoassay using ligand-linked fluorogenic enzyme substrates were developed by Burd and Wong (1977; 1979). In this assay (Fig. 4), a ligand derivative was covalently linked to a fluorogenic enzyme substrate to form a stable substrate-ligand (S-L) conjugate. The S-L competes with the analyte, L, for a limited concentration of antibody to L. Thus, through reactions I and II, L:Ab and S-L:Ab are formed. Because the free S-L conjugate is a fluorogenic substrate for the enzyme, E, whereas S-L bound to the antibody (S-L:Ab) is not, the absence of analyte (L) would allow the S-L to combine with the antibody to L so that the indicator reaction III would not take place. On the other hand, as the concentration of L increases, more S-L would remain free to act as substrates for E

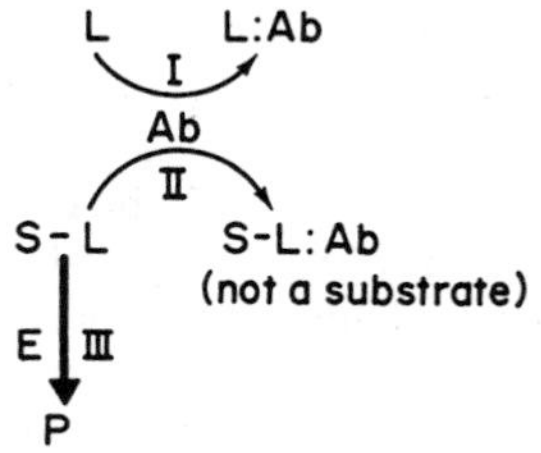

Fig. 4 Principle of separation-free EMI using fluorogenic enzyme substrate as the label.

and, hence, allowing more products to form by reaction III. Thus, the fluorogenic substrate-ligand conjugate (S-L) serves a dual role: (a) as a ligand analog which competes effectively for binding with antibody to the ligand, and (b) as an efficient substrate analog which allows the indicator reaction III to take place by forming products.

In this assay, rather than providing an amplification effect, the enzyme is used merely to distinguish and to quantify the proportion of free, unbound substrate-labeled ligand. The maximum signal obtainable is, of course, dictated by the concentration of free, unbound substrate-labeled ligand. The concentration of the latter is in turn determined by, and should be comparable to, the concentration of the ligand in the sample test. To get measurable signals, it is necessary to use a fluorogenic substrate in labeling the ligand and antigen.

Separation-free, fluorogenic enzyme substrate labeled assay, because of its current sensitivity limit (micromolar), has been used mostly for assaying therapeutic drugs of relatively high concentrations. The precision and accuracy of this method is comparable to that of gas and liquid chromatography, RIA and other EMI's (Ngo and Wong, 1985).

Using flurogenic enzyme substrate as a label, Ngo (1979, 1981) developed the first separation-free assay for macromolecule. Thus, human IgG was covalently labeled with a derivative of β-galactosylumbelliferone. The fluorogenic enzyme substrate labeled human IgG served both as a macromolecular substrate for β-galactosidase and as a labeled antigen which competed effectively with IgG from the serum for anti-human IgG (Fig. 4). When the IgG level in the test serum was low, antibodies combined with the labeled IgG forming three-dimensional lattice structures which made the β-galactosylumbelliferone residues less accessible to the enzyme and, therefore, less products formed via reaction III. On the other hand as the concentration of IgG in the serum increased, however, the large, three-dimensional lattice structures broke down into smaller units. Consequently, more β-galactosylumbelliferone residues became more accessible to the active site of the enzyme and, therefore, more products are formed.

Using this technique, human IgG at 0.5 μg/ml can be assayed in 20 minutes. The results correlate well with those obtained with radial immunodiffusion techniques which require a 24-hour delay in getting the results (Ngo, 1981).

Separation-free EMI based on antibody induced restriction on the conformation of apoenzyme labeled ligand

In this assay, an apoenzyme which has been covalently linked with several ligands (L-AE) competes with ligand analyte for a limited amount of antibody. When the concentration of the ligand analyte is low, reaction II (Fig. 5) will dominate. Consequently, more L-AE will combine with antibody (Ab) to form Ab:L-AE complex which is incapable to form an holoenzyme with FAD and leave less L-AE to combine with FAD (reaction III, in Fig. 5) to form enzymatically active L-AE:FAD. If the concentration of analyte ligand (L) was high, however, it would compete more successfully for the antibodies (reaction I) leaving more free L-AE to combine with FAD (reaction III), thereby forming more enzymatically active L-AE:FAD. Thus, the concentration of product formed (reaction IV) is directly proportional to the amount of the analyte ligand present.

Based on the principal of an antibody-induced conformational restriction of the apoenzyme labeled with ligands, Ngo and Lenhoff (1983) developed a separation-free EMI for DNP-aminocaproate using DNP conjugate apoglucose oxidase and FAD.

Under acid denaturing conditions, hologlucose oxidase covalently linked with 2,4-dinitrophenyl (DNP) was dissociated into flavin adenine dinucleotide (FAD) and DNP-apoglucose oxidase (DNP-AG). Both lacked catalytic activity. The activity was restored by combining FAD and DNP-AG at about pH 7. If, on the other hand, anti-DNP serum was preincumbated with the DNP-AG prior to the addition of FAD, activity was not restored. Furthermore, added DNP-aminocaproic acid counteracted the effects of the antibody in inhibiting the recombining of DNP-AG and FAD to form active enzyme. The anti-DNP serum probably prevented the DNP-AG from combining with FAD to form an active

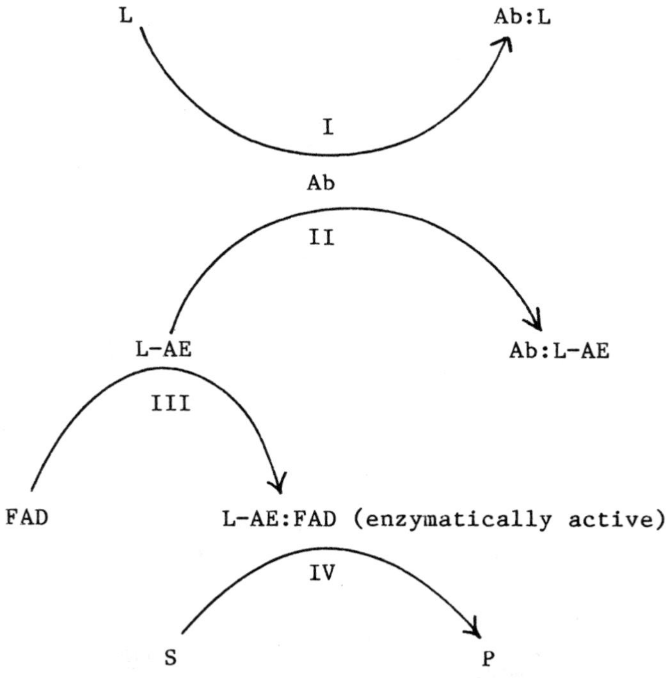

Fig. 5. Principle of separation-free EMI based on antibody induced conformational restriction.

holoenzyme by restricting the mobility of the polypedtide chain of DNP-AG from folding into a catalytically active conformation.

Separation-free EMI using the principle of enzyme channeling

When two enzymes, E_1 and E_2, catalyzing two consecutive reactions which transform the substrate (S) to P_1 and subsequently to P_2 according to the following equation:

$$S \xrightarrow{E_1} P_1 \xrightarrow{E_2} P_2$$

are brought together by, for example, co-immobilization on the same solid support, the initial rate of P_2 formation is greater than when the two enzymes, E_1 and E_2, are free and separated in a solution (Mosbach and Mattiasson, 1970).

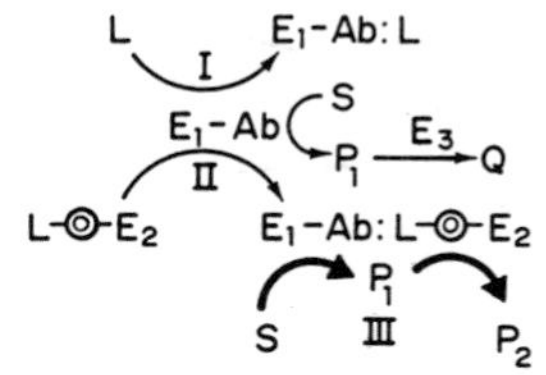

Fig. 6 Principle of separation-free EMI based on enzyme channeling.

In one version of the assay, the ligand antigen is coimmobilized with E_2, on fine beads (L-O-E_2) which can compete with L (reactions I and II, Fig. 6) for a limited and fixed number of antibodies which have been covalently linked to E_1 (E_1-Ab). The proportion of complexed E_1-Ab:L-O-E_2 to the uncomplexed enzyme species (i.e., E_1-Ab and L-O-E_2) determines the overall enhancement in the rate of formation of P_2 with the E_1-Ab:L-O-E_2 complex giving a faster overall rate than the two enzymes acting sequentially in their uncomplexed forms.

Thus, in the absence of L, reaction I will not occur and most of L-O-E_2 will be complexed via reaction II to form E_1-Ab:L-O-E_2. Consequently, the rate of P_2 formation from S is enhanced. Conversely, in the presence of high concentration of L, reaction I predominates and hence less E_1-Ab:L-O-E_2 is formed via reaction II. Therefore, increasing concentration of L (the analyte) reduces the extent of overall rate enhancement of reaction III. To minimize background reactions caused by free, uncomplexed E_1-Ab,L-O-E_2, and E_1-Ab:L (which can still catalyze the transformation of S to P_1 and to P_2), P_1 must be removed by adding a scavenger enzyme, E_3 to form Q. Furthermore, because the scavenger enzyme E_3 is sterically excluded from the E_1-Ab:L-O-E_2 complex, P_1 formed in the complex is much less readily converted to Q. Presumably, the presence of an unstirred layer around the bead E_1-Ab:L-O-E_2 slows down the rate of P_1 diffusing into the bulk solution.

Litman et al. (1980) demonstrated that using separation-free enzyme channeling immunoassay system, human IgG at ng/ml level can be quantified within 2 hours. The enzyme used in the channeling assay were hexokinase and glucose-6-phosphate dehydrogenase. Phosphoglucose isomerase and phosphofructokinase were the scavenger enzymes.

Separation-free EMI using liposome-entrapped enzyme

Liposome-entrapped EMI can be divided into two categories: (1) complement independent and (2) complement dependent systems. The former does not depend on components of complement system for lysis of the liposomal membrane while the latter requires a functionally active complement system.

Complement independent system: In this assay, the ligand is labeled with a cytolysin capable of lysing cell membrane or liposomal membrane. The ligand-cytolysin (L-C) serves, in addition, as a membrane piercing agent, also as an analog of the ligand which can compete with ligand analyte from the test sample. The lytic activity of L-C is, however, lost when it binds to an antibody specific for the ligand. The principle of the assay is schematically shown in Fig. 7. The covalent ligand-cytolysin conjugates (L-C) compete with ligands (L) for antibody (Ab) to that ligand to form the complex Ab:L-C (reaction II). The resultant Ab:L-C complex exhibits very little, if any, membrane lytic activity because the bound antibody presumably caused a steric hindrance preventing the cytolysin from contacting the membrane. Therefore, in the absence of L, reaction I would not take place and L-C and Ab would form Ab:L-C complex via reaction II with a concommitant loss of lytic activity. In the presence of high concentration of L, however, reaction I would prevail and there would be less antibody to complex with L-C leaving more L-C free to lyse the enzyme containing lipsome (reaction III) and therefore releasing more enzyme to generate more products (reaction IV). Thus the enzyme activity would be directly proportional to the amount of free L in the test sample.

Litchfield et al. (1984) developed a sensitive and rapid assay for digoxin using liposome-entrapped alkaline phosphatase and ouabain-melittin conjugate. Ouabain is an anolog digoxin and mellittin is a 26 amino acid lytic peptide from bee venom (Litchfield, 1984).

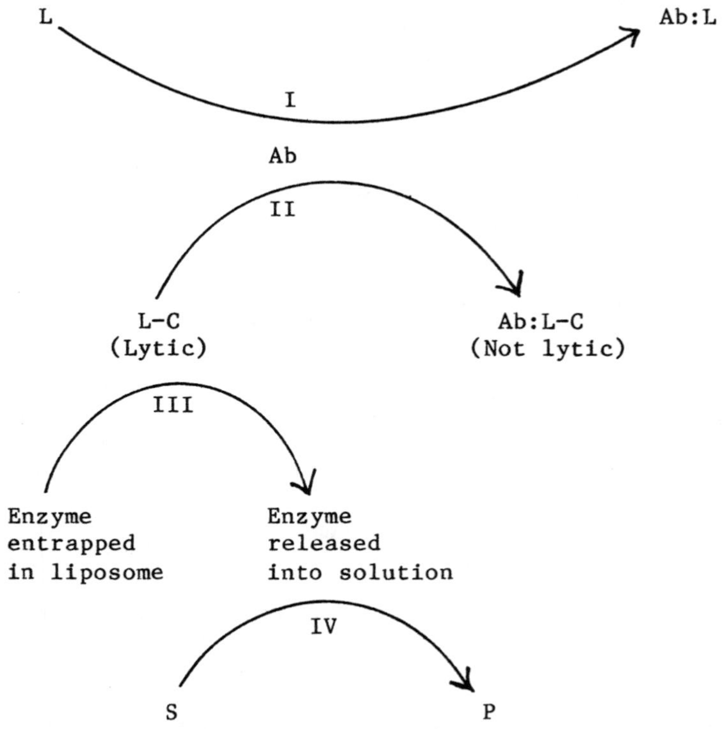

Fig. 7. Principle of separation-free EMI based on cytolysin labeled ligand and liposome-entrapped enzymes (complement independent system).

Complement dependent system: In this assay, the ligand is covalently linked to a polar lipid, such as phosphatidyl ethanolamine. This conjugate is then incorporated into the membrane of a liposome containing entrapped molecules. When an antibody specific for the ligand complexes with the liposome-bound ligand, the liposome becomes susceptible to lysis by the complements. For macromolecular antigens such as proteins, the liposomes with entrapped enzymes are formed before the antigens are covalently attached to the membrane on the surface of liposomes directly or via bifunctional crosslinkers. The principle of this assay is schematically presented in Fig. 8. The ligands which have been covalently bound to the liposomal membrane (L-Lip) compete with ligands (L) for antibody (Ab) to the ligand to form the complex Ab:L-Lip via reaction II. The resultant Ab:L-Lip is lysed, in the presence of complements (reactions III and IV) to release the entrapped enzymes. In the

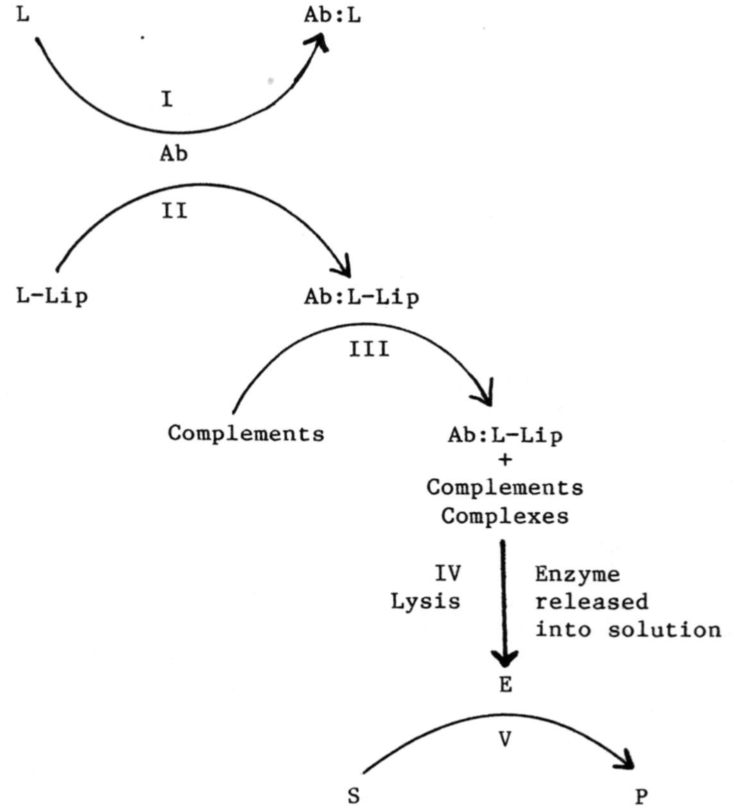

Fig. 8. Principle of separation-free EMI based on liposome labeled ligand and liposome-entrapped enzymes (complement dependent system).

absence of L, reaction I would not take place and L-Lip and Ab would form Ab:L-Lip via reaction II. Ab:L-Lip complex would subsequently be lysed by the complements to release the entrapped enzymes to catalyze the conversion of substrate to product (reaction V). In the presence of L, however, there would be a competition for the antibody leaving more L-Lip uncomplexed and therefore would not be lysed by the complements. Thus the released enzyme activity would be inversely proportional to the concentration of ligand analyte in the sample.

Separation-free EMI in reagent strip format

The convenience and perhaps the stability of separation-free EMI could be greatly increased if all reagents needed for carrying out the assay were put on a single solid phase polymeric matrix. All the user needs to do is to dip the strip into the sample solution for a given time and obtain the result by simply reading the strip with eyes or by using an instrument.

Glad and Grubb (1978) developed an immunocapillarymigration method for quantifying antigen. The method is based on the capillarymigration of an antigen in a porous matrix having immobilized antibodies in it. The capillarymigration of antigens is retarded by the binding of antigens to immobilized antibodies. As the concentration of antigen increases, more of the immobilized antibodies will be complexed by the antigens. This allows more antigens to migrate further before they are stopped by complexation to the immobilized antibodies. Therefore, the extent of migration is directly proportional to the amount of antigen in the sample. The area covered by the antigen due to capillary migration is revealed by introducing the strip to either fluorescence or enzyme lableled antibodies to the complexed antigens. Antigens such as C-reactive protein at 0.15 μg/ml can be quantified within 15 minutes.

By combining immunocapillary-migration (Glad and Grubb, 1978) and enzyme channeling techniques (Litman et al., 1980), Litman (1985) developed an enzyme chromatography method for quantitative measurement of anaylytes. The method uses: (1) antibodies immobilized on a paper strip, (2) labeled enzyme solution containing peroxidase labeled ligand and glucose oxidase, (3) substrate solution containing glucose and (4) chloronaphthol solution. The assay is performed by adding sample solution to labeled enzyme solution, then one end of the immobilized antibody paper strip is immersed into the solution which allows the solvent to migrate up the paper strip. After 10 minutes the paper strip is removed and is completely immersed in a substrate solution. After 5 minutes, the height of the colored zone is directly related to concentration of the ligand in the sample.

All reagents used in the fluorogenic enzyme substrate labeled immunoassay and prothetic group labeled enzyme immunoassay have also been successfully incorporated into reagent strips for dip and read type of assay (Greenquist et al., 1981; Tyhach et al., 1981).

SEPARATION-REQUIRED (HETEROGENEOUS) ENZYME MEDIATED IMMUNOASSAYS

Separation-required EMI was simultaneously developed by Van Weeman and Schuurs and by Engvall and Perlman in 1971. Most of the separation-required EMI's were based on assay principles which have been well developed in immunoassay using radioactive labels.

Separation-required competitive EMI

In this assay, the enzyme labeled ligand (L-E) competes with ligand (L) from the test sample for a limited amount of solid-phase immobilized antibody (I-Ab). After a brief incubation, the antibody-bound L-E is separated from the unbound L-E (Fig. 9). The antibody-found fraction is then assayed. The enzyme activity associated with the solid-phase is inversely related to the concentration of the test ligand.

The specificity of the assay depends, as does that of RIA's, on the specificity of the antibodies used. Using competitive method, picogram quantities of hormones, and other substances can be measured accurately (Bosh et al., 1978).

Separation-required inhibition EMI

In this assay (Fig. 10) immobilized ligands (I-L) compete with ligands (L) of the test sample for a fixed concentration of soluble enzyme labeled antibodies (E-Ab) (Reactions I and II). The enzyme activity in the I-L: E-Ab complexes will be inversely proportional to the concentration of the analyte (L) (Reaction III). In the absence of L, all E-Ab's bind to the I-L and the results would be maximal enzyme activity for the immobilized fraction.

The advantages of inhibition EMI using labeled antibody are: (i) Rapidity of the assay which involves only two incubations and one washing step; (ii) Stable enzyme-antibody conjugates can be prepared with standard procedures; (iii) It is

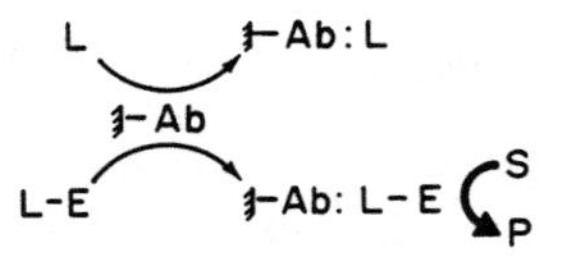

Fig. 9 Principle of separation-required EMI based on solid phase immobilized antibody (ǂ-Ab) enzyme labeled ligand (L-E).

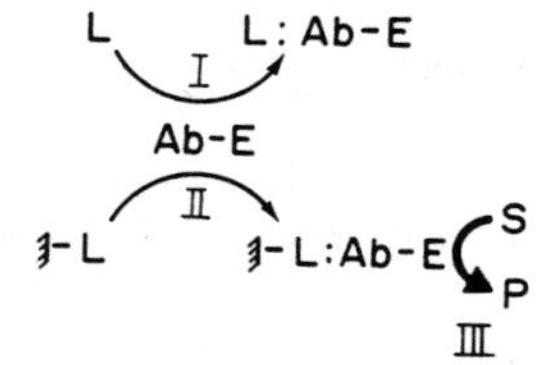

Fig. 10. Principle of separation-required inhibition EMI.

an accurate, reproducible and sensitive method. Human IgG at picomole levels can be quantified in less than 1.5 hours (Halliday and Wisdom, 1978).

<u>Separation-required sandwich EMI</u>

This method is used with antigens having multiple epitodes (Fig. 11). An excess of immobilized antibody is added to a test sample, after an incubation and washing, the enzyme labeled antibody is added. The enzyme activity remaining associated with immobilized antibody will be directly proportional to the amount of antigen in the test sample. Sandwich EMI does not involve competitions between labeled and unlabeled antigens, it is potentially a more sensitive technique than competitive EMI. Unlike competitive EMI, the development of a sandwich EMI has the advantage of not requiring purified antigens.

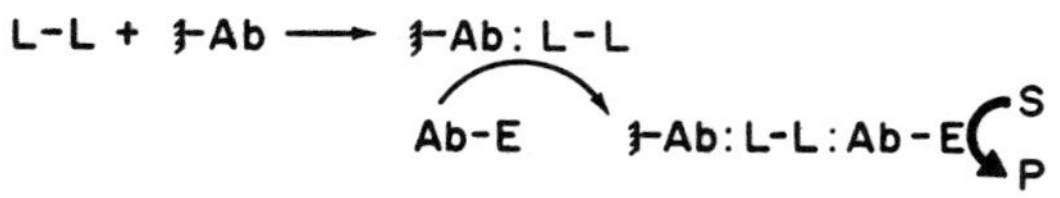

Fig. 11. Principle of separation-required sandwich EMI using immobilized antibody (⊣-Ab) and enzyme labeled antibody (Ab-E).

When sandwich EMI is developed by using polyclonal antibodies, unlike competitive EMI which takes only two incubations and one washing step, requires three incubations and two washing steps. When monoclonal antibodies directed toward different parts of the antigen are used in a sandwich EMI, however, the incubation and washing steps can be reduced to two and one respectively.

Sandwich EMI is well-suited for quantifying antigens with multiple antigenic determinants, such as antibodies, rheumatoid factors, polypeptide hormones, proteins and hepatitis B surface antigens. The results obtained with sandwich EMI are comparable to those obtained with those using radio-label in terms of precision, convenience and sensitivity. Macromolecular antigens at attomole levels have been quantified with this technique (Ishikawa et al., 1980).

Separation-required EMI using immunoenzymometric assay (IEMA)

This technique works with ligands having either single or multiple antigenic determinants (Fig. 12). An excess of enzyme labeled antibodies to a ligand (Ab-E) is added to the sample, after which an excess of immobilized ligand (I-L) is added. The immobilized ligands associated with the bound enzyme labeled antibodies are separated from the soluble ligand-enzyme labeled antibody complexes by a brief centrifugation. The enzyme activities of the soluble fraction should increase in proportion to the amount of ligand in the sample. Because of the use of excess reagents in IEMA, the analytical potentials of the system are pushed to its limit making it more sensitive than the competitive EMI.

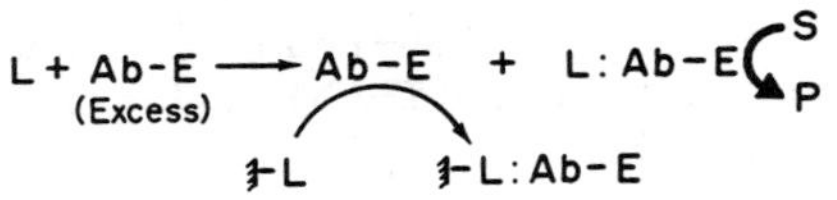

Fig. 12. Principle of separation-required EMI using immunoenzymometric assay (IEMA).

A new approach to IEMA to quantify hormones and drugs was described by Gnemmi et al. (1978). The method used an enzyme labeled anti-IgG antibody (called 2nd antibody, which is immunologically reactive to 1st antibody) and an antigen-specific antibody (called 1st antibody). The enzyme labeled 2nd antibody is reacted with 1st antibody to form an enzymic immunocomplex. In a typical assay, a sample is incubated with the preformed enzymic immuno-complex, then the unreacted enzymic immunocomplex is separated from the solution by binding to an immobilized solid phase antigen. The enzyme activity associated with the solid phase is inversely proportional to the antigen concentration in the sample.

The distinction between IEMA (Fig. 12) and inhibition EMI (Fig. 10) is the absence of a competition between analyte ligands and immobilized ligands for antibody binding sites in the IEMA. In IEMA, excess antibodies are used. While in inhibition EMI analyte ligands do compete with immobilized ligands for a limited amount of antibodies.

Separation-required indirect EMI

Indirect EMI is most frequently used for measuring serum levels of specific antibody produced in response to pathogens. In this assay (Fig. 13), ligands or antigens immobilized on a solid-phase are incubated with a test sample, after which the solid-phase ligands or antigens are washed and further incubated with enzyme labeled antispecies immunoglobulin antibodies (Ab_2-E). After separation of the soluble enzyme conjugates from those bound to immobilized solid-phase ligand-antibody complexes, the activity associated with these complexes is measured which is directly proportional to the concentration of antibody (Ab) in the test sample.

Indirect EMI has been used to detect and quantify antibody against bacterial and viral antigens, such as E. coli. plasmodium, trypanosoma, Vibrio cholerae, Typhus rickettsiae, hog cholera virus, Herpes simplex virus and Rubella virus (Halbert and Lin, 1985).

Separation-required EMI using tagged enzyme-ligand conjugate

The assay was based on the antibody masking the tag of a tagged enzyme-ligand conjugate. The principle of the assay is schematically presented in Fig. 14 and outlined as follows:

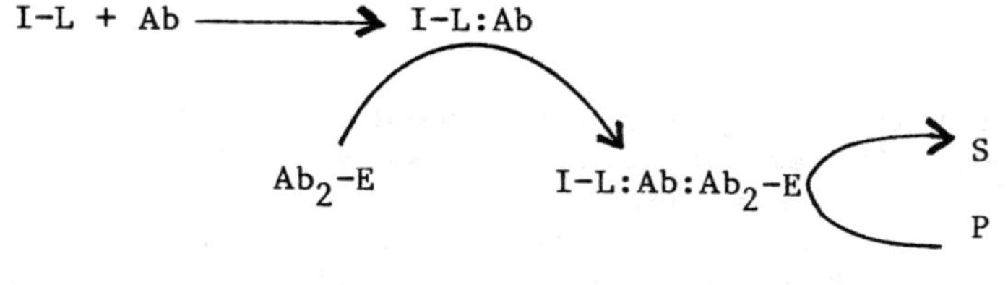

Fig. 13. Principle of separation-required indirect EMI. I-L: Immobilized ligand or antigen. Ab_2-E: Enzyme labeled anti-IgG.

(i) Competition in a homogeneous solution phase for antibody (Ab) by ligand (L) and ligand of tagged enzyme-ligand conjugate (L-E-T), as in reactions (a) and (b) of Fig. 14.

(ii) Separation by a brief centrifugation of antibody complexed tagged enzyme-ligand conjugate (Ab:L-E-T) from uncomplexed L-E-T by binding L-E-T to insolubilized receptor (R-I) to form L-E-T:R-I (reaction c). The binding between L-E-T and R-I occurs at a heterogeneous solid-liquid interphase. The T of Ab:L-E-T is not able to complex with receptor of R-I because of the presence of Ab.

(iii) Measuring the enzyme activity in either Ab:L-E-T of supernatant (reaction d), or L-E-T:R-I in the insoluble fraction (reaction e) can be achieved.

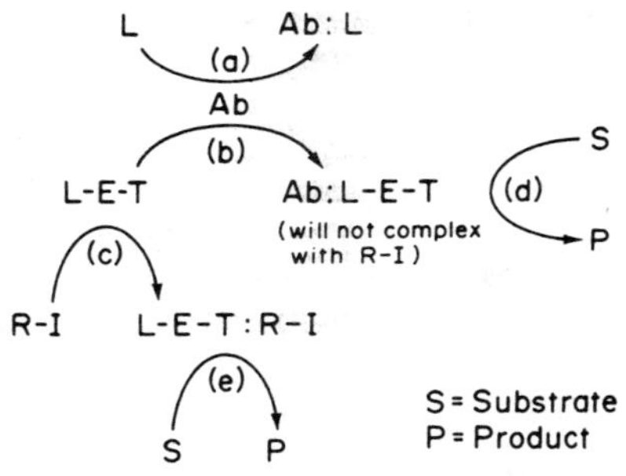

Fig. 14. Principle of separation-required EMI using tagged enzyme-ligand conjugate.

In this assay, a fixed concentration of L-E-T, Ab and R-I is used. The amount of free L-E-T available to complex R-I is dependent on the amount of L, because both L and L-E-T compete for Ab (reactions a and b). Thus, the lower the concentration of L, the more Ab are available to complex L-E-T (reaction b) so that less L-E-T is available to complex R-I. When L-E-T is complexed with Ab to form Ab:L-E-T by reaction b, it can no longer bind R-I, presumably because the Ab in Ab:L-E-T hinders physically the interaction between the tags and R-I.

Thus, it is obvious from Fig. 14 that a low concentration of L results in a high activity in the supernatant (reaction d) and low activity in the insoluble fraction (reaction e). On the other hand, more L will tie up more Ab (reaction a) freeing L-E-T so that it will complex with R-I (reaction c) resulting in higher activity in the insoluble fraction and less in the supernatant.

Using this assay, analyte at nanomolar levels can be measured in one hour (Ngo and Lenhoff, 1981).

Separation-required steric hindrance EMI

Castro and Monji developed a separation-required EMI based on a novel separation technique (1985). The assay uses enzyme labeled ligands and an affinity gel that is capable of binding the enzyme portion of the enzyme labeled ligand conjugates when the conjugates are free, i.e., not bound by the antibodies. The antibody bound enzyme labeled conjugate, however, does not bind the affinity gel, presumably because of a steric hindrance created by the antibody bound to the ligand of the enzyme labeled ligand conjugate that prevented the affinity gel to have effective interaction with the enzyme's active site. For this reason the assay is called Steric Hindrance Enzyme Immunoassay (SHEIA).

The principle of SHEIA is schematically shown in Fig. 15. The covalent enzyme labeled ligand (L-E) competed with ligand (L) from the sample for a limited amount of antibody (Ab). In the absence of L, reaction I would not take place and E-L and Ab would form Ab:L-E complex via reaction II. This complex would

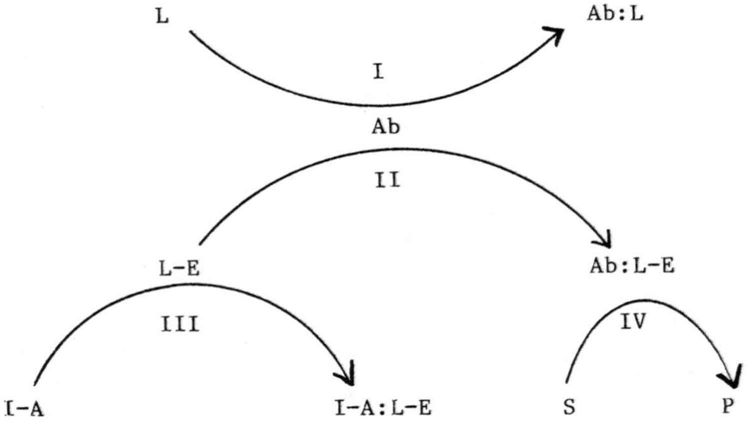

Fig. 15 Principle of SHEIA

not bind the affinity gel (I-A), therefore it would remain in supernatant phase with full enzymatic activity (reaction IV). In the presence of L, however, there would be a competition for the antibody leaving more L-E uncomplexed and free to bind to the affinity gel (I-A) and be removed from the supernatant (reaction III).

Using β-galactosidase as the enzyme label and agarose-6-aminocaproyl-β-D-galactosylamine as the affinity gel, sensitive assays for choriomammatropin and thyroxine have been developed (Castro and Monji, 1985).

Separation-required EMI for isoenzymes

Isoenzymes are structurally different but related enzymes catalyzing identical reactions. The composition and quantity of various isoenzymes, such as lactic dehydrogenases and creatine kinases change with pathological states of the body. For example, after a patient suffers from a myocardial infarction, the heart muscle derived lactic dehydrogenase and creatine kinase levels in the blood increase several folds. Isoenzymes are generally multisubunit enzymes composed of two or more types of monomeric subunits. Isoenzymes from different organs have different subunit compositions which are manifested in their electrophoretic mobilities. By using an antibody to specifically inhibit the enzyme activity expressed by one particular type of the subunit in a multi-subunit isoenzyme and to specifically separate and remove isoenzymes containing that particular type of sub-unit, the enzyme activity and quantity of the isoenzyme remaining in the solution, which is the disease marker, can be accurately measured (Wicks and Usategui-Gomez, 1985).

The principle of separation-required EMI for isoenzymes in its simplest form is diagramatically shown in Fig. 16. In this diagram, HH, HM and MM represent the three hypothetical isoenzymes from different organs and HH is the isoenzyme associated with, for example, a heart disease. The serum level of HH isoenzyme increase significantly at the onset of the disease.

The level of HH isoenzyme in a serum is quantified by incubating the serum with solid phase coupled antibody (I-Ab) that specifically inhibits the activity of M-subunit and binds to all M-subunit containing isoenzymes (Reactions I and II). After incubation and centrifugation to remove all M-subunit containing isoenzymes from the solution, the activity of HH isoenzyme is measured in order to obtain the quantity of HH isoenzyme in the serum (reaction III).

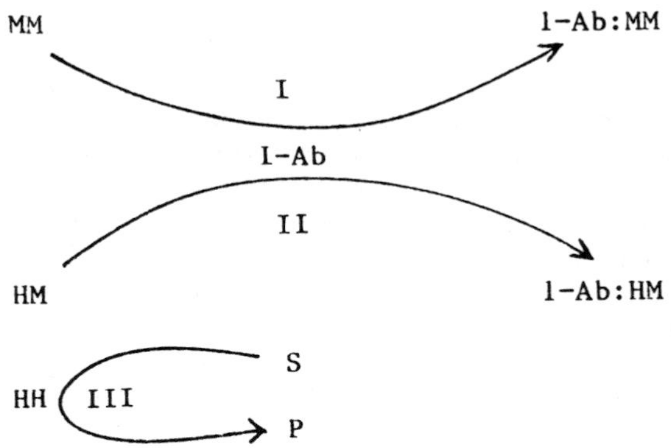

Fig. 16. Principle of separation-required EMI for isoenzymes.

Practical assays for cardiac isoenzymes, lactic dehydrogenase, LDH-1(H4) and creative kinase, CK-MB have been developed and gave good results which correlated well with electrophoretic methods (Wicks and Usategui-Gomez, 1985).

This method is only applicable to isoenzymes whose activity can be measured conveniently. In using this technique, patient serums should be handled with care so that the isoenzymes are not denatured. Preservation of the isoenzyme activity is crucial to the success of the assay.

<u>Amplification of enzyme activity</u>

The ultimate sensitivity of an immunoassay depends on the magnitude of the binding constant of a specific antibody. The greater the binding constant, the more sensitive the system would be (Yalow, 1980). In addition to an antibody with high binding constant, a sensitive label for easy detection is also needed for the development of a practical assay. Enzyme mediated labels are generally very sensitive labels because they provide amplification of weak signals through catalysis. A single molecule of the enzyme β-galactosidase has been measured (Rotman, 1961).

Several methods have been developed to further increase the sensitivity of enzyme labels. For example, the amplified enzyme-linked immunosorbent assay uses enzyme-antienzyme conjugates (Butler et al., 1985); the ABC system uses avidin-biotinylated enzymes and biotinylated antibodies complexes (Hsu, 1985) and enzyme polymers.

SUMMARY

Enzyme mediated labels (EML) are very versatile and sensitive labels for immunoassays. They have longer shelf-life, do not have the radiation related hazards and allow development of separation-free (homogeneous) enzyme mediated immuno-assays (EMI). Separation-free EMI's require less manipulation and are easier to automate. At the present state of development, however, they lack sensitivity. On the other hand, the separation-required (heterogeneous) EMI's are capable of providing great sensitivity and allow the development of an assay that can simultaneously assay two different analytes. The sensitivity of separation-required EMI's is best exemplified by a ferritin assay developed by Ishikawa et al. (1982) which is capable of measuring ferritin at 0.2 attomole, i.e., 2×10^{-19} moles.

Future developments in EMI are likely to involve: (1) more sensitive EML's, (2) more defined and controlled procedure for linking ligand to enzyme, (3) automated instruments and (4) dip-and-read solid-phase reagents.

REFERENCES

Adler, F. and Liu, C. T., 1971, Detection of morphine by hemaglutinatin-inhibition. J. Immunol. 106, 1684 - 1685.

Blecka, L. J., Shaffar, M. and Dworschack, R., 1983, Inhibitor enzyme immunoassays for quantitation of various haptens: A review, in: "Immunoenzymatic techniques," S. Avrameas, P. Druet, R. Mosseyeff and G. Feldman, eds., 207-214, Elsevier Science Publishers, Amsterdam.

Bosh, A. M. G., Van Hell, H., Brands, J. and Schuurs, A. H. W. M., 1978, Specificity, sensitivity and reproductibility of enzyme immunoassays, in: "Enzyme labelled immunoassay of hormones and drugs, S. B. Pal, ed., 175-187, Water de Gruyter, Berlin.

Braman, J. C., Broeze, R. J., Bowden, D. W., Myles, A., Fulton, T. R., Rising, M., Tjurston, J., Cole, F. and Vovis, G. F., 1984. Enzyme membrane immunoassay (EMIA), Biotechnology 2, 349-355.

Burd, J. F., Wong R. C., Feeney, J. E., Carrico, R. J. and Boguslaki, R. C., 1977, Homogeneous reactant-labeled fluorescent immunoassay for therapeutic drugs exemplified by gentamincin determintion in human serum. Clin. Chem. 23, 1402-1408.

Butler, J. E., Peterman, J. H. and Koertge, T. E., 1985. The amplified enzyme-linked immunosorbent assay (a-ELISA), This volume.

Cais, M., Dani, S., Eden, Y., Gandolfi, O., Horn, M., Isaacs, E. E., Josephy, Y., Saar, Y., Solvin, E. and Snarsky, L., 1977, Metalloimmunoassay. Nature, 270, 534-535.

Castro, A., and Monji, N., 1985, Steric hindrance enzyme immunoassay (SHEIA). This volume.

Chan, S. W., Tan, C. T. and Hsia, J. D., 1978, Spin membrane immunoassay: simplicity and specificity. J. Immunol. Meth. 21, 185-195.

Engval, E. and Perlmann, P., 1971, Enzyme-linked immunosorbent assay (ELISA). Quantitatie assay for immunoglobulin G. Immunochem. 8 871-874.

Finley, P. R., Williams, R. J. and Lichti, D. A., 1980, Evaluation of a new homogeneous enzyme inhibitor immunoassay of serum thyroxine with use of a bichromatic analyzer. Clin. Chem. 26, 1723-1726.

Gibbons, I., Skold, C., Rowley, G. L. and Ullman, E. F., 1980, Homogeneous enzyme immunoassay for proteins employs β-galactosidase, Anal. Biochem. 102-167.

Glad, C. and Grubb, A. O., 1978, Immunocapillary migration - a new method for immunochemical quantitation, Anal. Biochem. 85, 180-187.

Gnemmi, E., O'Sullivan, M. J., Chieregatti, G., Simmons, M., Simmonds, A., Bridges, J. W. and Mark, V., 1978, A senitive immunoenzymometric assay (IEMA) to quantitate hormones and drugs, in: "Enzyme labelled immunoassay of hormones and drugs," 29-41, S. B. Pal, ed., Walter de Gruyter, Berlin.

Gonnelli, M., Gabellieri, E., Montagnoli, G. and Felicioli, R., 1981, Complementing S-peptide as modulator in enzyme immunoassay, Biochem. biophys. Res. Commun., 103, 917-923.

Greenquist, A. C., Walter, B., and Li, T., 1981, Homogeneous fluorescent immunoassay with dry reagents. Clin. Chem. 27 1616-1617.

Haimovich, J. and Sela, M., 1969, Protein-bacteriophage conjugates: Application in detection of antibodies and antigens. Science, 164, 1279-1280.

Halbert, S. P. and Lin, T-M., 1985, Enzyme immunoassay of antibody. This volume.

Halliday, M. I. and Wisdom, G. B., 1978, A competitive enzyme-immunoassay using labelled antibody. FEBS Letters 96, 298-300.

Hsu, S-M., 1985, Immunoperoxidase techniques using the avidin-biotin system, This volume.

Ishikawa, E., Imagawa, M., Yoshitake, S., Niitsu, Y., Urushizaki, I., Inada, M., Imura, H., Kanazawa, R., Tachibana, S., Nakazawa, N., and Ogawa, H., 1982, Major factors limiting sensitivity of sandwich enzyme immunoassay for ferritin, immunoglubulin E, and thyroid-stimulating hormone, Ann. Chin. Biochem. 19, 379-384.

Ishikawa, E., Yoshitake, S., Endo, Y. and Ohtaki, S., 1980. Highly sensitive enzyme immunoassay of rabbit (anti-human IgG) IgG using human IgG-β-galactosidase conjugate, FEBS Letters III 353-355.

Jaklistsch, A., 1985. Separation-free enzyme immunoassay for haptens. This volume.

Kricka, L. J. and Carter, T. J. N., 1982, "Clinical and Biochemical Luminescence," Marcel Dekker, Inc., New York.

Leute, R. K., Ullman, E. F., Goldstein, A. and Herzenberg, L. A., 1972, Spin immunoassay techniques for determinination of morphine, Nature 236, 93-94.

Leuvering, J. H. W., Thal, P. J. H. M., Van der Waart, M. and Schuurs, A. H. W. M., 1980, Sol particle immunoassay (SPIA), J. Immunoassay, 1 77-91.

Litchfield, W. J., Freytag, J. W. and Adamich, M., 1981, Highly sensitive immunoassay based on use of liposome without complement, Clin. Chem. 30, 1441-1445.

Litman, D. J., 1985, Test strip immunoassay. This volume.

Litman, D. J., Hanlon, T. M. and Ullman, E. F., 1980, Enzyme channeling immunoassay: A new homogeneous enzyme immunoassay technique, Anal. Biochem. 136, 223-229.

Morris, D. L., Carrico, R. J., Ellis, P. B., Hornby, W. E., Schroeder, H. R., Ngo, T. T. and Boguslaski, R. C., 1980, Colorimetric immunoassays with flavin adenine dinucleotide as label, in: "Innovative approaches to clinical analytical chemistry," Lab. Professional Series, America Chemical Society.

Morris, D. L., Ellis, P. B., Carrico, R. J., Yeager, F. M., Schroeder, H. R., Albarella, J. P., Boguslaski, R. C., Horby, W. E. and Rawson, D., 1981, Flavin adanine dinucleotide as a label in homogeneous colorimetric immunoassays, Analyt. Chem., 53, 658-665.

Mosback, K. and Mattiasson, B., 1970, Matrix bound enzymes, Part 2, Studies on a matrix bound two enzyme system. Acta Chem. Scand., 24, 2093-2100.

Nakane, P., 1980, Future trends and application of immunoassays, in: Diagnostic Immunology, 73-77, P. W. Keitges and R. M. Nakamura, eds. College of American Pathologists, Illinois.

Ngo, T. T., 1983, Enzyme modulator mediated immunoassay (EMMIA), Int. J. Biochem. 15 583-590.

Ngo, T. T., 1985, Enzyme modulator as label in separation-free immunoassays: enzyme modulator mediated immunoassay (EMMIA). This volume.

Ngo, T. T., 1985, Prosthetic group labeled enzyme immunoassay. This volume.

Ngo, T. T., Bovaird, J.H. and Lenhoff, H.M., 1985, Separation-free amperometric enzyme immunoassay, Appl. Biochem. Biotechnol. 11, 63-70.

Ngo, T. T., Carrico, R. J. and Boguslaski, R. C., 1979, Homogeneous fluorescence immunoassay for protein using β-galactosyl-umbelliferone label, paper presented at 2nd International Conference on Diagnostic Immunology, New England College, Henniber, New Hampshire.

Ngo, T. T., Carrico, R. J., Boguslaski, R. C. and Burd, J. F., 1981, Homogeneous substrate-labeled fluorescent immunoassay for IgG in human serum. J. Immunol. Meth. 42, 93-103.

Ngo, T. T. and Lenhoff, H. M., 1980a, Enzyme modulators as tools for the development of homogeneous enzyme immunoassays. FEBS Lett., 116, 285-288.

Ngo, T. T. and Lenhoff, H. M., 1980, Amperometric determination of picamolar levels of flavin adenine dinucleotide by cyclic oxidation-reduction in apo-glucose oxidase system, Anal. Letters, 13, 1157-1165.

Ngo, T. T. and Tunnicliff, G., 1981, Inhibition of enzymic reactions by transition state analogs: An approach for drug design. Gen. Pharmac. 12, 129-138.

Ngo, T. T. and Lenhoff, H. M., 1981a, Recent advances in homogeneous and separation-free enzyme immunoassays. Appl. Biochem. Biotech. 6, 53-64.

Ngo, T. T. and Lenhoff, H. M., 1981b, New approach to hetrogeneous enzyme immunoassays using tagged enzyme-ligand conjugates. Biochem. Biophys. Res. Commun. 99, 495-503.

Ngo, T. T. and Lenhoff, H. M., 1982, Enzymes as versatile labels and signal amplifies for monitoring immunochemical reactions. Molec. Cell. Biochem. 44, 3-12.

Ngo, T. T. and Lenhoff, H. M., 1983, Antibody-induced conformational restriction as basis for new separation-free enzyme immunoassay, Biochem. Biophys. Res., 114, 1097-1103.

Ngo, T. T. and Wong, R. C., 1985, Fluorogenic enzyme substrate labeled immunoassays for haptens and macromolecules. This volume.

Rembaum, A. and Dreyer, W. J., 1980, Immunomicropheres: reagents for cell labeling and separqtion. Science 208, 364-368.

Rotman, B., 1961, Measurement of activity of single molecules of β-galactosidase, Proc. Nat. Acad. Sci., 47, 1981-1991.

Rowley, G. L., Rubenstein, K. E., Huisjen, J. and Ullman, E. F., 1975, Mechanism by which antibodies inhibit hapten-malate dehydrogenase conjugates, J. Biol. Chem. 250, 3759-3766.

Rubenstein, K. E., Schneider, R. S. and Ullman, E. F., 1972, "Homogeneous" enzyme immunoassay. New Immunochemical technique. Biochem. biophys. Res. Commun. 47, 846-851.

Schuurs, A. H. W. M. and Van Weemen, B. K., 1977, Enzyme immunoassay. Clinica chim. Acta 81, 1-40.

Serio, M. and Pazzagli, M., 1982, "Luminescent Assays," Raven Press, New York.

Simpson, J. S. A., Campbell, A. K., Ryall, M. E. T. and Woodhead, J. S., 1979, A stable chemiluminescent-labelled antibody for immunological assays, Nature, 279, 141-147.

Singer, S. J. and Schick, A. F., 1961, The properties of specific stains for electron microscopy prepared by the conjugation of antibody molecules with ferritin, J. Biophys, Biochem. Cytol., 9, 519-537.

Soini, E. and Hemmila, I., 1979, Fluoroimmunoassay: Present status and key problems, Clin. Chem. 25, 353-361.

Tyhach, R. J., Rupchock, P. A., Pendergrass, J. H., Shjold, A. C., Smith, P. J., Johnson, R. D., Albarrella, J. P. and Profitt, J. A., 1981. Adaptation of prosthetic-group-label homogeneous immunoassay to reagent-strip format, Chin. Chem., 27, 1499-1504.

Van Weemen, B. K. and Schuurs, A. H. W. M., 1971, Immunoassay using antigen enzyme conjugates, FEBS Letters, 15, 232-236.

Wicks, R. and Usategue-Gomez, M., 1985, Development of immunochemical enzyme assay for cardiac isoenzymes. This volume.

Wong, R. C., Burd, J. F., Carrico, R. J., Buckler, R. T., Thoma, J. and Boguslaski, R. C., 1979, Substrate-labeled fluorescent immunoassay for phenytoin in human serum; Clin. Chem., 25, 686-691.

Yalow, R. S., 1980, Radioimmunoassay, in: Ann. Rev. Biophys. Bioeng. (Edited by Mullins, L. J., Hagins, W. A., Newton, C. and Weber, G.) Vol. 9, pp. 327-345. Annual Reviews Inc., Palo Alto.

Yalow, R. S. and Berson, S. A., 1959, Assay of plasma insulin in human subjects by immunological methods, Nature, 184, 1648-1649.

SEPARATION-FREE ENZYME IMMUNOASSAY FOR HAPTENS

Anna Jaklitsch

SYVA Company
900 Arastradero Road
Palo Alto, California 94304

INTRODUCTION

Separation-free (homogeneous) enzyme immunoassays (Emit®) were first developed at the Syva Research Institute in the early 1970s. Their discovery was that binding of an anti-drug antibody to an enzyme labeled with the same drug, modulated (inhibited or activated) catalytic activity. Thus a competitive immunoassay can be developed in which drug in the sample and enzyme-labeled drug compete for antibody. Modulation of enzyme activity is directly related to the analyte concentration in the sample and is monitored without separation of antibody bound and unbound fractions. This principle was later extended to a variety of enzymes and also applied to the measurement of hormones (Ullman et al., 1979) and proteins (Gibbons et al., 1980).

The first commercial assays for detecting the presence of abused drugs in urine used lysozyme as the label (Rubenstein et al., 1972). Assays to monitor serum concentration of therapeutic agents and endogenous compounds were subsequently developed using glucose-6-phosphate dehydrogenase (G6PDH) (Rowley et al., 1976) or malate dehydrogenase (MDH) (Ullman et al., 1979). Most recently lysozyme derived assays are being converted to the G6PDH system which permits simplification and standardization of the Emit® assay protocol (Table 1).

This chapter reviews the mechanism of action of the separation-free (homogeneous) hapten assays. Table 2 lists the analytes to which the technology has been applied. Alternative technologies are also described. The development of assays to detect the presence of drug classes is discussed using the benzodiazepine and

Table 1. Standard EMIT Protocol[a]

1. Predilute 50 μl sample with 250 μl buffer[b].
2. Dilute 50 μl of diluted sample with 250 μl buffer.
3. Add 50 μl Reagent A[c] with 250 μl buffer.
4. Add 50 μl Reagent B[d] with 250 μl buffer.
5. Aspirate into the flow cell of a spectrophotometer (regulated to 30°C).
6. Read the change in absorbance (1 cm light path) at 340 nm between 15 sec and 45 sec.

[a]Drug abuse assays eliminate predilution.
[b]Buffer is 0.055 M Tris-HCl buffer pH 8.0, 0.9% NaCl, and 0.01% Triton X-100.
[c]Reagent A contains antibody, substrate, and cofactors including 0.066 M glucose-6-phosphate and 0.04 M NAD in 0.055 M Tris HCl buffer, pH 5.2, and preservatives.
[d]Reagent B contains enzyme conjugate, 0.055 M Tris at either pH 6.2 or 8.0, stabilizers, and preservatives.

qualitative tricyclic antidepressant assays as examples. The optimization of quantitative assays for therapeutic drug monitoring, including chemistry and data analysis, is illustrated by discussion of a representative class of assays, the aminoglycosides. Batch protocols required for measurement of low level analytes are also discussed and future applications of the technology explored.

MECHANISM

In the lysozyme-based immunoassays, a suspension of M. luteus is the substrate. Reduction in turbidity, monitored by the change in transmission at 436 nm, is observed when the cell wall of the bacteria is digested. Lysozyme covalently labeled with a drug derivative is inhibited by the binding of antibody, presumably due to steric exclusion of the substrate (Rubenstein, 1972). As drug in a sample competes with the enzyme-labeled drug for the antibody, there is less inhibition. Enzyme activity correlates with the concentration of drug introduced. The greater the amount of drug present, the greater the change in turbidity observed at 436 nm.

In the G6PDH-based immunoassays, drug derivatives of the enzyme are also used. Anti-drug antibody binding inhibits the enzyme drug conjugate by changing the enzyme's conformation. As in the lysozyme-based assay, sample, drug, and enzyme conjugate compete for anti-drug antibody and residual enzyme activity relates directly to the concentration of drug in the sample (Fig. 1A).

Table 2. EMIT Assays

A. EMIT SERUM THERAPEUTIC ASSAYS

Amikacin Assay*
Gentamicin Assay*
Netilmicin Assay
Tobramycin Assay*

Methotrexate Assay

Phenytoin Assay*
Phenobarbital Assay*
Primidone Assay**
Ethosuximide Assay
Carbamazepine Assay
Valproic Acid Assay

Digoxin Assay
Lidocaine Assay*
Procainamide Assay**
N-Acetylprocainamide Assay**
Quinidine Assay*
Disopyramide Assay

B. EMIT HORMONE ASSAYS

CFA Thyroxine Assay (for use with CentrifiChem Model 300 or 400)
ABA Thyroxine Assay (for use with Abbott Bichromatic Analyzer 100)
Manual Thyroxine Assay (for use with EMIT® Lab 6000 System)

Auto Cortisol Assay

C. EMIT SERUM TOXICOLOGY ASSAYS

Serum Benzodiazepine Assay
Serum Barbiturate Assay
Serum Tricyclic Antidepressants Assay
Acetaminophen Assay**

D. EMIT URINE TOXICOLOGY ASSAYS

Opiate Assay†
Amphetamine Assay†
Barbiturate Assay†
Methadone Assay†
Benzodiazepine Metabolite Assay†
Propoxyphene Assay
Cocaine Metabolite Assay
Phencyclidine Assay†
Cannabinoid Assay
Methaqualone Assay†

*Available in quantitative unit dose powder-formulated (Emit® Qst™ reagents.
**Available in Qst reagent form in 1985.
†Available in qualitative unit dose powder-formulated reagents.

Enzyme activity is conveniently measured by following production of NADH which absorbs light strongly at 340 nm (Rowley et al., 1976).

The thyroxine Emit® assay was developed with MDH as the label. MDH-T_4 conjugate shows an increased activity in the presence of anti-T_4 (Fig. 1B) (Ullman et al., 1979). Therefore the calibration curve in this assay shows decreasing enzyme activity as the

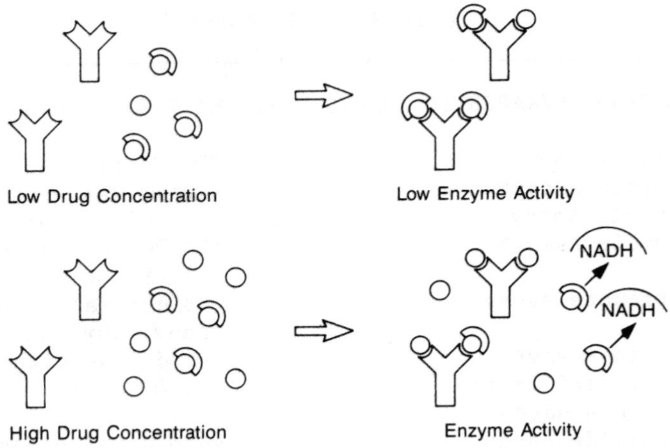

Fig. 1A. Mechanism of EMIT® Assays.

○ = Drug
= Enzyme labeled drug.

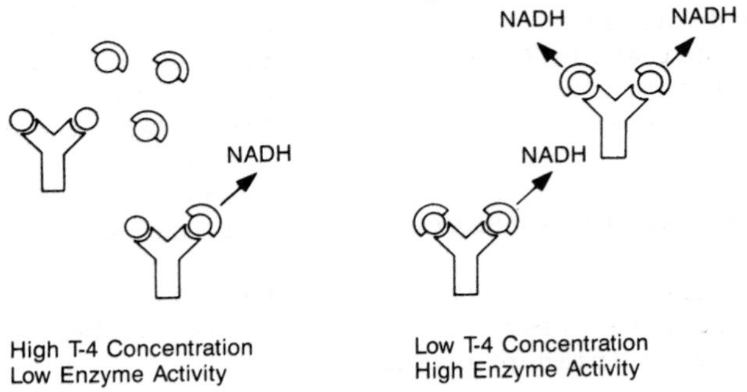

Fig. 1B. Mechanism of EMIT® Assays.

concentration of thyroxine from patients' sera increases (Jaklitsch et al., 1976).

DRUG ABUSE (URINE TOXICOLOGY) ASSAYS

Drug abuse assays have been set up to detect the drug rather than quantify its concentration. Each assay detects a class of structurally similar drugs and their metabolites. Fig. 2 shows the structural similarities among benzodiazepines commonly encountered. The Emit® assays are constructed so that any of these compounds will generate approximately the same response. For example, the benzodiazepine assay reports as positive 2 μg/ml flurazepam and

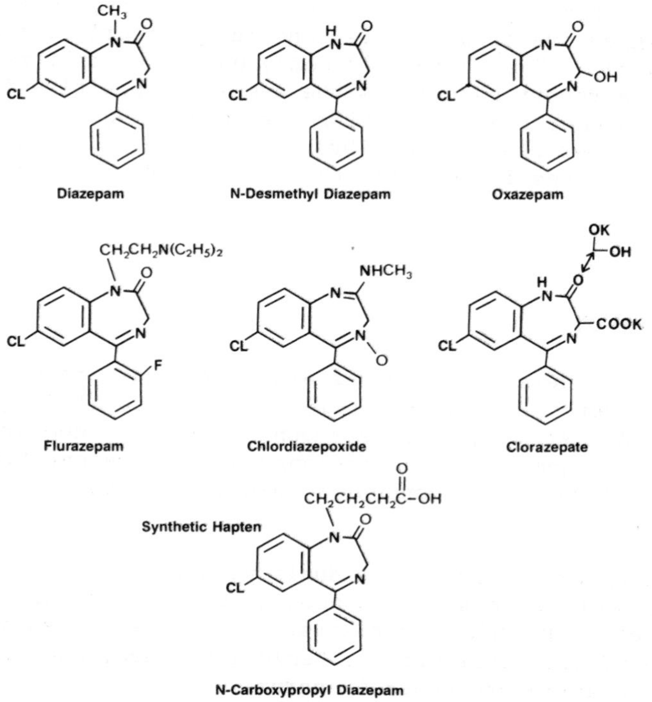

Fig. 2. Structural Similarities Among Benzodiazepines

desalkylflurazepam, its metabolite. This is achieved by coupling drug derivative to carrier protein by the functional group which differs among compounds of this class (Fig. 2). Thus antibodies have been raised to the part of benzodiazepine nucleus which is common to members of this drug class.

The response of a sample is compared to the response of a negative urine and a positive calibrator. The calibrator is prepared by adding a known quantity of a representative drug in the class to a negative urine pool. A sample that gives a change in absorbance value equal to or higher than the calibrator's value is interpreted to be positive, and a sample with a change in absorbance value lower than the calibrator's value is interpreted to be negative; either there is no drug in the sample, or it is present in a low concentration that is undetectable by the assay.

In addition to the assay configuration described above, these reagents were also developed in unit dose form. (Centofanti et al., 1981). The enzyme, antibody, substrate, buffers, and stabilizers are lyophilized separately and then blended together to produce the final product. The powder is then filled in one tube. Sample is added to one vial and calibrator to a second vial. Both are reconstituted with water, inverted vigorously, inserted in a custom-designed spectrophotometer and read directly through the tube for 90 sec. The unknown is compared to the calibrator automatically. A report card lists the sample as positive if the response is higher than the calibrator response or negative if the response is lower than the calibrator response. This assay system is designed for testing outside the laboratory environment.

SERUM TOXICOLOGY

Assays that detect a class of drug compounds are useful in the hospital emergency room when knowledge of the total drug load is desirable. An estimate of this amount is made by comparing the response of the sample to the response of a representative member of the drug class. The unknown is then quantified in total drug equivalents.

Collecting urine from individuals presenting in the hospital emergency room conscious or unconscious can be more difficult than obtaining serum from these patients. Therefore, class assays have been developed for barbiturates, benzodiazepines, and most recently the tricyclic antidepressants in serum.

For the antidepressant assays the response and number of the metabolites is substantial and their glucuronides are unavailable for evaluation. Only a qualitative estimation of the total tricyclics present in the sample can be made. A positive control can be assayed to indicate whether the sum of the parent drug and metabolite is within or above the therapeutic range. Figure 3 describes the response of metabolites in this assay.

The parent drug and demethylated metabolites show approximately equivalent responses. The assay was designed to achieve this result by obtaining antisera from sheep immunized with the carrier protein bovine serum albumin coupled to desipramine through an N-carboxypropyl linkage. The ring structure is exposed; therefore, the antisera produced minimize the differences in response to side chain substitution (Fig. 4).

A specific assay for acetaminophen was developed to define the need and to predict efficacy of the antidote required to treat life-threatening overdose cases. The elimination half-life can be calculated from measurements of samples drawn at two times several hours apart. Therefore, a quantitative Emit® assay was developed for this drug.

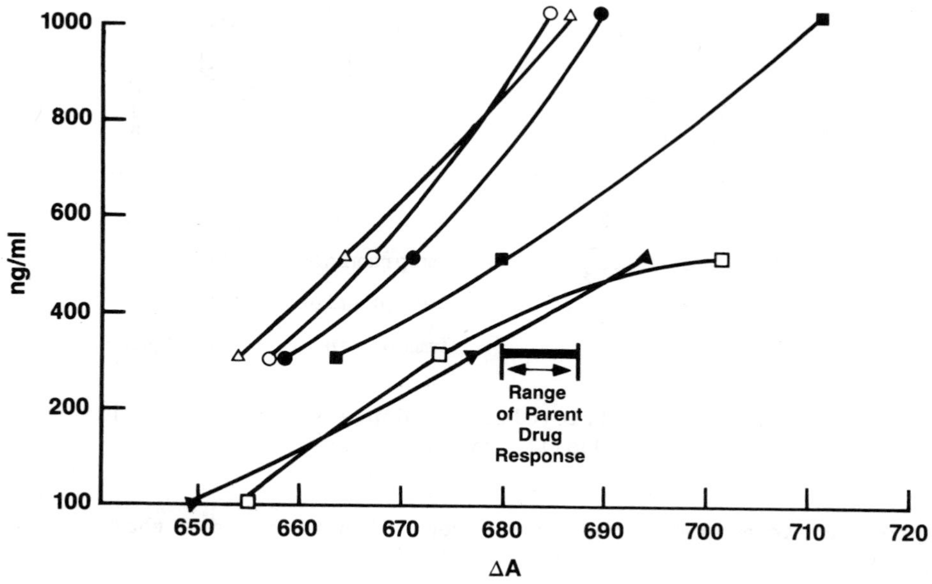

Fig. 3. Response of Structurally Related Compounds in the Tricyclic Toxicology Assay.

The response of each drug was measured in the toxicology assay using the standing Emit® protocol.

- ● 10-OH amitriptyline
- o 10-OH imipramine
- ▲ 2-OH imipramine
- △ 2-OH desipramine
- X chlorpromazine
- □ protriptyline, chlomipramine

THERAPEUTIC DRUG MONITORING

Purpose

The benefits of monitoring of drugs with narrow therapeutic ranges has become widely accepted. Although many inexpensive, accurate assays are available, Whiting et al., (1984) reported that only 42 of 145 theophylline samples (29%) from a survey quantified within the therapeutic range of 10-20 μg/ml, and even fewer phenytoin samples (18%) quantified within the therapeutic range. In a follow-up to this study, 70% of the theophylline samples and 45% of the phenytoin samples appeared within the therapeutic range when the attending physicians were rapidly apprised of the significance of the results and dosages were adjusted accordingly. Clearly communications between physicians and laboratories must be

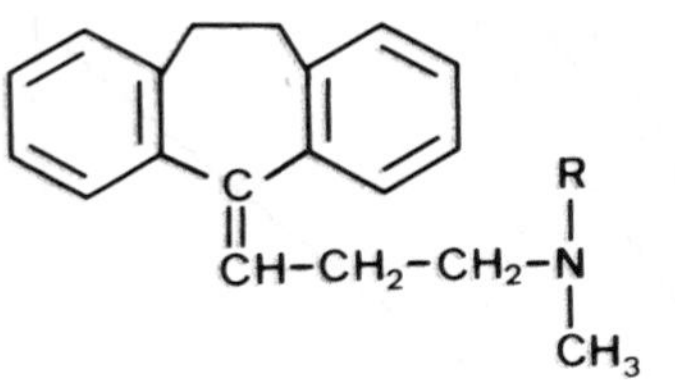

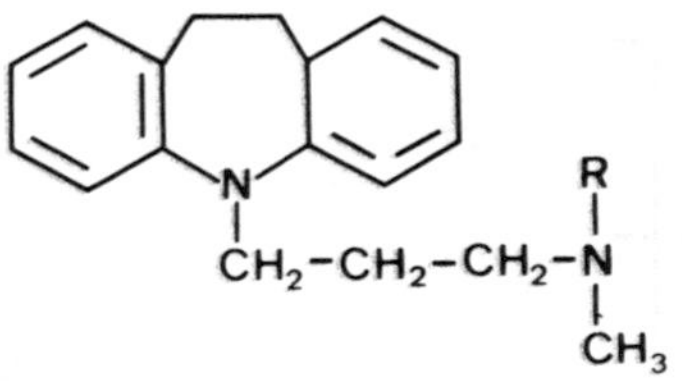

	R		R
Amitriptyline	- CH_3	Imipramine	- CH_3
Nortriptyline	- H	Desipramine	- H
		Immunogen	- $CH_2CH_2CH_2COOH$

Fig. 4. Structure of Tricyclic Antidepressants and the Immunogen Used for the Qualitative Tox Assay.

promoted to maximize the improvement of patient care that can be provided by therapeutic drug monitoring.

Protocol

Most Emit® assays are now developed with the same protocol on the semiautomated SYVA® Lab 5000 System (Table 1). The SYVA® AutoCarousel™ protocol differs only in the elimination of the preliminary dilution; 8 μl of serum is sampled directly. Numerous applications of these reagents have been developed for other instruments, such as the CentrifiChem (Dols and van Zanten, 1981), the IL Multistat (Iosefsohn, et al., 1981) and the Cobas Bio (Ou et al., 1981; Oellerich et al., 1982; Witebsky et al., 1983; and Hamlin and Sullivan, 1984). The ability of the user to adapt the chemistry to a variety of analyzers has resulted in widespread acceptance of the Emit® technology.

In addition to the applications, quantitative unit dose (Qst) formulations have been developed and assays now available are noted in Table 2. The Qst sampling module adds diluent, mixes, and automatically aspirates sample into the spectrophotometer. The quantitative result is available in about one minute.

DEVELOPMENT OF EMIT ASSAYS

Drug Haptens and Hapten Conjugates

We generally use one of two synthetic routes to prepare immunogens or enzyme drug conjugates. The most common procedure is similar to that described for theophylline (Chang et al., 1982) in which 3(carboxypropyl) theophylline is coupled to G6PDH by activating with N-hydroxysuccinimide in the presence of a condensing reagent such as dicyclohexyl carbodiimide. Because of their unique chemistry, aminoglycosides required a very different synthetic approach as described by Singh (1980) for the prototype assay gentamicin. Amino groups of G6PDH are modified with bromoacetyl glycine via an N-hydroxysuccinimide ester intermediate. The modified enzyme is then coupled to an aminoglycoside sulfhydryl derivative. The bromine is displaced and a covalent bond is formed between the drug and modified enzyme.

Assay Optimization

This section describes some of the challenges that arose during feasibility studies for a representative group of aminoglycosides. The structures of these compounds are shown in Figs. 5 and 6. Many interesting observations were made during assay development. Unique problems were encountered and resolved, although not all of the observations are fully explained. The relatively minor differences in structure among members of the gentamicin and tobramycin series provide an opportunity to explore response and specificity as a function of the aminoglycoside enzyme or antibody matched in an assay system. Properties of aminoglycosides that affect assay performance will also be described.

Gentamicin, Netilmicin, and Sisomicin

Figure 5 indicates that the structures of gentamicin, netilmicin, and sisomicin differ only in the appearance of a double bond between the 4' and 5' position in the pupurosamine ring and for netilmicin in the substitution of an ethylene group for a hydrogen at the number 1 nitrogen in the 2-deoxystreptamine ring.

These structures are so similar that a netilmicin assay was constructed using antisera raised against a gentamicin derivative-carrier protein conjugate and either netilmicin (unmatched) or gentamicin enzyme conjugate (matched). The difference in response between 1 and 7 μg/ml netilmicin calibrators is greater for the unmatched system (200 rate units) than for the matched system (125).

A sisomicin assay can also be constructed (Jaklitsch et al., 1980) using either of the above conjugates. The gentamicin antisera is about 30% more sensitive measuring 4-8 μg/ml of sisomicin, possibly because its immunogen is structurally less similar to sisomicin than netilmicin. Perhaps that structural difference favors drug over enzyme conjugate binding, hence more assay sensitivity is observed.

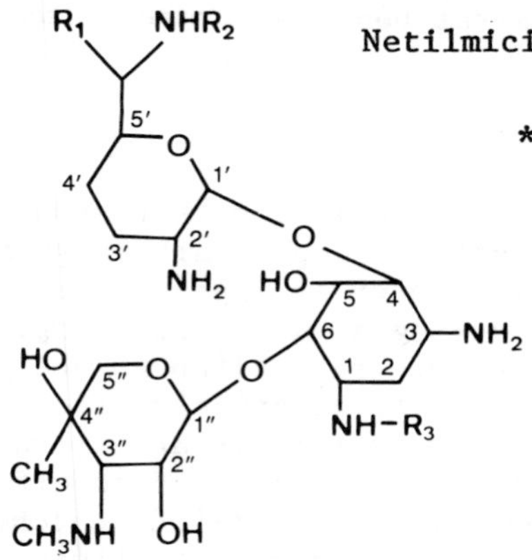

	R_1	R_2	R_3
Gentamicin C_1	CH_3	CH_3	H
Gentamicin C_2	CH_3	H	H
Gentamicin C_{1_a}	H	H	H
Sisomicin*	H	H	H
Netilmicin*	H	H	CH_2CH_3

*Double bond between 4' and 5'.

Fig. 5. Structure of Aminoglycoside Antibiotics.

	R_1	R_2	R_3	
Tobramycin	H_2	NH	H	
Kanamycin A	OH	OH	H	95%
Kanamycin B	OH_2	NH	H	5%
Amikacin	OH	OH	$HCOCOHCH_2CH_2NH_2$	

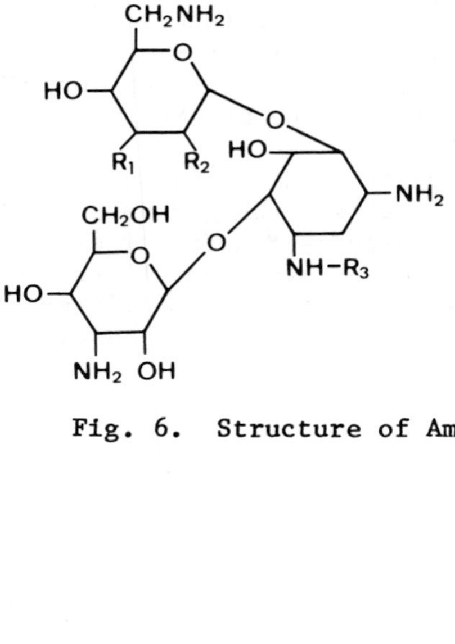

Fig. 6. Structure of Aminoglycoside Antibiotics.

The effect of sample and antibody incubation was studied using netilmicin reagents. The netilmicin curve in a fully matched system, i.e., netilmicin enzyme and antibody, increases at higher drug concentration when a 30 sec incubation is introduced into the

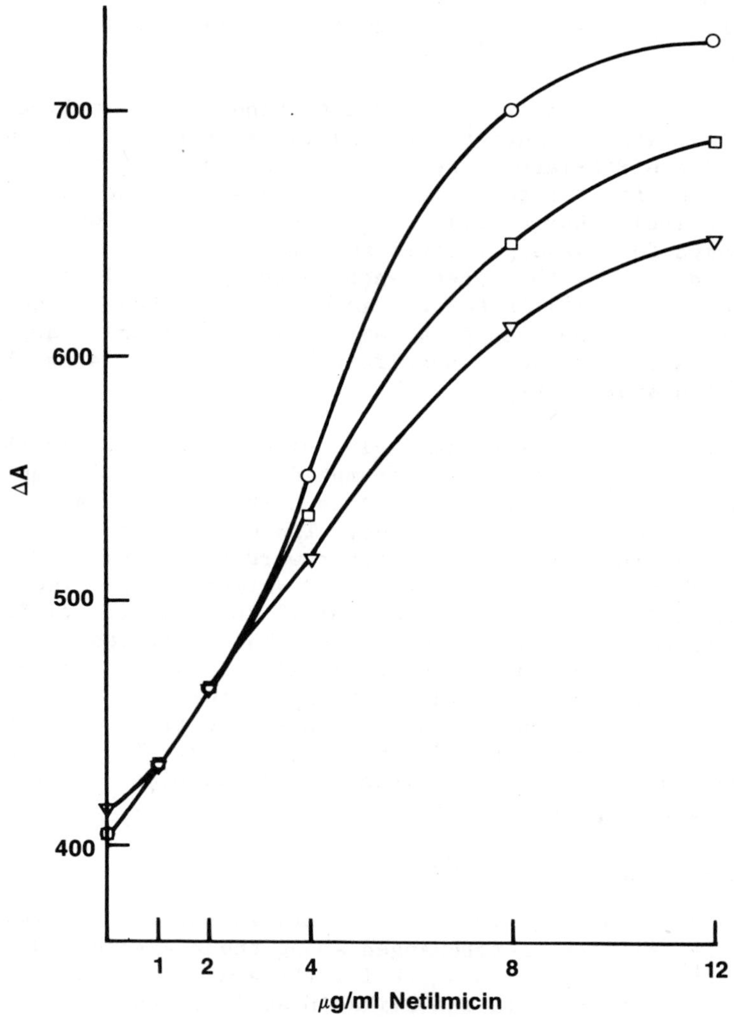

Fig. 7. Variation of Netilmicin Curve Shape with Antibody Incubation Time and Formulation.

○–○ Standard protocol: 30 sec incubation of sample and antibody.
□–□ No incubation.
▽–▽ No incubation, 1 μg/ml netilmicin was added to antibody.

protocol. Likewise the response is reduced and linearized when 1 μg/ml of drug is added to the antibody reagent. This phenomenon is illustrated in Fig. 7. Addition of drug to the antibody then becomes a technique by which curve shape can be experimentally manipulated. Results from the Emit® assay are quantitated in concentration units by comparing the response of an unknown to a calibration curve whose shape is defined by constants derived and validated at Syva.

The ability of aminoglycosides to bind anionic polymers such as dextran sulfate was investigated by the addition of dextran sulfate in serum to G6PDH-labeled gentamicin. As shown in Fig. 8, dextran sulfate inhibits the conjugate. From this data one can predict that negatively charged polymers, such as heparin, would interfere with assays for these polycationic compounds. Krogsted et al. (1982) did observe this phenomenon in the gentamicin assay and Matzke et al. (1984) in the tobramycin assay. Heparinized plasma therefore cannot be used as a sample matrix for Emit® assays of aminoglycosides and dextran-sulfate-treated serum cannot be used for a calibrator matrix.

Literature reports of aminoglycoside instability in the presence of β-lactam antibiotics prompted evaluation of sample collection and storage conditions since both β-lactams and aminoglycosides may be present in samples. Tindula's (1983) comprehensive study demonstrated that reactivity toward these compounds varies in the order: tobramycin > gentamicin > amikacin. He also found that penicillin, ampicillin, carbenicillin, and ticarcillin were the most reactive β-lactam compounds. The worst case was the tobramycin interaction with carbenicillin in which only 53% of the measured gentamicin remained after the mixture was stored at room temperature for 24 hours. Table 3 shows the time course of inactivation measured in the Emit® gentamicin assay. The conclusion from these studies is that aminoglycoside samples must be stored frozen immediately upon receipt in the laboratory.

<u>Tobramycin, Amikacin, and Kanamycin</u>

Figure 6 indicates the aminoglycosides structural similarities allow antisera to be interchanged among these assays. One objective of the development program for the amikacin assay was to evaluate both tobramycin and amikacin antisera and Fig. 9 shows that either tobramycin or amikacin reagents can be used in the assay. Although the tobramycin response curve is more linear and does not show time depending drug binding, the amikacin antisera was chosen for the Emit® amikacin assay for the following reasons. Occasionally patients are given amikacin, a broad spectrum antibiotic, after treatment with tobramycin proves ineffective. In uremic patients in whom clearance is reduced (Lockwood and Bower, 1973) and therefore half-life increased, the washout period can be

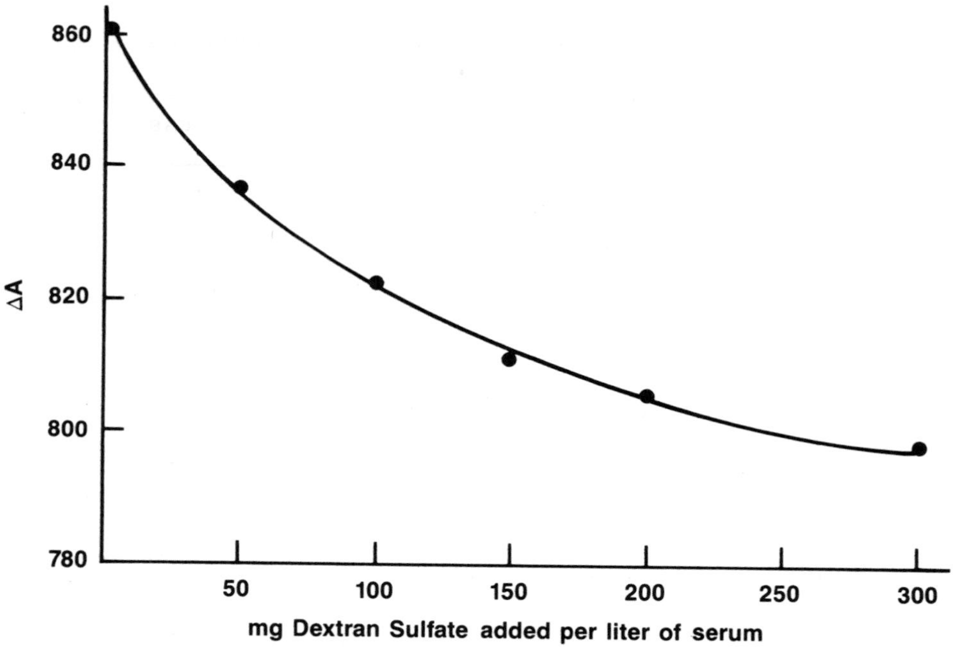

Fig. 8. Effect of Dextran Sulfate on G6PDH Enzyme Activity.

Dextran sulfate was added to raw sheep serum to form a precipitate with serum lipoproteins in the presence of calcium acetate. The precipitate is removed by centrifugation. The sheep serum was substituted for calibrator in the gentamicin assay. The change in NAD absorption at 340 nm is monitored at 30° in the standard protocol of 30 sec measurement after 15 sec incubation.

increased causing tobramycin to remain in the serum during the time when a sample would be drawn for amikacin quantitation. A substantial error can thus occur in the amikacin measurement using tobramycin antisera. To avoid this problem, the amikacin-matched enzyme and antisera pair were incorporated into the reagent formulation.

Figure 9 illustrates that variation in antibody-sample incubation time can strongly affect assay performance in the amikacin assay where variation in incubation time results in significant difference in response. To prevent potential compromise of assay precision, two approaches were studied. The simplest solution is to change the assay protocol so that the antibody is introduced into a mixture of enzyme and sample. The curve becomes more linear and the time dependence of binding is eliminated. We have also

Table 3. Loss of Gentamicin in the Presence of Carbenicillin over Time

Time	Gentamicin (μg/ml)	% Loss
0	7.8	
1 hr (R.T.)	7.2	10.2
15 hr (2-8°C)	5.7	26.9
1 wk (2-8°C)	3.4	56.4

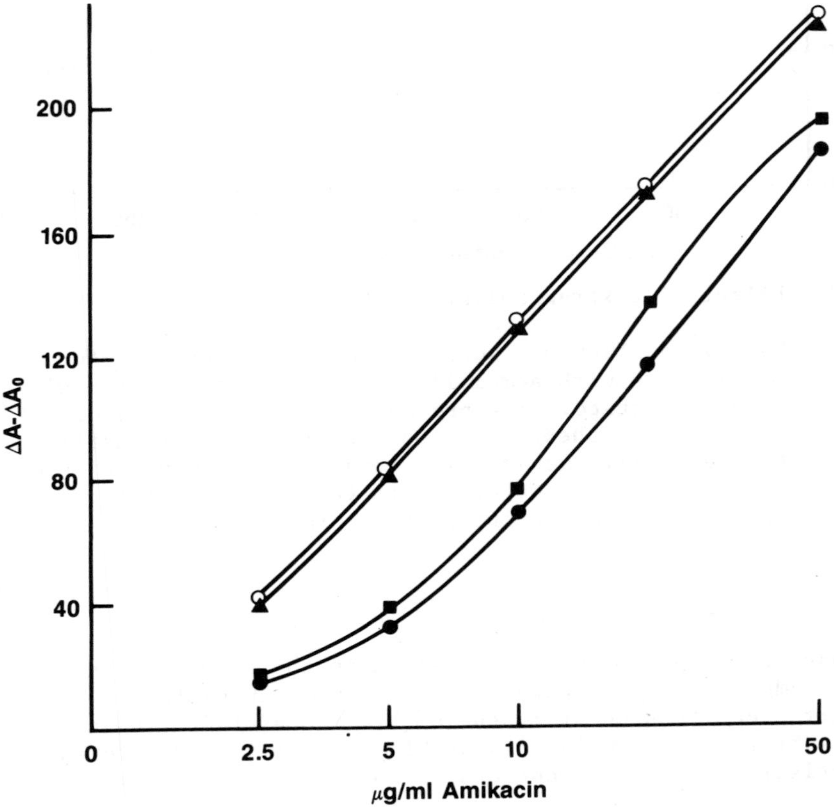

Fig. 9. The Amikacin Assay with Amikacin and Tobramycin Antisera.

The X01 amikacin conjugate was assayed in the standard Emit® protocol (Table 1) with tobramycin antisera.
○-○ 30 sec. sample and tobramycin antibody incubation.
▲-▲ No incubation with tobramycin antisera.
■-■ 30 sec. sample and amikacin antibody incubation.
●-● No incubation with amikacin antibody.

found that the time dependence is a function of the specific antibody used as shown in Table 4. The data indicates that 30 sec incubation does not change response using polyclonal or monoclonal antibody in the absence of drug. In the presence of 10 μg/ml amikacin only the polyclonal antisera shows time-dependent assay response. Therefore, by selecting the appropriate antibody it is possible to avoid the time effect.

A kanamycin assay can also be constructed using tobramycin, amikacin or kanamycin reagents as indicated in Fig. 10. The best response is achieved when the enzyme and antibody are not matched. The amikacin pair shows the poorest performance most likely because the antibody recognizes the amikacin side chain more than antigenic determinents shared with kanamycin which does not have a long side chain. Tobramycin is more like kanamycin than amikacin, however we can speculate that the tobramycin antibody may bind more tightly to the tobramycin conjugate than to free kanamycin and thus competition for the kanamycin drug is less favorable. Some evidence exists for this theory (Deleide et al., 1979). In the phenobarbital and phenytoin assays, the maximum enzyme inhibition achievable, which is related to the optimal assay response, occurs when the chain length of the hapten and immunogen derivatives are unequal.

Figure 11 shows the effect of antibody concentration on the kanamycin/tobramycin assay response. Because the response curves at different drug concentrations are not parallel, compromise between sensitivity at low and high kanamycin concentrations is made. The optimal load point would be 6 μl Ab/test for this formulation.

Performance

Performance of the Emit® aminoglycoside assays has been extensively documented in the literature. Recent studies include gentamicin (Matzke et al., 1982 and Radcliffe et al., 1981); tobramycin (Woo et al., 1982); and amikacin (Fukuchi et al., 1984). Within-run precision and day-to-day precision are generally < 10% and often < 5%. Correlation coefficients were > 0.9.

Reagent Stability

Many factors affect reagent stability including the quality of water used to reconstitute lyophilized reagents, storage conditions, and condition of the pipetting equipment used to sample or dilute the reagents. Several reports (Bach and Larsen, 1980; Gorsky, 1983; and Gorsky et al., 1983) indicate that standard curve stability can be extended as long as 16 days provided a new negative calibrator rate is measured and subtracted from the remaining calibrator rates. Calibration frequency can then be reduced. Reagents are provided in lyophilized form to enhance stability.

Table 4. Polyclonal vs. Monoclonal Antibody Protocol Effect.

Antibody		Sample and Antibody Incubation (ΔA) 0	Sample and Antibody Incubation (ΔA) 30 sec.	$\Delta A_{30}-\Delta A_0$
Polyclonal	Neg	460	462	2
	10 μg/ml Amikacin	695	713	18
4H6 Clone	Neg	421	424	3
	10 μg/ml Amikacin	765	769	4
12C6 Clone	Neg	417	423	6
	10 μg/ml Amikacin	785	787	2

ΔA = Change in absorbance in the standard protocol.
ΔA_{30} = Change in absorbance after 30 sec.
ΔA_0 = Change in absorbance of the background sample containing no drug (Neg).

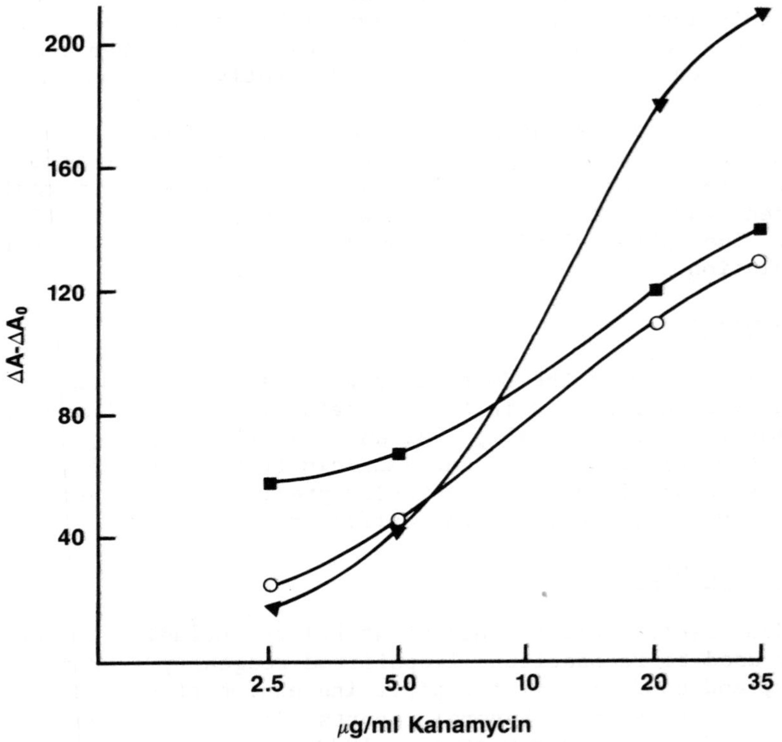

Fig. 10. Kanamycin Assay Calibration Curves.

■–■ Amikacin enzyme and antibody.
○–○ Tobramycin enzyme and antibody.
▼–▼ Kanamycin enzyme and tobramycin antibody.

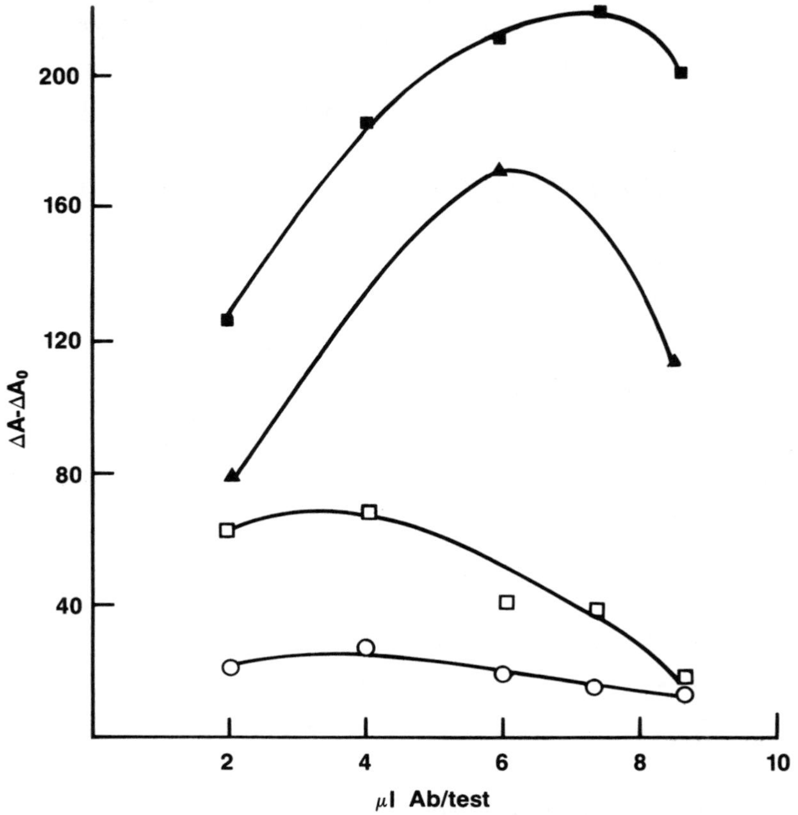

Fig. 11. Kanamycin Antibody Loading Study

Response as a function of antibody titer was studied at 2.5 μg/ml (○-○), 5 μg/ml (□-□), 20 μg/ml (▲-▲), and 35 μg/ml (■-■) kanamycin using the kanamycin enzyme conjugate and the tobramycin antisera.

Quantitation of reconstituted reagents is guaranteed for at least 12 weeks.

The quantitative unit dose (Qst) formulations for gentamicin and quinidine show standard curve stability for greater than one month (Table 5).

Data Handling

Data handling schemes have been devised over the years to provide accurate calibration and validation of calibration (Wellington,

Table 5. Control Quantitation over Time:
QST Quinidine and Gentamicin Assays

Assay Day / Control	Quinidine (Q)	Gentamicin (G)
Q / G	3.0 μg/ml	6.0 μg/ml
1	3.0	5.7
8	2.9	6.0
15	3.1	6.4
22	2.9	6.6
36	2.9	6.1
56 / 55	2.8	5.8
73 / 66	3.3	5.9
92 / 79	3.3	5.9
101 / 108	3.2	5.9
134 / 133	3.1	5.9
141	3.1	-
Total n*	28	25
$\overline{X}$	3.1	6.2
S.D.	0.28	0.31
C.V.%	7.4%	4.9%

*Multiple points run on more than 1 day.

1980). Several computer systems have become available to simplify these procedures using the HP 85 programmable desk top computer. Table 6 below lists the features available with each system. The Hewlett-Packard system was introduced to allow convenient software modification merely by exchanging tapes. The immediate advantage was the introduction of a a new calibration scheme that requires duplicate analyses of only the No. 1 and No. 5 calibrators, the lowest and highest calibrators provided for each assay. Utilizing these four data points and the mathematical constants derived at Syva describing the curve shape, a clinician can be assured that sample quantitation is equivalent to quantitation derived from a full standard curve in duplicate. The addition of control track allows a laboratory to monitor control quantitation more accurately and conveniently.

Batch Assays

To achieve the sensitivity required to measure low level analytes, sample, antibody incubation, and reaction time must be increased. Discrete sample analysis becomes inconvenient for large numbers of samples. Therefore batch assays have been developed for

Table 6. Data Handling Systems Available for SYVA Assays

System	Computer	Attributes
CP-5000	CP-5000	Compares calibrator (Neg, Neg. 1, 2, 3, 4, 5) response to standard curve derived from constants (linear least squares analysis) provided with each kit. If calibrators are within the tolerances specified, the curve is calculated from the nonlinear least squares regression analysis, i.e., derives curve shape directly from the data.
CP-5000 PLUS	CP-5000	Maintains nonlinear least squares option when constants are not available. Linear regression analysis is performed using constants supplied with each kit. The default to nonlinear analysis is eliminated. In addition, a reduced calibration scheme (1, 1, 5, 5) is optional for the user.
LP-6000	HP 85A	Provides the same data handling capabilities of the CP-5000 PLUS using an HP 85A computer. Adds Curve Track and Control Track options. Standard curve stability and control quantitation can be monitored within limits set by the user. New curve is run only when control(s) exceeds these limits.
LP-6500	HP 85B	Addition of electronic disk increases speed at which results are calculated. Increased memory consolidates data handling for all G6PDH assays on one tape.

digoxin, cortisol, and thyroxine. The digoxin and cortisol assays use G6PDH as the label and the thyroxine assay uses MDH as the label (Jaklitsch et al., 1976). Typically these assays require preparation of a diluted working antibody reagent, sample pretreatment to denature proteins allowing the drug/hormone to compete with the enzyme-labeled species for antibody binding, and 30 min measurement time. Reagents are added with a diluter or repeating dispenser and the optical density measurements are recorded manually although at least one automated protocol for digoxin and thyroxine on the Cobas has been published (Oellerich et al., 1982).

Alternative Technologies

In addition to the methodology whereby an enzyme is used as a label, several alternative homogeneous techniques have been reported. The coenzyme NAD can be labeled with haptens (Carrico et al., 1976). Interference from biological samples has been problematic in these assays. Cox and Buret (1982) report development of a gentamicin assay in which NAD is reacted with gentamicin through a carbodiimide intermediate. Antibody bound gentamicin cannot participate in a cycling reaction in which 6-phosphogluconate is measured by reaction with NADPH catalyzed by 6-phosphogluconate dehydrogenase. NADPH production increases as the concentration of gentamicin in the sample increases. However, a two-hour 30° incubation is required, followed by 3 min. in a boiling water bath to stop the reaction. The absorbance is recorded after an additional 30 min. incubation at 25°.

Ngo and Lenhoff (1980) reported an enzyme modulator mediated immunoassay in which an analyte labelled enzyme modulator alters activity of an indicator enzyme. Anti-analyte antibody bound to either analyte from the sample or labelled analyte modulates enzyme activity. The principle was demonstrated with an assay for 2,4 dinitrofluorobenzenelysine (DNP-lys). An anti-enzyme derivative functions as the modulator. DNP-lys was reacted with anti-horseradish peroxidase antibody (DNP-lys Anti-HRP). Anti-DNP antibody modulates the inhibition of horseradish peroxidase by the DNP-lys Anti-HRP conjugate.

Wu et al., (1982 and 1983) have reported quantitative, automated assays for phenobarbital, theophylline, tobramycin, and gentamicin using homogeneous immunoprecipitation. In this assay gentamicin from sample and a gentamicin-albumin conjugate compete for antibody binding. Precipitation of antibody/gentamicin-albumin complex is measurable after 3 min. incubation.

Future Directions

An emerging trend is the conversion of tedious batch protocols to more rapid convenient methods by introduction of more sensitive assay techniques or automation of the procedures required to measure analytes found in low concentration in biological fluids.

Applications of homogeneous enzyme immunoassays to the multitude of new, automated analyzers available is clearly the future direction for this technology. In the hospital laboratories the demand for walk-away automation is likely to increase as cost containment policies lead to staff reductions. These same policies are forcing a change in testing patterns from screening panels to individual tests selected from a large menu. Instruments capable of this "random access" feature are now available to the clinical laboratory.

Standardization costs can be modified either by reducing the number of calibrators or the frequency of calibration. Some of the approaches that can be investigated to achieve this objective are increasing reagent stability and/or development of mathematical models to predict the shape of the calibration curve or linearize it so that fewer assays are required for calibration.

SUMMARY

Separation-free homogeneous enzyme immunoassay for haptens is reviewed with emphasis on development of aminoglycoside assays. Alternative research techniques are described and future directions projected.

ACKNOWLEDGEMENTS

The author wishes to acknowledge the assistance of Teri Soares and Karen Lockman in preparation of this manuscript.

REFERENCES

Bach, P.R. and J.W. Larsen (1980). Stability of standard curves prepared for EMIT homogeneous enzyme immunoassay kits stored at room temperature after reconstitution, Clin. Chem. 26: 652-654.

Carrico, R.J., J.E. Christner, R.C. Boguslaski and K.K. Yeung (1976). A method for monitoring specific binding reactions with cofactor labeled ligands, Anal. Biochem. 72: 271-282.

Centofanti, J., A. Monte, D. Grandsaert, B. Chung, M. Ruggieri, R. Schneider, and D. Kabakoff (1981). The EMIT®-st™ drug detection system, J. Clin. Chem. Clin. Biochem. 19: 631.

Chang, J., S. Gotcher, J.B. Gushaw (1982). Homogeneous enzyme immunoassay for theophylline in serum and plasma, Clin. Chem. 28: 361-367.

Cox, C. and J. Buret (1982). Spectrophotometric homogeneous immunoassay of serum gentamicin with oxidized nicotinamide adenine dinucleotide as label, Anal. Chem. 54: 1862-1865.

Deleide, V., V. Dona, and R. Malvano (1979). Homogeneous enzyme-immunoassay for anticonvulsant drugs: Effects of hapten-enzyme bridge length, Clin. Chem. Acta 99: 195-201.

Dols, J.L. and A.P. van Zanten (1981). Gentamicin assay: comparison of an adapted EMIT method and an RIA method, Ann. Clin. Biochem. 18: 236-239.

Fukuchi, H., M. Yoshida, S. Tsukiai, T. Kitaura, and T. Konishi (1984). Comparison of enzyme immunoassay, radioimmunoassay, and microbiologic assay for amikacin in plasma, Am. J. Hosp. Pharm. 41: 690-693.

Gibbons, I., C. Skold, G.L. Rowley, and E.F. Ullman (1980). Homogeneous enzyme immunoassay for proteins employing β-galactosidase, Anal. Biochem. 102: 167.

Gorsky, J.E., P.R. Bach, and W. Wong (1983). Improved sensitivity of aminoglycoside enzyme immunoasays, Clin. Chem. 29: 1994-1995.

Gorsky, J.E. (1983). EMIT assay curve stability, Clin. Chem. 29: 1690.

Hamlin, C.R., and P.A. Sullivan (1984). Adaptation of EMIT reagents to the Cobas Bio centrifugal analyzer, Clin. Chem. 30: 314-315.

Iosefsohn, M., R.L. Boeckx, and J.M. Hicks (1981). Adaptation of the EMIT gentamicin and tobramycin procedure to the IL microcentrifugal analyzer, Ther. Drug Mon. 3: 365-370.

Jaklitsch, A.P., R.S. Schneider, R.J. Johannes, J.E. Lavine, and G.L. Rosenberg (1976). Homogeneous enzyme immunoassay for T_4 in serum, Clin. Chem. 22: 1185.

Jaklitsch, A.P., C.B. Gagne, D.S. Kabakoff, D.K. Leung, P. Singh, and W. Protzmann (1980). Measurement of sisomicin concentration in fermentation broths by homogeneous enzyme immmunoassay, Am. Chem. Soc. Meeting MBT 23, Las Vegas, Nev., American Chemical Society.

Krogsted, D.J., G.G. Granich, P.R. Murray, M.A. Pfaller, and R. Valdes (1982). Heparin interferes with the radioenzymatic and homogeneous enzyme immunoassays for aminoglycosides, Clin. Chem. 28: 1517-1521.

Lockwood, W.R. and J.D. Bower (1973). Tobramycin and gentamicin concentrations in the serum of anephric patients, Antimicrob. Agents Chemother. 3: 125-129.

Matzke, G.R., C. Gwizdala, J. Wery, D. Ferry, and R. Starnes (1982). Evaluation of three gentamicin serum assay techniques, Ther. Drug Mon. 4: 195-200.

Matzke, G.R., R. Piveral, C.E. Halstenson, and P.A. Abraham (1984). Heparin interferes with the tobramycin serum concentration determinations by EMIT, Drug Intell. Clin. Pharm. 18: 517-519.

Ngo, T.T. and H.M. Lenhoff (1980). Enzyme modulators as tools for the development of homogeneous enzyme immunoassays, FEBS Lett. 116: 285-288.

Oellerich, M., R. Haeckel, and H. Haindl (1982). Evaluation of EMIT adapted to the (Cobas) Bio centrifugal analyzer, J. Clin. Chem. Clin. Biochem. 20: 765-772.

Ou, C.N., V.L. Frawley, and G.J. Buffone (1981). Optimization of EMIT reagent system using a Cobas-Bio centrifugal analyzer, J. Anal. Toxicol. 5: 249-252.

Ratcliff, R.M., C. Mirelli, E. Moran, D. O'Leary, and R. White (1981). Comparison of five methods for the assay of serum gentamicin, Antimicrob. Agents Chemother. 19: 508-512.

Rowley, G.L., K.E. Rubenstein, S.T. Weber, and E.F. Ullman (1976). Mechanism by which antibodies inhibit hapten-glucose-6-phosphate dehydrogenase, Am. Chem. Soc. Meeting BIOL 151, Washington, D.C., American Chemical Society.

Rubenstein, K.E., R.S. Schneider, and E.F. Ullman (1972). "Homogeneous" enzyme immunoassay. A new immunochemical technique, Biochem. Biophys. Res. Commun. 47: 846-851.

Singh, P.S., D.K. Leung, G.L. Rowley, C. Gagne, and E.F. Ullman (1980). A method for controlled coupling of amino compounds to enzymes: a homogeneous enzyme immunoassay for gentamicin, Anal. Biochem. 104: 51-58.

Tindula, R.J., P.J. Ambrose, and A.F. Harralson (1983). Aminoglycoside inactivation by penicillins and cephalosporins and its impact on drug-level monitoring, Drug Intell. Clin. Pharm. 17: 906-908.

Ullman, E.F., R.A. Yoshida, J.I. Blakemore, W. Eimsted, R. Ernst, R. Leute, D.K. Leung, E.T. Maggio, R. Ronald, and B. Sheldon (1979). Mechanism of inhibition of malate dehydrogenase by thyroxine derivatives and reactivation by antibodies. Basis for a homogeneous enzyme immunoassay for thyroxine, Biochim. Biophys. Acta. 567: 66-74.

Wellington, D. (1981). Mathematical treatments for the analysis of enzyme-immunoassay data. In Enzyme-Immunoassay (E.T. Maggio, ed.) CRC Press, Boca Raton, Fla.

Whiting, B., A.W. Kelman, S.M. Bryson, F.H.M. Derkx, A.H. Thomson, G.H. Fotheringham, and S.E. Joel (1984). Clinical Pharmacokinetics: a comprehensive system for therapeutic drug monitoring and prescribing, Br. Med. J. 288: 541-545.

Witebsky, F.G., C.A. Sliva, S.T. Selepak, M.E. Ruddel, J.D. MacLowry, E.E. Johnson, and R.J. Elin (1983). Evaluation of four gentamicin and tobramycin assay procedures for clinical laboratories, J. Clin. Microbiol. 18: 890-894.

Woo, J., M.A. Longley, and D.C. Cannon (1982). Homogeneous enzyme immunoassay for tobramycin evaluated and compared with a radioimmunoassay, Clin. Chem. 28: 1370-1374.

Wu, J.W., C. Bunyagidj, S. Hoskin, S.M. Riebe, J. Aucker, K. White, and S.P. O'Neill (1983). Quantitation of haptens by homogeneous immunoprecipitation. 2. Centrifugal analysis for tobramycin, phenobarbital, and theophylline in serum, Clin. Chem. 29: 1540-1542.

Wu, J.W., S. Hoskin, S.M. Riebe, J.E. Gifford, and S.P. O'Neill (1982). Quantitation of haptens by homogeneous immunoprecipitation. 1. Automated analysis of gentamicin in serum, Clin. Chem. 28: 659-661.

ENZYME MODULATOR AS LABEL IN SEPARATION-FREE IMMUNOASSAYS: ENZYME MODULATOR MEDIATED IMMUNOASSAY (EMMIA)

T.T. Ngo

Department of Developmental and Cell Biology
University of California
Irvine, CA 92717

INTRODUCTION

Enzyme modulator mediated immunoassay (EMMIA) is a separation-free enzyme amplified immunoassay. The assay uses an enzyme modulator (M) as the tag to label a ligand analyte (L). Such stable covalent enzyme modulator-ligand conjugates (M-L) is capable of modulating the activity of an indicator enzyme by either causing a significant inhibition of the enzyme activity or a dramatic activation.

As an analog of the ligand, M-L is able to compete effectively with the ligand for a fixed number of antibodies to L. Furthermore, M-L can also lose its modulating activity when it combines with the antibody to L. In a competitive binding assay, the concentration of L determines the distribution of M-L between the antibody bound and unbound, free M-L. The unbound, free M-L is functional, i.e. capable of modulating the activity of the indicator enzyme, whereas the bound M-L is non-functional, i.e. not capable of modulating the enzyme activity (Ngo, 1983).

Enzyme modulators are defined as compounds capable of modifying the catalytic activity of an enzyme by either inhibiting or activating the enzyme activity. The enzyme activity can be modified by the modulator through either non-covalent, reversible or covalent, irreversible interactions. Non-covalent, reversible modulators include reversible enzyme inhibitors (Webb, 1963) activators (Wong, 1975), allosterics effectors (Stadtman, 1970), antibodies to some enzymes (Iwert et al., 1967; Cinader, 1976; Arnon, 1977) and tight binding inhibitors and transition-state analogs (Wolfenden, 1969; Wolfenden, 1972; Leinhard, 1973; Ngo and Tunnicliff, 1981). Covalent, irreversible modulators include active-site directed irreversible enzyme inhibitors (Baker, 1967; Shaw, 1980), k_{cat}, mechanism-based, suicide, enzyme-activated irreversible inhibitor (Rando, 1974; Abeles, 1978), and enzyme catalyzed post-translational enzyme modifications, e.g. enzymatic phospharylation or dephosphorylation of enzymes (Krebs, 1972).

Principle of enzyme modulator mediated immunoassay (EMMIA)

The ligand (L) of the test sample competes with M-L for a fixed and limited amount of antibody to L (Ab) (Scheme 1, reactions I and II). It can be seen in Scheme 1, when L is absent, reaction I would not take place, therefore, allowing M-L and Ab to combine through reaction II. The antibody bound M-L (Ab: ML) is unable to modify the activity of the indicator enzyme (E). As the concentration of analyte (L) increases, however, it would compete more successfully with M-L for Ab to L (reaction I dominates), leaving more M-L free to combine with more indicator enzymes (Scheme 1, reaction III) and thereby increasingly modifying the enzyme activity. Therefore, the amount of substrate (S) that is transformed by the enzyme into product (P) within a unit time (Scheme 1, reaction IV) will be accordingly modified. Depending on the properties of the modulator, the enzyme activity will be increased by an activator and decreased by an inhibitor. Thus, for EMMIA based on an enzyme activator, the standard curve will show increases in enzyme activity with increasing L concentrations. Whereas for EMMIA developed with an enzyme inhibitor, the standard curve will show decreasing enzyme activity with increasing L concentrations (Ngo, 1983).

MATERIALS AND METHODS

Strategies for developing EMMIA

The development of EMMIA begins by selecting a suitable enzyme which satisfies the following requirements: (1) enzyme activity can be easily measured; (2) high turnover rate, therefore providing a great signal amplification; (3) inexpensive, available in highly purified form and in large quantity; (4) stable under assay and storage conditions and (5) no such enzyme activity in test samples. If the enzyme is present in the test samples, such an enzyme can still be used, provided its activity is not expressed or can be supressed under assay conditions without affecting the indicator enzyme.

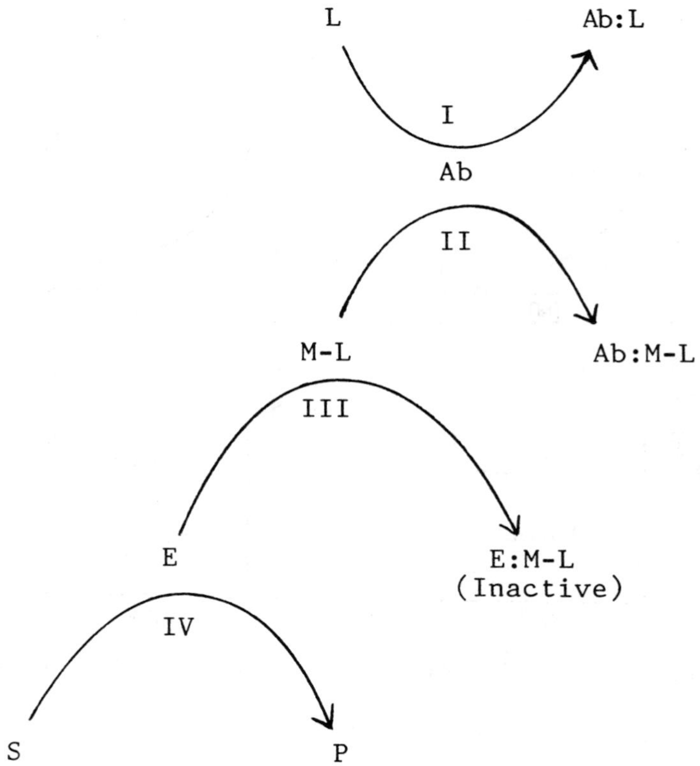

Scheme 1. Principle of an enzyme modulator mediated immunoassay (EMMIA): over-all reactions.

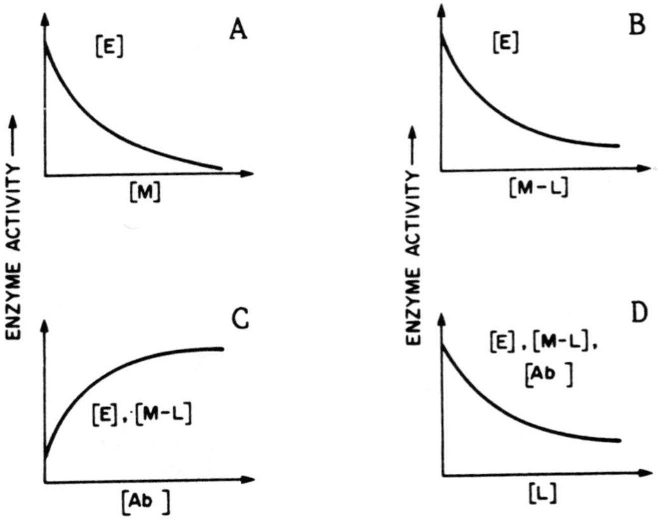

Scheme 2. Experimental stages in developing an EMMIA.

The next step in developing an EMMIA is the selection of a modulator having the following characteristics: (1) capable of modifing the activity of an indicator enzyme at low modulator concentrations. Examples of such modulators are enzyme inhibitors with high binding affinity, i.e. low Ki value; (2) absence of such a modulator activity in the test samples; (3) retention of substantial portion of modulator activity after covalent linking to ligands; (4) abrogation of the modifying activity of a modulator-ligand conjugate (M-L) upon binding with an antibody to L and (5) for irreversible modulators, the reaction must be rapid and specific to the indicator enzymes only and the covalent enzyme modulator adducts must be stable under the assay conditions and within the assay duration.

After the minimum detectable concentration of indicator enzyme has been determined, then the minimum concentration of modulator capable of causing at least 80% inhibition of the enzyme activity (assuming the modulator is an enzyme inhibitor) can be determined in experiments where varying amounts of modulator are mixed with a fixed concentration of enzyme (Scheme 2,A). From a similar experiment, the necessary concentration of modulator-ligand conjugate (M-L) can be determined. The modifying activity of M-L on the activity of the indicator enzyme is generally less than that of its virgin, unmodified counterpart, i.e. the unliganded modulator (Scheme 2,B). The next crucial question to be answered is whether the binding of a specific antibody to L to M-L can nullify or substantially decrease the modifying action of M-L. In other words, can the complexation of an L specific antibody with an M-L prevent M-L from inhibiting the activity of the indicator enzyme? By holding the concentrations of enzyme and M-L constant, the enzyme activity would be minimal in the absence of the specific antibody to L; as the concentration of the antibody increases, however, the activity of the indicator enzyme increases proportionally (Scheme 2,C). Once such phenomena were demonstrated, the establishment of a standard curve would be relatively simple. The standard curve can be obtained by keeping the concentrations of enzyme, M-L and Ab constant. The only variable is the concentration of the ligand, L (Scheme 2,D).

Preparation of modulator-ligand conjugates

In general, the ligand is directly linked to a modulator. So far, only enzyme inhibitors have been used as modulators. In EMMIA for a model ligand, 2,4-dinitrophenyllysine, Ngo and Lenhoff (1980) reacted 2,4-dinitrofluorobenzene with antibody to horseradish peroxidase (Ab_{pod}), the modulator. The resulting

2,4-dinitrophenylated anti-horseradish peroxidase (DNP-Ab_{pod}) is the M-L and the indicator enzyme is horseradish peroxidase. Finley et al. (1980) and Blecka et al. (1983) reported an EMMIA using an irreversible enzyme inhibitor as the modulator. The indicator enzyme was an acetylcholinesterase and a phosphonate inhibitor was used as the modulator. Thus in EMMIA for thyroxine, a covalent conjugate of thyroxine-phosphonate inhibitor was used. EMMIA for theophylline, using theophylline conjugated irreversible phosphonate inhibitor of acetylcholinesterase, has recently been developed (Blecka et al, 1983).

Place et al. (1983) on the other hand developed an EMMIA for thyroxine using a tight binding inhibitor of dihydrofolate reductase, methotrexate, as the modulator. Methotrexate was covalently linked through its γ-carboxyl of the glutamyl group to the α-amino of thyroxine by using isobutyl chloroformate mixed anhydride method. An EMMIA for diphenylhydantoin (Dilantin), an anti-convulsant drug, has recently been developed by using diphenylhydantoin linked avidin as the modulator-ligand conjugate and pyruvate carboxylase as the indicator enzyme (Bacquet and Twumasi, 1984). An N-hydroxysuccinimidyl ester of diphenylhydantoin-acetic acid was reacted with avidin to form a covalent avidin-diphenylhydantoin conjugate. During the coupling reaction 4-hydroxyazobenzene-2-carboxylic acid was included to block and therefore protect the biotin-binding sites from being reacted with the activated ligand. A novel EMMIA for thyroxine using complementing ribonuclease S-peptide as the modulator has been reported by Gonnelli et al. (1981). The S-peptide was obtained by enzymatic cleavage of the native ribonuclease. In this assay two thyroxinyl groups were covalently linked to one S-peptide chain.

RESULTS

Three distinct classes of modulators have been used in developing EMMIA: (1) high molecular weight (macromolecular) enzyme inhibitors (Ngo and Lenhoff, 1980; Ngo, 1983; Bacquet and Twumasi, 1984) (2) low molecular weight irreversible enzyme inhibitors (Finley et al., 1980; Blecka et al., 1983) and (3) low molecular weight reversible enzyme inhibitors (Gonnelli et al., 1981; Place et al., 1983). Examples for each of the aforementioned EMMIA are described below.

EMMIA using macromolecular inhibitors

The first EMMIA using a macromolecular enzyme inhibitor was developed by Ngo and Lenhoff (1980). Antibody to horseradish peroxidase was used as a specific and potent inhibitor of horseradish peroxidase. Many antibodies to enzymes are highly potent and specific enzyme inhibitors (Cinader, 1976; Arnon, 1977). There are several advantages in using a specific inhibitory antienzyme antibody as the enzyme modulator tag in developing an EMMIA: (1) because enzymes are antigenic macromolecules, antibodies to enzymes can usually be obtained by simply injecting the enzyme solution to animals; (2) highly specific and potent inhibitory monoclonal antibodies to enzymes can be obtained reproducibly by hybridoma technology (Kennett et al., 1980; Sevier et al., 1981); (3) analogs of the ligand analyte bearing either amino, carboxyl or sulfhydryl groups can be readily linked via carboxyl, amino, tyrosine or sulfhydryl groups of the antibody directly or indirectly through crosslinking reagents; (4) covalent linking of ligand analogs to an antibody does not significantly affect the immunological property of either the antibody or that of the ligand; (5) antibodies are stable chemicals under conditions normally encountered in bioanalytical laboratories and (6) small fragments of an antibody such as the Fab or $F(ab)_2$ fragments can be prepared readily.

The model EMMIA system developed for DNP-lysine measurement utilized an anti-horseradish peroxidase as the enzyme modulator and a horseradish peroxidase as the indicator enzyme (Ngo and Lenhoff, 1980). The modulator-ligand conjugate which consisted of a covalently-linked DNP-anti-horseradish peroxidase (DNP-Ab_{pod}), showed substantial inhibition on the horseradish peroxidase (Fig. 1). Furthermore, it was demonstrated that an anti-DNP antibody (Ab_{DNP}) was able to nullify the inhibition of enzyme activity brought about by DNP-Ab_{pod}. Figure 2 showed that in the presence of a constant amount of horseradish peroxidase and DNP-Ab_{pod}, the peroxidase activity increases with increasing amount of anti-DNP antibody. The standard curve for DNP-lysine was developed using constant amount of the indicator enzyme (horseradish peroxidase), antibody to DNP (Ab_{DNP}) and DNP labeled anti-horseradish peroxidase antibody (DNP-Ab_{pod}). Micromolar concentration of DNP-lysine was determined in less than 10 min (Fig. 3).

Avidin (molecular weight = 66,000), a naturally occurring protein that binds biotin tightly, also possesses advantages similar to those cited for an antibody. An EMMIA using avidin 5,5-diphenylhydantoin (DPH) conjugate as the inhibitory M-L and pyruvate carboxylase as the indicator enzyme has been developed for a separation-free enzyme immunoassay for 5,5-diphenylhydantoin (Bacquet and Twumasi, 1984). The avidin-DPH conjugate and DPH from test samples competed for a limited amount of anti-DPH antibodies. The complexation of anti-DPH antibodies with

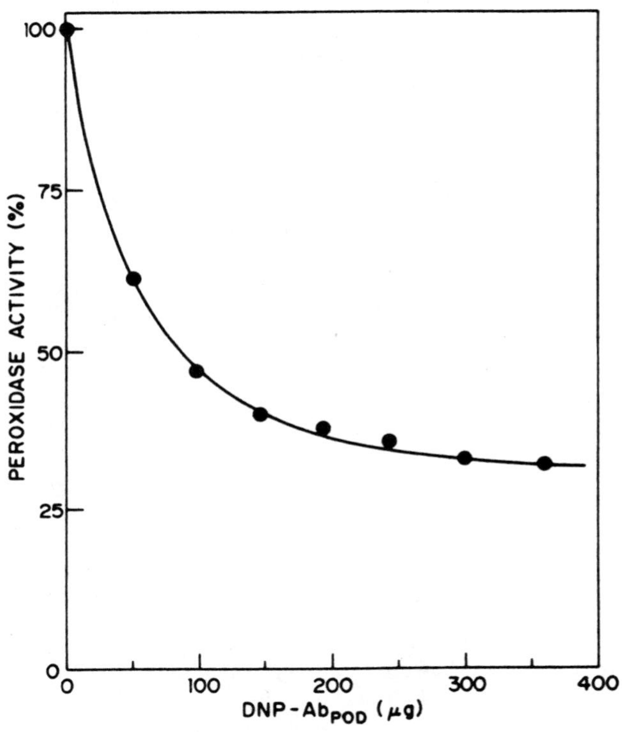

Fig. 1. The inhibition of horseradish peroxidase by DNP-labeled anti-horseradish peroxidase antibody (DNP-Ab$_{pod}$). (From Ngo and Lenhoff, 1980, with permission of the publisher.)

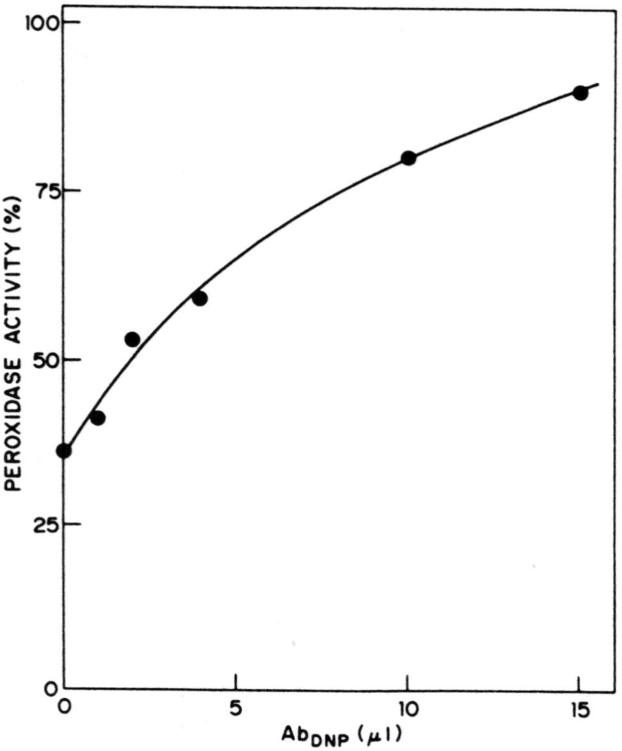

Fig. 2. Reversal by anti-DNP antibody (Ab_{DNP}) of the inhibition of horseradish peroxidase brought about by DNP-Ab_{pod}. (From Ngo and Lenhoff, 1980, with permission of the publisher.)

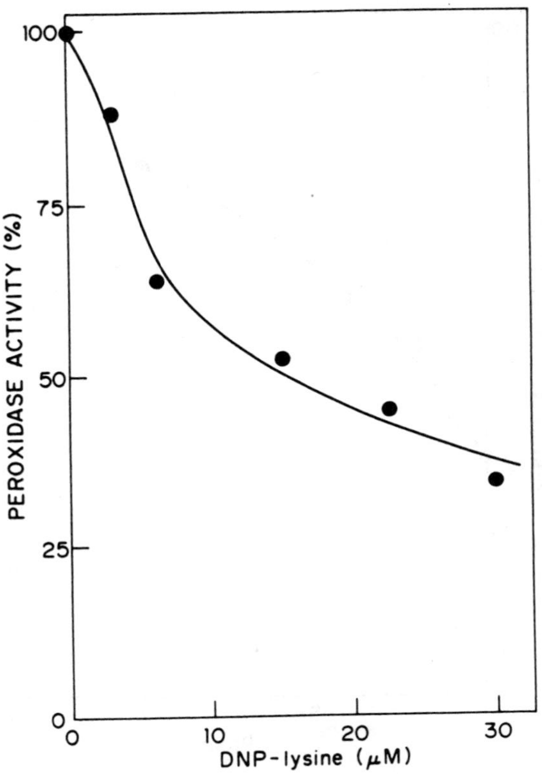

Fig. 3. Standard calibration curve for DNP-lysine using EMMIA. (From Ngo and Lenhoff, 9180, with permission of the publisher).

avidin-DPH conjugates either sterically or physically prevented the indicator enzyme from being inactivated by avidin-DPH conjugates. The presence of high concentration of DPH resulted in less avidin-DPH conjugate being complexed by anti-DPH antibodies. This, in turn, resulted in more indicator enzymes being inactivated by the free avidin-DPH conjugates.

EMMIA using low molecular weight reversible enzyme inhibitor

An EMMIA using methotrexate (molecular weight: 454.5) conjugated ligand has been developed (Place et al., 1983). The indicator enzyme in that EMMIA was dihydrofolate reductase (DHFR). Methotrexate (MTX) is a very potent reversible inhibitor of DHFR enzyme. Fifty percent of the enzyme is inhibited by nanomolar quantities of methotrexate. EMMIA for thyroxine (T_4) was developed using MTX labeled T_4 conjugate. The conjugate was prepared by the covalently linking γ-carboxyl group of the glutamyl residue of methotrexate to α-amino group of T_4. The MTX-T_4 conjugate is also a potent inhibitor of DHFR. MTX-T_4 conjugate, at 50nM was able to inhibit more than 80% of the enzyme activity. However, when MTX-T_4 was incubated with anti-T_4 antibody, it became less and less inhibitory with increasing concentrations of anti-T_4 antibody. Adding increasing concentrations of T_4 to solutions containing fixed concentrations of MTX-T_4, anti-T_4 antibody and DHFR resulted in decreasing enzyme activity.

The high affinity binding of MTX to dihydrofolate reductase is a very desirable and attractive feature of MTX-DHFR system for developing an EMMIA. The hydrophobic nature of the MTX due to the pteridine and aminobenzoyl rings, however, represent a serious drawback in using MTX as a modulator of an EMMIA. It was clearly shown that MTX-T_4 conjugate non-specifically bind to antibody to theophylline. As much as 50% of MTX-T_4 conjugate was bound non-specifically by anti-theophylline antibody (Place et al., 1983).

EMMIA using low molecular weight irreversible enzyme inhibitor

Finley et al. (1980) and Blecka et al. (1983) used a phosphonate derived, irreversible inhibitor of acetylcholinesterase as the modulator for developing EMMIA for thyroxine (T_4) and theophylline (Theo).

Both inhibitor-T_4 and inhibitor-Theo conjugates, when not bound to a specific antibody, were potent inhibitors of acetylcholinesterase. However, when these conjugates were bound to their respective antibodies, i.e. anti-T_4 and anti-Theo antibodies, they became non-inhibitory and therefore allowing the enzyme to hydrolyze the substrate acetyl-B-(methylthio)choline iodide. The product of enzymatic hydrolysis, β-(methylthio) choline, when reacted with 5,5'-dithiobis(2-nitrobenzoate), resulted in forming intensely yellow color of 5-thio-2-nitrobenzoate (λmax at 415 nM), Serum pseudocholinesterase was inhibited by adding N-methylorphenandrine sulfate to the assay solution. Using these irreversible inhibitor-T_4 and Theo conjugates, EMMIA's for T_4 and Theo were developed. The calibration curve for T_4 ranged from 58 - 154 nM and that for Theo ranged from 0 to 222 μM. The recovery of both T_4 and Theo, as measured by EMMIA, was almost 100%. The precision of both assays for T_4 and Theo was good, with inter-assay CV of less than 5% and intra-assay CV of less than 4%. The sensitivity of the assay was 2.56 nM for T_4 and 2.22 μM for Theo.

EMMIA based on S-peptide and S-protein of ribonuclease

It has been known for some time that ribonuclease, an enzyme with 124 amino acid residues, can be enzymatically cleaved into two dissociable inactive polypeptides (Richards and Vithayathil, 1959). One smaller peptide with 20 amino acid residues is called S-peptide and the larger peptide with 104 amino acid residues is termed S-protein. Neither S-peptide nor S-protein by itself has any enzymatic activity. However, 1:1 non-covalent complex of S-peptide and S-protein does have full ribonuclease activity. Gonnelli et al. (1981) made use of the two lysyl residues which were part of the S-peptide as functional groups for linking with T_4 derivatives. Unlike the native S-peptide which can provide full ribonuclease activity, the T_4 labeled S-peptide provides only up to 40% of the full enzyme activity.

The addition of anti-thyroxine antibody to a solution of T_4-S-peptide and S-protein, however, enhances the ribonuclease activity to almost 90% of the full enzyme activity. It appeared that the binding of anti-T_4 antibody with T_4-S-peptide conjugate formed a complex which provided a catalytically more favourable position to the amino acid residues involved in the expression of ribonuclease activity. The action of anti-T_4 antibody is specific and can be counteracted by free T_4.

By adding increasing amounts of T_4 to solutions containing a constant amount of S-protein, T_4-S-peptide and anti-T_4 antibody, decreasing ribonuclease activity was observed. A determination of T_4 can be made by using such a standard curve.

DISCUSSION

Enzyme modulators have been shown to be versatile labels for developing separation-free (homogeneous) enzyme immunoassays. The most useful modulators as labels in enzyme immunoassays must exhibit high affinity in binding to the indicator enzyme and can readily be linked to the ligands to form stable conjugates. Only those modulator-ligand conjugates with highest binding affinity toward the indicator enzymes will have the potential of making the enzyme immunoassay more sensitive.

Several different classes of enzyme modulators are available for developing enzyme immunoassays. For example:

1. Macromolecular modulators which include avidin, a glycoprotein, isolated from egg white. It consists of four identical sub units with combined molecular weight of 66,000. One avidin binds four biotin molecules. The biotin-avidin binding constant is $10^{15}M^{-1}$, which is an extremely high value (Green, 1975). Avidin is capable of inactivating a number of biotin-requiring enzymes and is potentially a good modulator for developing EMMIA. An EMMIA for 5,5-diphenylhydantoin, an anti-convulsant drug, using an avidin-drug conjugate has been developed by Bacquet and Twumasi (1984). The use of anti-enzyme antibody as a macromolecular inhibitor for EMMIA development has also been demonstrated by Ngo and Lenhoff (1980). The advantages of using anti-enzyme antibodies as enzyme modulators in EMMIA are the high affinity binding between the enzyme (antigen) and the anti-enzyme (antibody), and the ability of eliciting and obtaining inhibitory anti-enzyme antibodies for a number of enzymes (Cinader, 1977; Arnon, 1977). The selection and isolation of such antibodies are made easier by the advent of hybridoma technology (Kohler and Milstein, 1975) for producing monoclonal antibodies at commercial scales (Sevier et al., 1981).

2. Low molecular weight, reversible enzyme inhibitors such as methotrexate which is a very potent inhibitor of dihydrofolate reductase (Place et al., 1983). Although methotrexate binds dihydrofolate reductase with very high affinity, its use in developing a practical EMMIA is problematic. Methotrexate is a relatively hydrophobic molecule. It tends to bind to many macromolecules, particularly protein, non-specifically. The methotrexate-dihydrofolate reductase system is further complicated by the instability of the enzyme substrate.

3. Low molecular weight, irreversible enzyme inhibitors such as the phosphonate inhibitor of acetylcholinesterase prove to be good modualtors for EMMIA. This inhibitor has been conjugated to T_4 and theophylline and EMMIA's for T_4 and theophylline have been successfully developed. Both EMMIA's for T_4 and theophylline exhibited acceptable precision and sensitivity (Finley et al., 1980; Blecka et al., 1983).

4. Complementary peptide such as the ribonuclease S-peptide, which form a complementary complex with ribonuclease S-protein with concomittant generation of enzyme activity, has also been used as a modulator for developing EMMIA for T_4 (Gonnelli et al., 1981).

Potent and specific enzyme inhibitors, such as the transition state analogs (Lienhard, 1973; Wolfenden, 1969 and 1972) and the mechanism-based enzyme inhibitors, also known as k_{cat} inhibitors or as suicide enzyme inhibitors (Rando, 1974; Abeles, 1978) are potentially useful modulator-labels for EMMIA.

It is conceivable that some modulator-ligand conjugates can be mass-produced by genetically-engineered micro-organisms. For example, the gene for a single polypeptide chain containing codes for a polypeptide hormone and a ribonuclease S-peptide can be synthesized and incorporated into the genome of a micro-organism by recombitant DNA technology for a large scale production of the modulator ligand conjugate.

REFERENCES

Abeles, R.H., 1978, Suicide enzyme inactivators, in: "Enzyme-Activiated Irreversible Inhibitors," N. Seiler, M.J. Jung and J. Koch-Weser, eds., pp. 1-12., Elsevier North Holland Biomedical Press, Amsterdam.

Arnon, R., 1977, Immunochemistry of Iysozyme, in: "Immunochemistry of Enzymes and Their Antibodies," M.R.J. Galton, editor, pp. 1-28, Wiley, New York.

Bacquet, C and Twumasi, D.Y., 1984, A homogeneous enzyme immunoassay with avidin-ligand conjugate as the enzyme-modulator. Anal. Biochem., 136:487-490.

Baker, B.R., 1967, "Design of Active-Site-Directed Irreversible Enzyme Inhibitors," Wiley, New York.

Blecka, L.J., Shaffar, M., Dworschack, R., 1983, Inhibitor enzyme immunoassays for quantitation of various haptens: A review, in: "Immunoenzymatic techniques," S. Avrameas, P. Druet, R. Masseyeff and G. Feldman, eds., 207-214, Elsevier Science Publishers, Amsterdam.

Cinader, B., 1976, Enzyme-antibody interactions, in: Methods of Immunology and Immunochemistry, Chase M. & Williams C, editors, pp. 313-375. Academic Press, New York.

Finley, P.R., Williams, R.J. and Lichti, D.A., 1980, Evaluation of a new homogeneous enzyme inhibitor immunoassay of serum throxine with use of a bichromatic analyzer. Clin. Chem. 26:1723-1726.

Gonnelli, M., Gabellieri, E., Montagnoli, G. and Felicioli, R., 1981, Complementing S-peptide as modulator in enzyme immunoassay. Biochem. Biophys. Res. Commun. 103:917-923.

Green, N.M., 1975, Avidin, in: Advances in Protein Chemistry, C.B. Anfisen, J.T. Edsall and F.M. Richards, editors, pp. 85-133, Academic Press, New York.

Iwert, M.E., Nelson, N.S and Rust, J.H., 1967, Some immunologic properties of Jackbean urease and its antibody. Archs. Biochem. Biophys. 122:95-104.

Kennett, R.H., McKearn, J.J. and Bechtol, K.B., editors, 1980, Monoclonal Antibodies, Hybridoma: A New Dimension in Biological Analyses, Plenum Press, New York.

Kohler, G. and Milstein, C., 1975, Continuous cultures of fused cells secreting antibody of predefined specificity, Nature, 156:495-497.

Krebs, E.G., 1972, Protein kinases, in: Current Topics in Cellular Regulation, B.L. Horecker and E.R. Stadtman, editors, pp. 99-133, Academic Press, New York.

Ngo, T.T. (1983) Enzyme modulator mediated immunoassay (EMMIA), Int. J. Biochem., 15:583-590.

Ngo, T.T. and Lenhoff, H.M., 1980, Enzyme modulators as tools for the development of homogeneous enzyme immunoassays, FEBS Lett., 116:285-288.

Ngo, T.T. and Lenhoff, H.M., 1982, Enzymes as versatile labels and signal amplifiers for monitoring immunochemical reactions, Molec. Cell Biochem. 44:3-12.

Ngo, T.T. and Tunnicliff, G., 1981, Inhibition of enzymic reactions by transation state analogs: An approach for drug design, Gen. Pharmac. 12:129-138.

Place, M.A., Carrico, R.J., Yeager, F.M., Albarella, J.P. and Boguslaski, R.C., 1983, A colorimetric immunoassay based on enzyme inhibitor method, J. Immunol. Method, 61:209-216.

Richards, F.M. and Vithayathil, P.J., 1959, The preparation of subtilisin-modified ribonuclease and the separation of the peptide and protein components, J. Biol. Chem., 234:1459-1465.

Sevier, D.E., David, G.S., Martinis, J., Desmond, W.J., Bartholomew, R.M. and Wang, R., 1981, Monoclonal antibodies in clinical immunology, Clin. Chem., 27:1979-1806.

Shaw, E.N., 1980, Design of irreversible inhibitors, in: Enzyme Inhibitor as Drugs, M. Sandler , editor, pp. 24-42, University Park Press, Baltimore.

Stadtman, E.R., 1970, Mechanisms of enzyme regulation in metabolism, in: The Enzymes, P.D. Boyer, editor, Vol. 1, pp. 397-459, Academic Press, New York.

Webb, J.L., 1963, Enzyme and metabolic inhibitors, Vols 1-3, Academic Press, New York.

Wolfenden, R., 1969, Transition state analogs for enzyme catalysis, Nature, 223:704-705.

Wolfenden, R., 1972, Analog approaches to the structure of the transition state in enzyme reactions. Acc. Chem. Res., 5:10-18.

Wong, J.T.F., 1975, Kinetics of Enzyme Mechanisms, pp. 39-72, Academic Press, New York.

PROSTHETIC GROUP LABELED ENZYME IMMUNOASSAY

T.T. Ngo

Department of Developmental and Cell Biology
University of California
Irvine, CA 92717

INTRODUCTION

Background

Prosthetic group labeled enzyme immunoassay (PGLEIA) is a separation-free (homogeneous) enzyme mediated immunoassay. The assay depends on: (1) the ability of a stable covalent ligand-prosthetic group conjugate to activate an apoenzyme which by itself is enzymatically inactive; and (2) the ability of anti-ligand antibody to bind ligand-prosthetic group conjugate such that the bound ligand-prosthetic group is incapable of activating the apoenzyme.

The ligand-prosthetic group conjugate in PGLEIA serves two functional roles. First, it serves as a prosthetic group analog capable of activating the apoenzyme. Secondly, it serves as a ligand analog capable of competing with the ligand for ligand binding sites of the anti-ligand antibodies.

PGLEIA has been developed using apoglucose oxidase and ligand-labeled flavin adenine dinucleotide (L-FAD) pair because apoglucose oxidase binding FAD very tightly. A binding constant as high as 10^{10} M for apoglucose oxidase-FAD system has been reported (Okuda, et al., 1979). Furthermore, under some conditions apoglucose oxidase can be stabilized and for some analytes, stable covalent ligand-FAD conjugates can be prepared.

Principles of PGLEIA

A stable covalently linked ligand (L)-enzyme prosthetic group (FAD) conjugate, FAD-L, serves as a ligand analog capable of competing with the analyte ligand (L) for antibodies (Ab) specific to L (Scheme 1, reaction II). The formation of FAD-L:Ab complexes, through reaction II, prevents the FAD moiety of FAD-L to combine with apoglucose oxidase (AG) presumably due to steric hindrance imposed on it by the complex formation. Free FAD-L, i.e. not complexed by antibody, readily combine with AG to form enzymatically active hologlucose oxidase (HG) via reaction III. FAD-L serves as an effective substitute for the prosthetic group. In the absence of L, the analyte, reaction I would not occur; thus the antibody would combine with FAD-L (reaction II) making the FAD of FAD-L moiety incapable of combining with AG (reaction III). Conversely, as the concentration of L increases, it would be able to compete more successfully for Ab (reaction I) leaving more FAD-L free to combie with AG (reaction III), thereby increasing the amount of hologlucose oxidase to catalyze the oxidation of glucose (S) to gluconic acid (P) via rection IV (Ngo and Lenhoff, 1982).

Preparation of apoglucose oxidase

All procedures described for preparing apoglucose oxidase from the holoenzyme involved sulfuric acid dissociation step (Solomon and Levin, 1976; Massey and Mendelsohn, 1979; Ngo and Lenhoff, 1980a; Morris et al., 1981). The most satisfactory procedure (Ngo and Lenhoff, 1980; Morris et al., 1981) involved acidifying glucose oxidase in 30% glycerol solution at -5°C with 10% H_2SO_4 until the solution pH was approximately 1. After separating FAD from apoglucose oxidase by using a desalting gel, the apoenzyme was neutralized to pH 6.5 and then it was mixed with a power of activated charcoal to remove traces of adsorbed residual FAD. Finally the charcoal was removed from the apoenzyme by centrifugation and filtering the apoenzyme solution through a Millipore filter. This solution was stable for several months when stored at 4°C and in the presence of antimicrobial agents, such as 0.1% NaN_3.

Preparation of FAD-L conjugates

The general strategy for synthesizing FAD-L conjugate called for either a direct coupling of a reactive ligand analog to aminohexyl derivative of FAD, of a reactive carboxylic derivative of FAD to an amino derivative of the ligand, or by covalent linking of the ligand to aminohexyl-FAD by using a bifunctional crosslinking agent (Scheme 2).

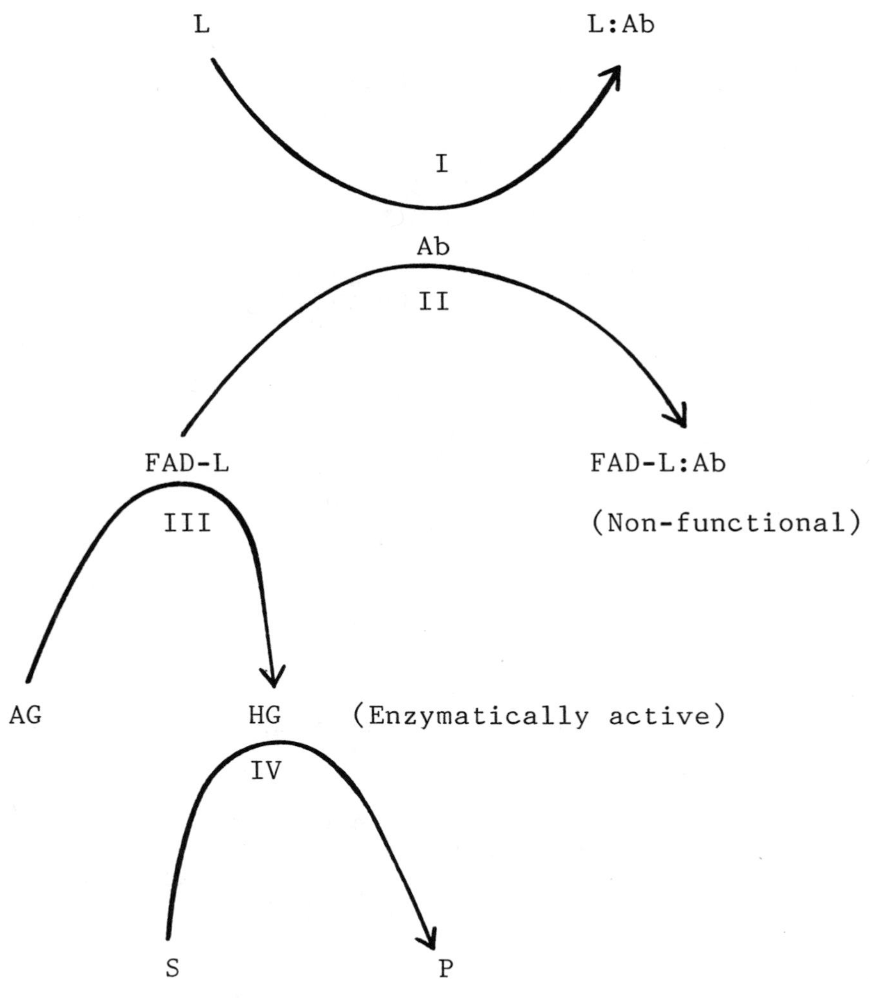
L
L:Ab
I
Ab
II
FAD-L
FAD-L:Ab
(Non-functional)
III
AG
HG
(Enzymatically active)
IV
S
P

Scheme 1

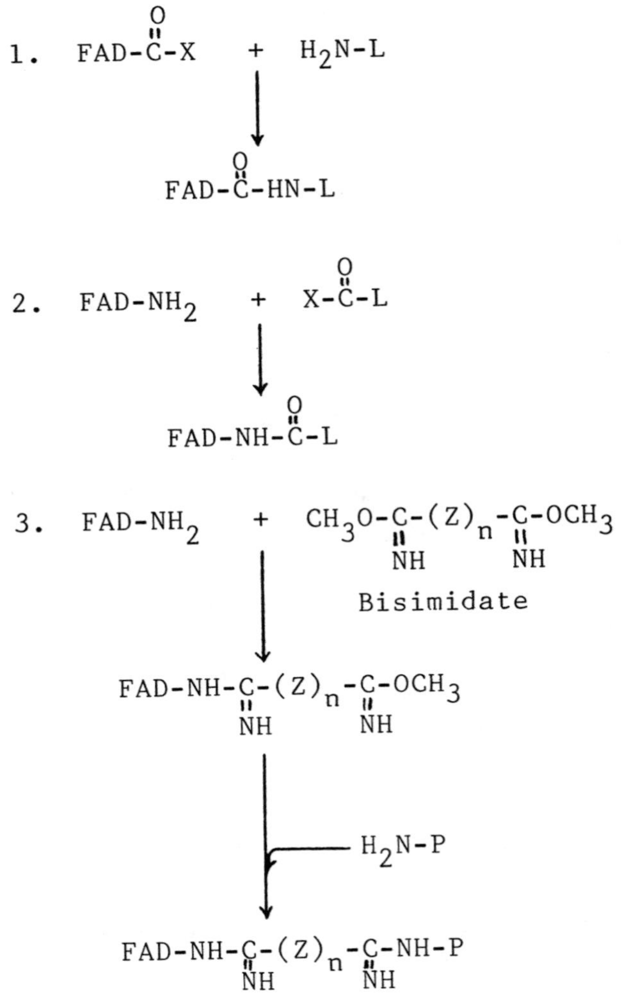

FAD = Flavin adenine dinucleotide
L = Analyte ligand
X = Activated group, e.g. chloride, ester of hydroxysuccinimide or p-nitrophenol
Z = Methylene group
P = Protein

Scheme 2. Routes for synthesis of FAD-L conjugates.

Synthesis of flavin N^6-(6-aminohexyl) adenine dinucleotide (aminohexyl-FAD)

Activation of N^6-(trifluoroacetamidohexyl) adenosine 5'-monophosphate with N,N'-carbonyldiimidazole yielded an imidazolide which was then coupled to riboflavin 5'-monophospate to form flavin N^6-(trifluoroacetamidohexyl) adenine dinucleotide. Alkaline hydrolysis of the latter compound yielded the desired aminohexyl-FAD (Morris et al., 1981).

Synthesis of FAD-theophylline

This was prepared by conjugating aminohexyl-FAD to 8-(3-carboxypropyl)-1,3-dimethylxanthine, a derivative of theophylline (Cook et al., 1976).

Synthesis of FAD-human immunoglobulin G (FAD-IgG)

This conjugate was prepared by crosslinking aminohexyl-FAD to human IgG using dimethylsuberimidate as the crosslinker. Typically, aminohexyl-FAD was first reacted with stoichiometric amount of dimethylsuberimidate in a pH 8-9 buffer for 5-10 minutes at room temperature, then a solution of human IgG buffered at pH 8-9 was added. The buffer used should be free of any amino group. The solution was continuously stirred at room temperature for 1-2 hours. After desalting chromatography and dialysis against phosphate buffer saline, the FAD-IgG was ready for use. FAD-IgG solution, in the presence of 0.1% sodium azide can be stored at 4°C for months (Ngo et al., 1981).

Measurement of glucose oxidase activity

Colorimetrically glucose oxidase can be measured by using horseradish peroxidase coupled rections which measured hydrogen peroxide produced by glucose oxidase. The more popular chromogens for H_2O_2 -- peroxidase reaction used aminoantipyrine and 2-ydroxy-3,5-dichlorobenzenesulfonic acid pair (Barham and Trinder, 1972). It was recently reported that the Ngo-Lenhoff (1980) peroxidase assay using 3-methyl-2-benzothiazolinone hydrazone and 3-(dimethylamino) benzoic acid pair was "superior to all other colorimetric methods for peroxidase detection" (Geoghegan et al., 1983).

RESULTS

Reconstitution of hologlucose oxidase from apoglucose oxidase and FAD-L

The first step in developing PGLEIA is to examine if the FAD-L conjugate is active as a prosthetic group of the apoenzyme, e.g. apoglucose oxidase. Both FAD-theophylline and FAD-IgG conjugates were very active as prosthetic groups of apoglucose oxidase. On molar basis, however, the conjugate was not as active as FAD in activating apoglucose oxidase.

Inhibition by antibody to L of the ability of FAD-L to activate apoglucose oxidase

The second step in PGLEIA development is to demonstrate the ability of ligand specific antibodies to inhibit FAD-L from activating apoglucose oxidase. Antibodies to theophylline and to human IgG, respectively, were able to specifically inhibit FAD-heophylline and FAD-IgG. As much as 95% of the total activating power of FAD-L can be inhibited when high concentration of specific antibody was used. Serum from either non-immunized animals or from animals immunized with other immunogens were not able to bring about any inhibition on the reconstruction of hologlucose oxidase activity from apoglucose oxidase and FAD-L. A typical relationship between the reconstitutable holoenzyme activity and the concentration of antibody is shown in Figure 1.

Reversing the inhibitory effect of antibodies on FAD-L activation of apoglucose oxidase -- competition of L and FAD-L for antibodies to L

Another prerequisite for a successful development of PGLEIA was the ability of ligand analyte (L) to counteract the inhibitory effect of antibodies to L on the activation of apoglucose oxidase by FAD-L, i.e. reconstitution of hologlucose activity. When solutions with increasing concentrations of analyte ligand, such as human IgG or, theophylline were mixed with soultions, each containing constant amounts of FAD-IgG or FAD-theophylline, antibodies to either human IgG or to theophylline and apoglucose oxidase, the final reconstitutable hologulase oxidase activity increases in direct proportion to the human IgG or theophylline concentrations (Morris et al., 1980, 1981). The amount of antibody used in most instances was enough to cause 50-80% inhibition of the maximal (100%) reconstitutable holoenzyme activity so that the background color absorbance was tolerable and yet efficient reversal of inhibition can be obtained by adding low concentrations of the ligand. Two protocols have been developed for PGLEIA: (1) the ligand L and FAD-L are simultaneously exposed to

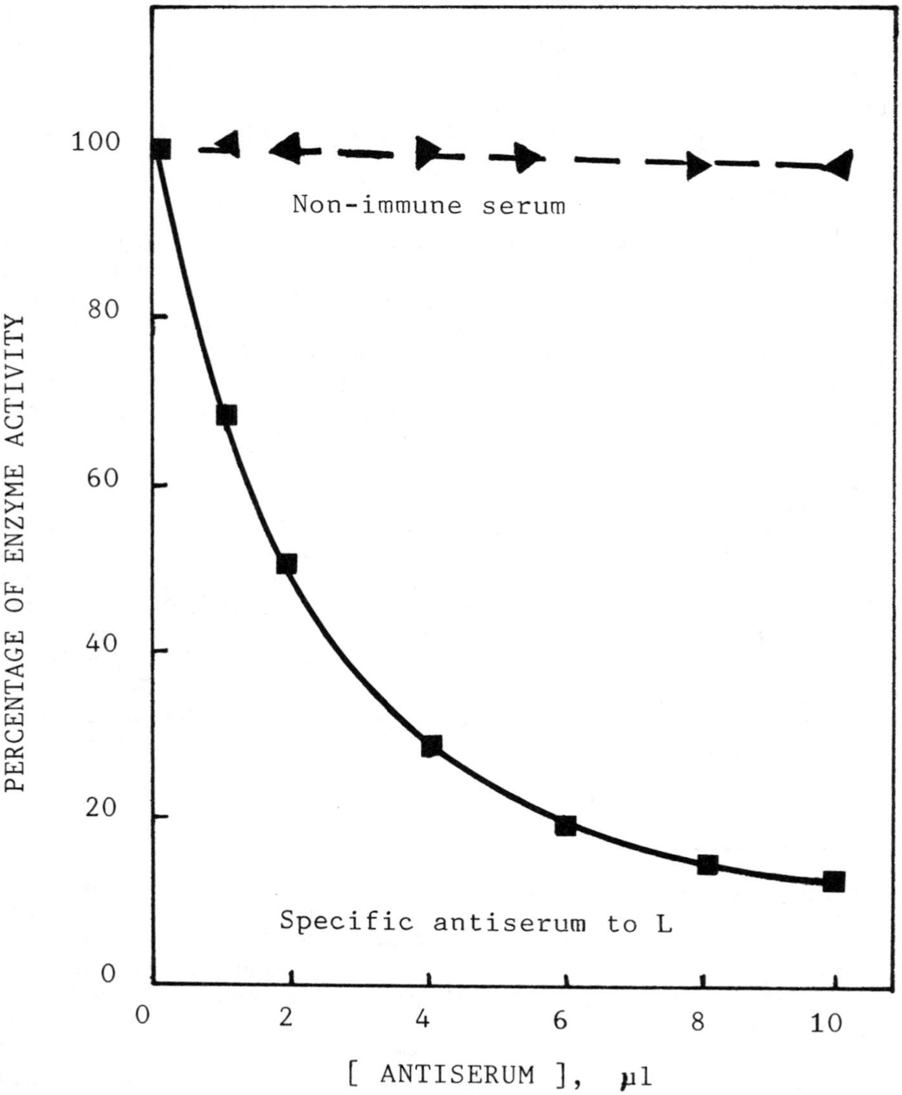

Fig. 1. Inhibition by antibody to L on the activation of apoglucose oxidase by FAD-L. The concentrations of apoglucose oxidase and FAD-L were kept constant.

both antibodies to L and apoglucose oxidase, and (2) ligand (L) and FAD-L are first mixed with antibodies to (L) and incubated for a period before adding apoglucose oxidase (Ngo et al., 1981) and measuring the reconstituted holoenzyme activity.

A typical standard curve of PGLEIA is shown in Figure 2. In the absence of ligand analyte (L), low reconstituted holenzyme activity was observed; generally only 20-50% of the maximal activity. As the concentrations of (L) increase, holoenzyme activity increases in proportion to (L) to a maximal activity.

PGLEIA on reagent-strip format

Tyhach et al. (1981) incorporated all the necessary reagents for PGLEIA on a filter paper. Small pieces of papers were then mounted on plastic handles to produce PGLEIA reagent-strips. Using the strip format, "dip and read" immunoassays for DNP-caproate, theophylline and phenytoin at micromolar levels have been successfully developed. The assay time was short, generally within 5 minutes after applying the analyte solution on the strip.

Precision and accuracy

The precision of PGLEIA is acceptable with average coefficients of variation of 1.2-7.5% and 7.6-8.4% for intra-assay and inter-assay, respectively. This level of precision is comparable to that obtained by using either enzyme labeled immunoassay or radioimmunoassay. The accuracy of PGLEIA is also within acceptable limits. Concentrations of ligand determined by PGLEIA agree well with those determined by EMIT (a homogeneous enzyme immunoassay of SYVA) for theophyline and by radial immunodiffusion method for human IgG. In all cases, correlation coefficients of > 0.90 were obtained.

Interfering substances

Compounds such as flavin mononucleotide (FMN), adenosine monophosphate (AMP), adenosine triphosphate (ATP), nicotinamide adenine dinucleotide (NAD) and free riboflavin, which occur naturally in biological systems, can competitively or non-competitively inhibit the recombination of apoenzyme with FAD (Friedman, 1963). It is prudent to rule out the possible presence of these inhibitors in the test samples. Exposure of FAD-L to acids, bases and light should be minimized (Friedman, 1963). Flavins, in the presence of dissolved oxygen, are well known to catalyze the oxidation of a wide range of substrates, e.g. amino acids, proteins, DNA, nucleotide and lipids (Hellis, 1982). The possibility of FAD oxidizing any component in PGLEIA should always be kept in mind.

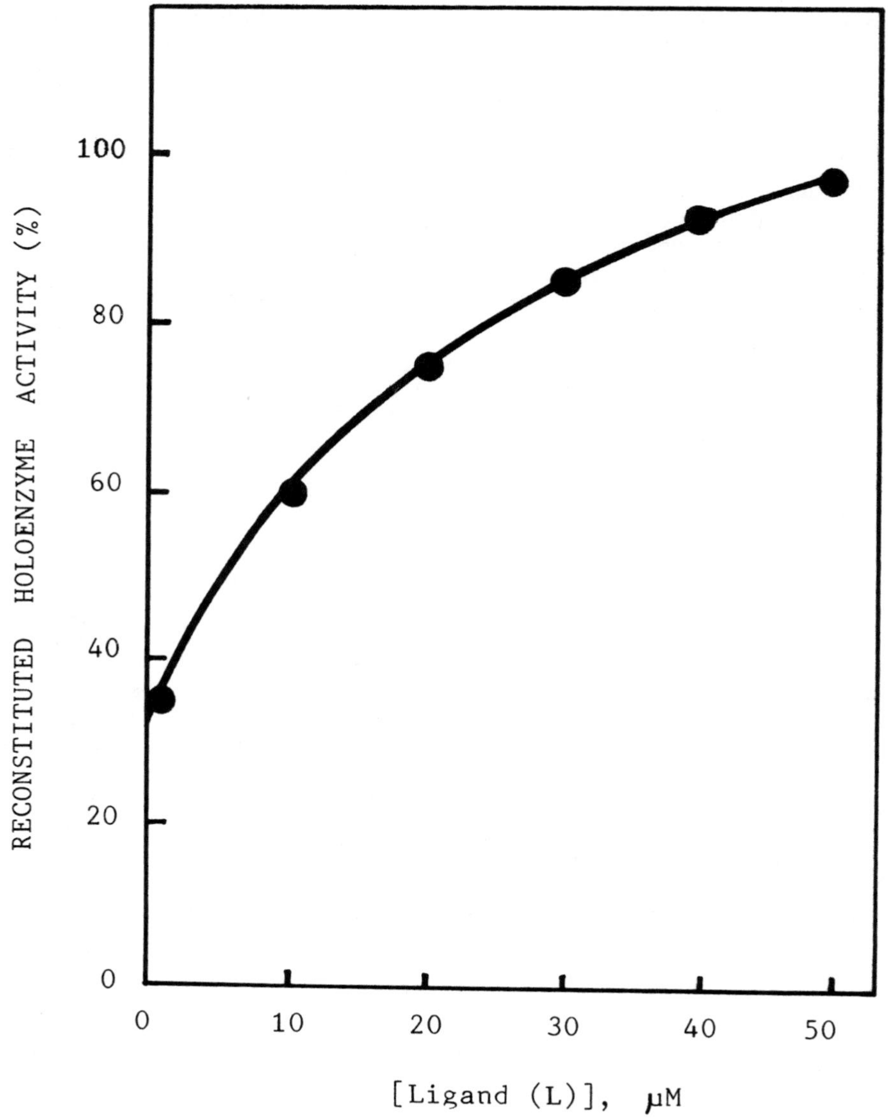

Fig. 2. Reversal by ligand (L) of the antibody induced inhibition — Standard curve for the ligand by PGLEIA. The concentrations of FAD-L, antibody to L and apoglucose oxidase were kept constant.

DISCUSSION

Prosthetic group labeled enzyme immunoassay (PGLEIA) is a separation-free colorimetric enzyme amplified immunoassay. Because no separation of the antibody bound and free, unbound prosthetic group labeled ligands is required, the assay is simple and convenient to perform. It is also more amenable to automation. PGLEIA can be performed in a rate or fixed time mode. Furthermore, PGLEIA works equally well with low molecular weight analyte, such as theophylline (m.w. = 180) or with high molecular weight analyte, such as human immunoglobulin G (m.w. = 160,000). PGLEIA is potentially a sensitive technique, because the apoglucose oxidase-bound FAD-L undergoes cyclic oxidation and reduction reactions during the production of H_2O_2 and gluconic acid from catalytic oxidation of glucose. When the substrates, i.e. glucose and oxygen, are present in saturating amounts, the cyclic oxidation-reduction of small amounts of the apoenzyme bound FAD-L is capable of producing large quantity H_2O_2 which is detected colorimetrically via sentive peroxidase coupled reaction.

SUMMARY

A separation-free colorimetric enzyme immunoassay using a prosthetic group, FAD, as the label is described. A ligand analyte (L) covalently linked to FAD forms a stable FAD-L conjugate. Such a conjugate can compete with L for antibodies specific to L. In the absence of L, most of the FAD-L conjugate is bound by antibody; when bound, the FAD-L conjugate is incapable of further binding with apoglucose oxidase to form the enzymatically active hologlucose oxidase. When L is present at high concentration, however, L will compete more successfuly with FAD-L for the antibodies. As a result, more FAD-L will be unbound, which is in proportion to the concentration of the ligand. The unbound FAD-L can, therefore, serve as a prosthetic group analog to combine with apoglucose oxidase to form hologlucose oxidase. Consequently, the reconstituted hologlucose oxidase activity increases with increasing concentrations of the ligand.

"Dip and read" immunoassay reagent-strip based on PGLEIA has also been developed for determining DNP-caproate, theophylline and phenytoin.

REFERENCES

Barham, D. and P. Trinder, 1972. An improved color reagent for the detrmination of blood glucose by the oxidase system. Analyst, 97, 142-145.

Cook, C.E., M.E. Twine, M. Myers, E. Amerson, J.A. Kepler, and G.F. Taylor, 1976. Theophylline radioimmunoassay: synthesis of antigen and characterization of antiserum. Res. Comm. Chem. Pathol. Pharmacol. 13, 497-504.

Friedman, H.C., 1963. Flavin-adenine dinucleotide. In: Methods of Enzymatic Analysis, H.U. Bergmeyer, ed., Verlag Chemie, Weinheim, Vol. 4, pp. 2182-2185.

Geoghegan, W.D., M.F. Struve, and R.E. Jordan, 1983. Adaptation of Ngo-Lenhoff peroxidase assay for solid phase ELISA. J. Immunol. Methods 60, 61-68.

Hellis, P.F., 1982. The photophysical and photochemical properties of flavins (isoalloxazines). Chem. Soc. Rev. 11, 15-39.

Massey, V. and L.D. Mendelsohn, 1979. Immobilized glucose oxidase and D-amino acid oxidase: a convenient method for the purification of flavin adenine dinucleotide and its analogs. Anal. Biochem. 95, 156-159.

Morris, D.L., R.J. Carrico, P.B. Ellis, W.E. Hornby, H.R. Schroeder, T.T. Ngo and R.C. Boguslaski, 1980, Colorimetric immunoassays with flavin adenine dinucleotide as label, in: "Innovative approaches to clinical analytical chemistry," Lab Professional Series, American Chemical Society.

Morris, D.L., P.B. Ellis, R.J. Carrico, F.M. Yeager, H.R. Schroeder, J.P. Albarella, R.C. Boguslaski, W.E. Hornby, and D. Rawson, 1981. Flavin adenine dinucleotide as a label in homogeneous colorimetric immunoassays. Anal. Chem. 53, 658-665.

Ngo, T.T. et al. (1981). Reference 22 in Anal. Chem. 53, 1981, 658-665 and reference 12 in Applied Biochem. Biotech. 7, 1982, 401-414.

Ngo, T.T. and H.M. Lenhoff, 1980a, Amperometric determination of picomolar levels of flavin adenine dinucleotide by cyclic oxidation-reduction in apo-glucose oxidase system. Anal. Letters 13, 1157-1165.

Ngo, T.T. and H.M. Lenhoff, 1980, A sensitive and versatile chromogenic assay for peroxidase and peroxidase-coupled reactions. Anal. Biochem. 105, 389-397.

Ngo, T.T. and H.M. Lenhoff, 1981, Recent advances in homogeneous and separation-free enzyme immunoassays. Applied Biochem. Biotechnol. 6, 53-64.

Ngo, T.T. and H.M. Lenhoff, 1982, Enzymes as versatile labels and signal amplifiers for monitoring immunochemical reactions. Mol. Cell. Biochem. 44, 3-12.

Okuda, J., J. Nagamine and K. Yagi, 1979, Exchange of free and bound coenzyme of flavin enzymes studied with [^{14}C]FAD. Biochim. Biophys. Acta 566, 245-252.

Solomon, B. and Y. Levin, 1976, Flavin-protein interaction in bound glucose oxidase. J. Solid-Phase Biochem. 1, 159-171.

Tyhach, R.J., P.A. Rupchock, J.H. Pendergrass, A.C. Skjold, P.J. Smith, R.D. Johnson, J.P. Albarella, and J.A. Profitt, 1981, Adaptation of prosthetic-group-label homogeneous immunoassay to reagent-strip format. Clin. Chem. 27 (1981), 1499-1504.

FLUOROGENIC SUBSTRATE LABELED SEPARATION-FREE ENZYME MEDIATED IMMUNOASSAYS FOR HAPTENS AND MACROMOLECULES

That T. Ngo and Raphael C. Wong

Department of Developmental and Cell Biology
University of California
Irvine, California 92717 USA

INTRODUCTION

Background

Fluorogenic substrate labeled separation-free enzyme mediated immunoassay, hereafter abbreviated as FSIA (Boguslaski et al., 1980; Burd, 1981) (it is also known as Substrate-labeled Fluorescent Immunoassay) uses a modified fluorogenic enzyme substrate as a label to form a stable covalent substrate-analyte ligand conjugate. This is in contrast to most enzyme mediated immunoassays which use an enzyme rather than an enzyme substrate as the label (Ngo and Lenhoff, 1981 and 1982). There are two critical prerequisites that must be satisfied before a fluorogenic substrate can be used as a label in FSIA development: (1) the modified fluorogenic substrate after covalent conjugation to the analyte ligand must retain its function as an enzyme substrate and (2) the substrate-analyte ligand conjugate upon binding to an antibody specific to the ligand must not be able to serve as an enzyme substrate. By definition, a fluorogenic enzyme substrate should not fluoresce at the wavelength used to monitor the assay; however, its enzymatic product should exhibit strong fluorescence at the appropriate monitoring wavelengths. The substrate-analyte ligand conjugate in FSIA serves as a modified enzyme substrate capable of binding to the active-site of an enzyme and being transformed into a detectable product. Furthermore, it serves as an analog of the analyte ligand capable of competing with the analyte ligand for ligand binding sites of the anti-ligand antibodies.

Several practical FSIA's for measuring therapeutic drugs and immunoglobulins have been developed using a derivative of umbelli-

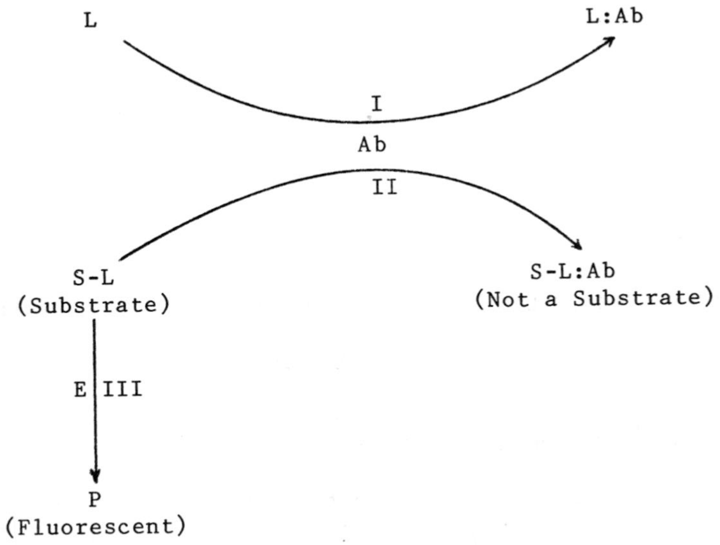

Scheme 1. Principles of FSIA

feryl-β-galactoside which serves as a fluorogenic substrate for *E. coli* β-galactosidase.

Principles of FSIA

A stable, covalently linked fluorogenic enzyme substrate-analyte ligand conjugate (S-L) competes with the analyte ligand (L) for a limited concentration of antibody to L (Ab). Thus, through reactions I and II (Scheme 1), L:Ab and S-L:Ab complexes are formed. Since the free, unbound S-L conjugate is a good fluorogenic substrate for the enzyme, E, whereas S-L bound to the antibody (S-L:Ab) is not, the absence of analyte ligand (L) would allow the S-L to combine with Ab so that indicator reaction would not take place and very little or no fluorescence would be observed. On the other hand, as the concentration of analyte ligand (L) increases, more S-L would remain free to act as fluorogenic substrate for the enzyme (E) and, hence, yielding more fluorescent enzymatic products through reaction III. Thus, the standard curve of a FSIA would show an increasing fluorescent intensity with increasing concentrations of the analyte ligand.

MATERIALS AND METHODS

Enzymes

The most widely used enzyme in FSIA is *E. coli* β-galactosidase. This enzyme is very stable showing no significant loss of activity over a twelve month period when stored at 4°C in the presence of antibacterial agents. Nucleotide pyrophosphatase which catalyzes the hydrolysis of flavin adenine dinucleotide to flavin mononucleotide and adenosine monophosphate, has also been used in one FSIA system.

Synthesis of Carboxy Derivative of Umbelliferyl-β-D-galactoside

The synthesis of β-[7-(3-carboxycoumarinoxy)]-D-galactoside (Fig. 1, V), a carboxy derivative of umbelliferyl-β-D-galactoside was achieved by reacting 3-carboethoxy-7-hydroxycoumarin (Fig 1, III) with 2,3,4,6-tetraacetyl-α-D-galactosyl bromide (Fig. 1, IV) according to a procedure used by Leaback (1965) for preparing methylumbelliferyl-β-D-galactoside. 3-Carboethoxy-7-hydroxycoumarin (Fig. 1, III) was prepared using a Knoevenagel condensation of 2,4-dihydroxybenzaldehyde (Fig. 1, I) with diethylmalonate (Fig. 1, II) in acetic acid, benzene and piperidine (Cope et al., 1941).

Synthesis of Fluorogenic Enzyme Substrate-analyte Ligand Conjugate (S-L)

The general strategy for synthesizing S-L called for either a direct coupling of the carboxy-activated β-[7-(3-carboxycoumarinoxyl)]-D-galactoside (abbreviated as β-CCG) (Fig. 1, V) to the amino group of the analyte ligand or indirect coupling of β-CCG through a spacer and/or a bifunctional crosslinking agent to the analyte ligand. The analyte ligand can be a low molecular weight substance, such as therapeutic drugs (M.W.< 1000) or a macromolecule such as human IgG or IgM (M.W.> 150,000).

Synthesis of S-L by Direct Coupling of β-CCG to an Amino Group Containing Analyte Ligand

Several conjugates of β-CCG labeled aminoglycoside antibiotics have been prepared by direct coupling of β-CCG to the amino group of the antibiotics (Fig. 2) using a water soluble condensing agent, 1-ethyl-3-(3-dimethylaminopropyl)-carbodiimide (Burd et al., 1977). Alternatively, the carboxy group β-CCG can be activated first by reacting β-CCG with N-hydroxysuccinimide in dry polar organic solvent, using N,N'-dicyclohexylcarbodiimide as the condensing agent (Fig 3). The N-hydroxysuccinimide ester of β-CCG can be coupled directly to amino group of tobramycin (Burd et al., 1978) and amikacin (Thompson and Burd, 1980). The carboxy group of β-CCG can also be activated by forming a mixed anhydride with ethyl chloroformate. Such a

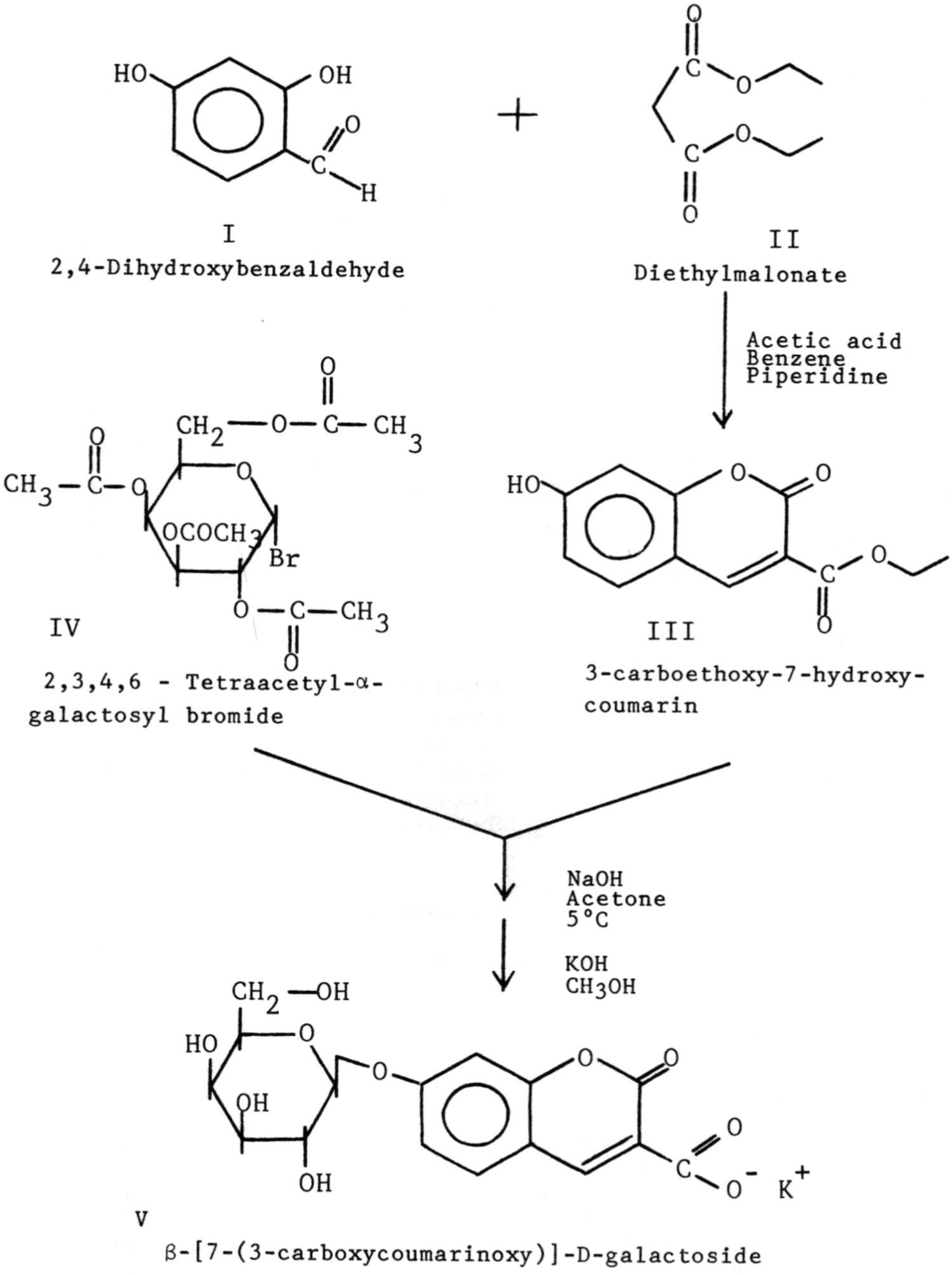

Fig. 1. Synthesis of β-[7-(3-Carboxycoumarinoxy)]-D-galactoside

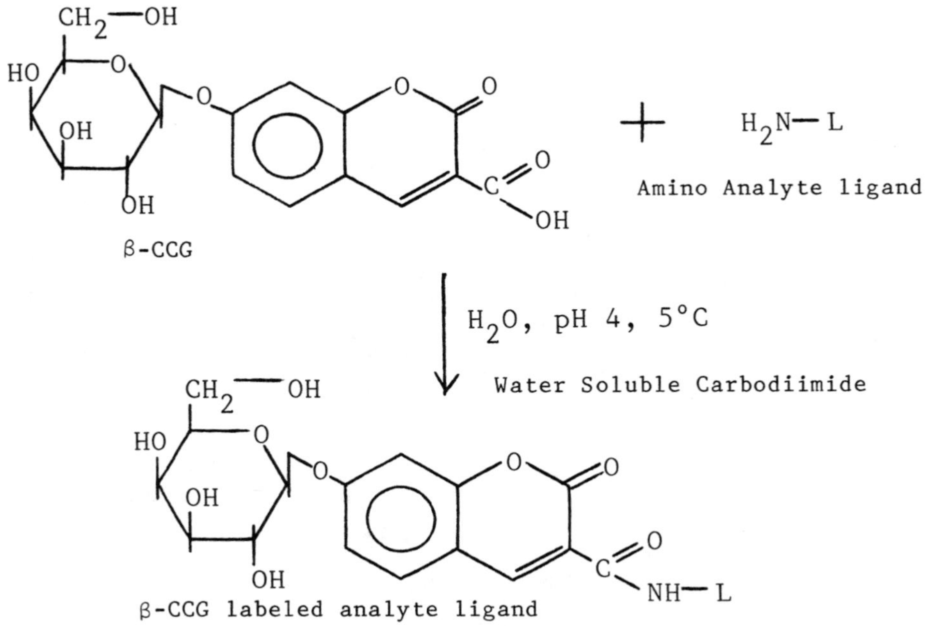

Fig 2. - Synthesis of β-CCG labeled analyte ligand using water soluble carbodiimide

mixed anhydride of β-CCG has been reacted with 5-[2-(4-aminobutoxy-phenyl]-5-phenylhydantoin, an amino group derivative of 5,5-diphenylhydantoin (Wong et al., 1979).

<u>Indirect Coupling of β-CCG to Macromolecules</u>

In order to allow a more facile interaction between the active site of the enzyme and the β-CCG bound to a macromolecule, a spacer was introduced between β-CCG and the macromolecule so that the β-CCG moiety is well extended into the solution surrounding the macromolecule. The spacer molecule, 1,6-hexanediamine (Fig. 4, II) was first linked to β-CCG (Fig. 4,I) by using 1-ethyl-3(3-dimethylaminopropyl) carbodiimide as the condensing agents to form 6-(7-β-galactosylcoumarin-3-carboxamido) hexylamine (β-GCHA). Then, the substrate derivative was reacted with an equal molar quantity of a homobifunctional crosslinking reagent (Fig. 4, III) such as dimethyladipimidate for a short time before being introduced to a solution of macromolecule (Fig. 4, IV). The substrate labeled macromolecule was separated from the unreacted substrate derivatives by a gel chromatographic method. (Ngo et al., 1981)

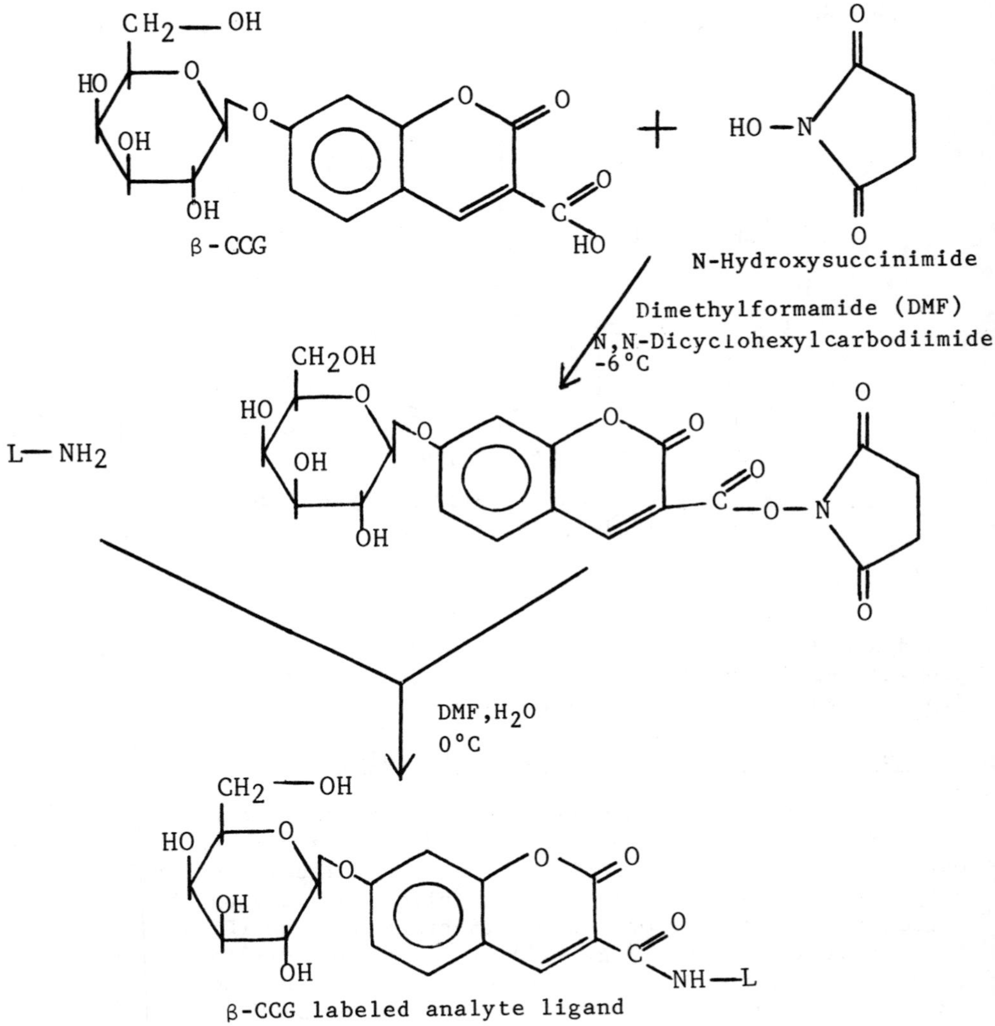

Fig 3. Synthesis of β-CCG labeled analyte ligand via an intermediary N-hydroxysuccinimide ester of β-CCG.

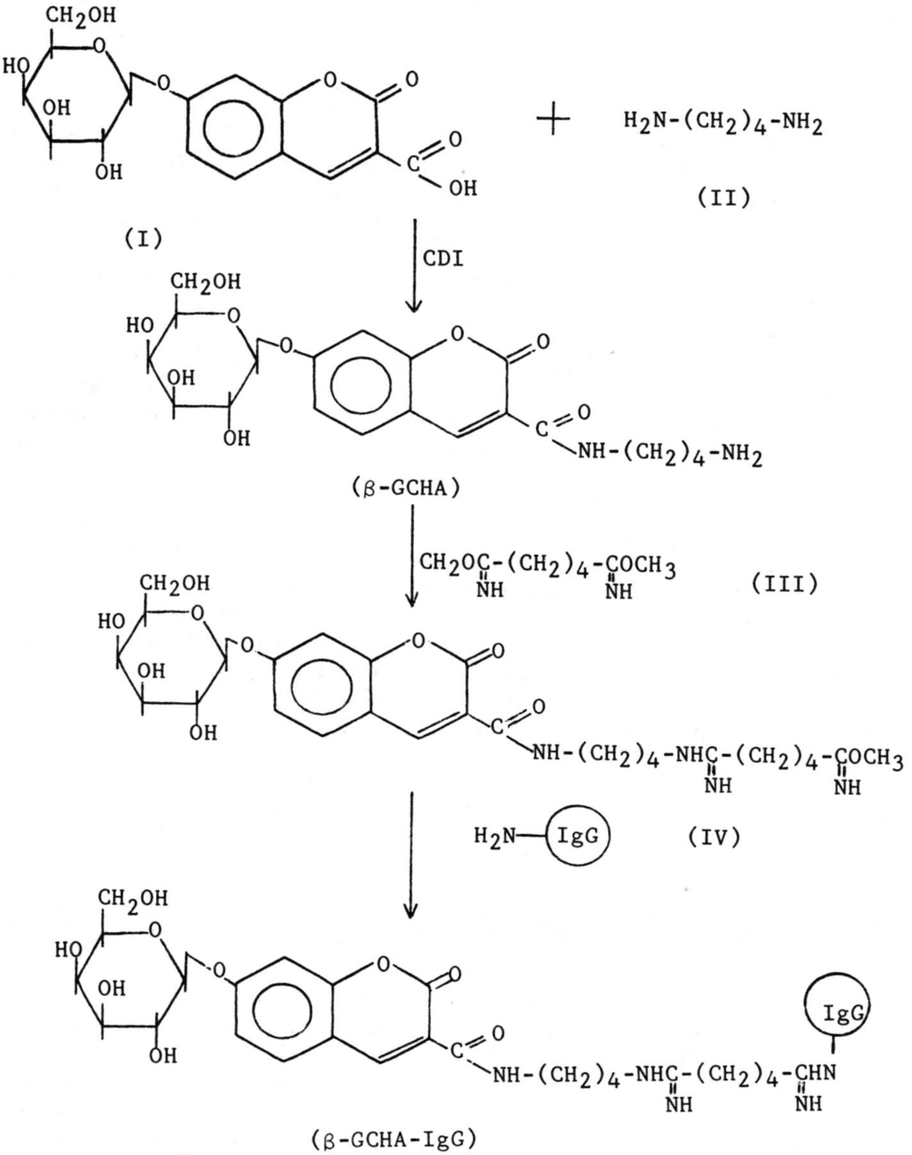

Fig 4. Schematic representation of the labeling of IgG with β-GCHA

Typical Protocols for Conducting FSIA

There are two typical assay protocols for carrying out analyte measuring using FSIA. The first protocol is used for measuring low molecular weight analyte ligands (haptens). The second protocol is for measuring high molecular weight analyte ligands (macromolecules).

(A) FSIA protocol for haptens (Scheme II.a): The test sample was first diluted (1:50) with 2.5 ml of 50 mM Bicine buffer, pH 8.2 containing 0.1% sodium azide as a preservative. The diluted sample (50 µl) was added to 1.0 ml solution containing the anti-analyte ligand antibody and the enzyme, *E. coli* β-galactosidase. After mixing, 50 µl fluorogenic enzyme substrate-analyte ligand conjugate (S-L) along with 0.5 ml buffer was added to initiate the reaction. The reaction was allowed to proceed at room temperature (≏22°C) for 5 minutes to several hours, but generally for 20 to 30 minutes. In addition to the above fixed-time method, FSIA can also be performed, based on similar protocol by using an initial rate method. The above mentioned volumes for various reagents are recommended by the manufacturer's product inserts, but other volumes have also been used.

(B) FSIA protocol for macromolecules: (Scheme II.b) The test sample was first diluted with the same Bicine buffer and 100 µl of the diluted sample was mixed with 3 ml fluorogenic enzyme substrate-analyte ligand conjugate (S-L). Then 100 µl of the diluted antibody solution was added to the above solution and mixed well. Finally, the enzyme solution was added to initiate the fluorogenic reaction. After 20-30 minutes at room temperature, the fluorescence of the solution was measured.

In both protocols, the solution fluorescence was measured using excitation and emission wavelength of 400 and 450 nm, respectively.

RESULTS

Absorption Spectra of Fluorogenic Enzyme Substrate-analyte Ligand Conjugate Before and After Enzymatic Hydrolysis

(A) β-CCG-hapten conjugates: The absorption spectra of β-CCG labeled haptens such as, β-CCG-gentamicin, β-CCG-tobramycin, β-CCG-amikacin, β-CCG-phenytoin etc. showed a characteristic absorption maximum at around 343 nm (Fig. 5). The absorption spectra of the enzyme treated β-CCG-labeled hapten conjugates, i.e., after enzymatic cleavage of β-galactosyl group from β-CCG labeled haptens, showed a diminished absorption at 343 nm, but a new and more intense absorption peak at 403 nm (approximately 1.5 times of absorption at 343 nm) which is a characteristic absorption of a dissociated (ionized) coumarin.(Fig. 5)

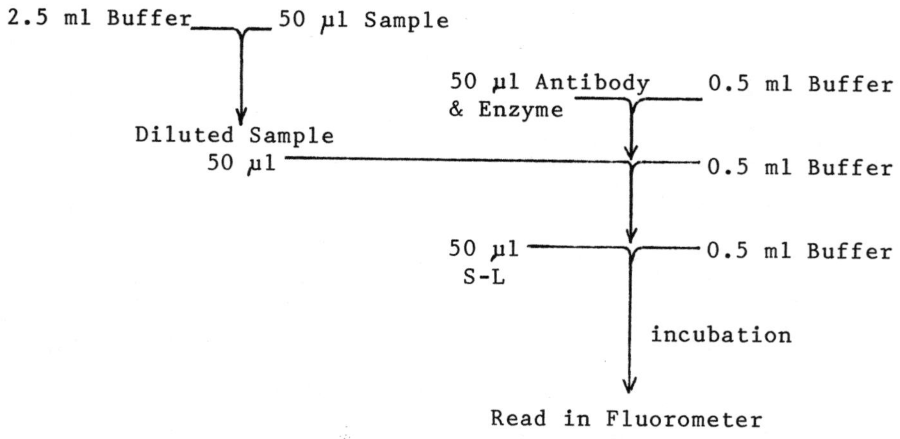

Scheme II.a FSIA protocol for haptens

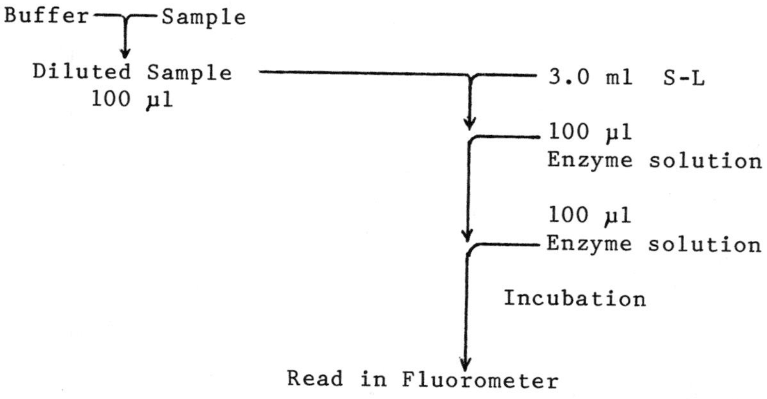

Scheme II.b FSIA protocol for Macromolecules

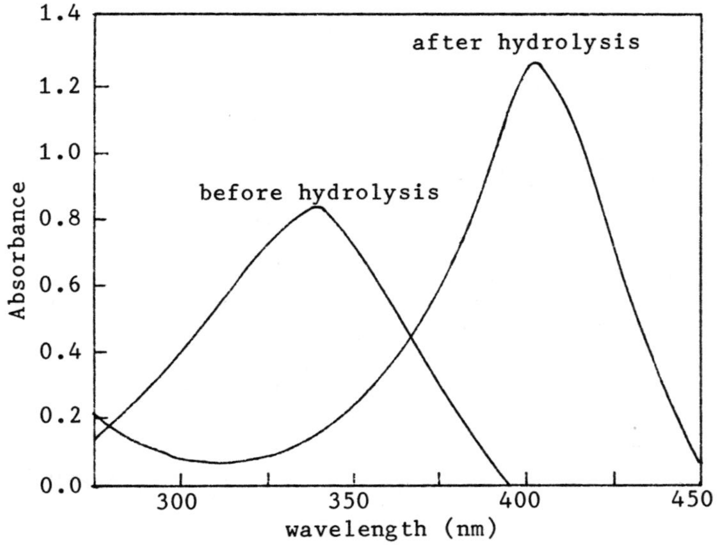

Fig 5. Absorption spectra of βCCG-phenytoin before and after β-galactosidase hydrolysis in pH 8.5 bicine buffer.

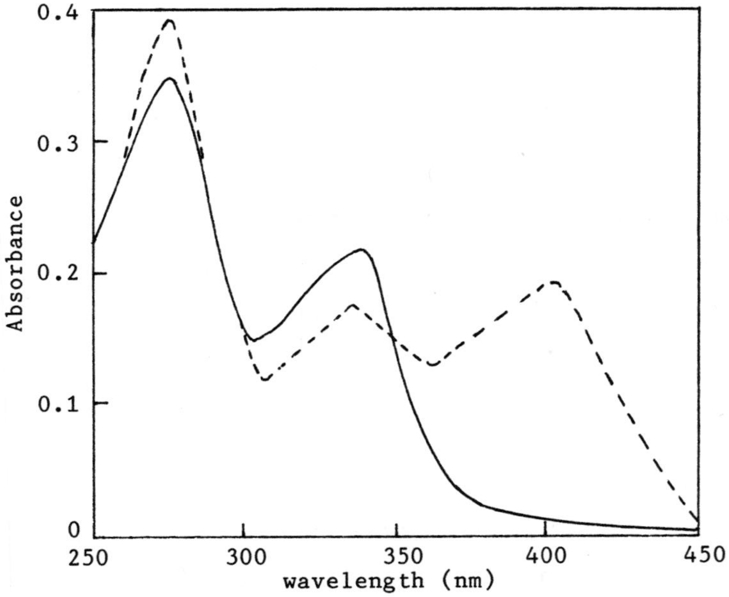

Fig 6. Absorption spectra of β-GCHA-IgG in 50 mM Bicine buffer pH 8.2 before (solid line) and after (broken line) treatment with 0.5 units of β-galactosidase for 30 min at room temperature.

(B) β-GCHA-macromolecular conjugates: Two β-CCG-macromolecular conjugates have been prepared, i.e., β-GCHA-human IgG and β-GCHA human IgM. The absorption spectra of these conjugates showed absorption maxima at 278 and 340 nm. The 278 nm absorption is due mainly to the absorption of aromatic amino acid residues of the immunoglobulins, while the 340 nm absorption is the characteristic absorption of a galactosyl-umbelliferone derivative (Fig. 6). The intensities of these two absorption and extinction coefficients of the immunoglobulin and β-GCHA at the corresponding wavelengths were used to calculate molar ratios of β-GCHA coupled per immunoglobulin. Routinely, 6-7 moles of β-GCHA were found per mole of monomeric immunoglobulin unit. Upon treatment of the β-GCHA-macromolecule conjugates with the enzyme, β-galactosidase, the absorbance at 340 nm decreases, with a concomittant increase in absorbance at 400 nm (Fig. 6). Approximately 50% of the bound β-GCHA was hydrolyzable by the enzyme.

Fluorescence Spectra of Fluorogenic Enzyme Substrate-analyte Ligand Conjugate Before and After Enzymatic Hydrolysis

The fluorescence spectra of both β-CCG-haptens and β-GCHA-macromolecules showed that wavelengths for maximal excitation and emission were, respectively, at 365 and 405 nm (Figs. 7a,b). However, after enzymatic hydrolysis, i.e., removal of β-galactosyl groups, the excitation and emission maxima for both β-CCG-haptens and β-GCHA-macromolecules shifted to 405 and 450 nm, respectively. Furthermore, the emission intensity increased many fold (Figs. 7a, b). Under the assay conditions, using 400 nm excitation and 450 emission wavelengths, neither β-CCG-haptens nor β-GCHA-macromolecules exhibits any significant fluorescence before the enzymatic leavage of the linked β-galactosyl groups.

Titration of Analyte Ligand Specific Antibody with Fluorogenic Enzyme Substrate-analyte Ligand Conjugate

The covalent linking of fluorogenic enzyme substrates, β-CCG or β-GCHA to analogs of analyte ligand does not impair the immunologic properties of the ligand, i.e., the specific antibody binds to the β-CCG or β-GCHA labeled ligand as well as it does to the native ligand. When antibody to an analyte ligand was titrated with fixed levels of fluorogenic enzyme substrate-analyte ligand conjugate and the enzyme, the fluorescence of the solution, as a result of the enzyme action, decreased with increasing concentrations of the antibody. Normal, non-immune immunoglobulin has little or no effect on the solution fluorescence (Fig. 8). It is clear from Fig. 8 that greater than 90% of total fluorescence can be suppressed upon binding of the fluorogenic conjugate by a specific antibody.

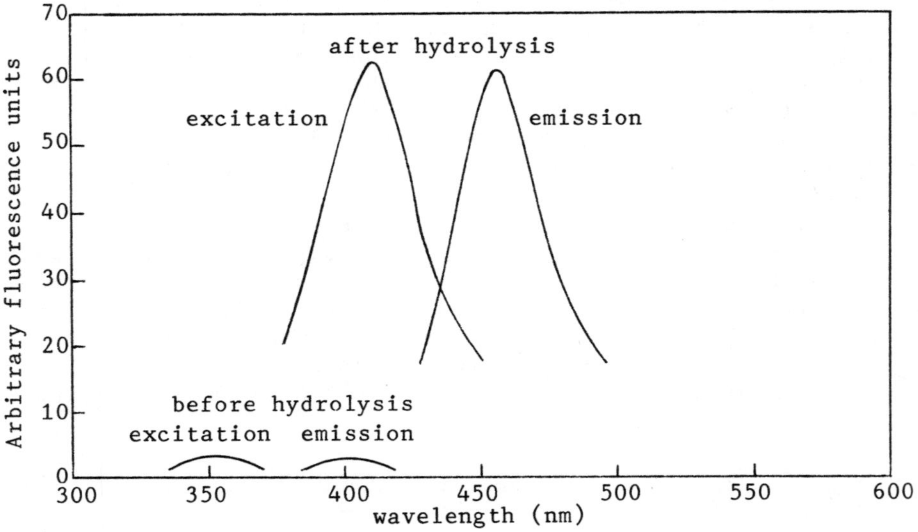

Fig 7a. Fluorescence spectra of β-CCG-phenytoin in 50 mM Bicine buffer pH 8.2 before and after hydrolysis with β-galactosidase.

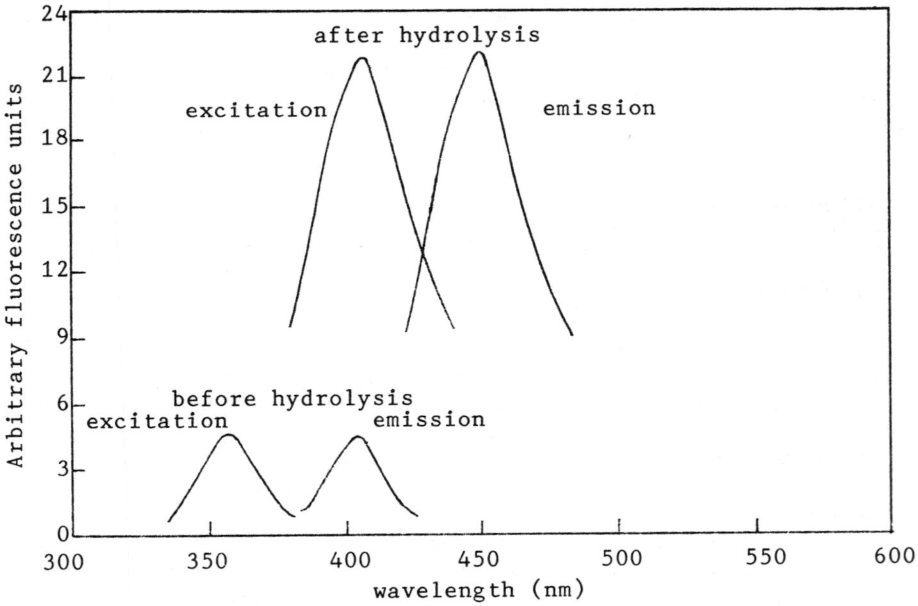

Fig 7b. Fluorescence spectra of β-GCHA-IgG before and after hydrolysis with β-galactosidase in 50 mM Bicine buffer pH 8.2.

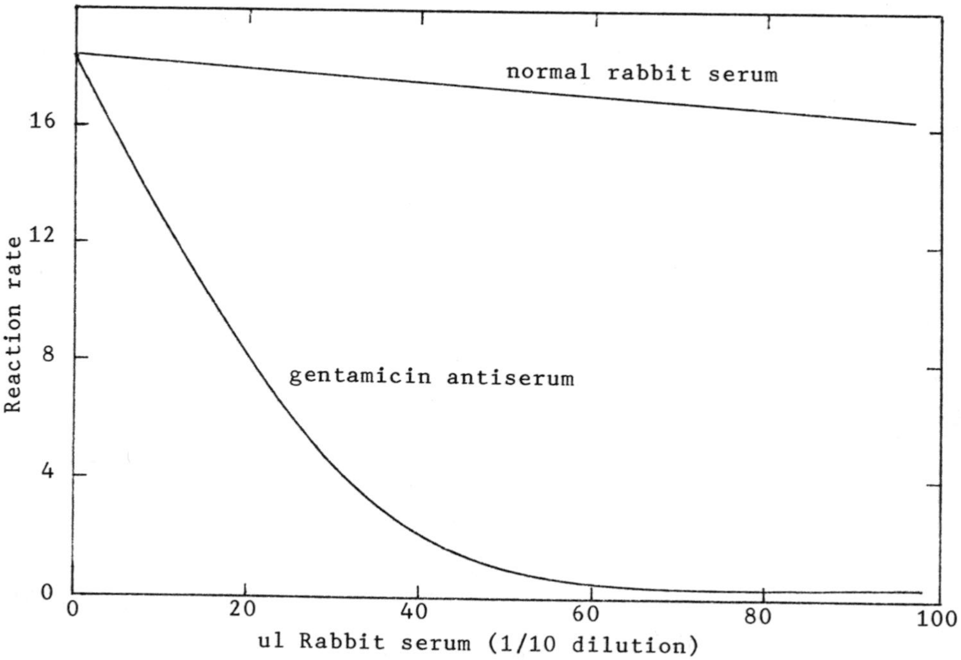

Fig 8. Titration of antiserum to gentamicin with β-CCG labeled gentamicin analog - sisomicin. Various amounts of a 10-fold dilution of rabbit gentamicin antiserum or normal rabbit serum were added to 2.0 ml of 50 mM Bicine buffer containing 50 ng of β-galactosidase. Reaction rate was then determined when 50 ul of β-CCG sisomicin was added.

Competitive Binding Reactions - The Standard Curves

In FSIA, the concentration of an analyte ligand in a test sample is generally measured by comparing fluorescence generated in the cuvet containing the test sample to the fluorescence of a series of cuvets containing standard solutions of precisely known amount of the analyte ligand. Thus in FSIA, the concentration of antibody specific to the analyte ligand, the enzyme and the fluorogenic enzyme substrate-analyte ligand conjugate are purposely kept constant. The only intended variable is the concentration of the test analyte ligand. The amount of antibody is usually sufficient to diminish 70 to 90% of the total fluorescence.

Figure 9 showed a typical standard curve obtained by an initial rate method for gentamicin measurement by FSIA (Burd et al, 1977). The standard curve for human IgG, a macromolecule, using a fixed-time FSIA is shown in Fig. 10 (Ngo et al, 1981).

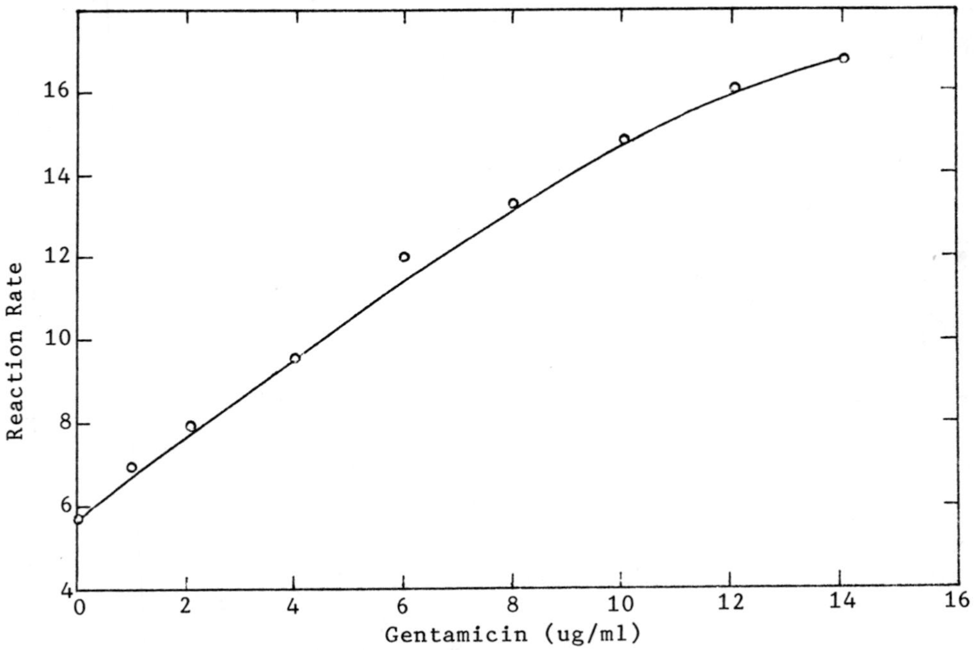

Fig 9. Typical FSIA standard curve obtained by an initial rate method as examplified by gentamicin.

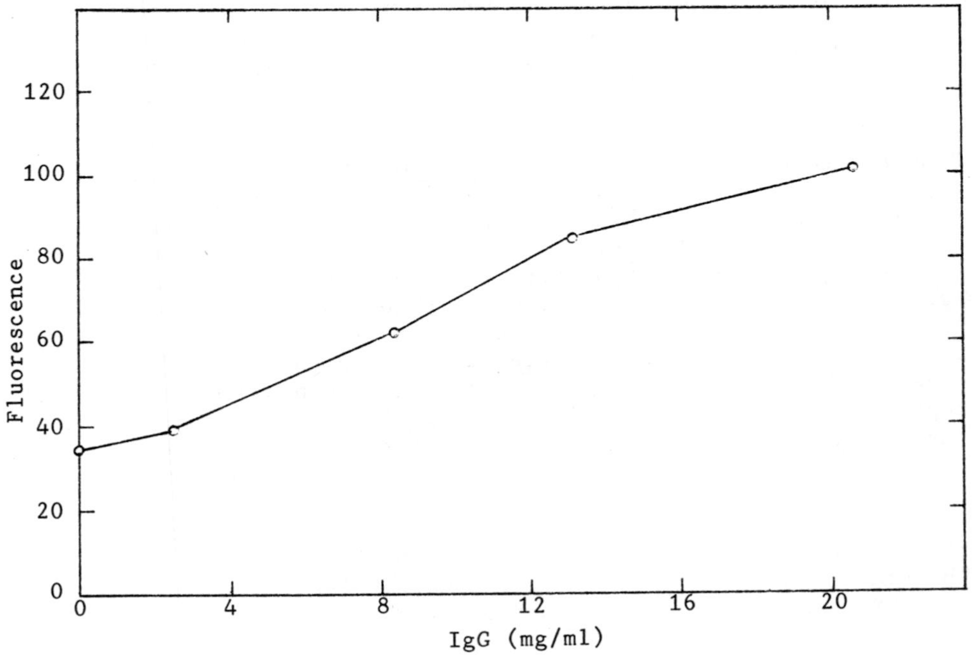

Fig 10. Standard curve for IgG using a fixed time method.

Precision of FSIA

The precision of FSIA for haptens and macromolecules is shown in Table I. The standard deviation and coefficient of variation for both intra-day and inter-day comparison are within an acceptable limit for immunoassays.

Table 1. Precision of FSIA

	Analyte	Intra-day			Inter-day		
A. Haptens	Conc ug/ml	Mean ug/ml	S.D. ug/ml	C.V. %	Mean ug/ml	S.D. ug/ml	C.V. %
Gentamicin (Burd et al,77)	2.0	1.7	0.6	35.3	1.8	0.3	16.7
	6.0	6.0	0.3	5.0	6.0	0.5	8.3
	10.0	9.7	0.5	5.1	10.1	0.7	6.9
Tobramycin (Burd et al,78)	2.0	2.0	0.1	5.0	2.0	0.3	15.0
	6.0	6.0	0.2	3.3	6.0	0.4	6.7
	10.0	10.0	0.2	2.0	10.0	0.6	6.0
Amikacin (Thompson et al, 80)	5.0	4.7	0.3	6.9	4.7	0.2	4.7
	15.0	15.8	0.4	2.4	15.8	0.2	1.2
	30.0	31.7	0.7	2.3	31.7	1.0	3.2
Kanamycin (deCastro et al, 84)				<3.8			
Phenytoin (Wong et al,79)	5.0	4.8	0.25	5.2	5.1	0.65	12.7
	15.0	15.3	0.59	3.9	15.3	1.09	7.1
	25.0	25.5	1.41	5.5	23.7	1.04	4.4
(Wong et al,80)	5.0	4.7	0.25	5.3			
	15.0	15.4	0.37	2.4			
	25.0	25.2	0.81	3.2			
Phenobarbital (Krausz et al, 80)	50.0	50.9	2.72	4.5	50.9	2.96	5.8
	20.0	20.2	0.81	4.0	20.2	0.75	3.7
	10.0	9.7	0.44	4.6	9.7	0.36	3.7
Carbamazepine (Li et al,82)	4.0	4.1	0.2	5.5	3.5	0.1	3.5
	12.0	12.1	0.2	1.6	11.6	0.2	1.9
	16.0	15.8	0.5	2.9	15.8	0.4	2.3

(continued)

Table 1. Precision of FSIA (Continued)

	Analyte	Intra-day			Inter-day		
	Conc	Mean	S.D.	C.V.	Mean	S.D.	C.V.
A. Haptens	ug/ml	ug/ml	ug/ml	%	ug/ml	ug/ml	%
Primidone	4.0			<6			<6
(Johnson et al							
81)	12.0			<6			<6
	16.0			<6			<6
Valproic Acid	40.0			6.2			
(Feinstein et al							
82)	80.0			6.3			
	120.0			3.5			
Ethosuximide							
(Pahuski et al,				2.8-4.5			3.2-4.5
82)							
Theophylline	5.0	4.7	0.13	2.8	4.7	0.21	4.5
(Li et al,81)	15.0	15.6	0.33	2.1	15.2	0.34	2.3
	25.0	24.6	0.26	1.0	24.0	0.63	2.6
Caffeine				5.7-9.9			
(Benovic et al,82)							
Solid Phase	5.0		0.42	8.4		0.42	8.4
Theophylline	15.0		0.57	3.8		0.56	3.7
(Walter et al,							
83)	25.0		0.90	3.6		0.90	3.6
Quinidine	4.0			3.3			1.2
(Csiszar et al,							
81)	6.0			4.4			1.2
	8.0			6.9			4.4
Disopyramide	1.5	1.4		4.3			
(Lima et al,84)	3.0	3.1		1.9			
	7.0	7.2		2.9			
Dibekacin	2.0	1.8	0.08	4.6	1.9	0.12	6.5
(Place et al,83)	6.0	6.1	0.08	1.3	6.1	0.06	1.1
	14.0	14.3	0.34	2.4	14.0	0.24	1.7
B. Macromolecules							
IgG		2.3	0.41	18.3	2.2	0.27	12.2
(Ngo et al,81)		10.3	0.45	4.4	10.2	0.64	6.3
		20.7	1.62	7.8	20.4	1.07	5.2

Table 1. Precision of FSIA (Continued)

	Analyte Conc ug/ml	Intra-day Mean ug/ml	Intra-day S.D. ug/ml	Intra-day C.V. %	Inter-day Mean ug/ml	Inter-day S.D. ug/ml	Inter-day C.V. %
B. Macromolecules							
IgM (Worah et al, 81)		0.6	1.10	15.9	0.6	0.09	14.9
		2.1	0.15	7.1	1.5	0.15	9.8
		3.1	0.22	7.0	3.0	0.30	10.0

Analytical Recovery and Effect of Sample Dilution

Experiments on analytical recovery and sample dilution are important in assessing matrix effect on the precision and accuracy of FSIA. Table 2 showed that, within experimental errors, the recovery was quantitative. Similarly, the effect of sample dilution did not significantly alter the results of FSIA (Table 3).

Table 2. Analytical Recovery of FSIA

	No. of Sample	Recovery % Range	Recovery % Mean
Gentamicin (Burd et al, 77)	18	87.7 - 109.0	97.9
Amikacin (Thompson et al, 80)		93.0 - 113.0	
Phenytoin (Wong et al, 79)	15	97.2 - 123.2	107.8
Phenobarbital (Krausz et al, 82)	15	92.5 - 108.0	100.4
Theophylline (Li et al, 81)	20	91.3 - 110.0	103.0
Dibekacin (Place et al, 83)	1		100.6
IgG (Ngo et al, 81)	6	96.0 - 107.0	101.2
IgM (Worah et al, 81)	4	91.2 - 105.0	99.4

Table 3. Effect of Sample Dilution on FSIA

	No. of Determinations	Percentage of the Actual Value Obtained with Dilution After Multiplied by Dilution Factor
Gentamicin (Burd et al, 77)	15	97.2
Phenytoin (Wong et al, 79)	15	94.2
Theophylline (Li et al, 81)	15	102.5
IgG (Ngo et al, 81)	17	95.3
IgM (Worah et al, 81)	26	98.5

Accuracy

The accuracy of results obtained by using FSIA has been compared with those obtained by using microbiological testing, gas chromatography, high performance liquid chromatography, 'EMIT', radioimmunoassay (RIA) and radial immunodiffusion (RID). For example, correlation between FSIA and EMIT determinations of phenytoin, a hapten, in clinical serum samples have been shown (Fig. 11) (Wong et al, 1979). Similarly, a good correlation between FSIA and RID has been obtained for human IgG, a macromolecule (Fig. 12) (Ngo et al, 1981). Table 4 summarized the accuracy of FSIA for haptens and macromolecules in terms of the correlation coefficient, r, standard error of estimate, S.E., the correlation equation, the reference methods used and the number of samples (n).

Table 4. Accuracy of FSIA

A. Haptens	Reference	n	Equation	r	S.E.
Gentamicin (Burd et al, 77)	RIA	66	y=1.07x-0.19	0.94	0.66
Tobramycin (Burd et al, 78)	RIA	53	y=0.88x+0.20	0.99	0.41
	Microbio.	53	y=0.99x+0.40	0.97	0.71
Amikacin (Thompson et al, 80)	RIA	93	y=0.95x+0.15	0.99	1.55

Table 4. Accuracy of FSIA (continued)

A. Haptens	Reference	n	Equation	r	S.E.
Kanamycin	RIA			0.99	
(deCastro et al, 84)	HPLC			0.98	
Phenytoin	GC	136	y=1.01x+0.80	0.98	1.57
(Wong et al, 79)	EMIT	103	y=0.95x+0.65	0.97	1.75
(Wong et al, 80)	GC	45	y=1.06x-0.45	0.99	1.17
	HPLC	37	y=0.96x-0.25	0.99	1.35
	EMIT	67	y=0.99x-0.68	0.97	2.26
	RIA	34	y=1.01x-1.69	0.98	2.57
Phenobarbital	EMIT	102	y=1.11x-2.71	0.97	2.32
(Krausz et al, 80)					
Carbamazepine	HPLC	45	y=1.10x+1.21	0.97	0.77
(Li et al, 82)	GC	56	y=1.01x-0.30	0.95	0.99
	EMIT	53	y=1.07x-0.82	0.98	0.51
Primidone	EMIT	97	y=1.11x-0.59	0.98	
(Johnson et al, 81)	GC	97	y=0.94x-0.34	0.98	
Valproic Acid	GC		y=1.08x-2.5	0.99	
(Pahuski et al, 83)					
Ethosuximide	GC		y=1.04x-1.19	0.93	
(Pahuski et al, 83)					
Theophylline	GC	85	y=0.97x+0.82	0.97	1.74
(Li et al, 81)	EMIT	92	y=1.06x-1.18	0.99	1.60
Caffeine	HPLC	26	y=0.95x-0.02	0.94	0.83
(Benovic, 82)					
Solid Phase Theophylline					
(Greenquist et al, 81)	EMIT	50	y=1.06x-1.31	0.99	2.07
(Walter et al, 83)	HPLC	30	y=1.02x+0.09	0.98	1.72
Quinidine	EMIT	84	y=1.04x-0.07	0.97	0.36
(Csiszar et al, 81)					
Disopyramide	HPLC	67	y=1.00x+0.19	0.95	
(Lima et al, 84)					
Dibekacin	HPLC	151	y=0.94x+0.24	0.99	0.36
(Place et al, 83)					

(continued)

Table 4. Accuracy of FSIA (continued)

B. Macromolecules	Reference	n	Equation	r	S.E.
IgG (Ngo et al, 81)	RID	125	y=0.81x+1.14	0.94	1.02
IgM (Worah et al, 81)	RID	87	y=0.96s+9.60	0.96	0.22

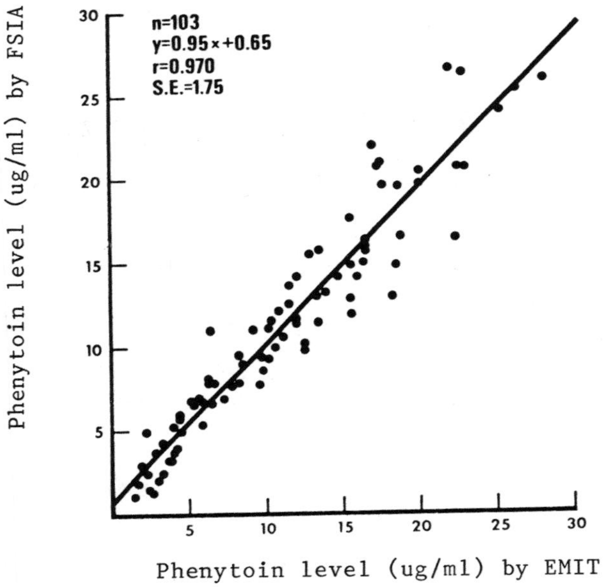

Fig 11. Correlation between FSIA and EMIT determinations of Phenytoin in clinical serum samples.

DISCUSSION

Fluorogenic enzyme substrate has been shown to be a good label for separation-free enzyme mediated immunoassays. An example is the carboxylic derivative of umbelliferyl-β-galactoside, β-7-(3-carboxy-coumarinoxy)-D-galactoside (β-CCG) as a fluorogenic substrate for E. coli β-galactosidase. This substrate has been used to develop a number of separation-free immunoassays for both haptens and macromolecules (Tables 1 and 4). Because of its simplicity, FSIA is a useful and attractive immunochemical method for measuring analytes that occur at relatively high concentrations (in micromolar levels).

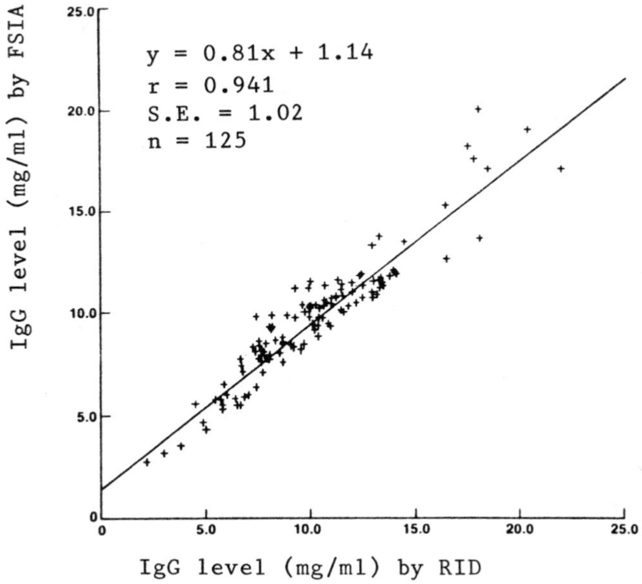

Fig 12. Comparison of IgG levels measured in human sera by RID and FSIA

There are two critical steps in FSIA development. First, the enzyme should be able to transform the covalent fluorogenic enzyme substrate analyte ligand conjugates into fluorescent products at an acceptable rate. Secondly, after anti-analyte antibody binds to the fluorogenic enzyme substrate-analyte ligand conjugate, the conjugate should no longer serve as a substrate. Consequently, no fluorescent product will be produced by the enzyme from the antibody bound conjugate. Once these two criteria have been fulfilled, the development of a useable FSIA is almost certain.

Results obtained from FSIA for therapeutic drugs and immunoglobulins showed that the precision, analytical recovery and accuracy of these FSIA are within the accepted limits for immunoassay systems (Tables 1-4).

Solid-phase reagent strips for serum theophylline measurement based on FSIA has been developed (Walter et al, 1983). They incorporated all the essential chemistries of FSIA onto paper. The reagents were then dried on the paper to provide a one-step immunoassay after appropriate dilution of the sample. Two assay formats were developed for solid-phase FSIA. They were based on: (1) competitive binding reactions between labeled and unlabeled ligands and (2) displacement of antibody-bound labeled ligand by unlabeled ligand. The displacement concept has been used by Weetall and Odstrchel (1976) in developing a highly reproducible and sensitive radioimmunoassay for T4 that required only a single pipetting step. In the competitive binding format, the solid-phase strip was prepared by a two-step impregnation of the paper. The first impregnating solution contained a specific antibody and β-galactosidase in a buffer. The second impregnating solution consisted of the fluorogenic enzyme substrate-analyte ligand conjugate in acetone. In between impregnations, the paper was dried at 50°C. In the displacement format a solution containing the fluorogenic enzyme substrate-analyte ligand conjugate was first incubated with the anti-ligand antibody for 15 minutes before the enzyme was added to the solution. After an additional 15 minutes, the solution was applied on the paper and the paper was dried at 50°C for 15 minutes.

The assay using either format consists of applying up to 70 µl of diluted sample on the paper and the increase in fluorescence on the paper within a pre-determined time (3-15 min) was recorded. The fluorescence intensity will increase with increasing concentrations of the analyte ligand.

The precision and accuracy of solid-phase reagent strip FSIA are comparable to that of solution-phase FSIA (Greenquist et al, 1981 and Walter et al, 1983) (Table 1 and 4).

In connection with solid-phase FSIA, it should be noted that Terumo Corporation of Japan (1983) has simplified the FSIA procedure by incorporating the method into a multi-layer immobilized filter test strip (see Chapter by Tim A. Kelly in this volume and Japan Patent No. 58,150,861. Immunoassay for the Determination of Drugs).

The major limitations of FSIA are: (1) Triamterene (or 6-phenyl 2,4,7-triamino pteridine), a diuretic drug, is a fluorescent compound with spectral characteristics which are similar to those of umbelliferone derivatives. Hence, presence of this drug in patient sera being monitored by FSIA may give a false elevation of the analyte concentration. (Ames Product Insert for Gentamicin FSIA, 1983). (2) The need of extremely pure analyte ligand or its derivative in preparing the fluorogenic enzyme substrate-analyte ligand conjugate. While this may not be a problem for drugs or haptenic analyte, it is a problematic and expensive proposition when the ligand is a macromolecule or an antigen. The situation is magnified if the antigen is scarsely available. The fluorogenic enzyme substrate-analyte ligand conjugates prepared from impure analyte ligand will result in high fluorescence background, because the labeled impurities will not be recognized by the antibody and therefore will continuously serve as fluorogenic substrates and yield fluorescent products. The consequence of impure conjugate is a decrease in the assay sensitivity. (3) The quenching of the fluorescence of 3-carboethoxy-7-hydroxy-coumarin by the covalently linked analyte ligand derivative may significantly decrease the sensitivity of the assay. This problem can be solved by attaching the analyte ligand to a site which, after enzymatic hydrolysis, is structurally not linked to the fluorescent product. Thus, freeing the fluorescent products from any quenching effect that might have been caused by the analyte ligand. For example, Li and Burd (1981) covalently linked a derivative of theophylline (Theo) to the adenine moeity of flavin adenine dinucleotide, (FAD) to form a Theo-FAD conjugate. Theo-FAD is a fluorogenic substrate for nucleotide pyrophosphatase. The enzymatic hydrolysis of Theo-FAD yields Theo-AMP and flavin mononucleotide (FMN). However, such an enzymatic hydrolysis is prevented if the Theo-FAD conjugate is bound by anti-theophylline antibody. The fluorescence intensity of the isoalloxazine group in FMN is about 9 times stronger than in FAD. The quenching of the fluorescence of isoalloxazine moiety in FAD is brought about by intramolecular contact of the isoalloxazine ring with the adenine ring of the AMP. Using Theo-FAD, anti-theophylline antibody and nucleotide pyrophosphatase system, an FSIA for theophylline was demonstrated. (4) Unlike most enzyme mediated immunoassays (Ngo and Lenhoff, 1982) which use an enzyme as label, FSIA, on the other hand, uses an enzyme substrate as the label. Consequently, the amplifying power of the enzyme is not manifested in FSIA. It is, therefore, of necessity to use a fluorogenic substrate label instead of a chromogenic substrate label. The role of enzyme in FSIA is merely to provide a means to quantify the proportion of the antibody bound and unbound fluorogenic enzyme substrate-analyte ligand conjugates. Without the participation of the amplifying power of an enzyme, the sensitivity of a FSIA is limited by the quantity of the fluorogenic enzyme substrate-analyte ligand conjugate that is introduced in an assay. The concentration of the conjugate is, in turn, controlled by the concentration of the analyte ligand.

ACKNOWLEDGEMENT

Figures 5 - 12 are adapted from the authors' publications and with the permission of the publishers.

REFERENCES

Benovic, J.L., Cary, L.C., Li, T.M., Hatch, R.P., and Burd, J.F., 1982, Substrate-labeled fluorescent immunoassay for caffeine in human serum, Clin Chem, 28, 1666.

Boguslaski, R.C., Li, T.M., Benovic, J.L., Ngo, T.T., Burd, J.F. and Carrico, R.C., 1980, Substrate labeled homogeneous fluorescent immunoassays for haptens and proteins, in: "Immunoassays: Clinical Laboratory Techniques for the 1980's," R.M. Nakamura, W.R. Dito, E.S. Tucker III, ed., Alan Liss, New York, pp. 45-64.

Burd, J.F., Wong, R.C., Feeney, J.E., Carrico, R.J. and Boguslaski, R.C., 1977, Homogeneous reactant labeled fluorescent immunoassay for therapeutic drugs exemplified by gentamicin determination in human serum, Clin Chem, 23, 1402-1408.

Burd, J.F., Carrico, R.J., Kramer, H.M. and Denning, C.E., 1978, Homogeneous substrate-labeled fluorescent immunoassay for determining tobramycin concentrations in human serum, in: "Enzyme Labelled Immunoassay of Hormones and Drugs," S.B. Pal, ed., Walter de Gruyter and Company, Berlin-New York, pp. 387-403.

Burd, J.F., 1981, The homogeneous substrate-labeled fluorescent immunoassay, in: "Methods in Enzymology," Volume 74, J.J. Langone and H.V. Vunakis, ed., Academic Press, New York, PP. 79-87.

Cope, A.C., Hofmann, C.M., Wyckoff, C. and Hardenbergh, E., 1941, Condensation reactions. II. alkyldene cyanoacetic and malonic esters, J Am Chem Soc, 63, 3452.

Csiszar, L., Li, T.M., Benovic, J.L. Buckler, R.T. and Burd, J.F., 1981, Substrate-labeled fluorescent immunoassay for quinidine, Clin Chem, 27, 1087.

deCastro, A.F., Lam, C.T., Place, J., Parker, D. and Patel, C., 1984, Kanamycin concentration in serum using a substrate-labeled fluorescent immunoassay, Clin Chem, 30, 1027.

Feinstein, H., Hovav, H., Fridlender, B., Inbar, D. and Buckler, R. T., 1982, Substrate-labeled fluorescent immunoassay (SLFIA) for valproic acid, Clin Chem, 28, 1665.

Greenquist, A.C., Walter, B. and Li, T.M., 1981, Homogeneous fluorescent immunoassay with dry reagents, Clin Chem, 27, 1614-1617.

Johnson, P.K., Messenger, L.J., Krausz, L.M., Buckler, R.T. and Burd, J.F., 1981, Substrate-labeled fluorescent immunoassay for primidone, Clin Chem, 27, 1093.

Krausz, L.M., Hitz, J.B., Buckler, R.T. and Burd, J.F., 1980, Substrate-labeled fluorescent immunoassay for phenobarbital, Ther Drug Monit, 2, 261-272.

Leaback, D.H., 1965, Preparation of methylumbelliferyl-β-D-galactoside, Clin Chim Acta, 12, 658.

Li, T.M., Benovic, J.L., Buckler, R.T. and Burd, J.F., 1981, Homogeneous substrate-labeled fluorescent immunoassay for theophylline in serum, Clin Chem, 27, 22-26.

Li, T.M. and Burd, J.F., 1981, Enzymic hydrolysis of intramolecular complexes for monitoring theophylline in homogeneous competitive protein-binding reactions, Biochem Biophys Res Comm, 31, 1157-1165.

Li, T.M., Miller, J.E., Ward, F.E. and Burd, J.F., 1982, Homogeneous substrate-labeled fluorescent immunoassay for carbamazepine, Epilepsia, 23, 391-398.

Lima, J.J., Shields, B.J., Howell, L.H. and Mackichan, J.J., 1984, Evaluation of fluorescence immunoassay for total and unbound serum concentrations of disopyramide, Ther Drug Monit, 6, 203-210.

Ngo, T.T., Carrico, R.J., Boguslaski, R.C. and Burd, J.F., 1981, Homogeneous substrate-labeled fluorescent immunoassay for IgG in human serum, J Immunolog Methods, 42, 93-104.

Ngo, T.T. and Lenhoff, H.M., 1981, Recent advances in homogeneous and separation-free enzyme immunoassays, Applied Biochem Biotechnol, 6, 53-64.

Ngo, T.T. and Lenhoff, H.M., 1982, Enzymes as versatile labels and signal amplifiers for monitoring immunochemical reactions, Mol Cell Biochem, 44, 3-12.

Pahuski, E.E., Hixson, C.S., Petrozolin, A.K., Hatch, R.P. and Li, T.M., 1982, Homogeneous substrate-labeled fluorescent immunoassay (SLFIA) for ethosuximide in human serum, Clin Chem, 28, 1664.

Pahuski, E.E., Manzuk, D.M., Maurer, J.L. and Li, T.M., 1983, Total automation of the SLFIA's for ethosuximide and valproic acid, Clin Chem, 29, 1236.

Place, J.D. and Thompson, S.G., 1983, Substrate-labeled fluorescent immunoassay for measuring dibekacin concentrations in serum and plasma, Antimicrob Agents Chemother, 24, 240-245.

Terumo Corp., 1983, Sept. 7, Japan Patent No. 58,150,861, Immunoassay for the determination of drugs.

Thompson, S.G. and Burd, J.F., 1980, Substrate-labeled fluorescent immunoassay for amikacin in human serum, Antimicrob Agents Chemother, 18, 264-268.

Walter, B., Greenquist, A.C. and Howard, III, W.E., 1983, Solid-phase strips for detection of therapeutic drugs in serum by substrate-labeled fluorescent immunoassay, Anal Chem, 55, 873-878.

Weetall, H.H. and Odstrchel, G., 1976, A biased solid-phase radioimmunoassay for thyroxine (T_4) using T_4-^{125}I presaturated antibodies, J Solid-phase Biochem, 1, 241-245.

Wong, R.C., Burd, J.F., Carrico, R.J., Buckler, R.T., Thoma, J. and Boguslaski, R.C., 1979, Substrate-labeled fluorescent immunoassay for phenytoin in human serum, Clin Chem, 25, 686-691.

Wong, R.C., George, R., Yeung, R. and Burd, J.F., 1980, A comparison of serum phenytoin determination by the substrate-labeled fluorescent immunoassay with gas chromatography, liquid chromatography, radioimmunoassay and 'EMIT', Clin Chim Acta, 100, 65-69.

Worah, D., Yeung, K.K., Ward, F.E. and Carrico, R.J., 1981, A homogeneous fluorescent immunoassay for human immunoglobulin M., Clin Chem, 27, 673-677.

IMMUNOASSAYS BASED ON CONFORMATIONAL RESTRICTION OF THE ENZYME LABEL INDUCED BY THE ANTIBODY

T.T. Ngo and H.M. Lenhoff

Department of Developmental and Cell Biology
University of California, Irvine, CA 92717

INTRODUCTION

We describe here an approach to separation-free enzyme immunoassays based on the ability of specific antibodies to ligands to restrict the conformational mobility of an apoenzyme that has already been labeled with that ligand, to fold into a catalytically active conformation.

In our model system (Scheme I), the holoenzyme is glucose oxidase; it requires flavine adenine dinucleotide (FAD) as its prosthetic group in order to function. Before using the enzyme in this immunoassay, however, it is necessary to label it covalently with the ligand to be tested [step (1)]. In our model system we treated the enzyme with 2,4-dinitrofluorobenzene in order to label it with 2,4-dinitrophenyl groups. The resultant enzyme was still active, and has to be active for this assay to work.

Next, in step (2) we removed the FAD from the labeled enzyme by an acid denaturation process (Morris et al., 1981; Ngo and Lenhoff, 1980a), and obtained labeled apoglucose oxidase by gel chromatography. This labeled apoglucose oxidase, however, was enzymatically inactive and could undergo one of two reactions. On one hand it could combine with added FAD to reconstitute the active enzyme [step (3)]. On the other hand, the bound ligand on the labeled apoenzyme could combine with antibody toward that ligand [step (4)] to form an antibody:apoenzyme complex that is incapable of reconstituting the enzymatically active form of the labeled enzyme on the addition of FAD [step (5)]. Presumably the antibody restricted the mobility of the polypeptide chain of the labeled apoenzyme, and thereby prevented it from folding into a catalytically active conformation on the addition of FAD as it did in the absence of antibody in step (3). Hence, based on such an antibody-induced conformational restriction of the labeled apoenzyme, we developed a separation-free (homogeneous) enzyme immunoassy called AICREIA (Ngo and Lenhoff, 1983).

With the above rationale in mind, to develop AICREIA we needed to demonstrate that we could: (a) combine apoglucose

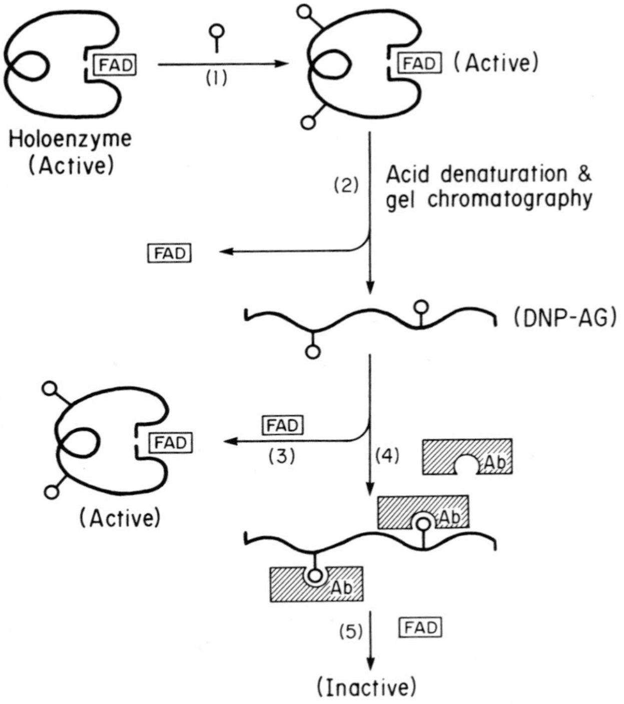

Scheme 1. Rational and experimental processes involved in developing AICREIA. Holoenzyme: glucose oxidase; ♀ or DNP: 2,4-dinitrophenyl groups; DNP-AG: denatured apoglucose oxidase labeled with 2,4-dinitrophenyl groups; and Ab: antibody to DNP-group.

oxidase (AG) labeled with dinitrophenol groups (DNP) with FAD to give enzyme activity; (b) demonstrate that the level of the enzyme activity formed by the above reaction is directly dependent on the concentration of FAD binding sites on the AG; (c) prevent the reconstitution of glucose oxidase activity from DNP-AG and FAD by adding serum containing antibodies for DNP (anti-DNP); and (d) demonstrate that a free ligand having DNP groups could compete with DNP-AG for anti-DNP. We describe the results of such a study and the overall reaction scheme for a model system, thereby demonstrating that the concept of AICREIA is practical to perform in the laboratory.

MATERIALS AND METHODS

Chemicals and suppliers: 2,4-dinitrofluorobenzene, flavin adenine dinucleotide, horseradish and N-2,4-dinitrophenyl-E-aminocaproic acid (DNP-ACA) (Sigma Chemical Co.); 3-dimethylaminobenzoic acid (DMAB) and 3-methyl-2-benzothiazolinone hydrazone (MBTH) (Aldrich Chemical Co.); rabbit anti-DNP serum and highly purified glucose oxidase from *Aspergillus niger* (Miles Labs, Inc.).

Glucose oxidase labeled with 2,4-dinitrophenyl groups (DNP) was made by reacting glucose oxidase with 2,4-dinitrofluorobenzene (Johnson et al., 1980).

Removal of FAD from labeled hologlucose oxidase by acid denaturation was accomplished by treating the labeled enzyme with sulfuric acid (Morris et al., 1981; Ngo and Lenhoff, 1980a).

Assay for glucose oxidase was carried out with a horseradish peroxidase coupled reaction using DMAB,MBTH and H_2O_2 as substrates (Ngo and Lenhoff, 1980b).

RESULTS

The FAD of DNP-labeled hologlucose oxidase treated with sulfuric acid (Ngo and Lenhoff, 1980a; Morris et al., 1981) dissociated from the holoenzyme, and was separated from the denatured DNP-labeled apoglucose oxidase (DNP-AG) by gel permeation chromatography on a desalting gel. The enzyme activity of the acid denatured DNP-AG was restored by incubating it with FAD at around pH 7 (see also Johnson et al., 1980). By increasing the concentration of FAD, we observed an increasing amount of glucose oxidase activity until a maximal level of reconstitutable enzyme activity was achieved (Fig. 1). From these experiments we calculated the concentration of FAD binding sites. The level of reconstitutable glucose oxidase activity is directly dependent on the concentration FAD binding sites (Fig. 2).

Addition of increasing amounts of rabbit anti-DNP serum to solutions containing a fixed concentration of acid denatured DNP-AG, and preincubating them at 25°C for 30 min, gave decreasing levels of reconstitutable glucose oxidase activity being formed on the addition of excess FAD (Fig. 3). In contrast, when anti-DNP serum was added to DNP-labeled hologlucose oxidase which was not previously treated with sulfuric acid or to DNP-labeled hologlucose oxidase which was reconstituted from DNP-AG and FAD, we observed no change in enzyme activity upon adding such anti-serum.

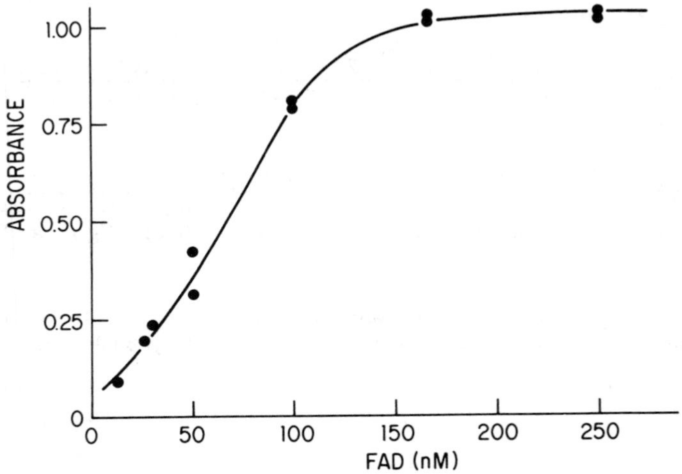

Fig. 1. Reconstitution of glucose oxidase activity from DNP-AG and FAD. Solutions (10 μℓ) containing varying concentrations of FAD (25-500 nM) in 0.1 M sodium phosphate pH 6.5 were added to 10 μℓ DNP-AG and incubated at 25° for 30 min. Assay solution (4 ml) was added to each of the above solutions and incubated at 25° for another 5 min before the absorbance (at 590 nm) was read. The assay solution was made up in 0.1 M sodium phosphate pH 6.5. It contained 75 mM β-D-glucose, 3 mM DMAB, 0.075 mM MBTH, 31.25 nM horseradish peroxidase and 37.5 mM NaCl.

To solutions containing different concentrations of DNP-ACA, the following sequence of materials were added and mixed separately: a fixed concentration of DNP-AG and of anti-DNP serum, and excess FAD and excess enzyme assay solution. We found that the enzyme activity increased with increasing concentrations of DNP-ACA (Fig. 4). Presumably the DNP-ACA competed with the DNP group of DNP-AG for the anti-DNP for antibodies; hence, there were less antibodies to complex with DNP-AG. As a result, the free, uncomplexed DNP-AG presumably folded into the right conformation allowing it to bind FAD and therefore become enzymatically active.

DISCUSSION

Separation-free enzyme immunoassays usually employ an anti-ligand antibody to modulate the activity of a covalently linked: (i) ligand-enzyme conjugate (Rubenstein et al., 1972); (ii) ligand-substrate or ligand-prosthetic group conjugate (Burd et al., 1977; Ngo et al., 1981; Morris et al., 1981); and

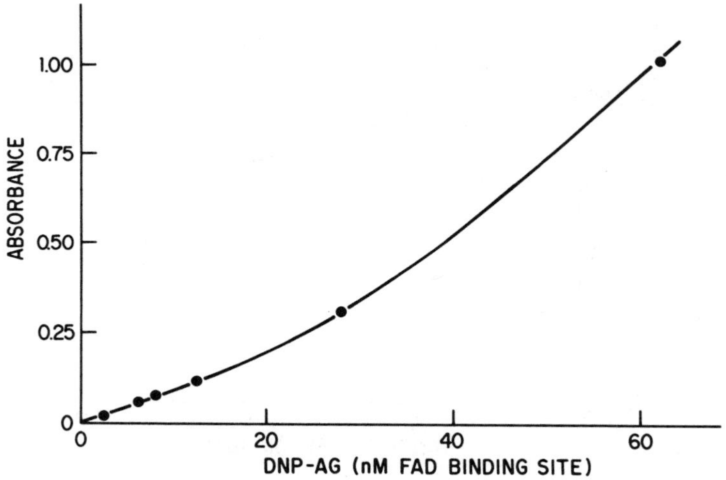

Fig. 2. Relationship between reconstitutable glucose oxidase activity and the concentration DNP-AG. Solutions of 10 μℓ containing varying amounts of DNP-AG, expressed in terms of FAD binding sites, were added to 10 μℓ of 1000 nM FAD (saturating concentration, see Fig. 1), and incubated at 25° for 15 min; then 4 ml of assay solution (composition described in legend of Fig. 1) were added and the solutions were incubated at 25° for 15 min before measuring their absorbance at 590 nm.

(iii) ligand-enzyme inhibitor conjugate (Ngo and Lenhoff, 1980c).

We describe here a different approach to separation-free enzyme immunoassays. This approach, called AICREIA, is based on the ability of specific anti-ligand antibodies to restrict the

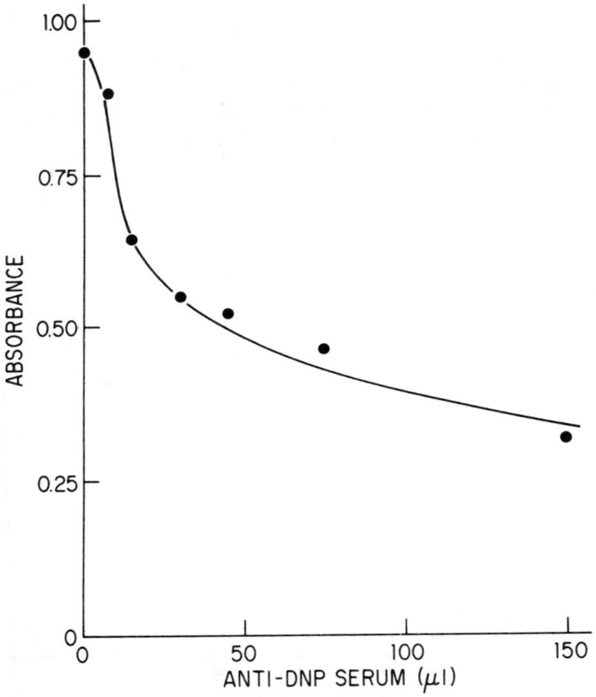

Fig. 3. Prevention of reconstitution of glucose oxidase activity from DNP-AG and FAD by anti-DNP serum. Varying amounts of anti-DNP serum were added to 50 μℓ DNP-AG (25 nM of FAD binding sites). The solutions were adjusted to 200 μℓ with 0.1 M sodium phosphate, pH 6.5, and incubated at 25° for 30 min. To these solutions were added 100 μℓ 10 μM FAD, the solutions were incubated at 25° for another 30 min. Glucose oxidase activity was measured as described in legend to Fig. 1.

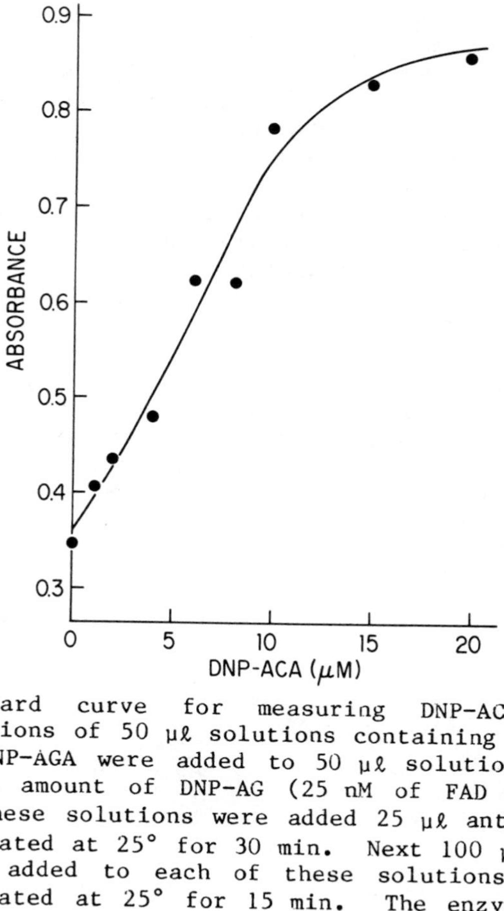

Fig. 4. Standard curve for measuring DNP-ACA by AICREIA. Solutions of 50 μℓ solutions containing various amounts of DNP-AGA were added to 50 μℓ solutions containing a fixed amount of DNP-AG (25 nM of FAD binding sites). To these solutions were added 25 μℓ anti-DNP serum and incubated at 25° for 30 min. Next 100 μℓ of 100 nM FAD were added to each of these solutions and they were incubated at 25° for 15 min. The enzyme activity was assayed by adding 4 ml of assay solution to the above solutions and incubating them at 25° for 15 min before taking the absorbance.

conformational mobility of a ligand labeled protein to fold into a catalytically active conformation.

The rationale and experimental stages for developing ACREIA have been described in the Introduction and are depicted in Scheme 1.

The overall process of AICREIA as described in this chapter is summarized in Scheme 2. The success of AICREIA depended upon the experiments summarized in Figures 1-4. The results in Figures 1 and 2 showed in extremely acidic conditions DNP-labeled glucose oxidase holoenzyme was denatured and dissociated into FAD and DNP-AG. Neither FAD nor DNP-AG possessed significant enzyme activity. The combining of FAD to DNP-AG at about pH, 7 however, restored enzyme activity. Such a restoration of activity was prevented if DNP-AG was preincubated with anti-DNP serum prior to adding FAD (Fig. 3). Addition of free DNP in the form of DNP-ACA to DNP-AG counteracted the inhibitory action of anti-DNP serum with a concomitant increase in enzyme activity (Fig. 4).

Hence, following Scheme 2, in the absence of the analyte (DNP-ACA), reaction (a) would not occur and DNP-AG and the antibody (Ab) would combine [reaction (b)] leaving less DNP-AG to combine with FAD [reaction (c)], thereby forming more enzymatically active DNP-AG:FAD. Thus, the amount of combined product formed [reaction (D)] is proportional to the amount of

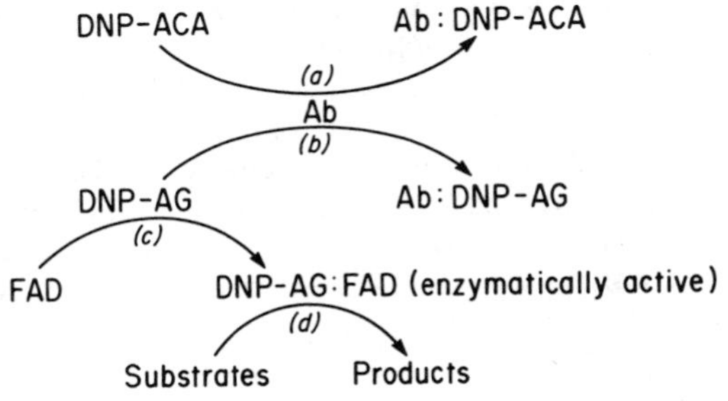

Scheme 2. Principles of AICREIA as applied to detecting DNP-ACA as analyte. DNP-ACA:2,4-dinitrophenyl-ε-aminocaproic acid. Other symbols are explained in lengend to Scheme 1.

analyte present. The results presented in Figures 3-4 bear out this new type of separation-free (homogeneous) enzyme immunoassay.

REFERENCES

Burd, J.F., R.C. Wong, J.E. Feeney, R.J. Carrico, and R.C. Boguslaski (1977). Homogeneous reactant-labeled fluorescent immunoassay for theraupetic drugs exemplified by gentamicin determination in human serum. Clin. Chem. 23: 1402-1408.

Johnson, Jr., R.B., R.M. Libby, and R.M. Nakamura (1980). Comparison of glucose oxidase and peroxidase as labels for antibody in enzyme-linked immunosorbent assay. J. Immunoassay 1: 27-37.

Morris, D.L., P.B. Ellis, R.J. Carrico, F.M. Yeager, H.R. Schroeder, J.P. Albarella, R.C. Boguslaski, W.E. Hornby, and D. Rawson (1981). Flavin adenine dinucleotide as a label in homogeneous colorimetric immunoassays. Anal. Chem. 53: 658-665.

Ngo, T.T. and H.M. Lenhoff (1980a). Amperometric determination of picomolar levels of flavin adenine dinucleotide by cyclic oxidation-reduction in apo-glucose oxidase system. Anal. Letters 13: 1157-1165.

Ngo, T.T. and H.M. Lenhoff (1980b). A sensitive and versatile chromogenic assay for peroxidase and peroxidase-coupled reactions. Anal. Biochem. 105: 389-397.

Ngo, T.T. and H.M. Lenhoff (1980c). Enzyme modulators as tools for the development of homogeneous enzyme immunoassays. FEBS Letters 116: 285-288.

Ngo, T.T. and H.M. Lenhoff (1983). Antibody-induced conformational restriction as basis for new separation-free enzyme immunoassay. Biochem. Biophys. Res. Comm. 114: 1097-1103.

Ngo, T.T., R.J. Carrico, R.C. Boguslaski, and J.F. Burd (1981). Homogeneous substrate-labeled fluorescent immunoassay for IgG in human serum. J. Immunol. Methods 42: 93-103.

Rubenstein, K.E., R.S. Schneider, and E.F. Ullman (1972). "Homogeneous" enzyme immunoassay: a new immunochemical technique. Biochem. Biophys. Res. Comm. 47: 846-851.

NONSEPARATION ENZYME IMMUNOASSAYS FOR MACROMOLECULES

Ian Gibbons

Syva Research Institute
900 Arastradero Road
Palo Alto, California 94304

INTRODUCTION

Two classes of homogeneous enzyme immunoassays for macromolecules have been developed at the Syva Research Institute. The first class uses β-galactosidase together with a macromolecular substrate (Gibbons et al., 1980). The second is based on the phenomenon of "enzyme channeling" (Litman et al., 1980) involving two or more enzymes. This chapter will discuss the principles, characteristics, performance and clinical utility of both types of assay.

Several previous publications document the preparation and characterization of the reagents so the experimental section will be restricted to a description of assay protocols and results.

ASSAYS USING β-GALACTOSIDASE AND MACROMOLECULAR SUBSTRATES

Three assays for serum proteins will be described: a colorimetric enzyme inhibition immunoassay (Gibbons et al., 1980), a turbidometric enzyme activation immunometric assay (Gibbons et al., 1981) and a more sensitive fluorometric enzyme inhibition immunoassay (Armenta et al., submitted).

Colorimetric Enzyme Inhibition Immunoassay

In deciding how to configure a homogeneous enzyme immunoassay for macromolecules, the size of the antigen was an obvious concern. Antibody binding to enzyme labeled antigen would not be expected to have any effect on enzyme activity if the substrate were small because the bulky antigen would not transmit any effect to the enzyme.

If the substrate were large, however, antibody binding should cause inhibition by steric exclusion of the substrate, thus permitting an enzyme inhibition immunoassay to be set up.

Although many enzymes have macromolecular substrates, they were rejected as enzyme labels in favor of β-galactosidase. β-Galactosidase from *E. coli* is commercially available, stable, has a high turnover number with conveniently measured chromogenic substrates

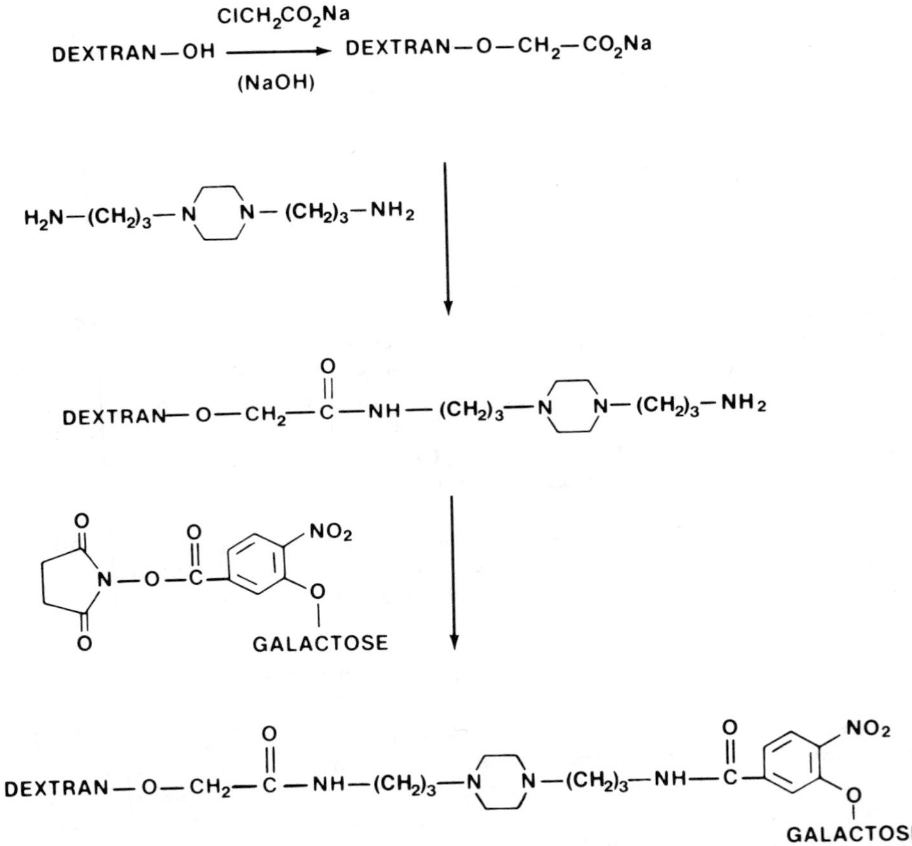

Fig. 1. Preparation of a chromogenic macromolecular substrate for β-galactosidase. The chemistry was carried out in aqueous solution as described by Skold (1981). Dextrans with average molecular weights ranging from 40,000 to 180,000 were used. The degree of substitution was about one galactoside per five glucose residues.

and is readily coupled to proteins without loss of activity. Moreover, no interfering activity was found in human serum at the pH range where the *E. coli* enzyme is active.

No macromolecular substrate being available for β-galactosidase, Skold (1981) prepared a substrate polymer by attaching many molecules of *o*-nitrophenylgalactoside, a good chromogenic substrate, to dextran (as shown in Fig. 1). The conjugate, after optimization of the linking chemistry, was an excellent substrate giving a maximum velocity about half that of the parent small substrate (Skold and Gibbons, unpublished observations).

Using this macromolecular substrate, Gibbons et al. (1980) established an immunoassay for serum proteins. The assay reagents and principle are illustrated in Fig. 2.

By combining the macromolecular substrate with antibody to make a single reagent, a very rapid, two reagent assay protocol is possible. Fig. 3 shows such a protocol for a commercially available version of the assay used to measure serum C-reactive protein, which is a sensitive indicator of inflammation and tissue necrosis (Pepys, 1981). This assay was designed for use with an automated analyser but can also be carried out manually. Table 1 shows that the performance of the C-reactive protein assay is comparable with that of other standard methods. The assay range is from 1-16 mg/dl. Within this range, precision is about 5-10%. The major advantage of this enzyme inhibition assay is its adaptability to automated analysers equipped with standard spectrophotometric detectors.

Table 1. Performance of the EMIT® Assay for Serum C-Reactive Protein Compared with Standard Methods[a]

Reference Method	Number of Samples	Correlation Coefficient	Slope of Regression Line	Intercept mg/dl
Radial Immuno-diffusion	75	0.99	0.98	0.13
Nephelometry	85	0.99	1.05	-0.08
Pocket Electro-immunodiffusion	70	0.96	1.03	-0.88

[a]Source: Information supplied with the EMIT C-reactive protein assay, December 1982. Syva Company, Palo Alto, California 94304.

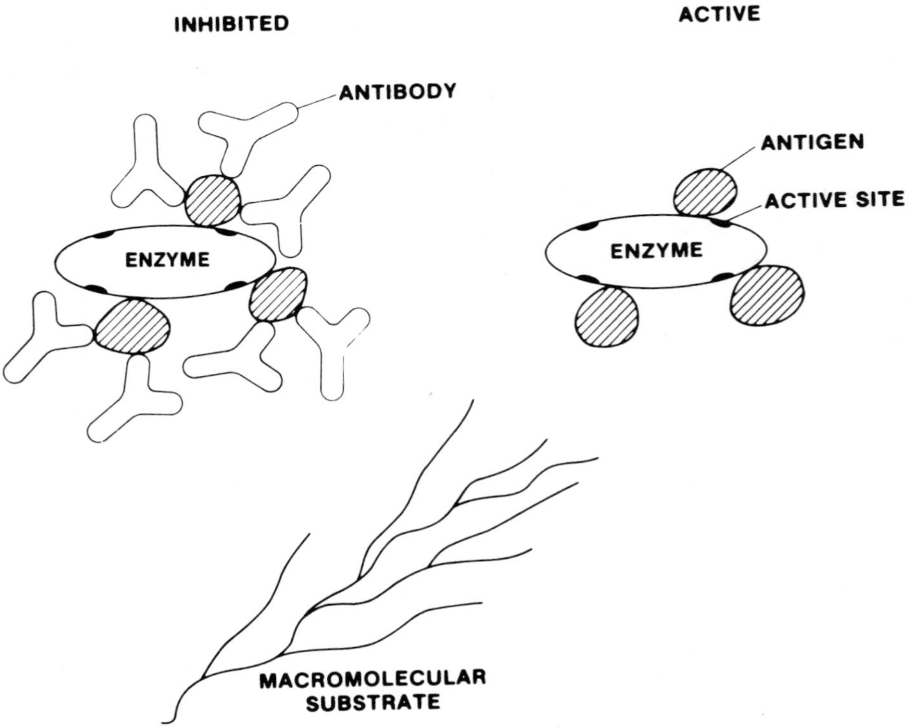

Fig. 2. Reagents and assay principles of the colorimetric enzyme inhibition assay. β-galactosidase labeled protein antigen is constructed so that each molecule of enzyme has several antigens attached (Gibbons et al., 1980). The macromolecular substrate (see Fig. 1) is made from dextran, a flexible linear branched polymer, which can easily reach the active sites of the enzyme. The binding of antibodies to the enzyme labeled antigen inhibits enzyme activity (up to 95%) by causing a steric block. In an immunoassay, the sample is incubated with antibody and enzyme labeled antigen either simultaneously or sequentially. Sample antigen binds to antibody, leaving less antibody to bind to and inhibit the enzyme labeled antigen. The resulting activity of the enzyme-labeled antigen is therefore directly related to the sample antigen concentration.

1. Dilute 8.3 µl of the serum sample with 300 µl buffer.

2. Add 50 µl reagent A and 250 µl buffer.

3. Add 50 µl reagent B and 250 µl buffer.

4. Aspirate into the flow cell of a spectrophotometer (regulated to 37°) and read the change in absorbance (1 cm light path) at 405 nm between 15 sec and 45 sec.

Reagent A:	Anti C-reactive protein and chromogenic macromolecular substrate
Reagent B:	β-galactosidase labeled C-reactive protein
Buffer:	Imidazole - HCl, pH 7.1

Fig. 3. Colorimetric enzyme inhibition immunoassay for C-reactive protein in serum:protocol.

Turbidometric Enzyme Activation Immunometric Assay

The assay described above requires highly purified antigen to make the enzyme labeled reagent. Unfortunately, it is often difficult to purify and stabilize serum proteins and other macromolecules of clinical interest. To circumvent this problem, Gibbons et al. (1981) looked for ways to convert the above assay into an immunometric method (i.e., one which uses labeled antibody, cf Miles and Hales, 1968). Fig. 4 illustrates how this was achieved.

Before describing the assay in detail, it will be helpful to look at a model experiment designed to study the role of charge in the colorimetric enzyme inhibition assay. β-galactosidase-labeled human IgG was titrated with anti-IgG or with anti-IgG which had been succinylated to various extracts to increase its negative charge. Enzyme activity was then determined with the chromogenic macromolecular substrate. As shown in Fig. 5, native antibody inhibited the activity by as much as 85%. In contrast, lightly succinylated antibody (24 groups/molecule) caused essentially no effect and heavily succinylated (44 groups/molecule) antibody caused increased enzyme activity. Obviously, the effect of antibody binding depends on the charge of the antibody. Native antibody causes

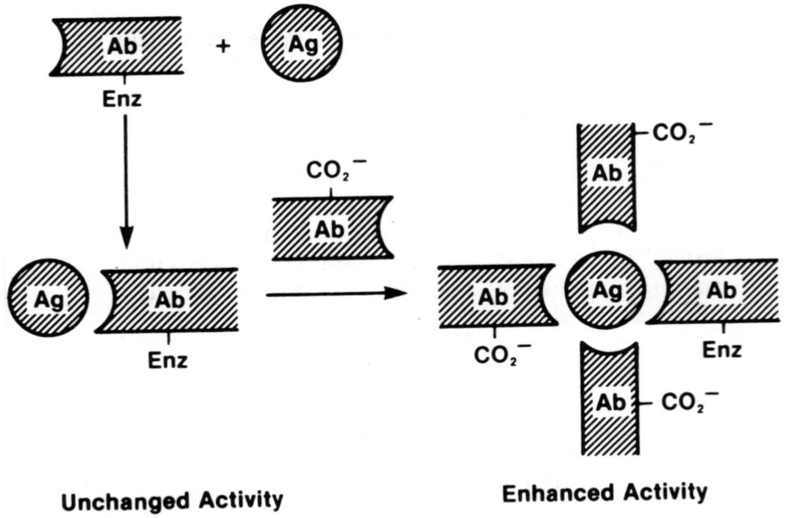

Fig. 4. Turbidometric enzyme enhancement assay for polyvalent antigens. Polyvalent antigen in the sample is reacted successively with β-galactosidase labeled antibody and antibody modified so as to be highly negatively charged (e.g., by reaction with succinic anhydride). In the resulting complex, the enzyme is brought into a negatively charged environment. When assayed with the polycationic macromolecular substrate (Fig. 1), the complexed enzyme conjugate exhibits enhanced activity relative to non-complexed enzyme conjugate or enzyme conjugate complexed with antigen alone.

inhibition of enzyme activity due to steric exclusion. (This phenomenon is the basis of the assay described above.) Two properties of the macromolecular substrate are presumably responsible for the activation seen with the highly negatively charged antibody. The substrate is a polycation by virtue of the positively charged linking groups connecting dextran and to the *o*-nitrophenylgalactoside substituents. Furthermore, dextran has a flexible backbone. Because of these two properties, charge-charge interaction can overcome the steric exclusion seen with native antibody.

The activation effect has been shown to result not from an increase in turnover of substrate but from a change in the nature of the product (Gibbons et al., 1981). β-Galactosidase, when in a negatively charged environment (e.g., when negatively charged anti-

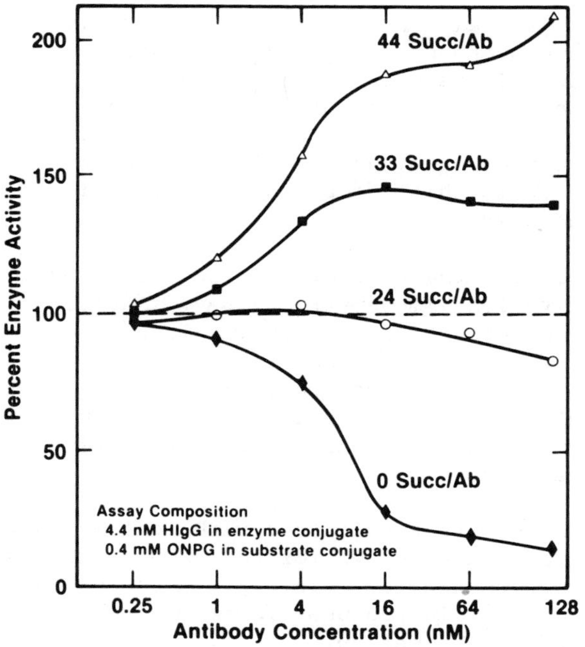

Fig. 5. Effects of native and negatively charged antibody on the activity of enzyme labeled antigen. Human IgG (0.76 μg) labeled with β-galactosidase (0.50 μg) was incubated in 0.1 ml of buffer for 3 h at room temperature with sheep anti-human IgG carrying the indicated number of succinyl groups. Enzyme activity was then measured by adding 0.4 μ moles of chromogenic macromolecular substrate (Dextran MW = 40,000) in 0.9 ml buffer and recording the change in absorbance at 420 nm for 30 seconds at 37°. The final concentration of antibody is indicated. Reagents were made as described by Gibbons et al. (1980, 1981) and were dissolved in 10 mM sodium phosphate pH 7.0 containing 0.13 M NaCl, 5 mM NaN_3, 1 mM magnesium acetate and 1 mg rabbit serum albumin per ml.

This figure is reproduced from Ullman (1981), by permission of Elsevier/North Holland Biomedical Press.

body is bound to enzyme labeled antigen), acts on the macromolecular substrate by hydrolyzing essentially all the substrate substituents on each substrate polymer which binds to the enzyme. When this occurs, the product forms a second aqueous phase which is seen as a turbid suspension of small droplets. In contrast, when the enzyme is not bound to negatively charged antibody, hydrolysis of macromolecular substrate proceeds by a random mechanism in which a soluble (non-turbid) product is released from the enzyme after one or only a few hydrolytic events. By measuring the increase in turbidity of the reaction mixture due to the enzyme reaction, it is then possible to detect the binding of negatively charged antibody to enzyme labeled antigen. It is this effect which we exploit in the immunometric activation assay.

Fig. 6 exemplifies the type of response obtained in the turbidometric enzyme activation immunometric assay. Antigen alone caused almost no change in the activity of enzyme labeled antibody. When excess succinylated (negatively charged) antibody was added, however, more than twofold activation was seen. The assay response is biphasic because, at very high antigen levels, the succinylated antibody is consumed and not enough remains to activate enzyme labeled antibody:antigen complex.

The assay can be performed either with long incubations or in a "mix and read" fashion as shown for the case of an assay for C-reactive protein in human serum (Fig. 7). Although the rapid protocol involves some loss in response for any given antigen level, it is better suited to automation.

When results (y) from the incubation protocol (Fig. 7) were compared with those obtained by radial immunodiffusion (x) in a clinical study on 17 patients' samples, the regression analysis was as follows: $y = 1.04x - 2.1$, $r = 0.96$; S.E. = 7.1 μg/ml. The coefficient of variation of replicate analyses was 5%. The assay clearly quantifies C-reactive protein levels in clinical samples satisfactorily.

Fluorometric Enzyme Inhibition Immunoassay

The sensitivity of the spectrophotometric assays outlined above is sufficient to quantify serum protein concentrations down to about 10 μg/ml. While many serum proteins of clinical interest are found at concentrations higher than 10 μg/ml, there are others for which much greater sensitivity is needed.

By using a fluorogenic rather than a chromogenic macromolecular substrate, a one thousand-fold improvement in the sensitivity of the enzyme inhibition immunoassay has been achieved (Armenta et al., submitted). Umbelliferone galactoside was substituted for *o*-nitrophenylgalactoside in the dextran-linked substrate giving the structure shown in Fig. 8. Using this fluorogenic substrate and a conventional fluorometer, concentrations of β-galactosidase that are one thousand times lower can be determined precisely (±1%) in a short (30 second) read time compared with spectrophotometric measurements with the chromogenic macromolecular substrate.

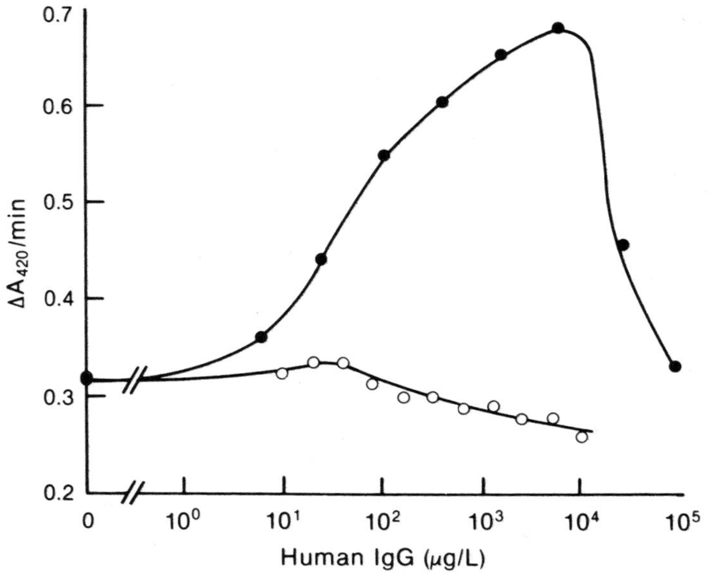

Fig. 6. Effect of succinylated antibody on the activity of a complex of human IgG (antigen) with a β-galactosidase-antibody conjugate. Anti-human IgG labeled with 0.5 μg of β-galactosidase was incubated for 1 h with human IgG in 0.3 ml of buffer at room temperature. A further 0.35 ml of buffer either with (·) or without (o) 0.17 mg succinylated antibody (45 groups/molecule) followed by 0.4 μ mole of chromogenic macromolecular substrate (dextran molecular weight 40,000) was added and the rate of change of apparent absorbance recorded over 30 s at 37°. Human IgG concentrations are those in the final assay mixture. The reagents were described by Gibbons et al. (1980, 1981) and were dissolved in the same solvent as in Fig. 5.

This figure is reprinted from Gibbons et al. (1981) with permission of the American Association for Clinical Chemistry.

Ferritin was selected as a representative low concentration serum analyte. Serum levels are of clinical interest since iron deficiency anemia results in lowered concentrations of ferritin (Worwood, 1979).

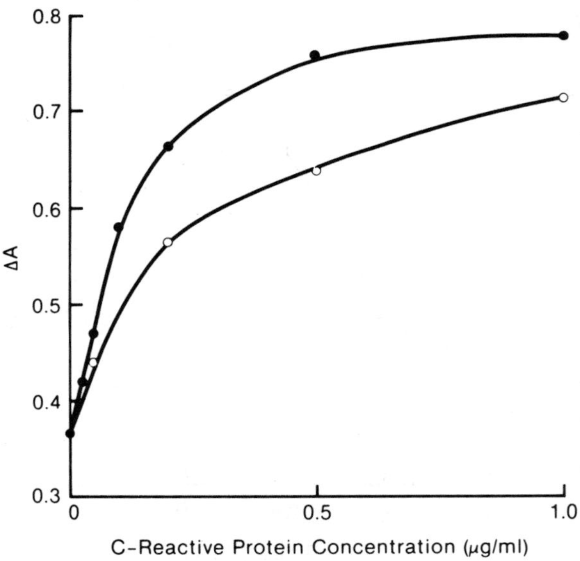

Fig. 7. Turbidometric enzyme activation immunometric assay for C-reactive protein in serum. Anti-human C-reactive protein labeled with 2 μg of β-galactosidase was incubated with human C-reactive protein in 0.7 ml of buffer for 2 h (•) or 15 s (o), after which 0.5 ml of a solution containing a mixture of succinylated anti-human C-reactive protein (0.12 mg sheep IgG) and chromogenic macromolecular substrate (dextran molecular weight 40,000) (1.6 μ mole) was added. Change in apparent absorbance over 20 seconds was immediately measured at 37°.

Reagents were made according to Gibbons et al. (1981) and were dissolved in the same solvent as in Fig. 5 except that the concentration of salts was increased by 15%.

This figure is reprinted from Gibbons et al. (1981) with permission of the American Association for Clinical Chemistry.

As in the colorimetric enzyme inhibition immunoassay, the antigen is labeled with enzyme. Three β-galactosidase molecules were attached to each ferritin. The conjugate activity with fluorogenic macromolecular substrates was inhibited by incubation with goat antiferritin (Fig. 9). Using saturating levels of

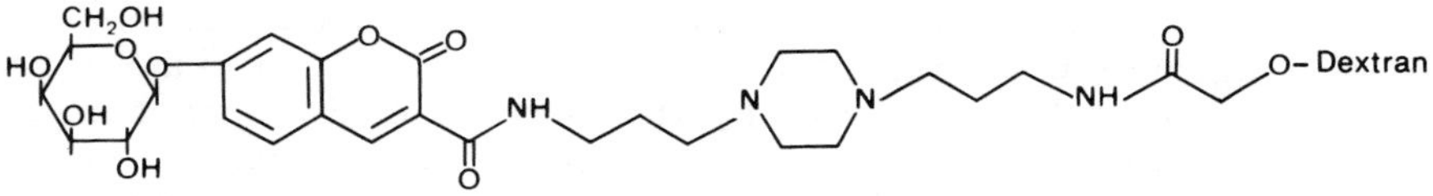

Fig. 8. Structure of a fluorogenic macromolecular substrate for β-galactosidase.

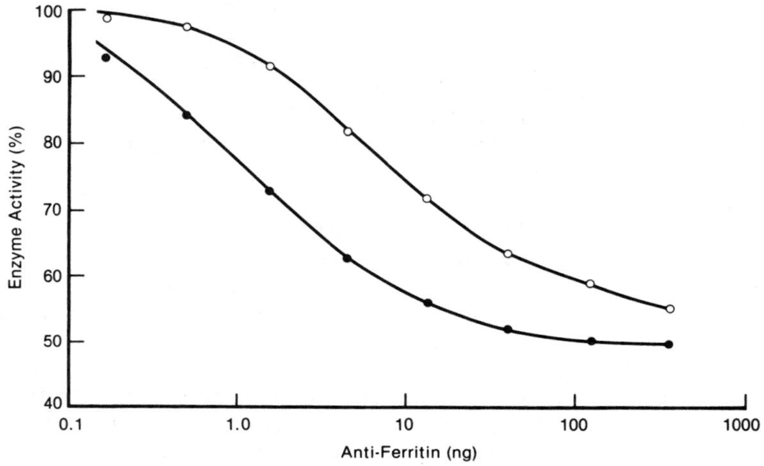

Fig. 9. Inhibition of β-galactosidase labeled ferritin by anti-ferritin. β-galactosidase (1 ng) coupled to ferritin was incubated in 0.2 ml of buffer for 2.5 h at room temperature with goat anti-ferritin. Enzyme activity was determined after adding 0.7 ml of a solution containing 40 n moles fluorogenic macromolecular substrate either with (•) or without (o) 1 μg rabbit anti-goat IgG. Fluorescence was measured using a Varian Fluorochrome fluorimeter equipped with a 405 nm excitation filter and a 450 nm emission filter with 10 nm band widths. Change in fluorescence over 30 seconds at 37° was recorded. Reagents were as described by Armenta et al. (submitted). All reagents were dissolved in 17.5 m$\underline{M}$ sodium phosphate pH 7.0 containing 0.23 $\underline{M}$ NaCl, 8.8 m$\underline{M}$ NaN_3, 1.75 m$\underline{M}$ magnesium acetate and 5 mg rabbit serum albumin/ml.

This figure is from Armenta et al. (submitted) and is reprinted by permission.

antibody, about 50% inhibition was seen. Addition of excess anti-goat IgG together with the fluorogenic macromolecular substrate served to greatly increase the inhibition caused by subsaturating amounts of antiferritin (Fig. 9). There was only marginally greater inhibition of high levels of antiferritin, however. Presumably, the "second" antibody acts by binding to antiferritin and thus increasing the steric block to macromolecular substrate.

To test the feasibility of a competitive immunoassay, serum samples containing ferritin were incubated with enzyme labeled ferritin and antiferritin prior to addition of substrate and anti-goat IgG. Results similar to those in Fig. 10 were obtained, indicating that the assay was sensitive enough to quantify clinically relevant ferritin concentrations (10-500 ng/ml). There were, however, severe problems with many serum samples which could not be quantified because they contain an inhibitor of β-galactosidase. In some cases the inhibitor has been shown to be antibody to β-galactosidase which inhibits activity by blocking access of macromolecular substrate to the enzyme.

Two techniques to eliminate serum interference in the ferritin immunoassay have been developed. Firstly, the enzyme ferritin conjugate was "protected" from antienzyme by coupling many albumin molecules all over the enzyme. Interestingly, this can be done without reducing enzyme activity.* The albumin appears to prevent binding of antienzyme. Secondly, an inactive β-galactosidase derivative which retains much of the native antigenic character of the enzyme was made. Large amounts (in great excess over the enzyme labeled ferritin) of the inactive enzyme can be added to the ferritin immunoassay so as to bind to antienzyme in the samples. Using both these methods, serum interference in the ferritin immunoassay can be essentially eliminated.

The assay response shown in Fig. 10 was obtained with the albumin-covered enzyme conjugate and the inactive enzyme additive. Ferritin can be quantified in the range 10-500 ng/ml. The results of the assay (y) correlate well with those from a commercial radioimmunoassay (x). Regression analysis gave: $y = 0.80x - 5.2$; $r = 0.97$, $N = 31$; S.E.E. = 8.3 ng/ml. In this comparison, patients' sera including several strongly inhibitory sera were determined and the results were not discrepant.

*Steric inhibition of enzyme activity with the fluorogenic macrosubstrate is strongly influenced by electrostatic interactions. Thus, substrate (which is strongly positively charged) is readily blocked by binding of antibody, which is almost neutral, to enzyme labeled ferritin but is not blocked by covalently bound albumin which carries a negative charge.

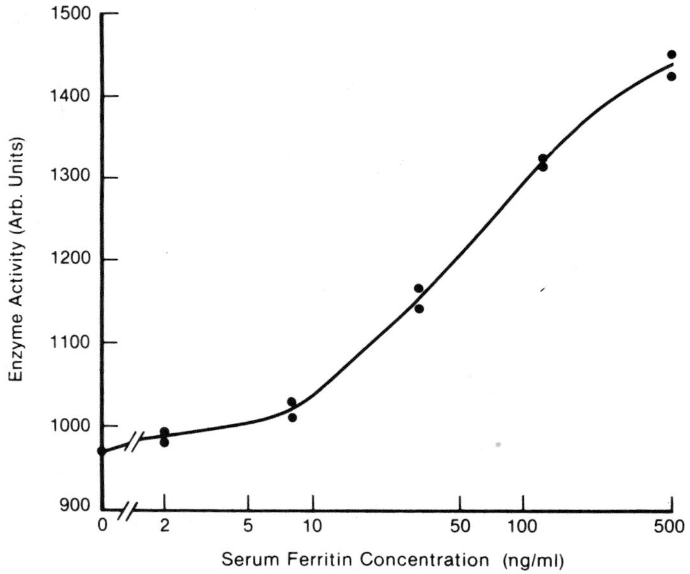

Fig. 10. Fluorogenic enzyme inhibition immunoassay for serum ferritin. Serum calibrators with known ferritin content were made by adding pure human liver ferritin to ferritin-free sera. Samples (10 μl) were incubated for 2 h at room temperature with 1 ng β-galactosidase coupled to ferritin and albumin and 10 μg inactive β-galactosidase in a final volume of 0.12 ml. Enzyme activity was determined fluorometrically (as in Fig. 9) at 37° for 30 seconds immediately after addition of 0.75 ml of a solution containing 0.11 μ moles of fluorogenic macromolecular substrate and 10 μl of anti-goat IgG. Reagents were as described in Armenta et al. (submitted). All reagents were made up in the buffer described in Fig. 9.

This figure is from Armenta et al. (submitted) and is reprinted by permission.

Discussion

The three assays described above all depend on strong interactions between antibody and a polymeric substrate. Native antibodies are almost neutral in charge and on binding to enzyme labeled antigen block access of macromolecular substrate and therefore inhibit the enzyme activity. As would be expected, both increasing the size of the macromolecular substrate and bulking the antibody with a second (anti-immunoglobulin) antibody increases the inhibition (Crowl et al., 1980).

Both the chromogenic and the fluorogenic macromolecular substrates are highly charged cationic polymers. After attaching many negative charges to antibodies, the interaction between antibody and macromolecular substrate becomes attractive. Coulombic forces overcome the steric exclusion. According to the charge on the antibody, either inhibition or activation assays can be set up with either substrate.

The assays I present here give a cross-section of the possibilities which have been shown to be practical for clinical purposes. All of them were designed to have a short (< 1 min) measurement of enzyme activity so that high throughput can be achieved with automated instrumentation. The fluorometric enzyme inhibition immunoassay requires a prolonged (about 1 hour) incubation prior to reading the enzyme activity, but this does not compromise throughput since assays can be batched. The reagents used in these assays are stable for many weeks, both in solution and after lyophilization, provided appropriate buffering agents, salts, carrier proteins (e.g., albumin) and antimicrobial agents (e.g., sodium azide) are present.

Two significant problems have emerged. Firstly, the activity of enzyme with macromolecular substrate is very sensitive to the ionic composition of the medium. The ionic composition of the reagents must therefore be very carefully controlled. Precise pipetting of reagents and samples is also crucial so that the composition of the final reaction medium is kept constant. Variations between serum samples do not cause any difficulties since there is very little difference in ionic content between sera and the sample represents a maximum of only 1% of the final reaction volume. Secondly, the presence in serum of factors (e.g., antibody) which bind to β-galactosidase presented a problem in assays where a prolonged incubation is needed. Effective counter measures have been worked out, however.

Each of the assays described here has been shown to be applicable to several analytes. It appears that essentially any protein antigen can be measured in serum using these methods. The sensitivity limits are about 10 μg/ml for the assays using the chromogenic substrate and 10 ng/ml when the fluorogenic substrate is used. I have restricted discussion to the measurement of antigens but obviously these methods can be configured to measure antibodies (see, for example, Fig. 6 and Fig. 7 in Gibbons et al., 1981).

There are some related techniques which exploit steric exclusion of a macromolecular substrate from an enzyme in nonseparation immunoassays. Morita and Woodburn (1978) used β-amylase labeled antigen to set up an assay for Staphylococcal enterotoxin B. The enzyme activity measurement was elaborate and would limit the value of the method for routine clinical use. Ngo et al. (1981) introduced a technique using a protein antigen labeled with a fluorogenic substrate for β-galactosidase. They could measure serum antigen concentrations by allowing the substrate labeled antigen and sample antigen to compete for antibody and then determining the extent of substrate hydrolysis caused by adding excess β-galactosidase.

ENZYME CHANNELING IMMUNOASSAY

Enzyme channeling is a phenomenon which occurs when two (or more) enzymes catalyzing sequential reaction (Scheme 1) are together in a microenvironment where restricted diffusion causes the local concentration of product 1 to exceed that in the bulk phase. Product 2 is then made more rapidly than when the enzymes are free in solution or when only one enzyme is in the microenvironment.

Scheme 1

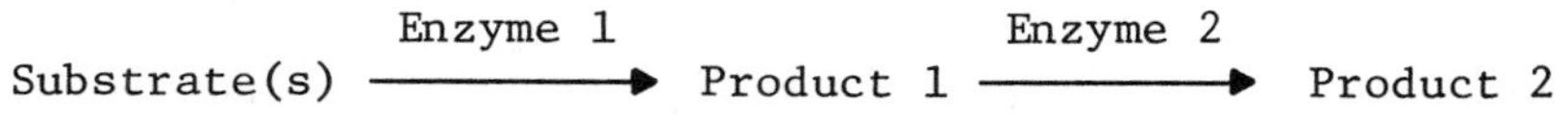

In this section I will discuss nonseparation enzyme channeling immunoassays in which immunological binding brings two suitable enzymes together within a suitable microenvironment. Several enzyme channeling immunoassays have been described previously using a variety of enzyme pairs and microenvironments (Litman et al., 1980; Ullman et al., 1983). Ullman et al. (1983) have thoroughly reviewed the theory of enzyme channeling as applied to immunoassays so I will only briefly deal with theoretical considerations here. Instead, I have chosen to present two assays which illustrate the scope, potential, and limitations of enzyme channeling-based methods. These assays are classified (following Ullman et al., 1983) according to which enzyme constitutes the label.

First Enzyme as Label

Litman et al. (1980) introduced an immunoassay method using hexokinase and glucose-6-phosphate dehydrogenase as the enzyme pair. The reactions catalyzed by this pair are shown in Scheme 2.

Scheme 2

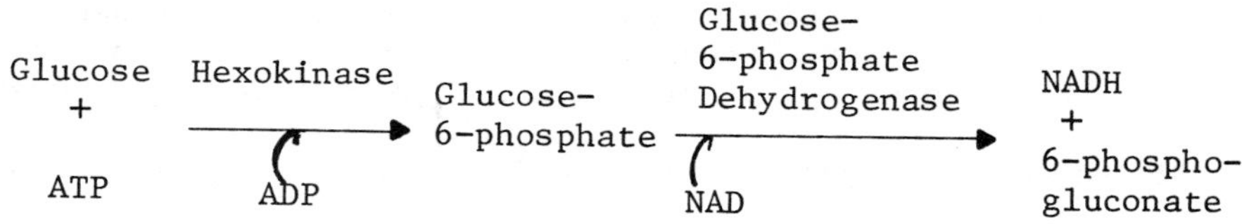

The assay reagents were glucose-6-phosphate dehydrogenase co-immobilized with antigen (human IgG) onto agarose beads and hexokinase labeled anti-human IgG. In the immunoassay, these reagents are mixed with samples and competition between sample antigen and immobilized antigen determines the amount of hexokinase labeled antibody which binds to the beads. When substrates (glucose, ATP and NAD) are added, the activity of the enzyme pair, as measured by the rate of production of NADH, is diminished according to the antigen concentration in the sample (Fig. 11). A concentration of 1 ng IgG/ml in the final assay medium was detected. A theoretical issue limits the sensitivity of this method. During the enzyme activity measurement, the concentration of glucose-6-phosphate is initially lower than its Km for glucose-6-phosphate dehydrogenase. Thus, as glucose-6-phosphate accumulates, the rate of the glucose-6-phosphate dehydrogenase reaction accelerates. When hexokinase is bound to the beads (no antigen in sample), the glucose-6-phosphate concentration inside the beads rapidly increases because of the small volume of the beads and restricted diffusion. In contrast, when no hexokinase is bound to the beads (saturating antigen in sample), the glucose-6-phosphate concentration increases more slowly. Eventually, however, regardless of the location of the hexokinase, the glucose-6-phosphate concentration exceeds Km and the rate of NADH production reaches the same maximum value. The consequence of this is that antigen modulation of the rate of NADH production is only seen for a short time after adding substrates.

This limitation can be obviated by adding a "scavenger" of glucose-6-phosphate to the bulk medium. A combination of phosphoglucose isomerase, phosphofructokinase, and ATP scavenges glucose-6-phosphate by converting it to fructose 1,6-diphosphate. When this is done, most of the glucose-6-phosphate produced by hexokinase in the bulk phase is destroyed but that made on the beads is not. The kinetics of NADH production become linear and antigen modulated signal increases indefinitely with time.

The channeling phenomenon depends on restricted diffusion in the agarose beads. The "efficiency" of channeling can be expressed as the chance that a molecule of glucose-6-phosphate made within a bead will react with glucose-6-phosphate dehydrogenase before leaving the bead. The greatest channeling efficiency (0.65), which is close to the theoretical maximum (1.0), is achieved in large beads (20 μ diameter).

The rate of diffusion from the bulk medium into the beads determines "background" signal which is not modulated by antigen. By adding a high concentration of sucrose to the medium (as was done in Fig. 11) to increase viscosity and thus decrease the rate of diffusion of glucose-6-phosphate, the background rate is reduced.

The assay for human IgG presented in Fig. 11 clearly estab-

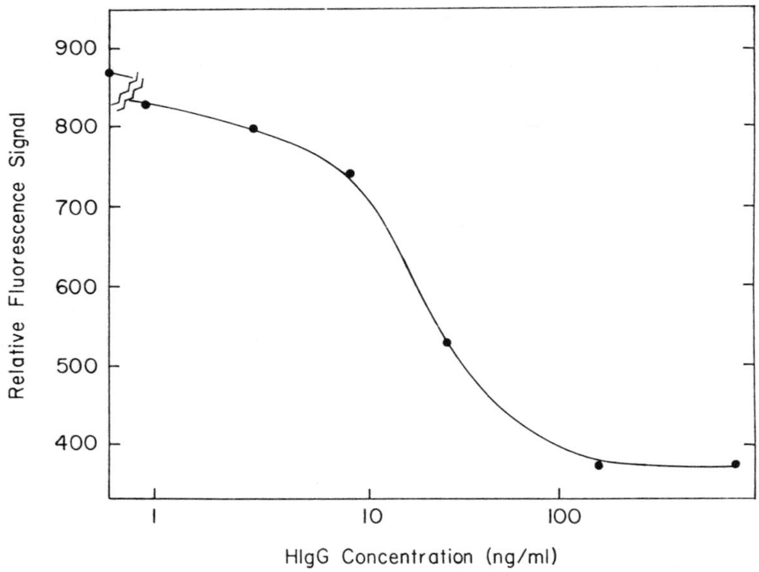

Fig. 11. Enzyme channeling immunoassay for human IgG. A suspension of agarose (1 μl beads coupled to 0.15 IU of glucose-6-phosphate dehydrogenase and 0.22 μg human IgG) was incubated with hexokinase labeled antibody (0.12 IU) and human IgG in a final volume of 0.14 ml buffer (50 mM bicine pH 8.4, containing 100 mM KCl, 0.2% bovine serum albumin and 0.05% sodium azide) for 2 h at room temperature. The mixture was diluted with 1 ml of a solution containing 50 mM bicine pH 8.4, 100 mM KCl, 6 mM $MgCl_2$, 3 mM ATP, 3 mM NAD^+, 40 mM glucose and 40% glycerol, vortex mixed and aspirated into the flow cell of a modified Varian Fluorochrome® fluorimeter equipped with a 2 mm diameter flow cell thermoregulated at 37°. The rate of NADH production was monitored for 1 minute at 450 nm with 340 nm excitation.

Reagents were made as described by Litman et al. (1980). The final bead concentration (about 100,000 per ml) caused little visible turbidity.

This figure is from Ullman and Litman (1980) and appears by permission.

lished the potential for homogeneous enzyme channeling immunoassay as a convenient and sensitive nonseparation method. Practical constraints, however, have precluded the development of the method as a clinically useful assay. In particular, there is a conflict between the need for large beads to support efficient channeling and the requirement for reproducible dispensing of the bead reagent which is problematic with large beads.

Second Enzyme as Label

In theory, this configuration is more desirable since the antigen modulated activity gives rise to a continuously increasing signal.

In studying this configuration, we chose a different enzyme pair (shown in Scheme 3). Horseradish peroxidase is an attractive enzyme label because of its high turnover with a variety of chromogenic substrates. Furthermore, an excellent scavenger of hydrogen peroxide (Product 1) is available in the form of catalase (Scheme 3).

Scheme 3

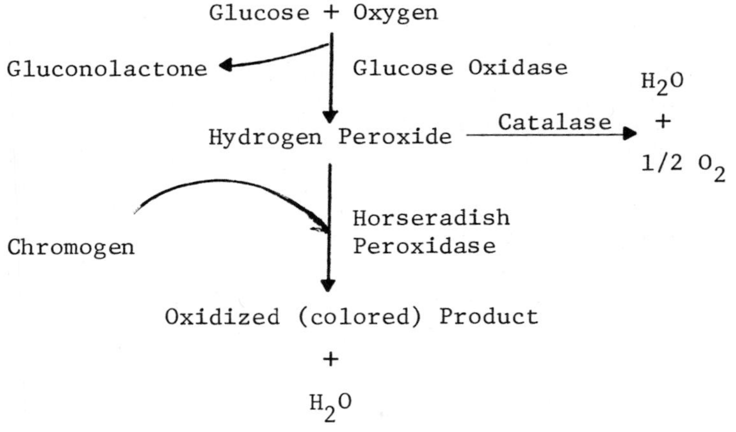

We set out to solve the problems inherent in dispensing particulate reagents by designing the assay with soluble reagents which combine in the assay to make a particulate microenvironment suitable for channeling. Fig. 12 illustrates how this was achieved. To minimize the assay time, we chose to use labeled antibodies and to drive the binding reactions with the sample antigen using excess reagents (cf, radio-immunometric assay, Miles and Hales, 1968).

As a model analyte "polyribose phosphate," the capsular antigen of *Haemophilus influenzae* was selected. Polyribose phosphate is a large heterogeneous polymer with repeating units of ribose, ribitol

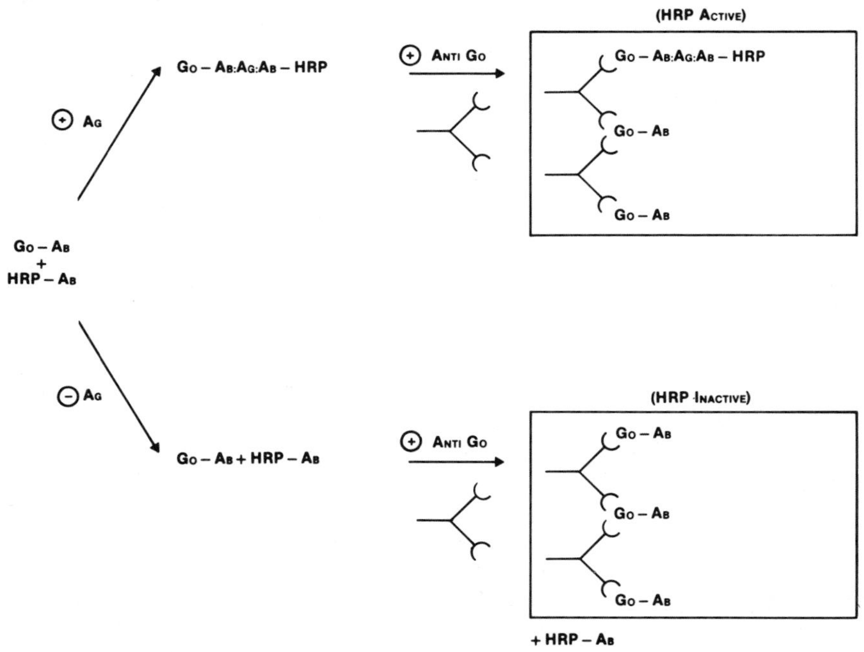

Fig. 12. Soluble reagent enzyme channeling immunoassay. Antigen (Ag) binds to glucose oxidase labeled antibody (GO-Ab) and HRP-labeled antibody (HRP-Ab) to form a soluble complex. Enough anti-glucose oxidase (noninhibitory) is added to precipitate all the glucose oxidase labeled antibody together with any bound antigen and horseradish peroxidase labeled antibody. When substrates (glucose, oxygen, and a chromogen) are added, peroxidase bound to the immune precipitate participates in a channeled reaction with the product of glucose oxidase (hydrogen peroxide) to give a colored product. Peroxidase outside the immune precipitate is inactive because any hydrogen peroxide diffusing from the precipitate is destroyed by catalase, which is present in large excess.

and phosphate (Crisel et al., 1975). Results of a soluble reagent channeling immunoassay for polyribose phosphate are shown in Fig. 13. The assay is very sensitive with a detection limit of about 5 pg. As anticipated, the assay protocol can be completed quickly. The reaction between antigen and the two enzyme-labeled antibody reagents occurs virtually instantaneously. Incubation for 15 minutes prior to adding substrates is needed for the precipitation reaction. Shaking during the incubation promotes the growth of small particles of immune precipitate. Optimum channeling is

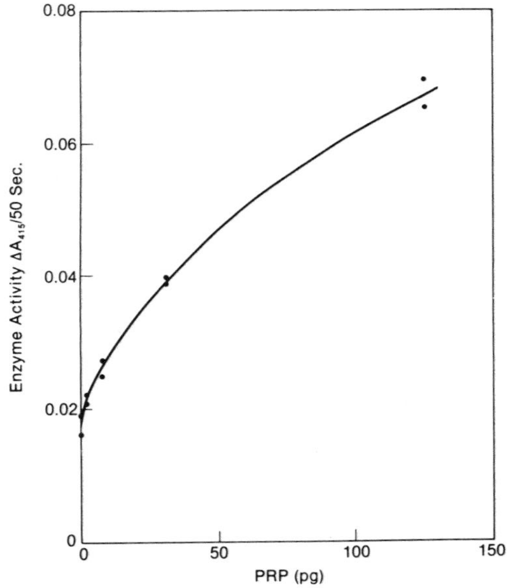

Fig. 13. Soluble reagent immunometric enzyme channeling immunoassay for polyribose phosphate. Polyribose phosphate samples were dissolved in human serum. The samples (10 µl) were mixed with 10 µl of a reagent containing 0.55 µg of horseradish peroxidase labeled anti-polyribose phosphate and 0.20 µg of glucose oxidase labeled anti-polyribose phosphate with 20 µl of a reagent containing anti-glucose oxidase (0.22 mg protein/ml), 5% polyethylene glycol and 0.2% Tween®-20. After incubation at room temperature for 15 m with gentle orbital shaking, 0.96 ml of a reagent containing 250 mM glucose, 10 mM 2,2'-Azinodi-(3-ethylbenzthiazoline - 6-sulfonic acid) [the chromogen] and 500 µg catalase/ml was added. The change in absorbance at 415 nm was immediately measured at 37° over a 50 second period. Reagents were as described by Gibbons et al. (in press). All reagents were made up in 10 mM sodium phosphate pH 7.2 containing 150 mM sodium chloride and 0.1% ovalbumin.

This figure is from Gibbons et al. (in press) and is reprinted by permission.

obtained when the particles reach about 5-10 µ.

Despite the huge excess of HRP labeled antibody over antigen (approximately 10^4 fold on a molar basis at the highest antigen level), the background due to unbound HRP is low. This is because of effective suppression of background by catalase. As may be seen in Fig. 14, background is reduced as the catalase concentration is

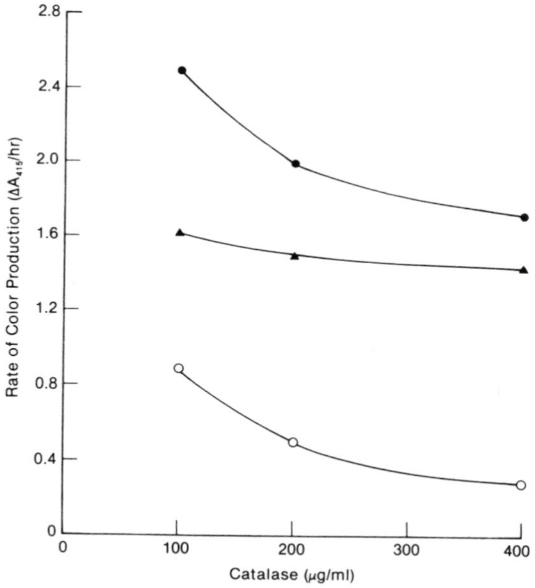

Fig. 14. Effect of catalase on background and antigen specific signal in the soluble reagent enzyme channeling immunoassay. Reagents (Gibbons et al., in press) were made up in 10 m$\underline{M}$ sodium phosphate pH 7.2 containing 150 m$\underline{M}$ sodium chloride and ovalbumin at 1% (reagents A and B) or 0.1% (reagent C). Reagent A contained 120 µg horseradish peroxidase conjugated to anti-polyribose phosphate and 200 µg glucose oxidase conjugated to anti-polyribose phosphate per ml. Reagent B contained anti-glucose oxidase diluted to 0.44 mg/ml. Reagent C contained 284 m$\underline{M}$ glucose, 12 m$\underline{M}$ 2,2'-Azinodi-(3-ethylbenzthiazoline - 6-sulfonic acid) and sufficient catalase to give the indicated concentration in the final assay medium. Aliquots of water (10 µl) containing either 1 ng (•) or 0 ng (o) of polyribose phosphate were combined with 10 µl of reagent A and 10 µl of reagent B in the wells of a microtiter plate. The plate was covered and shaken at 37° for 90 minutes. Reagent C (220 µl) was added and the absorbance of the wells recorded immediately and after a further incubation at 37° for 30 minutes with shaking. The difference in these two readings is presented. The data indicated (Δ) represents the difference between the rates with and without antigen.

This figure is from Gibbons et al. (1984) and is reprinted by permission.

increased, whereas antigen specific signal is essentially unchanged.

The soluble reagent assay is the most sensitive homogeneous enzyme channeling immunoassay available. It approximates the sensitivity achieved by the most sensitive ELISA. The method can detect not only soluble antigens like polyribose phosphate but whole microorganisms such as *Chlamydia trachomatis* (Gibbons et al., 1984). As yet, the method has not been applied to clinical samples. The convenience of the method will make it attractive provided interferences from clinical samples can be overcome.

REFERENCES

Armenta, R., Tarnowski, T., Gibbons, I., and Ullman, E. F., Improved sensitivity in homogeneous enzyme immunoassays using a fluorogenic macromolecular substrate, Anal. Biochem., submitted.

Crisel, R. M., Baker, R. S., and Dorman, D. E., 1975, Capsular polymer of Haemophilus influenzae, Type b, J. Biol. Chem. 250:4926.

Crowl, C. P., Gibbons, I., and Schneider, R. S., 1980, Recent advances in homogeneous enzyme immunoassays for haptens and proteins, in: "Immunoassays: clinical laboratory techniques for the 1980's," R. Nakamura, ed., Alan R. Liss, New York.

Gibbons, I., Armenta, R., DiNello, R. K., and Ullman, E. F., Non-separation enzyme channeling immunometric assay, in: "Immobilized enzymes and cells," a volume of Methods in Enzymology, K. Mosbach, ed., Academic Press, New York, in press.

Gibbons, I., DiNello, R. K., Greenburg, R. R., Olson, J., and Ullman, E. F., 1984, Sensitive homogeneous enzyme immunoassays for microbial antigens, in "Rapid detection and identification of infectious agents," D. J. Kingsbury and S. Falkow, ed., Academic Press, New York (in press).

Gibbons, I., Hanlon, T. M., Skold, C. N., Russell, M. E. and Ullman, E. F., 1981, Enzyme enhancement immunoassay: a homogeneous assay for polyvalent ligands and antibodies, Clin. Chem., 27:1602.

Gibbons, I., Skold, C., Rowley, G. L., and Ullman, E. F., 1980, Homogeneous enzyme immunoassay for proteins employing β-galactosidase, Anal. Biochem., 102:167.

Litman, D. J., Hanlon, T. M., and Ullman, E. F., 1980, Enzyme channeling immunoassay: a new homogeneous enzyme immunoassay technique, Anal. Biochem., 106:223.

Miles, L. E. M. and Hales, C. N., 1968, Labeled antibodies and immunological assay systems, Nature, 219:186.

Morita, T. N. and Woodburn, M. J., 1978, Homogeneous enzyme immunoassay for Staphylococcal enterotoxin B, Infection & Immunity, 21:666.

Ngo, T. T., Carrico, R. J., Boguslaski, R. C., and Burd, J. F., 1981, Homogeneous substrate-labeled fluorescent immunoassay for IgG in human serum, J. Immunol. Methods, 42:93.

Pepys, M. B., 1981, C-reactive protein fifty years on, Lancet, 1:653.

Skold, C. N., 1981, U.S. Patent 4,268,663.

Ullman, E. F., 1981, Homogeneous enzyme immunoassay techniques for proteins, in: "Monoclonal antibodies and developments in immunoassays," A. Albertini and R. Ekins, ed., Elsevier/North Holland Biomedical Press, Amsterdam.

Ullman, E. F., Gibbons, I., Litman, D. J., Weng, L., and DiNello, R., 1983, Highly sensitive enzyme channeling immunoassay for macromolecules, in "Immunoenzymatic Techniques," S. Avrameas et al., ed., Elsevier Science Publishers B.V., Amsterdam.

Ullman, E. F., Gibbons, I., Weng, L., DiNello, R., Stiso, S. N., and Litman, D. J., 1983, Homogeneous immunoassays and immunometric assays employing enzyme channeling, in: "Diagnostic Immunology: Technology Assessment and Quality Assurance," (CAP Conference, 1983), R. M. Nakamura and J. H. Rippey, ed., College of American Pathologists, Skokie, Illinois.

Ullman, E. F. and Litman, D. J., 1980, Enzyme channeling immunoassay. A new homogeneous enzyme immunoassay technique, in: "Enzyme Engineering," vol. 5, H. H. Weetall and G. P. Royer, ed., Plenum Publishing Corporation, New York.

Worwood, M., 1979, Serum ferritin, CRC Critical Reviews in Clinical Laboratory Sciences, 10:171.

LIPOSOME-ENTRAPPED ENZYME MEDIATED IMMUNOASSAYS

William J. Litchfield and J. William Freytag

Diagnostic & Bioresearch Systems Division
E. I. duPont de Nemours & Co., Inc.
Glasgow Research Laboratory, Wilmington, DE 19898

INTRODUCTION

Liposomes, first described by Bangham, et al.(1965), are membrane vesicles that can form spontaneously by suspending a variety of lipids in water. These vesicles can be prepared as large multilamellar structures containing many internal aqueous compartments or as smaller unilamellar structures with only one internal compartment. If prepared in the presence of marker molecules, such as soluble enzymes, ions or flurophores, the markers are entrapped within the aqueous spaces surrounded by bilayer lipid membranes. In addition, liposome membranes can be labeled with either antigens or antibodies, and they can be lysed by a variety of agents.

Since Kataoka et al.(1971) initially reported a homogeneous immunoassay using antibody and complement to lyse liposomes sensitized with lipid A, at least 15 different papers have been published on similar immunoassay designs (fig.1). These have included assays for measuring specific antibodies (Urema, 1972, and Rosenquist, 1977), IgG (Braman, 1984), hormones (Hsia, 1978, Tan, 1981, and Braman, 1984), and therapeutic drugs such as theophylline (Haga, 1980 and 1981). In all of these cases, complement was required for vesicle lysis, and the membranes were specifically prepared with either covalently attached haptens or naturally occurring lipid antigens.

Recently, we have reported a new immunoassay design using digoxin as a model analyte (fig.4) which does not require complement or the use of labeled liposomes (Litchfield, 1984, and Freytag, 1984). Instead, novel hapten-cytolysin conjugates, such

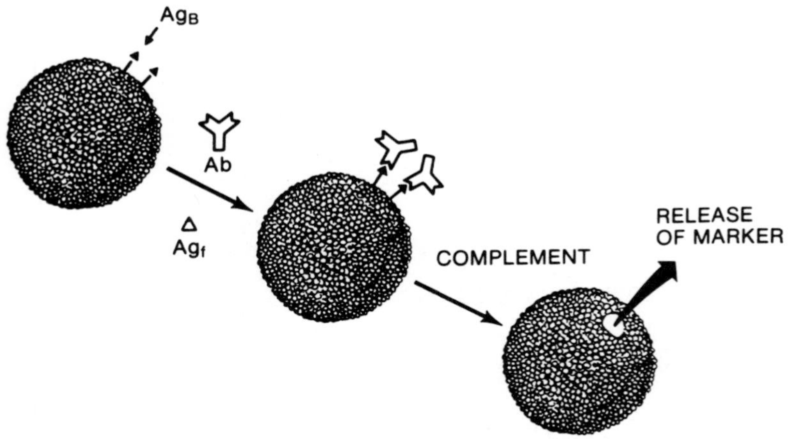

Fig. 1: Schematic representation of a complement based liposome immunoassay. Free antigen (AG_f) competes with bound antigen (Ab_B) for antibody binding. Surface bound antibody activates complement resulting in vesicle lysis.

as ouabain-melittin, are used which retain their lytic activity until bound by an antibody specific for the hapten. This liposome-entrapped enzyme mediated immunoassay appears to have significant advantages over other assays requiring complement.

METHODS

Liposomes of varying size and charge can be made by using a variety of phospholipids that differ in the head group attached to the phosphate, by incorporating cholesterol into the membrane or by using surfactants such as dodecyl phosphate or stearylamine. They can range in size from a few nanometers to a few micrometers; however, for most immunoassay designs, it's desirable to have large, single compartment vesicles that can release significant quantities of entrapped marker. Large unilamellar vesicles of 0.1 to 1 micron can be formed in a number of ways, such as by infusing a solution of phospholipids in ether into warm water (Deamer, 1976) and by a procedure called reverse-phase evaporation (Szoka, 1978). In the latter technique, phospholipids are first dissolved in an organic solvent and then mixed with an aqueous buffer to produce an emulsion. The organic solvent is subsequently removed under reduced pressure, and the liposomes form. In the work described herein, another technique known as detergent dilution, or dialysis, was employed following a modification of the procedure reported by Mimms et al. (1981).

In the detergent dialysis protocol, egg lecithin (24 mg), cholesterol (4.2 mg), and a detergent (3.75 mL or 200 mmol octyl glucoside/L of water) were cosolubilized into mixed micelles which were then dialyzed against 50 mmol/L Tris buffer, pH 7.8. Since octyl glucoside has a high critical micelle concentration, it was easily removed within a few hours of dialysis at room temperature. When alkaline phosphatase as a marker was included in the mixed micelle preparation, liposomes were obtained with up to 40% of the enzyme entrapped. The trapped marker was then separated from any free enzyme by passing the lipid vesicles over a Sepharose 4B-Cl column equilibrated with the Tris buffer. By monitoring absorbance (or turbidity) at 280 nm, the fractions containing liposomes were identified, and by monitoring enzyme activity at 410 nm with p-nitrophenyl phosphate, the fractions containing alkaline phosphatase were found. The activity of enzyme in the liposomes could not be seen until a lytic detergent such as 1mL/L Brij-58 was added to lyse the vesicles.

To demonstrate the immunoassay design using hapten-cytolysin conjugates, a low molecular weight conjugate was synthesized using melittin, a 26 amino acid lytic peptide from bee venom, and ouabain, an anolog of digoxin. Other cytolytic agents, such as cytotoxin I and II from Naja naja oxiana venom, were also studied but found to be less lytic than melittin. During the coupling, ouabain was first treated with excess sodium periodate to open up its sugar moiety and then reacted with melittin before reduction with cyanoborohydride (Litchfield, 1984). Three ouabains per melittin were typically observed in the conjugate preparations by measuring binding with [^{3}H]ouabain.

The ouabain-melittin conjugate was tested for lytic activity at 37°C by adding it to 2 mL suspensions of liposomes in Tris buffer that also contained 2 mmol/L of p-nitrophenyl phosphate. By monitoring absorbance at 410 nm, the amount of released enzyme was determined and related to the concentration of lytic agent present. Ouabain-mellitin was surprisingly found to be slightly more lytic than melittin, while as expected, ouabain alone was not lytic. In the homogeneous immunoassay for digoxin, various amounts of the free drug were incubated for 5 min at 37°C with antibody to digoxin (affinity > 10^9L/mol), ouabain-melittin conjugate, enzyme substrate, and Tris buffer before the liposomes were added.

DISCUSSION

With complement based immunoassays (as illustrated in fig.1), antigens such as theophylline or thyroxin are first immobilized, or pre-bound, on the surface of liposomes containing marker

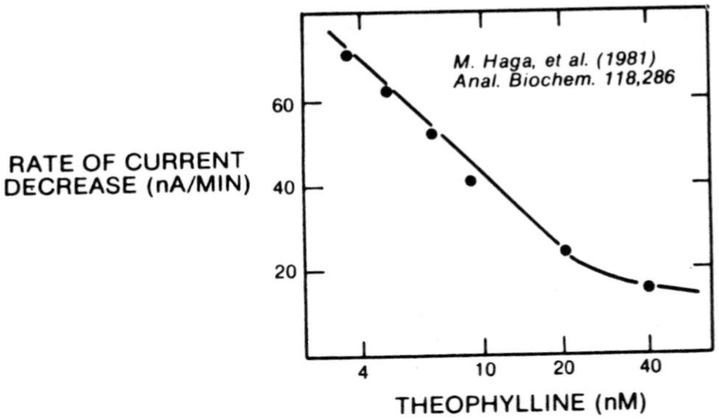

Fig. 2: Dose curve for theophylline using a complement based liposome immunoassay. The release of horseradish peroxidase from the liposomes was measured with an oxygen electrode.

enzymes. Free antigen from the sample competes with bound antigen for antibody binding, and two bound antibodies in close proximity are needed for activation of the complement cascade. Complement proteins punch relatively large holes in the vesicle membranes, allowing for either marker enzymes to escape or substrate molecules to enter.

Haga et al.(1981) published a very good example of this complement dependent immunoassay design using multilamellar liposomes prepared with a theophylline-phosphatidyl ethanolamine conjugate. Entrapped horseradish peroxidase was released after antibody, complement and liposomes were mixed together, and the released enzyme was measured using an oxygen electrode. Free theophylline from the samples prevented antibody binding and thus the lytic action of complement. The rate of oxygen consumption was therefore decreased, resulting in an impressive dose curve over the 4 to 40 nanomolar range of free drug (fig. 2). Not only did this demonstrate good sensitivity, but it did so with an inexpensive electrode system for detection. Braman et al.(1984) also reported encouraging results with a complement based system for spectrophotometrically measuring thyroxin and IgG in serum samples. Using liposomes coated with the particular antigens and containing alkaline phosphatase, they demonstrated reasonably good correlation of values obtained with both assays to those found by radioimmunoassays. In their work, the enzyme activity unmasked by complement appeared to remain within the liposomes to which the substrate became permeable.

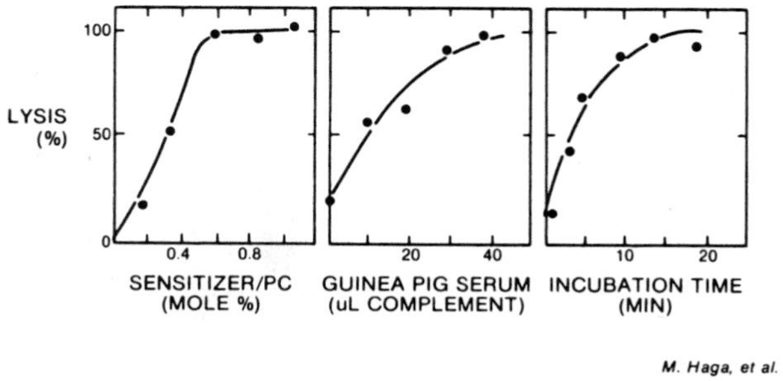

Fig. 3: Optimization of a complement based liposome immunoassay. Effects on liposome lysis are shown for varying ratios of antigen sensitizer to phosphatidyl choline (left), varying volumes of complement added (middle), and varying incubation times of the liposomes with complement.

The main disadvantages with complement based immunoassay designs are threefold. First, the liposomes are specially prepared with critical amounts of sensitizing drug on the vesicle surface (1 theophylline for every 2 phosphatidyl cholines as shown in fig. 3) which causes problems in terms of using this procedure for measuring other analytes. Each additional assay requires a new liposome preparation that is laborious to both prepare and stabilize. Secondly, a relatively large amount of guinea pig complement, which is both expensive and difficult to stabilize, is needed to obtain vesicle lysis. And lastly, a considerable amount of time, up to 20 minutes or more, is required to achieve assay results. This is not attractive when compared to other homogeneous assays for analytes like theophylline, thyroxin, and IgG that can be completed in a matter of 5 min or less.

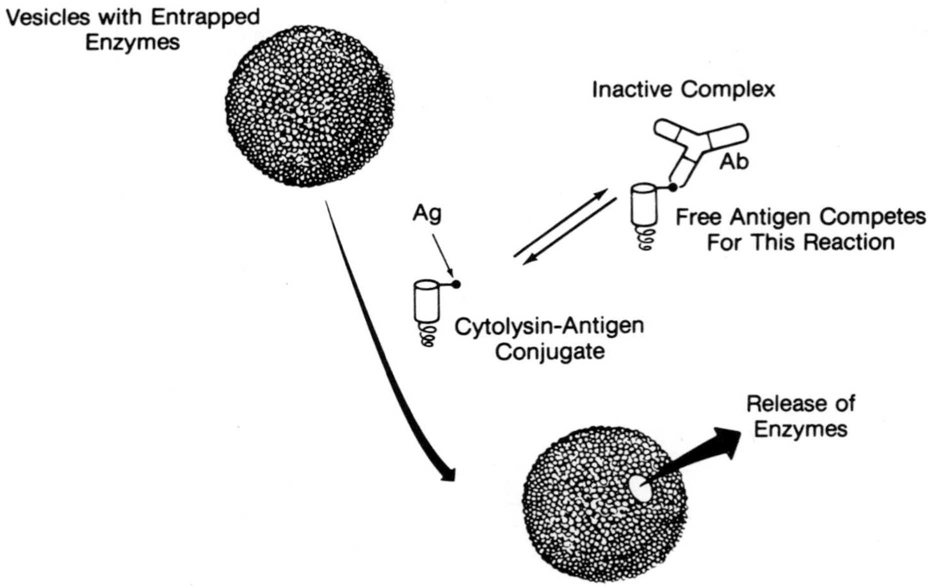

Fig. 4: Schematic representation of a liposome immunoassay independent of complement. Free antigen competes with an analyte-cytolysin conjugate for antibody binding. The analyte-cytolysin conjugate, if unbound by antibody, is able to lyse liposomes.

To get around these disadvantages, the cytolysin based immunoassay (illustrated in fig. 4) was designed using unilamellar liposomes prepared by detergent dialysis. This approach did not require the attachment of a drug to the liposome, did not require complement, and gave rapid lysis of the vesicles. The liposomes used in this work were stable against release, or unmasking, of alkaline phosphatase activity for almost a year when stored refrigerated under an argon atmosphere, but were not stable if cholesterol was removed from the preparation. By electron microscopy with negative staining, the average vesicle size was roughly 0.2 microns. Addition of 20 nmol/L ouabain-melittin conjugate to these vesicles immediately promoted 100% lysis which was totally inhibited by pre-incubating the conjugate with excess antibody to digoxin (fig. 5). By adding digoxin standards to the pre-incubation solution, the dose curve shown in fig. 6 was obtained.

An important advantage of this design is that one liposome preparation can be used as a reagent with different hapten-cytolysin or analyte-cytolysin conjugates. In fact, the same liposomes used here were also employed with a biotin-melittin conjugate to measure free biotin (Litchfield, 1984). With more work, we expect that a number of immunoassays for therapeutic drugs

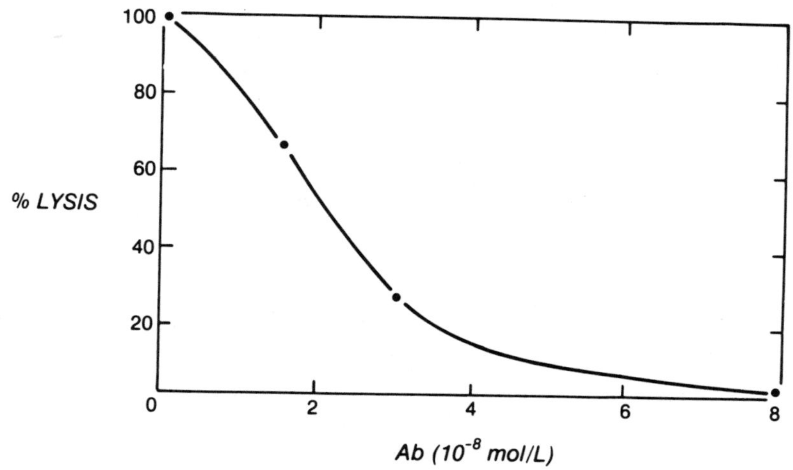

Fig. 5: Effect of specific antibody on lytic activity of an analyte-cytolysin conjugate. Various amounts of immunopurified anti-digoxin antibodies were incubated with 20 nmol/L ouabain-melittin conjugate for 5 min at 37°C prior to addition of liposomes.

could be developed using the same liposomes with different analyte-melittin conjugates that are relatively easy to prepare. For analytes at extremely low concentrations (generally below 1 nanomolar), however, this approach would require more potent cytolysins. Thirty-eight different cytolytic agents have been classified by Thelestam (1979), and many of these may be more lytic than melittin when used in this system.

As with all liposome-entrapped enzyme mediated immunoassays, this approach is limited by the affinity of antibodies employed, and it must be scaled up in terms of both liposome preparation and assay characterization before any claims about its clinical utility can be made. To date, there are no commercially available methods using liposomes with entrapped enzymes mainly because of problems associated with the availability of raw materials and the preparation of stable vesicles on a large scale. While little is also known about agents in serum that could potentially interfere with these assays, we predict that the above concerns will be overcome and that this technology will expand considerably in the future.

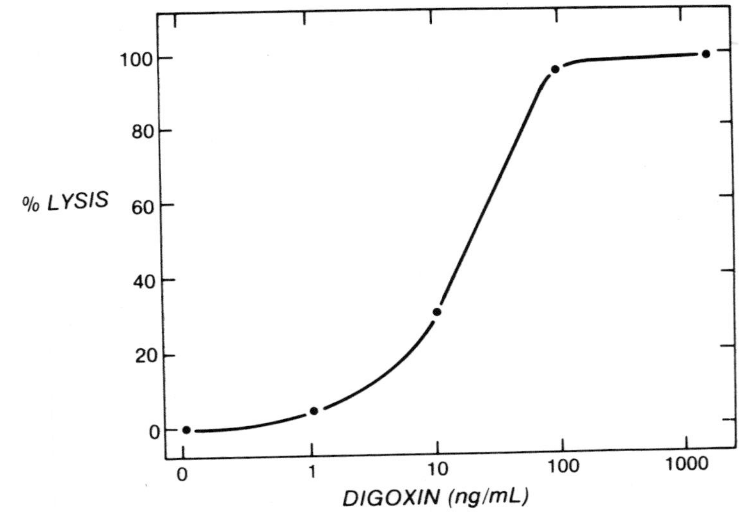

Fig. 6: Homogeneous immunoassay for digoxin using liposomes with an analyte-cytolysin conjugate. Various amounts of free digoxin in the presence of 20 nmol/L ouabain-melittin conjugate and 78 nmol/L antibody were incubated for 5 min at 37°C. After this, liposomes were added, and the enzyme activity released from the vesicles was monitored at 410 nm.

ACKNOWLEDGEMENTS

Figs. 2 and 3 from Haga et al. (1981) were redrawn with permission from the copyright owners, Academic Press Inc. Figs. 4 and 6 from Litchfield et al. (1984) were copyrighted by Clinical Chemistry and used with permission.

REFERENCES

Bangham, A. D., Standish, M. M., and Watkins, J. C. (1965). Diffusion of univalent ions across the lamellae of swollen phospholipids. J. Mol. Biol. 13, 238.

Braman, J. C., Broeze, R. J., Bowden, D. W., Myles, A., Fulton, T. R., Rising, M., Thurston, J., Cole, F. X., and Vavis, G. F. (1984). Enzyme membrane immunoassay (EMIA). Bio/Technology 1, 349.

Deamer, D. W. and Bangham, A. D. (1976). Large volume liposomes by an ether vaporization method. Biochem. Biophys. Acta 443, 629.

Freytag, J. W. and Litchfield, W. J. (1984). Liposome mediated immunoassays for small haptens (digoxin) independent of complement. J. Immunol. Methods 70, 133.

Haga, M., Itagaki, H., Sugawara, S., and Okano, T. (1980). Liposome immunosensor for theophylline. Biochem. Biophys. Res. Commun. 95, 187.

Haga, M., Sugawara, S., and Itagaki, H. (1981). Drug sensor: Liposome immunosensor for theophylline. Anal. Biochem. 118, 286.

Hsia, J. C. and Tan, C. T. (1978). Membrane immunoassay: Principle and applications of spin membrane immunoassay. Ann. NY Acad. Sci. 308, 139

Kataoka, T., Inoue, K., Galanos, C., and Kinsky, S. C. (1971). Detection and specificity of lipid A antibodies using liposomes sensitized with lipid A and bacterial lipopolysaccharides. Eur. J. Biochem. 24, 123.

Litchfield, W. J., Freytag, J. W., and Adamich, M. (1984). Highly sensitive immunoassays based on use of liposomes without complement. Clin. Chem. 30, 1441.

Mimms, L. T., Zampighi, and Nozaki, Y. (1981). Phospholipid vesicle formation and transmembrane protein incorporation using octyl glucoside. Biochemistry 20, 833.

Rosenquist, E. and Vistnes, A. I. (1977). Immune lysis of spin loaded liposomes incorporating cardiolipin: A new sensitive method for detecting anticardiolipin antibodies in syphilis serology. J. Immunol. Methods 15, 147.

Szoka, F. and Papahadjopoulos, D. (1978). A new procedure for preparation of liposomes with large internal aqueous space and high capture by reverse phase evaporation. Proc. Natl. Acad. Sci. USA 75, 4194.

Tan, C. T., Chan, S. W., and Hsia, J. C. (1981). Membrane immunoassay: A spin membrane immunoassay for thyroxine. Methods Enzymol. 74, 152.

Thelestam, M. and Mollby, R. (1979). Classification of microbial, plant and animal cytolysins based on their membrane-damaging effects on human fibroblasts. Biochem. Biophys. Acta 557, 156.

Uremura, K. and Kinsky, S. C. (1972). Active vs. passive sensitization of liposomes toward antibody and complement by dinitrophenylated derivatives of phosphatidyl ethanolamine. Biochemistry 11, 4085.

TEST STRIP ENZYME IMMUNOASSAY

David J. Litman

Syntex Medical Diagnostics Division
3221 Porter Drive
Palo Alto, California 94304

INTRODUCTION

Enzyme immunoassays (EIA) have become one of the most widely disseminated and broadly applied group of analytical techniques for biochemical analysis. Laboratory based EIA methods have found important applications in the areas of clinical chemistry, veterinary medicine, microbiology, and human fertility. The great versatility of these methods, for both the identification and quantitation of clinically relevant antigens, is illustrated by example throughout this volume and is largely attributable to the stability and amplification power of enzyme labels.

Some of the most exciting new applications for enzyme immunoassays are in the area of rapid, on-site testing. Methods that can provide accurate, real time results would be highly useful for STAT analysis in the hospital, therapeutic drug monitoring in the physician's office, and self-assessment in the home. The key requirements for practical non-laboratory methods are convenience, protocol simplicity, speed, and reliability. Ideally, these on-site methods should require minimal instrumentation, calibration, and operator expertise.

Enzyme based test strip immunoassays have a number of properties which are consistent with these requirements and which make them particularly well suited for non-laboratory applications. Probably their most important property, is the ability to use a wide variety of chromogenic substrates. In contrast to fluorescent or radioactive labels, the products of enzyme reactions can be easily detected, either visually or with very simple, portable instrumentation. The test strip itself, of course, is a very convenient vehicle for assay reagents and can be used to simplify sample handling and separation steps.

This chapter will focus primarily on the properties and characteristics of several representative test strip immunoassay methods developed in our laboratory. The methods have in common the use of an enzyme channelling detection system and a cellulose paper based immunoreactive support. A more detailed account of the experimental conditions and performance characterizations can be found in the original publications (Litman et al., 1980; Litman et al., 1983; Chen et al., 1984; Zuk et al., 1985).

Signal Generating Systems

Test strip immunoassays can be based on either heterogeneous (separation requiring) or homogeneous (non-separation) enzyme detection systems. Heterogeneous strip methods, configured similar to solid phase ELISA techniques (Engvall & Perlmann, 1971), can be highly sensitive but usually require bound/free separations, multistep protocols, and separate packaging of the enzyme and substrate reagents. Much simpler protocols can be obtained by applying non-separation immunoassay techniques to test strip formats. In order to provide convenient read-out systems, we have adapted many of the principles of the homogeneous enzyme channelling immunoassay technique (Litman et al., 1980) to the immunochemical strips described in this chapter.

Enzyme channelling immunoassays use a pair of enzymes that participate in sequential reactions such that the product of the first enzyme serves as the substrate for the second (Figure 1). The catalytic efficiency of the coupled reaction is greatest when both enzymes are in close physical proximity either on the surface of a particle in solution or in a multienzyme aggregate (Mosbach and Mattiasson, 1970). In such cases, high local concentrations of the intermediate product can accumulate and promote rapid turnover by the second enzyme. The enzyme pair used in this work consists of glucose oxidase and horseradish peroxidase.

Immunoassays based on enzyme channelling can exist in a variety of different configurations but all share the common feature that an immune binding event is used to partition one of the members of the channelling pair (usually the second enzyme) into a microenvironment containing the complementary enzyme. Only when both enzymes are sequestered together on the surface does maximum catalysis occur. Since unbound conjugate experiences a very low concentration of intermediate product, it contributes minimally to the assay signal and the need for bound/free separations is therefore obviated. The characteristics of enzyme channelling detection have been recently reviewed (Ullman et al., 1983) and applications to protein assays are specifically discussed in this volume (Gibbons, Chapter III, 1985).

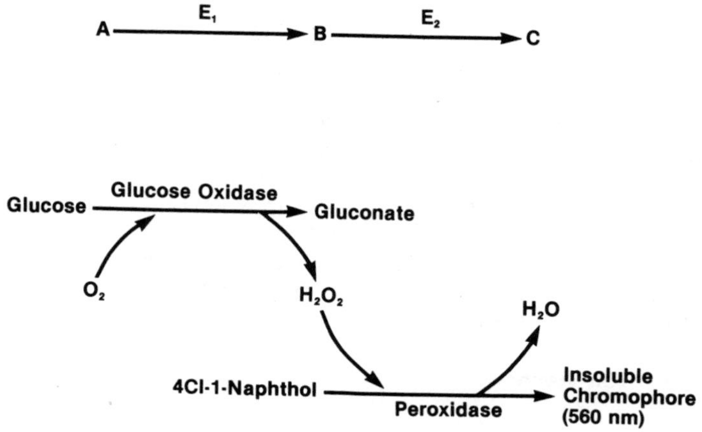

Figure 1 **Sequential enzyme reactions catalyzed by the glucose oxidase/horseradish peroxidase channelling pair.** (Reprinted by permission, from Litman et al., 1983)

Test Strip Supports

Ideally, the test strip immunoreactive surface should be highly uniform, exhibit minimal non-specific interactions, and provide a convenient means for controllable immobilization. We have chosen to prepare test strip supports by covalent immobilization to chromatography grade cellulose papers because they have a number of desirable chemical and physical properties. In addition to their hydrophilicity and minimal non-specific binding, cellulose papers can be made with a relatively high degree of uniformity and are available in a variety of thicknesses and densities. Enzymes and antibodies can be immobilized onto cellulose at high capacity by a variety of well understood methods (Lilly, 1976) to provide supports with excellent stability and immunoactivity.

MATERIALS AND METHODS

The preparation of some of the key assay components and representative assay protocols are described in this section. The reader is referred to several previous publications for a more detailed account of the reagent characterizations and experimental procedures (see above).

Materials

Horseradish peroxidase (Type VI), glucose oxidase (Type V), Triton QS-44, s-acetylmercaptosuccinic anhydride, N-ethyl maleimide, and cysteine were obtained from Sigma Chemical Company, St. Louis, MO 63178. 4-chloronaphthol, sodium meta-periodate, 2,2'oxy-bis(ethylamine), and p-nitrophenylchloroformate were obtained from Aldrich Chemical Co., Milwaukee, WI 53233. 1,1'carbonyldiimidazole was from Polysciences, Warrington, PA 18976 and bovine albumin was from Miles Laboratories, Elkhart, IN 46515. Chromatography grade filter paper (grades 1C, 31ET Chrom) were from Whatman Inc., Clifton, NJ 07014.

Immobilization Methods

Test strip supports were prepared by covalently coupling aqueous solutions of antibody and enzymes to chemically activated chromatography paper. Whatman 1C paper was used for the qualitative assays (morphine and choriogonadotropin) and 31ET paper was used for the quantitative immunochromatographic method (theophylline). A variety of conventional immobilization reagents, such as sodium meta-periodate, 1,1'carbonyldiimidazole (CDI), and p-nitrophenylchloroformate were used to prepare the activated paper with similar results. A representative protocol, using CDI as an example, is described below:

Qualitative Supports. Incubate filter paper circles (Whatman 1C, 12.5 cm) for 2 hr in 0.5 L 0.2 M CDI (22°C, in dichloromethane), and dry under nitrogen for 30 min to active paper. Incubate the activated paper in 15 mL protein coupling solution (per liter; 0.1 M sodium phosphate, 0.2 M NaCl, pH 7.0, 200-2000 mg IgG antibody, 200 mg glucose oxidase) for 4 hr with agitation, wash with phosphate buffer (pH 7.0), immerse in preservative solution (per liter; 0.1 M phosphate, pH 7.0, 150 g sucrose, 2 g BSA), and dry under reduced pressure. Papers typically contain from 5-100 $\mu g/cm^2$ protein and are stable for up to one year (desiccated).

Quantitative Supports. Prepare essentially as described above. Use 31ET paper stock and omit glucose oxidase from the protein solution.

Conjugate Preparation

Hapten conjugates of horseradish peroxidase were prepared in a two-stage procedure as described previously (Litman, 1983). First, an amine derivative of HRP was prepared by oxidizing the native enzyme with sodium metaperiodate followed by reductive amination in the presence of 2,2'oxy-bis(ethylamine). The amine HRP was then reacted with activated esters of the haptens of interest to yield conjugates with up to 14 haptens/enzyme (95% immunoreactivity, > 65% enzyme activity).

Antibody-peroxidase conjugates were prepared as described previously (Chen et al., 1983) by a modification of the thiol/maleimide procedure of Yoshitake et al (1979).

Reflectance Measurements

The color intensity on the test strip was measured with a Macbeth reflectance spectrophotometer (Model MS2000; Macbeth, Newburgh, NY 12550) and is reported as color difference units (CDU). Moist, unreacted test strips gave a background reading of 3-4 CDU.

Assay Protocols and Reagent Composition

Morphine Assay. To perform the assay, immerse test strips (10 $\mu g/cm^2$ antimorphine IgG, and 4 mIU/cm^2 glucose oxidase) into 2 mL of sample (buffer or urine) and incubate for 1 min. Transfer the strips into 2 mL of developer solution (per liter; 0.1 mol sodium phosphate, 0.2 mol sodium chloride, 2 g BSA, 300 mg 4-chloronaphthol, 50 mmol glucose, 0.25 g Triton QS-44, and 300 mg peroxidase-morphine conjugate), agitate for 3-5 sec, and incubate 10 min. Remove the strips, blot to remove excess liquid, and measure the color intensity of each pad using the Macbeth reflectance spectrophotometer. It should be noted that the assay is substantially independent of the volume of sample or developer above the minimum necessary to cover the test strip surfaces. All steps performed at room temperature.

HCG Ultrasound Assay. Combine 1 mL of sample (urine or buffer) with 100 uL of conjugate reagent (0.05 mol per liter; sodium phosphate, 0.1 mol sodium chloride, 0.25 g Triton QS-44, 2.5 mg antibody-peroxidase conjugate, pH 7.2), immerse the test strip, sonicate for 5 min, transfer to a developer solution (0.05 mol per liter; sodium phosphate, 0.1 mol sodium chloride, 2 g BSA, 300 mg 4-chloronaphthol, 0.25 g Triton QS-44, pH 6.5), incubate for 10 min at room temperature, and measure the reflectance as described above.

Enzyme Immunochromatographic Method. Add 10-20 µL of sample (buffer, serum, whole blood) to 1 mL of enzyme reagent (0.1 mol per liter; sodium phosphate, 0.2 mol sodium chloride, 0.2-1.0 mg theophylline-peroxidase conjugate, 100 mg glucose oxidase, 2 g non-immune sheep IgG, pH 7.0). Insert one end of the 4 x 90 mm test strip, containing ~ 30 $\mu g/cm^2$ immobilized anti-theophylline, into the enzyme reagent and allow the solution to wick up the entire length of the strip. When the wicking step is complete (~ 10 min), transfer the strip to the developer solution (0.01 M sodium phosphate, 0.02 M sodium chloride, 0.05% Triton QS-44, 400 µg/mL 4-chloronaphthol, 0.05 M glucose, 2 mg/ml BSA, pH 7.0). After 5 min, a visible blue color develops on the strip in the form of a color bar or rocket. The height of the color bar is measured visually and is proportional to the analyte concentration. Quantitation is derived from a predetermined standard curve which relates color front height to theophylline concentration.

Other Methods

Native protein concentrations were determined from the absorptivities at 280 nm for immunoglobulins ($\varepsilon = 2.16 \times 10^5 L \cdot mol^{-1} \cdot cm^{-1}$) and at 403 nm for peroxidase ($\varepsilon = 1 \times 10^5 L \cdot mol^{-1} \cdot cm^{-1}$). Conjugate concentrations were determined by the Lowry et al. method (1951). For comparative studies, morphine and theophylline were determined using the EMIT® Opiate and EMIT® Theophylline Assays (Syva Co., Palo Alto, CA 94304), and HCG was determined using the Tandem-E HCG immunoenzymetric assay (Hybritech, Inc., San Diego, CA 92121). Monoclonal and polyclonal antisera were obtained by standard immunological methods and were purified by sodium sulfate precipitation.

INTERNALLY REFERENCED QUALITATIVE TEST STRIP

Qualitative, non-laboratory methods have a number of important diagnostic applications in which the clinically relevant parameter is the presence or absence of a specific analyte at or above a defined concentration. Some examples of these applications would include assays for abused drugs, infectious agents, and pregnancy related proteins. These methods should be rapid, environmentally insensitive, and require no special expertise or precise manipulations on the part of the operator. An enzyme channelling test strip for morphine, which uses a novel internal reference system to compensate for environmental factors, is described below.

Assay Configuration and Protocol

As shown in Figure 2, the test strip active surface consists of glucose oxidase and antibody to morphine co-immobilized onto a

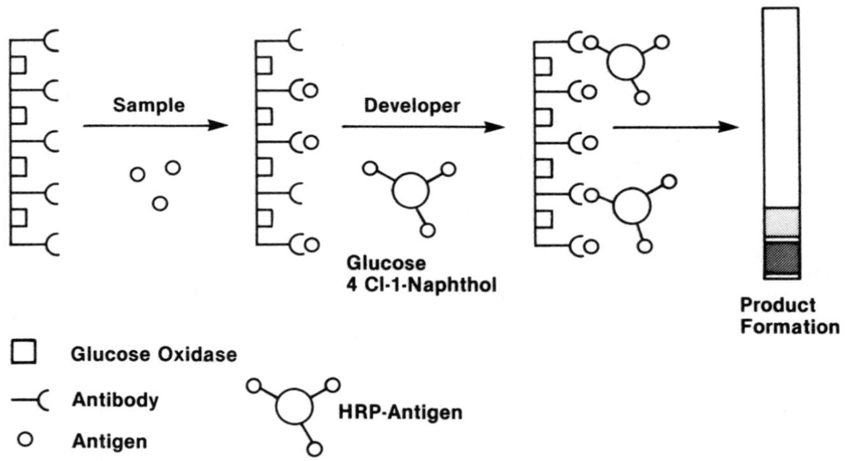

Figure 2 Schematic representation of the enzyme channelling immunochemical test strip.
(Reprinted by permission, from Litman et al., 1983)

cellulose paper support. The first step in an assay is to immerse the strip in a liquid sample. The antibody sites on the surface are filled in proportion to the sample antigen concentration. The strip is then transferred to a developer solution which contains a peroxidase-morphine conjugate, glucose, and a chromogenic substrate for peroxidase (4-chloronaphthol). During the developer incubation, conjugate binds to vacant antibody sites on the strip and uses hydrogen peroxide, generated by the immobilized glucose oxidase, to oxidize 4-chloronaphthol to an insoluble blue product which adheres to the surface of the strip. Within ten minutes the strip can be read, either visually or with a reflectance meter.

As expected for a competitive assay, the color generated on the strip is inversely related to the sample antigen concentration. Bound/free separations and wash steps are obviated by the channelling detection system, and biological fluids (urine, saliva) can be assayed without pretreatment or dilution.

Internal Reference System

Although enzyme labels have many properties which make them eminently well suited for laboratory immunoassay applications, the requirements for non-laboratory tests are more rigorous. Enzymes have finite thermal stability, time and temperature dependent catalysis, and susceptibility to sample interference. Laboratory methods commonly use temperature controlled cells, precise timing, and sample pretreatments to compensate for these deficiencies. We have addressed these problems in the strip format by devising an internal reference pad to normalize for changes in catalytic activity. The reference pad, which consists of co-immobilized glucose oxidase and anti-peroxidase, differs from the anti-morphine indicator pad only in the specificity of the antibody. The color reactions on both surfaces would therefore be expected to be similarly affected by temperature, sample interference, and enzyme activity.

By adjusting the anti-peroxidase concentration, the color intensity of the reference pad can be matched to the color intensity of the indicator pad at the detection limit of the assay. The assay cut-off limit is then defined by the ratio of the indicator color intensity (I) to the reference color intensity (R). If the indicator color is greater than the reference ($I/R > 1.0$) the test is scored negative, if the indicator color is less than or equal to the reference ($I/R \leq 1.0$), the test is scored positive.

Antibody and Glucose Oxidase Concentration

The rate of color formation on the test strip surface is a function of the concentration of the immobilized antibody and glucose oxidase, both of which can be varied by controlling the concentration of protein during the immobilization reaction. To study the effects of antibody concentration on color intensity, we immobilized test strips with varying amounts of anti-morphine at constant glucose oxidase activity (4 mU/cm^2) and used them to assay negative urine samples (Figure 3). The color increased monotonically with increasing antibody concentration until a plateau was reached. The saturation at high antibody site concentrations is most probably due to the rate limiting diffusion of conjugate to the test strip surface. Steric or covalent inactivation of the antibody during coupling is considered unlikely since it was established, in separate experiments, that the extent of ^3H-morphine binding to the paper was linearly related to the mass of immobilized antibody (data not shown).

The function of the glucose oxidase on the test strip is to generate high local concentrations of hydrogen peroxide to drive the peroxidase catalyzed oxidation of 4-chloronaphthol. The kinetics of color generation on the strip would therefore be expected to increase with increasing glucose oxidase activity until the concentration of peroxide on the strip substantially

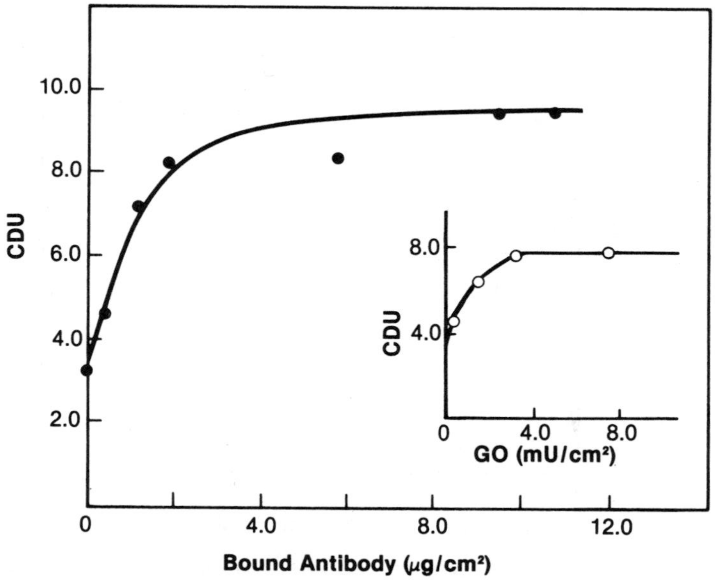

Figure 3 **Effect of immobilized antibody and glucose oxidase.** Test strips containing increasing amounts of antibody or glucose oxidase (inset) were used to assay a negative urine sample as described in Materials and Methods. (Reprinted by permission, from Litman et al., 1983)

exceeds the Km for peroxidase. Results consistent with this were obtained by measuring the color produced by test strips containing increasing amounts of glucose oxidase activity (Figure 3, inset). The color intensity increased with increasing glucose oxidase activity until a plateau was reached at ~ 4 mU/cm^2.

Antigen Binding Kinetics

The kinetics of morphine binding to immobilized antibody was studied as a function of morphine concentration (Figure 4). Both the binding kinetics and the equilibrium concentration of bound morphine were directly related to the drug concentration. At concentrations of morphine exceeding 100 ng/mL the binding was rapid with greater than 80% saturation achieved in the first minute. As expected, lower concentrations of morphine were accompanied by slower kinetics and reduced binding. Since the analyte was in large excess at all concentrations tested, the differences in the equilibrium binding are probably related to the antibody avidity and mass action considerations. The kinetics of conjugate

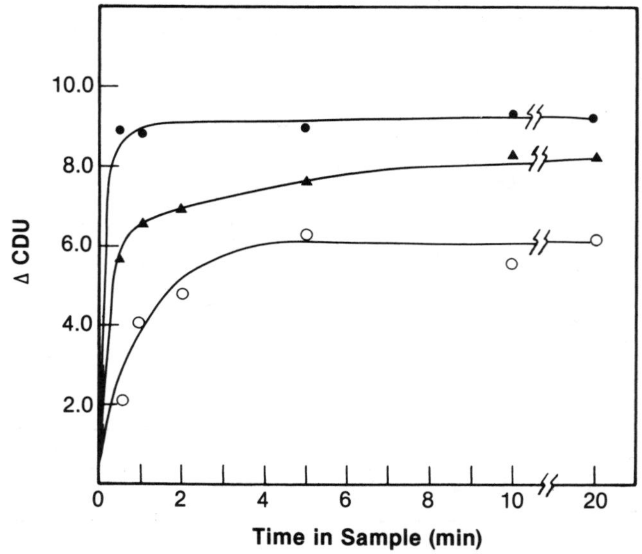

Figure 4 Dependence of assay response on duration of sample incubation.
Data are represented as the difference in color intensity (Δ CDU) produced on strips incubated in urine containing no morphine as compared with those incubated in urine supplemented with 10 (0), 100 (▲), or 500 (●) µg of morphine per liter. Negative samples gave an average color of 20.2 CDU. (Reprinted by permission, from Litman et al., 1983)

and macromolecular antigen binding to test strip surfaces were found to be similarly dependent on concentration (data not shown).

Morphine Dose Response Curves

The color intensities of the anti-morphine indicator and anti-peroxidase reference pads as a function of sample morphine concentration are shown in Figure 5. The indicator intensity was inversely related to the analyte concentration over the range from ~5 to 1000 ng/mL. The sensitivity limit was approximately 5-10 ng/mL with 50% modulation of the immunospecific signal occuring at ~ 70 ng/mL. As anticipated, the color response on the reference surface was independent of morphine concentration over the range tested. The dose response curve can quite adequately be described by the ratio of the indicator to the reference as shown. It is also easily seen that the cut-off limit of the assay, defined as the concentration of analyte at which the color intensity of the indicator equals that of the reference, can be adjusted by raising

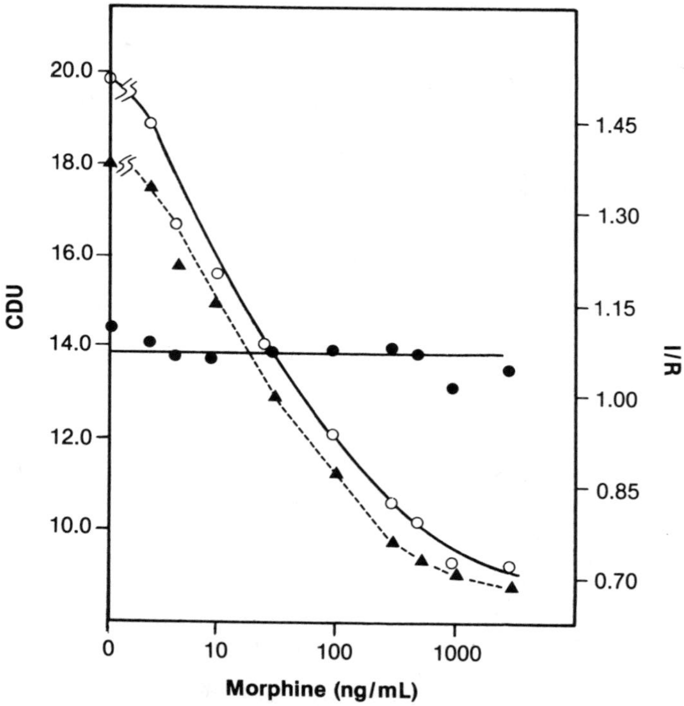

Figure 5 **Morphine Standard Curve.**
Pooled normal urine, supplemented with morphine as indicated, was assayed as described in Materials and Methods. Color intensity of anti-morphine indicator pad (O) and anti-peroxidase reference pad (●) as well as the I/R ratio (▲) are plotted. (Reprinted by permission, from Litman et al., 1983)

or lowering the color value of the reference. Assays can thus be tailored to the different ranges of analyte concentrations. Like other competitive methods, the inherent sensitivity of the test strip assay is directly related to the antibody affinity and inversely related to the antibody concentration (unpublished results).

Effect of Development Time and Temperature

The effect of development time on the color response of the anti-morphine indicator and anti-peroxidase reference reactions was studied for positive (20 µg/mL) and negative urine samples (Figure 6). The data show a similar trend in color intensity as a function of time for both the indicator and reference reactions. The qualitative relationship between the two also remained the

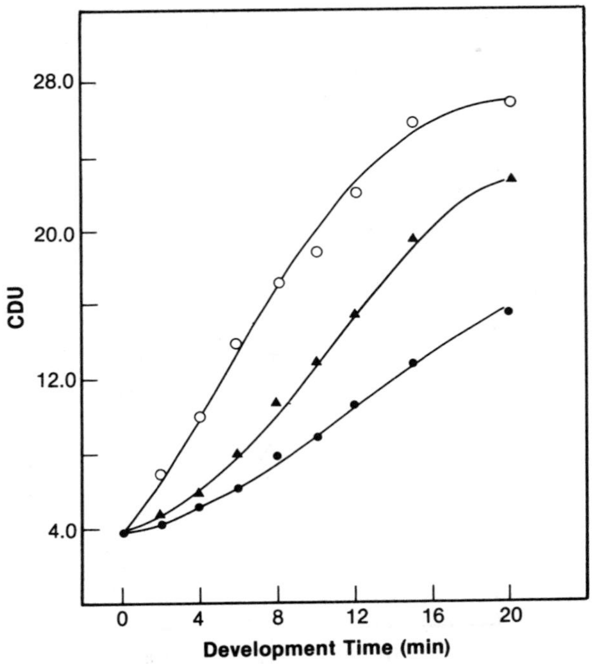

Figure 6 **Color formation as a function of development time.** Test strips pre-incubated in negative or positive (20 μg/mL) urine were developed at 25°C for the times indicated. (O) negative indicator pads, (●) positive indicator pads, (▲) reference pads. (Reprinted by permission, from Litman et al., 1983)

same over the time course studied; negative indicators were always greater than the reference ($I/R > 1.0$) and positive indicators were always less than the reference ($I/R \leqslant 1.0$).

The effect of temperature on color formation was similarly studied (Figure 7). The color intensity of both the indicator and reference reactions increased in parallel with temperature over the range from 0 to 50°C. The influence of temperature on the indicator to reference ratio (I/R) for positive and negative urine samples was much less pronounced (inset). I/R ratios accurately distinguished positive ($I/R \leqslant 1.0$) from negative ($I/R > 1.0$) samples over the entire range studied.

Sample Interference and Matrix Effects

Ascorbate is a common inhibitor of peroxidase which is present in biological fluids at concentrations that vary widely

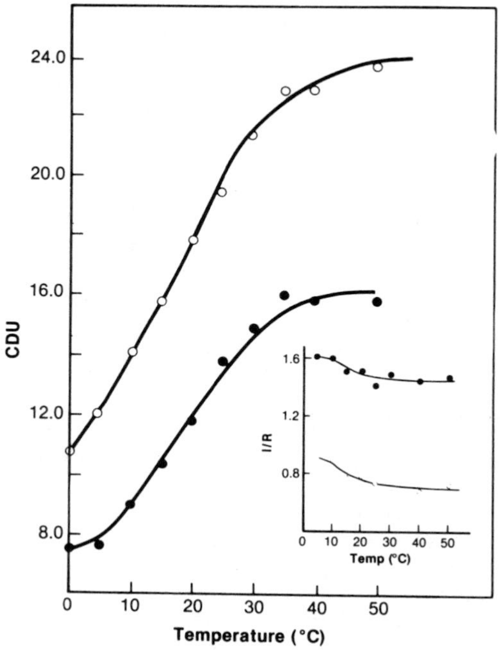

Figure 7 **Color formation as a function of development temperature.**
Indicator pads (O) and reference pads (●) were pre-incubated in morphine-free urine, followed by 10-min incubation in developer at the indicator temperatures. Inset: indicator-to-reference ratios (I/R) for dual-pad strips incubated in negative urine (●) and urine containing 20 mg/L morphine (O), followed by 10-min incubation in developer at the indicated temperatures.

from individual to individual. Ascorbate interference of the indicator and reference color reactions was studied as a model for sample interference effects (Figure 8). Although the color intensity of both test strip reactions was inhibited by ascorbate in a concentration dependent manner, the effect on the I/R ratio (Figure 8b) was almost negligible. At ascorbate concentrations that reduced the absolute color by nearly 50%, the I/R ratio still provided clear discrimination between positive and negative samples.

Sample matrix effects were investigated further by assaying individual urine samples before and after spiking with 500 ng/mL morphine. The results are shown as histograms of color intensity (Figure 9a) and I/R ratios (Figure 9b). Several negative samples, subsequently found to contain high levels of ascorbate, were

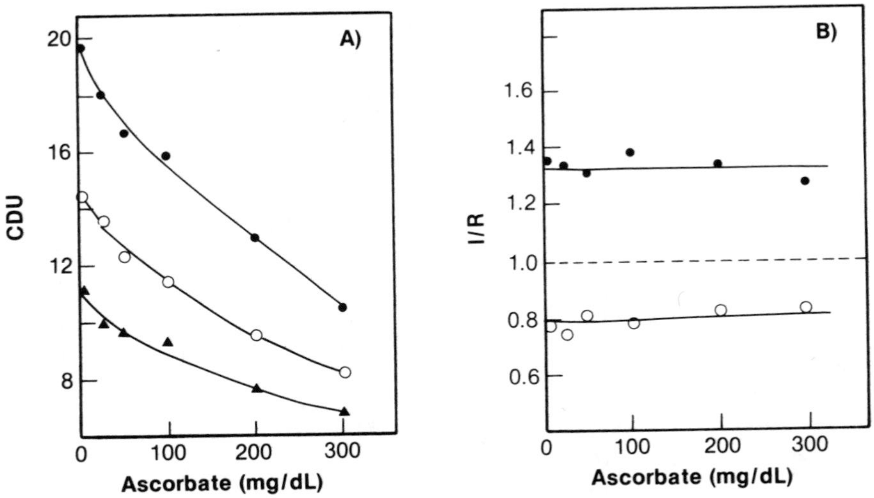

Figure 8 **Ascorbate mediated sample interference.** Test strips were incubated in pooled negative and positive (500 mg/L) urine supplemented with ascorbate, as indicated, and developed as described in Materials and Methods. A) color intensity of negative indicator (●), reference (O), and positive indicator (▲) reactions. B) I/R ratios obtained for negative (●) and positive (O) urine samples. (Reprinted by permission, from Litman et al., 1983)

scored as false positives when measured by absolute color but were correctly identified as negatives when measured by the I/R ratio. A close comparison of the two histograms clearly shows that the use of the I/R ratio eliminates the overlap between the positive and negative populations and decreases the variability within each population.

Assay Reproducability and Accuracy

The data presented in Figure 9 give an indication of the reproducibility of the test strip method. Standard deviations, in color response, are significantly larger for sample-to-sample variability (1.8 CDU) than for strip-to-strip variability (0.6 CDU). As discussed above, inter-sample variability can be greatly reduced by using the indicator/reference ratio (I/R) to score the tests results. In Figure 9, for example, using the I/R ratio

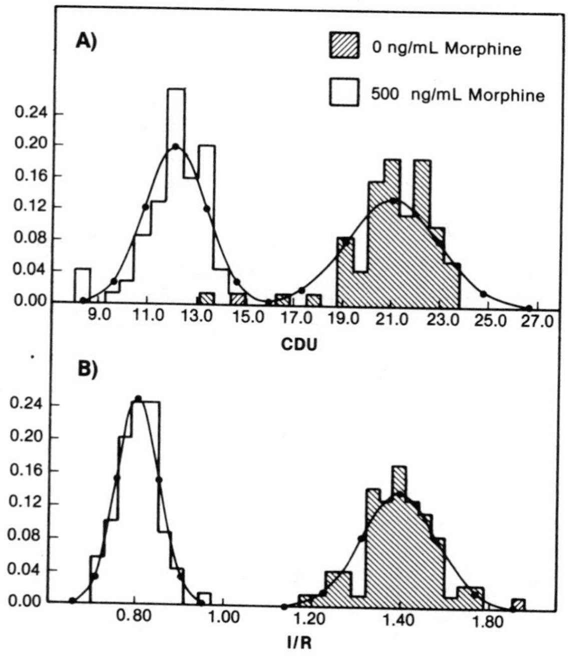

Figure 9 **Assay response distribution for individual urine samples supplemented with zero or 500 mg/L morphine.** Data are presented in color-difference units of the indicator pad (A) and as the I/R ratio (B). Ordinate = fraction of total samples. (Reprinted by permission, from Litman et al., 1983)

results in a separation of the mean positive and negative signals by ~ 9 standard deviations, compared to ~ 6 standard deviations using absolute color values.

The clinical accuracy of the method was assessed by its ability to correlate with established laboratory methods. Urine samples (n=71), obtained from patients at a drug clinic, were assayed by the test strip method and by the Syva EMIT® Opiate Assay (Schneider et al., 1973). The detection limit in the test strip assay was set at ~ 30 ng/mL morphine. Of the 53 positives by the test strip method, 52 (98%) were confirmed by EMIT. There were no false negatives. The one apparent false positive by the test strip method gave an I/R ratio equivalent to 40-60 ng/mL morphine, which is considerably below the detection limit of the confirmation technique. Although more clinical studies are needed, the present results show excellent correlation with established methods and demonstrate the clinical accuracy of the method.

Discussion

The internally referenced test strip fulfills many of the requirements for a useful non-laboratory qualitative test. The homogeneous detection system eliminates the need for separation steps and permits a convenient presentation of the reagents in the form of one dry strip and one liquid developer. The internal reference concept compensates for many environmental factors and renders the assay relatively insensitive to temperature fluctuations, errors in timing, enzyme instability, and sample interference. The method should be suitable for use by untrained personnel by virtue of the fact that it involves no pipetting, precise timing, or liquid measurements.

ULTRASOUND ACCELERATED IMMUNOASSAY

The rate limiting step in many solid phase immunoassays is related to the slow kinetics of macromolecular antigen and enzyme conjugate binding to the immobilized phase. Restricted diffusion and mass transport across the liquid/solid interface are most likely responsible for the slow binding kinetics. We have investigated the use of ultrasound to accelerate macromolecular antigen binding to immobilized antibody in a qualitative test strip immunoassay for human choriogonadotropin, HCG (Chen et al, 1984). HCG is two subunit (α,β) glycoprotein hormone and was chosen as a model protein analyte because it is a highly useful pregnancy indicator.

Assay Configuration and Protocol

We configured the test strip assay as a two-site sandwich immunoassay (Maiolini and Masseyeff, 1975) with a coupled enzyme detection system (Figure 10). The test strip indicator surface consisted of glucose oxidase and monospecific antibody to the α-subunit of HCG co-immobilized onto Whatman 1C chromatography paper. A peroxidase conjugate of monoclonal anti-β HCG was used in solution as the immunospecific signal generator.

The assay protocol involved mixing the sample and enzyme conjugate, adding the strip, incubating 5 min in a bath sonicator, and developing the color reactions for 10 min (see Materials and Methods). During the first incubation, antigen binds simultaneously to the solid phase antibody and the antibody enzyme conjugate. Color formation during the development step is analagous to that described for the morphine assay above. In contrast to the morphine assay, which is competitive, the color generated in two site sandwich assay configurations is proportional to the antigen concentration.

Effect of Ultrasound on HCG Binding to Immobilized Antibody:

We first investigated the effect of ultrasound on the rate of HCG binding to the test strip surface as a function of acoustic power (Figure 11). Test strips were sonicated for 10 minutes in the presence of 10 U/L ^{125}I-HCG (1 ng/mL) at the power levels indicated. The amount of ^{125}I-HCG bound was found to be directly related to ascoustic power.

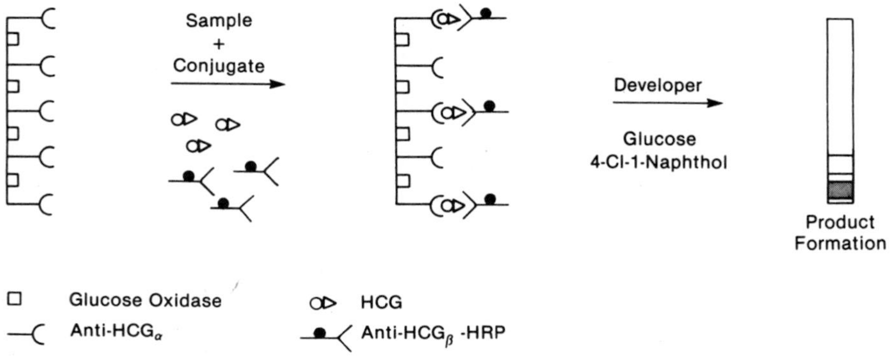

Figure 10 Two site sandwich immunoassay for HCG. (Reprinted by permission, from Chen et al., 1984)

The effect of ultrasound on the binding kinetics was further studied as a function of antigen concentration. Test strips were incubated in 1, 10, 100, or 1000 U/L 125-I HCG solutions (± sonication) for the times indicated. The data obtained at 1 U/L HCG (~ 0.1 ng/mL) are representative of the entire data set and clearly show that ultrasound can dramatically accelerate the binding of antigen to the test strip surface (Figure 12). In the presence of ultrasound, nearly all of the immunoreactive HCG is bound within the first 20 minutes, whereas in the absence of ultrasound, saturation is not achieved even after 150 hours. As shown in Table 1, the binding rate constants, with or without ultrasound, were apparently first order and were essentially

TABLE 1

EFFECT OF SONICATION ON RATE CONSTANTS OF BINDING[a]

HCG (IU/L)	k, $s^{-1} \times 10^6$ Unsonicated	k, $s^{-1} \times 10^6$ Sonicated
1	5.3	2100
10	4.4	1500
100	5.3	4500
1000	6.1	2500
Mean (SD)	5.3 (0.7)	2700 (1300)

a) Rate constants determined from plots of $\ell n\ (C_t/C_o)$

TABLE 2

EFFECT OF TREATMENTS ON BINDING RATE CONSTANTS[a]

Treatment	k, $s^{-1} \times 10^6$
None	5
Pre-sonication	5
Heat (43°C)	8
Vortex-Mixing	300
Sonication	2700

a) Rate constants determined with 100 IU/L HCG.

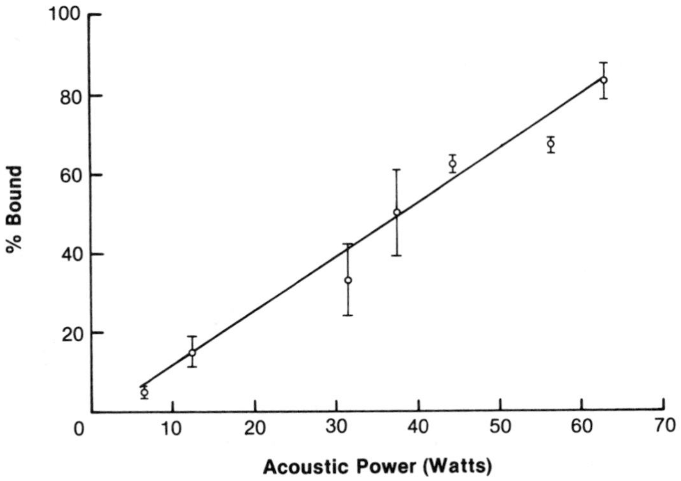

Figure 11 Effect of sonicator power on binding of ^{125}I-labeled HCG to immobilized antibody.
Power was varied by varying the voltage to the sonicator. Bars represent the range of replicte determinations. (Reprinted by permission, from Chen et al., 1984)

independent of antigen concentration. Overall, ultrasound resulted in a ~ 500 fold acceleration in rate of antigen binding. Similar results were observed for conjugate binding to surfaces pre-bound with HCG (data not shown).

Solid Phase Reactions and Comparative Studies

The rate of antigen and conjugate binding to antibody in solution was studied at reactant concentrations similar to those used in the test strip assay. In the solution phase, we could observe no effect of ultrasound on the rate or amount of antigen binding (Chen et al., 1984). Attempts to duplicate the sonication effect by other more common thermal or mechanical treatments were largely unsuccessful (Table 2). Performing the binding reactions at elevated temperature had only marginal effect on the binding rate and pre-sonication of the test strip, prior to antigen binding, resulted in no significant change in the binding rate compared to the unsonicated control. Vigorous mechanical agitation

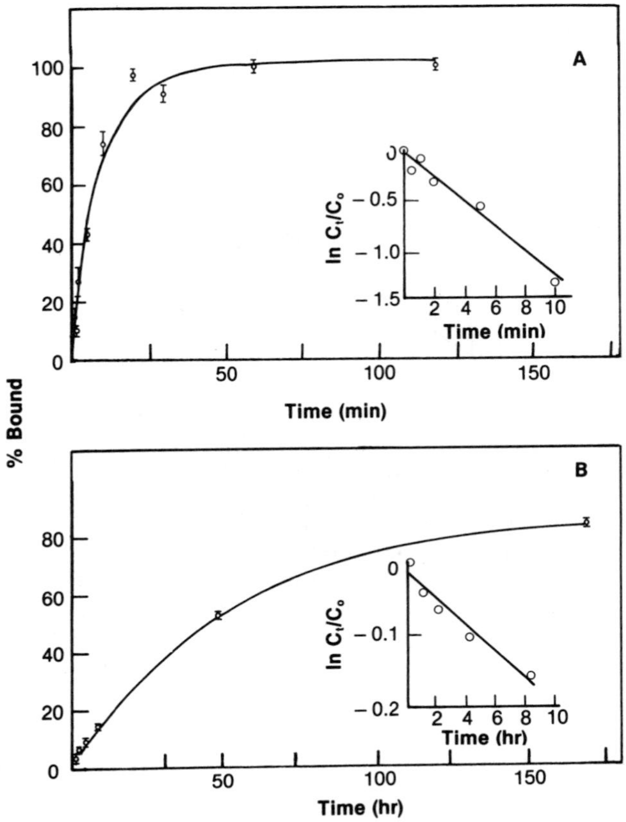

Figure 12 Kinetics of ^{125}I-labeled HCG binding to immobilized antibody test strips in the presence (A) and absence (B) of ultrasound.
Bars represent the range of replicate determinations. Insets show the plot of ln (C_t/C_o) vs t, used to derive the pseudo-first-order rate constants. (Reprinted by permission, from Chen et al., 1984)

accelerated the binding kinetics by a factor of about 50, but was still an order of magnitude less effective than ultrasound mixing.

Dose Response and Clinical Performance

The assay response (color intensity) as a function of HCG concentration was determined using pooled male urines supplemented

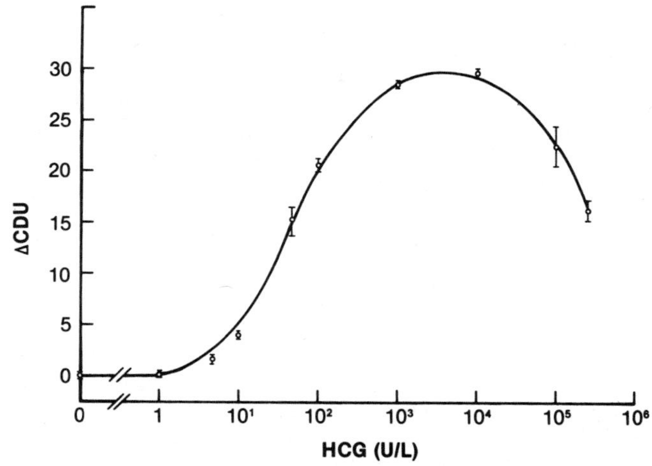

Figure 13 Standard curve for HCG.
Pooled urine from normal men, supplemented with purified HCG as indicated, was assayed as described in Materials and Methods. Bars indicate the range of replicate determinations. (Reprinted by permission, from Chen et al., 1984)

with purified hormone (Figure 13). The assay had a lower limit of detection of ~ 10 mIU/mL and exhibited a "high dose hook effect" at concentrations in excess of 1000 mIU/mL. The loss of signal at high HCG concentrations is related to the limited capacity of the support and the large molar excess of sample antigen over enzyme conjugate in the reaction mixture. The biphasicity in this type of assay can be alleviated by increasing the conjugate concentration or by removing the excess antigen before adding conjugate. Despite the hook effect, the assay provides clear discrimination between negative and positive samples even at HCG concentrations as high as 2.5×10^5 mIU/mL.

A preliminary evaluation of the clinical accuracy of the ultrasound test strip was obtained by assaying a small population of individual urine samples (11 normal males, 65 confirmed pregnant women) and comparing the results with those obtained with a commercially available ELISA method (Tandem-E®, Hybritech, Inc.). The detection limit of both assays was set at 25 mIU/mL. Excellent correlation between the methods was obtained. Both methods identified the male urines as negative and 55 of the 65 female urines as positive. The ten female urine samples, scored negative by both methods, presumably contain less than 25 mIU/mL HCG. Nine of the ten apparent false negative were collected before the expected onset of menses, and the tenth was obtained from a women with an ectopic pregnancy.

Discussion

Taken as a whole, the results in this section support the idea that ultrasound acts to accelerate the mass transfer of proteins across the liquid/solid interface. Similar effects of ultrasound on heterogeneous catalysis have been observed in immobilized enzyme systems (Berezin et al., 1977). The possibility that ultrasound irreversibly alters the accessibility or affinity of the immobilized antibody is unlikely, but transient perturbations of the support matrix, which could temporarily relieve steric hindrance, cannot be excluded. The most satisfactory explanation for the effect of ultrasound is that it creates highly efficient microscopic mixing by the combined effects of collapse cavitation and liquid streaming. This type of mixing is thought to reduce the thickness of the unstirred layer of solvent, which surrounds surfaces in solution, and thereby enhances diffusion and interfacial mass transfer (Borisov and Stantnikov, 1966). The effect should be quite general and might be useful for other solid phase binding assays such as microtiter plate ELISA, radioimmunoassay, DNA probes, and receptor assays.

ENZYME IMMUNOCHROMATOGRAPHY - NON-INSTRUMENTAL QUANTITATIVE ANALYSIS

It is clear that a non-instrumental quantitative method would have a number of highly useful applications in areas of low volume testing where portability and speed are most important. This objective has proven quite difficult to achieve with most enzyme immunoassays in which the immunospecific signal is related to the activity of the enzyme label. Since enzyme reactions are time and temperature dependent, assays based on the measurement of activity usually require external calibration, precise timing, and temperature control. This, in addition to the difficulty of visually measuring small changes in absorbance or reflectance, has necessitated the use of relatively sophisticated instrumentation.

We have developed a fundamentally different approach in which the immunospecific signal is related to the position of the enzyme label on the test strip surface, rather than to the enzyme activity (Zuk et al., 1985). As discussed below, this enyzme immuno-

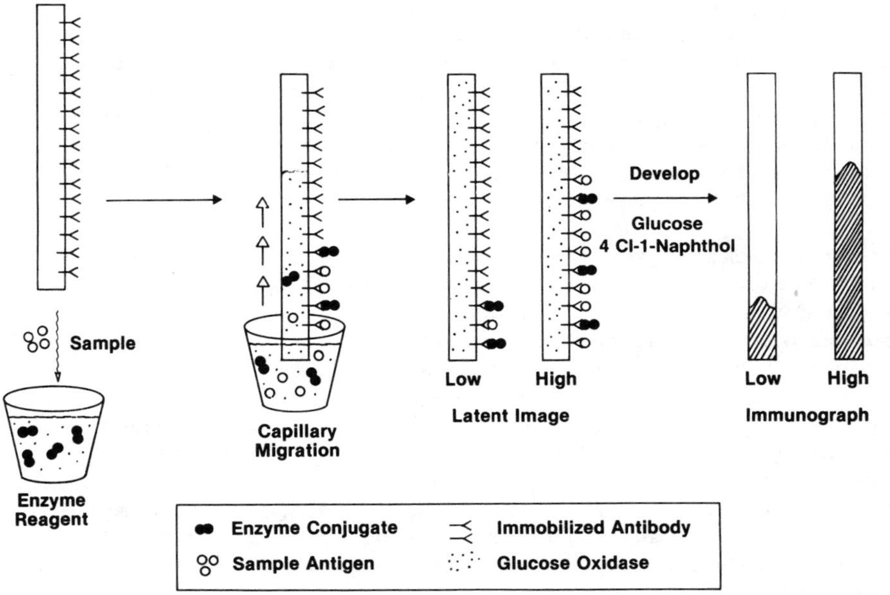

Figure 14 Schematic representation of the enzyme immunochromatographic test strip assay.
(Reprinted by permission from Zuk et al., 1985)

chromatographic method is relatively insensitive to temperature, assay timing, and sample effects. The characteristics of a visual whole blood assay for theophylline are described.

Assay Principle and Protocol

Enzyme immunochromatography combines many features of enzyme channelling with those of immunocapillary migration (Glad and Grubb, 1978) to provide a method in which antigen can be visually quantitated by the height of the color front on a strip of immunoreactive paper. The assay consists of three main components: 1) a dry paper test strip containing immobilized antibody, 2) a liquid enzyme reagent containing glucose oxidase and a peroxidase-theophylline conjugate, and 3) a developer solution containing the enzyme substrates, glucose and 4-chloronaphthol (Figure 14).

To perform an assay, sample antigen is added to the enzyme reagent and mixed briefly. Next, one end of the immobilized antibody strip is inserted into the enzyme reagent/antigen mixture and the liquid components are allowed to wick up the length of the strip by capillary action. During the wicking step the strip absorbs a finite and reproducible volume of solution. As the antigen and enzyme conjugate migrate past the immobilized antibody, they are immunospecifically bound to the strip. Glucose oxidase, also present during wicking, becomes evenly distributed throughout the strip. The capillary migration step is complete within 10 minutes and results in a latent image of antigen and conjugate on the strip. The height to which the label migrates is directly related to the analyte concentration.

After wicking, the strip is completely immersed in the developer reagent. Hydrogen peroxide, generated by the glucose oxidase catalyzed reaction of glucose with O_2, is used by the immunobound peroxidase conjugate to oxidize 4-chloronaphthol to an insoluble blue product which adheres to the surface of the strip. Although hydrogen peroxide is formed uniformly throughout the strip, visible product is formed only in the antigen bound region where the peroxidase is localized. The use of the coupled enzyme detection system avoids the need for relatively unstable peroxide containing developer solutions. Color development is complete within 5 min and results in a color zone or rocket with a sharp front at a height related to the sample antigen concentration (Figure 15).

Front Sharpness and Edge Effects

It is important for non-instrumental methods to provide a clear and unambiguous readout for optimal accuracy and reproducibility. In the immunochromatographic method, the uniformity and sharpness of the color front are essential for precision. One of the major factors that influences front quality is the manner in which the paper is cut. Papers cut with a straight edge give rise to a meniscus shaped front (Figure 15, panel A) presumably due to faster capillary flow at the paper edge. Serration of the paper edge, to increase the distance of capillary flow, can eliminate the meniscus and even reverse the effect to give rocket shaped fronts (Figure 15, panel B). The degree to which the rocket is formed can be controlled by adjusting the amplitude of the serrations and can yield fronts which are easily readable and reproducible for a given antigen concentration.

Effect of Antibody Concentration on Assay Response

The height to which a given concentration of antigen and conjugate migrate is a function of the surface density of immobilized antigen and the rate of binding of the migrating antigen relative to the rate of wicking. The effect of the immobilized antibody concentration on the migration front height was studied as a function of theophylline concentration (Figure 16). For any

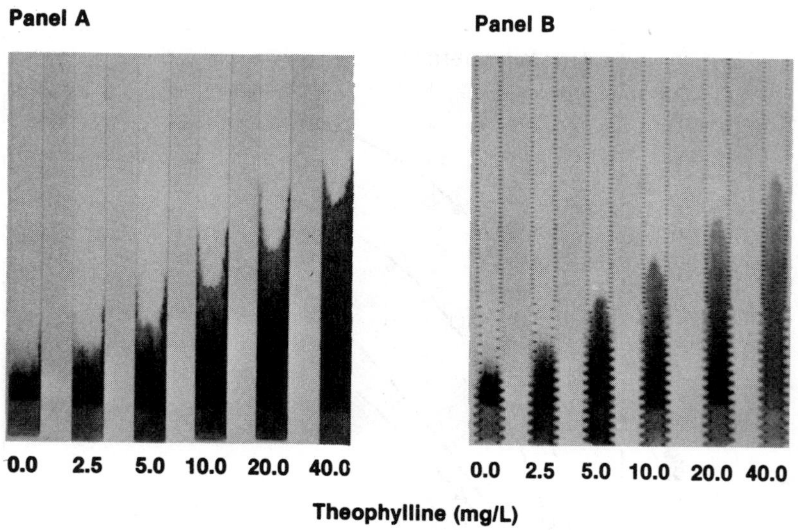

Figure 15 Color front sharpness and edge effects. Theophylline assays were run on 12.5 μL of serum calibrators in 1 mL of enzyme reagent as described in Materials and Methods. Panel A, test strips cut with straight edges; Panel B, test strips cut with serrated edges. Concentrations cited in this and subsequent figures represent the theophylline concentration in the initial, undiluted sample. (Reprinted by permission, from Zuk et al., 1985)

given concentration of immobilized antibody, the migration height increases in a non-linear monotonic manner with increasing antigen concentration. For any given concentration of analyte, the migration heights are inversely related to the immobilized antibody density. The predictable effect of antibody concentration on assay response, which is similar to that seen in other competitive assays, suggests methods of optimizing for different ranges of analyte concentration by appropriate antibody loading.

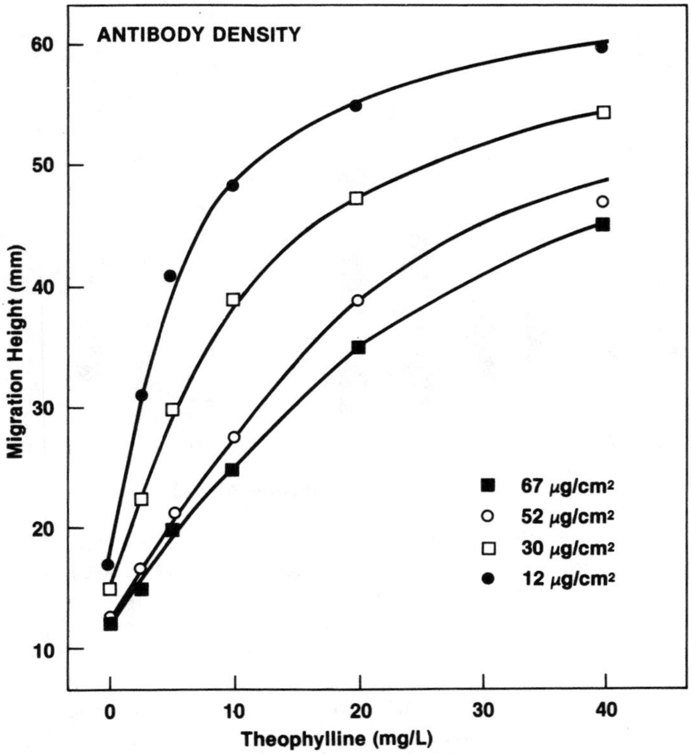

Figure 16 Effect of antibody density on color front migration. Test strips with different amounts of immobilized antibody were used to assay pooled normal plasma supplemented with theophylline as indicated. Assays were performed as described in Materials and Methods using 20 µL of plasma in 1 mL of enzyme reagent. 12 µg/cm^2 (●), 30 µg/cm^2 (□), 52 µg/cm^2 (O), 67 µg/cm^2 (■). (Reprinted by permission, from Zuk et al., 1985)

Effect of Conjugate Concentration

We investigated the effect of the theophylline-peroxidase conjugate concentration and hapten/enzyme ratio on the assay response to theophylline (Figure 17 A, B). The standard curves for theophylline were largely independent of conjugate concentration over a four-fold range and were similarly unaffected by the number of haptens coupled to peroxidase. The data is consistent with the fact that the lowest antigen concentration in the wicking solution is in significant excess (17x) over the highest conjugate concentration used. Thus, the conjugate acts primarily as a

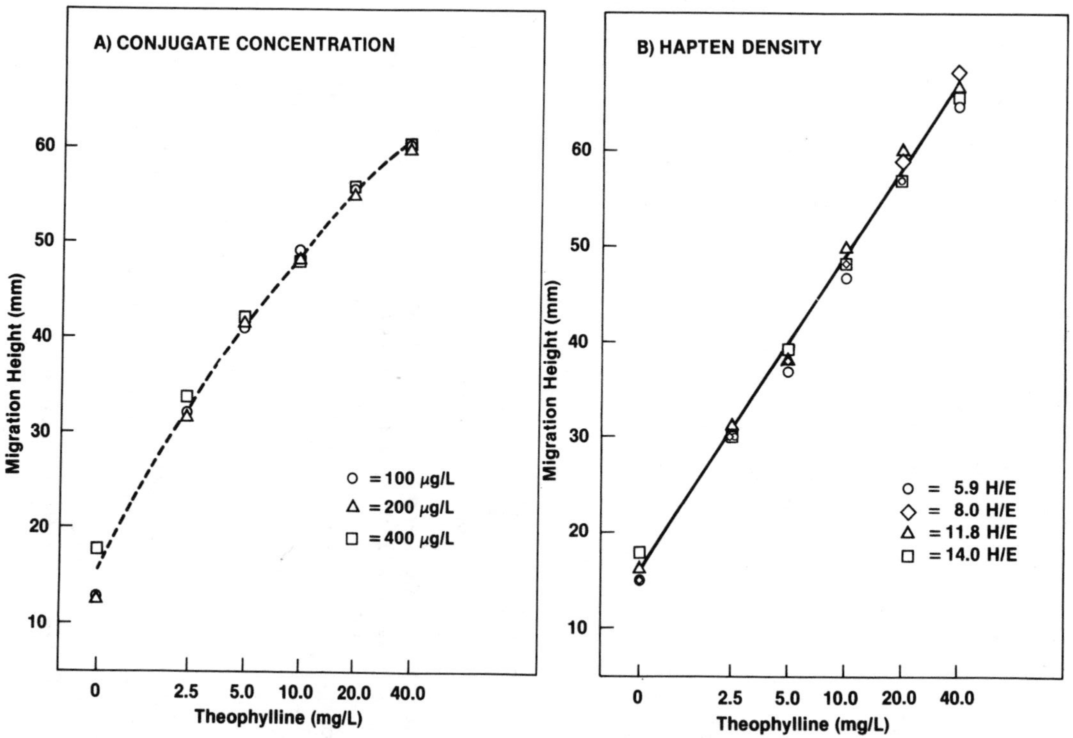

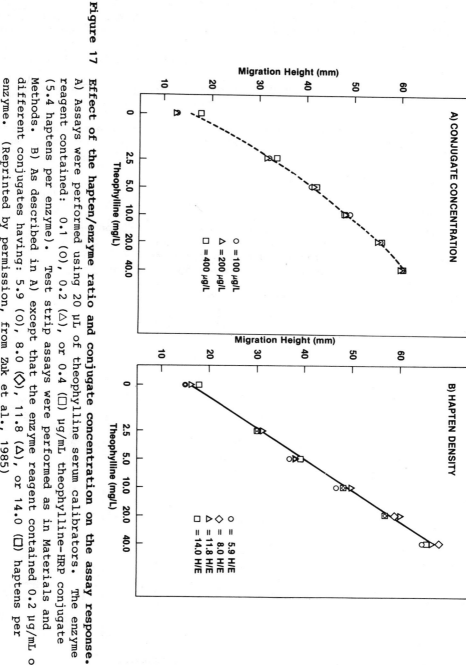

Figure 17 **Effect of the hapten/enzyme ratio and conjugate concentration on the assay response.** A) Assays were performed using 20 µL of theophylline serum calibrators. The enzyme reagent contained: 0.1 (○), 0.2 (△), or 0.4 (□) µg/mL theophylline-HRP conjugate (5.4 haptens per enzyme). Test strip assays were performed as in Materials and Methods. B) As described in A) except that the enzyme reagent contained 0.2 µg/mL of different conjugates having: 5.9 (○), 8.0 (◇), 11.8 (△), or 14.0 (□) haptens per enzyme. (Reprinted by permission, from Zuk et al., 1985)

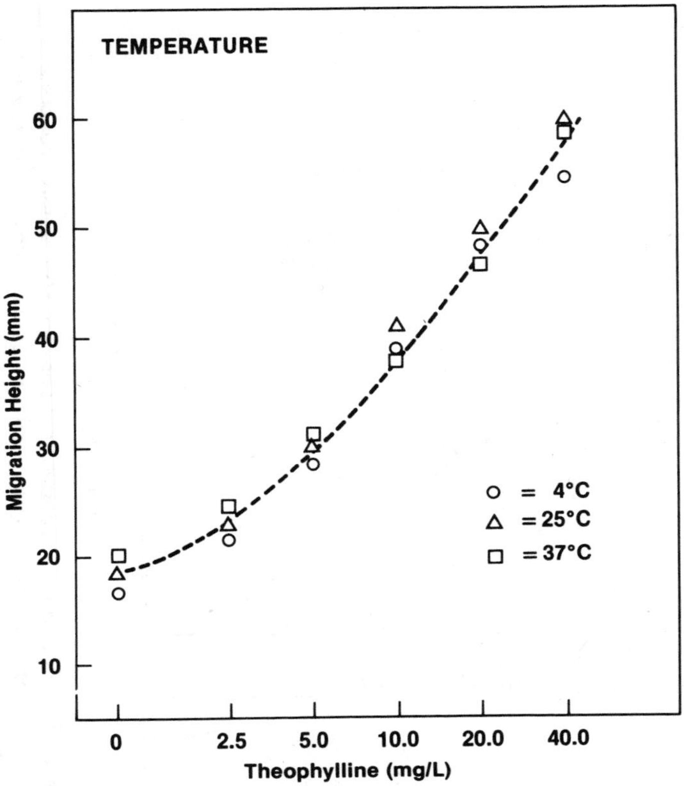

Figure 18 Effect of temperature on assay response. 20 µL of theophylline serum calibrators were assayed at 4°C, 24°C, and 37°C, as described in Materials and Methods. (Reprinted by permission, from Zuk et al., 1985)

tracer for bound antigen and migration heights are affected by conjugate concentration only when negative calibrators are assayed.

Effect of Incubation Time and Temperature

Immunoassays for non-laboratory applications should require little or no calibration and be capable of giving reliable results over a wide range of environmental conditions. The immunochromatographic method would be expected to be minimally influenced by incubation timing and temperature since quantitation does not rely on the precise measurement of enzyme activity. Results consistent with this hypothesis were obtained by running theophylline dose response curves at three different temperatures (Figure 18) and observing that the migration heights were not dramatically affected by temperature over the range from 4 to 37°C. In separate experiments (data not shown), test strips were sequentially incu-

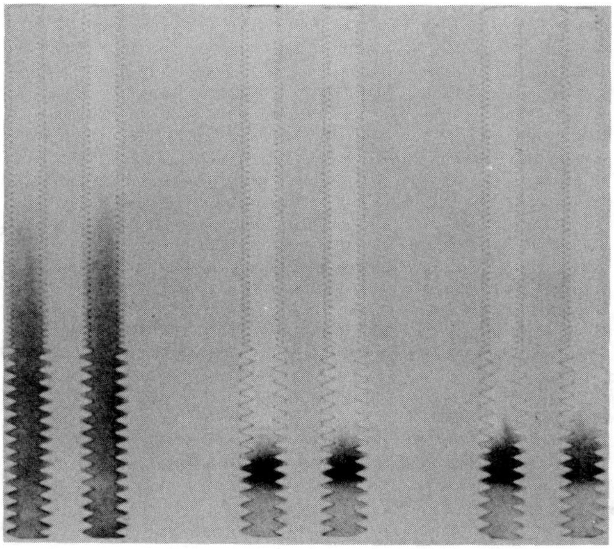

Figure 19 Red cell trapping by agglutinating agents. The distribution of red cells is shown on test strips after wicking 2% whole blood in enzyme reagent. No additions; (lanes 1 and 2), 20 µg/mL wheat germ agglutinin; (lanes 3 and 4), or 20 µg/mL sheep anti-human RBC; (lanes 5 and 6). (Reprinted by permission, from Zuk et al., 1985)

bated in the enzyme reagent and the developer solution, for varying times. Although no differences in sample quantitation were observed with up to a 30 minute incubation in either reagent, longer incubations resulted in diffuse color fronts and a reduction in precision, presumably due to the combined effects of diffusion and solvent evaporation.

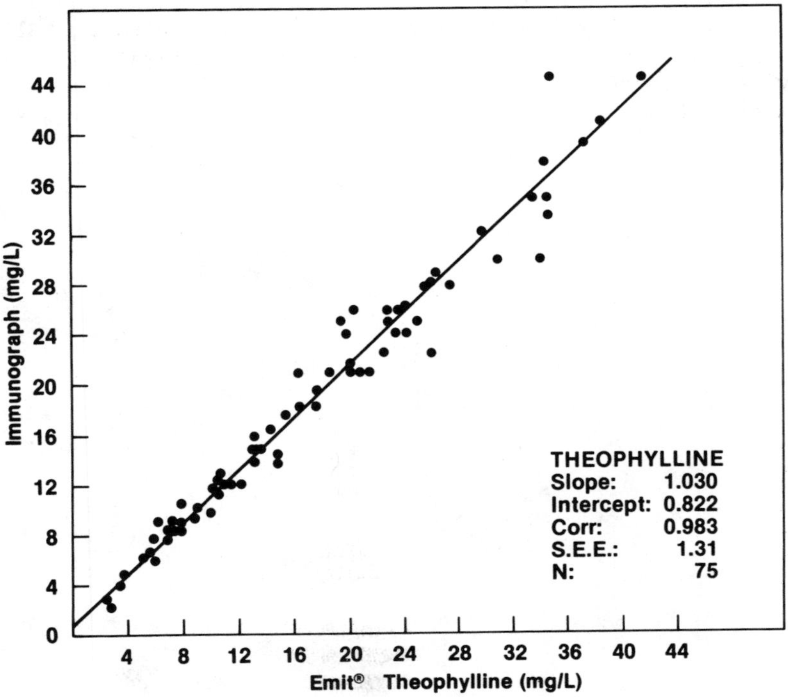

Figure 20 Sample correlation study.
12.5 µL aliquots of serum samples, obtained from asthma patients, were assayed by the enzyme immunochromatographic method (immunograph) and by the EMIT® enzyme immunoassay. (Reprinted by permission, from Zuk et al., 1985)

Whole Blood Analysis and Sample Matrix Effects

In order to minimize sample handling and provide a simplified assay protocol, methods of using fresh whole blood instead of serum or plasma were explored. We observed that red cells remained intact when suspended in enzyme reagent, but that they formed a diffuse zone obscuring approximately 50% of the strip when subjected to capillary migration (Figure 19, lanes 1 & 2). Red cell migration was prevented by including aggregating agents, such as antibody to red cells (Figure 19, lanes 3 & 4) or wheat germ agglutinin (Figure 19, lanes 5 & 6), in the enzyme reagent. The paper fibers could then effectively trap the agglutinated red cells in a tight band at the bottom 10 mm of the strip.

We assessed whether the sample matrix had any effect on the assay response by performing tests on theophylline calibrators prepared in buffer, normal serum, or fresh whole blood (Table 3). The results showed no significant differences in the height of the color fronts as a function of sample composition. Despite the well known partitioning of drugs into erythrocytes (Mitenko and Ogilvie, 1973), all of the theophylline in the enzyme reagent, regardless of sample matrix, is quantitated in the assay. In separate experiments, Zuk et al., (1985) have shown that there is a rapid and complete re-equilibration of the red cell bound drug within one minute of addition to the enzyme reagent.

The ability to use whole blood in a non-instrumental method offers several advantages over traditional testing methods which are especially relevant for the physician's office and emergency room testing. The analysis time is reduced by eliminating the need to prepare serum or plasma, the protocol is simplified, and the patient is less inconvenienced because capillary blood from a finger puncture can be used rather than venous blood.

Standard Curve Stability

It would be highly desirable to eliminate the need for frequent external calibration by fixing the standard curve of the immunochromatographic test at the time of manufacture. To do this requires that the assay components, most critically the immobilized antibody, be stable and quantitate accurately over the entire shelf life of the test. Standard curve stability was investigated by periodically assaying a set of theophylline calibrators with test strips stored desiccated at 4 and 37°C. No consistent drift in migration heights, either at 4°C or elevated temperature, was observed over a period of 21 weeks.

Clinical Performance

Like other solid phase competitive methods, the precision of the immunochromatographic method is largely dependent on a uniform distribution of antibody (or antigen) on the solid phase surface. The data shown in Table 3 are indicative of the uniformity routinely obtained with this technique. Standard deviations, in migration height, are generally less than 1 mm throughout the calibration range. Under assay conditions, this degree of uniformity translates into coefficients of variation (in concentration) ranging from 5-8% (Table 4). The analytical recovery of theophylline spiked into serum or whole blood ranged from 95-106% and was not significantly effected by high levels of endogenous substances such as hemoglobin, triglycerides, or bilirubin (Zuk et al., 1985). The accuracy of the test strip method was assessed by comparison with theophylline results obtained using the EMIT® enzyme

TABLE 3

EFFECT OF SAMPLE MATRIX ON MIGRATION HEIGHT[a]

Theophylline Concentration		Buffer	Plasma	Blood
0.0 mg/L				
Migration (mm)		18.4	19.1	18.8
SD	0.8	0.6	0.5	
% CV	4.2	3.0	2.9	
2.5 mg/L				
Migration (mm)		30.7	31.0	31.1
SD	0.6	0.6	1.1	
% CV	2.0	2.1	3.5	
5.0 mg/L				
Migration (mm)		36.2	36.7	36.1
SD	0.7	0.7	0.9	
% CV	1.6	1.5	1.6	
10.0 mg/L				
Migration (mm)		43.5	43.7	43.1
SD	0.7	0.7	0.7	
% CV	2.0	1.8	2.4	
20.0 mg/L				
Migration (mm)		52.3	52.9	52.7
SD	1.0	0.6	0.9	
% CV	1.8	1.2	1.6	
35.0 mg/L				
Migration (mm)		60.0	60.1	60.2
SD	0.7	1.0	0.8	
% CV	1.1	1.6	1.4	

a) Samples were prepared by spiking buffer (0.1 mol/L sodium phosphate, 0.2 mol/L NaCl, pH 7.0), pooled normal plasma, or pooled normal whole blood with a 1.0 mol/L solution of theophylline to the concentrations indicated. After a 1 hour equilibration period, 12.5 L aliquots were added to 1 mL of enzyme reagent. Data represent the average of ten replicates assays performed as described in Materials and Methods.

TABLE 4

ASSAY PRECISION AND REPRODUCIBILITY

A. Within-Run Precision[a]

	Theophylline (mg/L)		
	2.5	10.0	35.0
Serum (N = 20)			
Mean (mg/L)	2.4	10.1	33.4
SD	0.1	0.5	1.7
% CV	4.9	5.1	5.2
Whole Blood (N = 20)			
Mean (mg/L)	2.5	10.5	32.9
SD	0.2	0.7	2.7
% CV	7.1	6.9	8.1

B. Between-Run Precision[b]

	Patient Serum Theophylline		
	1	2	3
Mean (mg/L)	6.9	15.3	29.1
SD	0.4	1.2	1.5
% CV	5.8	7.9	5.0

a) Pooled normal serum or whole blood samples were spiked with theophylline at the concentrations indicated. Data represent twenty replicate assays on each sample.

b) Three individual patient serum samples were assayed, in replicates of four, on five consecutive days. The data represent the average of the five determinations.

immunoassay as a reference (Figure 20). Excellent correlation between the two methods was obtained ($Y = 1.03\ X + 0.82$, $R = 0.98$, $SEE = 1.3$, $n = 75$, where Y is the immunochromatograph result).

Discussion

Quantitative test strips can be constructed based on conventional homogeneous or heterogeneous enzyme immunoassay principles. In either case, instruments are usually used to meet the stringent requirements for temperature control, accurate timing, and calibration. For protocol simplicity, speed, and convenience, adaptations of homogeneous methods to strip formats are clearly the preferred route. Both Greenquist et al. (1981) and Tyhach et al. (1981) have demonstrated elegant conversions of the substrate labeled immunoassay and the apoenyzme reactivation immunoassay, respectively, to reagent strip formats. (Please refer to Chapters V & VI in this volume for a detailed description of these immunoassay methods.) In both examples, the strip served as an inert carrier for the dry immunoassay reagents. When a defined amount of diluted patient serum was added to the strip, the reagents were reconstituted and the immunochemical reactions took place within the strip volume. After a short incubation the color generated on the strips was measured using a bench top reflectance meter. These examples first demonstrated rapid, quantitative, minimal instrumentation methods for monitoring serum or plasma therapeutic drug levels.

The immunochromatographic method adds further convenience to non-laboratory testing by permitting visual quantitation and whole blood analysis. These features are made possible by virtue of a unique assay configuration in which the immunospecific signal is related to the chromatographic position of the enzyme label rather than its activity.

SUMMARY AND OUTLOOK

The test strip immunoassay methods discussed in this chapter are representative of a rapidly evolving class of minimal instrumentation techniques for on-site analysis. Enzyme based immunochemical test strips can provide simple, rapid, and highly sensitive assays for low molecular weight drug analytes as well as for large macromolecular protein antigens. Both qualitative and quantitative methods can be designed to address a wide spectrum of analytical and clinical applications. The current trend towards simple distributed testing is exemplified by the immunochromatographic method in which minimal user expertise is required to obtain accurate quantitation in a visual whole blood assay for therapeutic drugs. We believe that this trend towards simplicity and convenience will continue and that future developments will focus on further reducing operator manipulations, assay time, and susceptibility to environmental influences.

The rapid technological advances in test strip immunoassay methods have opened the possibility for practical and convenient on-site testing in many non-traditional testing environments such as the physician's office, the home and the workplace. These methods have the potential to improve the quality of human health care by providing rapid and immediate answers to clinical questions and by enabling physicians to conveniently individualize drug therapy and monitor the status of their patients. Applications in the areas of home testing, animal husbandry, law enforcement, and agribusiness are also easily envisioned and would be expected to provide a wide spectrum of benefits from family planning to improved industrial safety.

REFERENCES

Berezin, I.V., Klibanov, A.M., Samokhin et al., G.P., 1977, Mechanosensitive and sound-sensitive enzymatic systems as chemical amplifiers of weak signals, in: "Biomedical Applications of Immobilized Enzymes and Proteins," 2, T.M.S. Chang, ed., Plenum Press, New York, pp 237-251.

Borisov, Y.Y. and Statnikov, Y.G., 1966, The measurement of a boundry layer thickness in accoustical field, Akust Zh 12, 372-373.

Chen, R., Weng, L., Sizto, N.C., Osorio, B., Hsu, C.J., Rodgers, R., and Litman, D.J., 1984, Ultrasound accelerated immunoassay, as exemplified by enzyme immunoassay of choriogonadotropin. Clin Chem 30, 1446-1451.

Engvall, E., and Perlmann, P., 1971, Enzyme linked immunosorbent assay (ELISA). Quantitative assay of immunoglobulin G, Immunochemistry 8, 871-874.

Glad, C. and Grubb, A.O., 1978, Immunocapillary migration - A new method for immunochemical quantitation, Anal Biochem 85, 180-187.

Greenquist, A.C., Walter, B., and Li, T., 1981, Homogeneous fluorescent immunoassay with dry reagents, Clin Chem 27, 1614-1617.

Litman, D.J., Hanlon, T.M., and Ullman, E.F., 1980, Enzyme channelling immunoassay: A new homogeneous enzyme immunoassay technique, Anal Biochem 136, 223-229.

Lilly, M.D.. 1976, Enzymes Immobilized to Cellulose, in: "Methods of Enzymology," Volume XLIV: Immobilized Enzymes, Klaus Mosback, ed., Academic Press, New York, pp. 46-53.

Litman, D.J., Lee, R.H., Jeong, H.J., Tom, H.K., Stiso, S.N., Sizto, N.C., and Ullman, E.F., 1983, An internally referenced test strip immunoassay for morphine, Clin Chem 29, 1598-1603.

Lowry, O.H., Rosebrough, N.H., Farr, A.L., Randall, A.J., 1951, Protein measurements with the folin phenol reagent, J Biol Chem 193, 265-275.

Maiolini, R. and Masseyeff, R., 1975, A sandwich method of enzyme immunoassay. I. Application to rat and human alphafetoprotein, J Immunol Methods 8, 223-234.

Mitenko, P.A. and Ogilvie, R.I., 1973, Pharmacokinetics of intravenous theophylline, Clin Pharmacol Ther 14, 509-513.

Mosbach, K. and Mattiasson, B., 1970, Matrix bound enzymes. Part 2. Studies on a matrix bound two enzyme system, Acta Chem Scand 24, 2093-2100.

Schneider, R.S., Linquist, P., Wong, E.T., 1973, Homogeneous enzyme immunoassay for opiates in urine, Clin Chem 19, 821-825.

Tyhach, R.J., Rupchock, P.A., Pendergrass, J.H., Skold, A.C., Smith, P.J., Johnson, R.D., Albarella, J.P., Profitt, J.A, 1981, Adaptation of prosthetic-group label homogeneous immunoassay to reagent strip format, Clin Chem 27, 1499-1504.

Ullman, E.F., Gibbons, I., Weng, L., Dinello, R., Stiso, S.N., and Litman, D.J., 1983, Homogeneous immunoassays and immunometric assays employing enzyme channelling, in: "Diagnostic Immunology: Technology Assessment," (CAP Conference 1983), R.M. Nakamura & J.H. Rippey, eds., College of American Pathologists.

Yoshitake, S., Yamada, Y., Ishikawa, E., Masseyeff, R., 1979, Conjugation of glucose oxidase from aspergillus niger and rabbit antibodies using N-hydroxysuccinimide ester of N-(4-carboxycyclohexylmethyl)maleimide, Eur J Biochem 101, 395-399.

Zuk, R.F., Ginsberg, V.K., Houts, T., Rabbie, J., Merrick, H., Ullman, E.F., Fisher, M.M., Sizto, C.C., Stiso, S.N., Litman, D.J., 1985, Enzyme immunochromatography - A non-instrumental, quantitative immunoassay method. Submitted to Clin Chem.

SEPARATION-FREE ENZYME FLUORESCENCE IMMUNOASSAY BY CONTINUOUS FLOW INJECTION ANALYSIS

Tim A. Kelly

Bartels Immunodiagnostic Supplies, Inc.
P. O. Box 3093
Bellevue, WA 98009

INTRODUCTION

Enzyme immunoassays (EIAs) are now known to be a viable alternative to the more established radioimmunoassay techniques. The improved sensitivity along with the inherently better speed, stability, and cost have resulted in the commercial production of several EIA systems (Monroe, 1984). Most systems have colorimetrically monitored the indicating enzyme reaction. With the appropriate considerations, a fluorescence monitor can further enhance the sensitivity and selectivity of such methods without introducing significant sacrifices in the ease, speed, or cost of analysis (Kelly and Christian, 1982).

Several variations and parameters can be considered in utilizing a fluorescence system (Soini and Hemmila, 1979). Both the excitation and emission wavelengths can be selected to avoid interferences and enhance the signal-to-noise ratio. Compounds that fluoresce, can be derivatized to fluoresce, or extinguish or quench the fluorescence of other molecules can be used. Solution conditions such as temperature, pH, solvent, ionic strength, and viscosity can be optimized for fluorescence as well as enzyme performance. Also such instrumental techniques as polarization and time-resolution can aid in the design of a particular system.

The automation of EIAs present in several of the new commercial systems adds a greatly needed quality to these tests (Monroe, 1984). Not only is the speed of these procedures increased but the ease and precision also benefit. Flow injection analysis can also serve in this capacity. Most easily applied to homogeneous EIAs, with the advent of certain modifications flow injection analysis could actually provide greater improvements for heterogeneous systems.

Flow injection analysis is the rapid reproducible injection of a liquid sample plug into a continuously laminar flowing stream (without air segmentation) where mixing occurs by radial diffusion (Ruzicka and Hansen, 1981). The stream can be prepared to contain reagents or the reagents can be immobilized along the path of the stream. Detection by appropriate means occurs at some fixed point along the stream and response peaks appear that are proportional in height to the analyte concentration. The parameters of tube length, tube diameter, flow rate, and manifold design can be adjusted to provide the amount of sample dispersion best suited to the particular test while minimizing the analysis time, sample size, and reagent volume. With autosamplers and microprocessor-controlled operations and data reduction the process may be entirely automated.

This chapter describes an homogeneous EIA where a new fluorescence detection scheme is utilized and the application of flow injection analysis is first made (Kelly and Christian, 1982). The enzyme, horseradish peroxidase (HRP), is conjugated to the antibody, anti-human IgG (Ab). HRP catalyzes the oxidation of leuco-diacetyldichlorofluorescein (LDADCF, Fig. 1.) by hydrogen peroxide which produces the highly fluorescent dichlorofluorescein (DCF). The binding of the labeled antibody to the antigen, IgG (Ag), partially inhibits the activity of the enzyme. This is noted as a decrease in the fluorescence intensity produced by the indicator reaction. This is represented in Fig. 2.

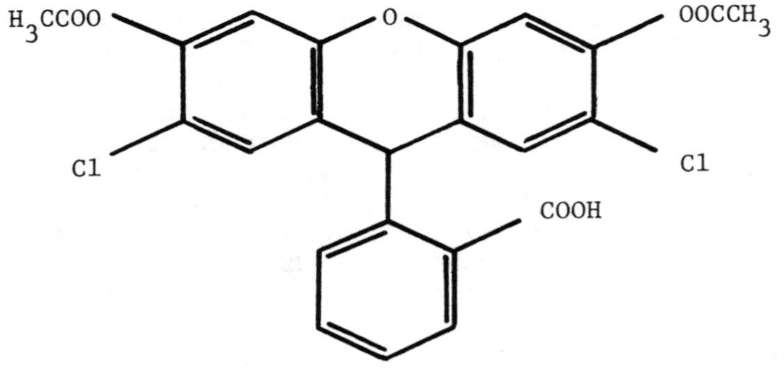

Fig. 1. Structure of Leuco-diacetyldichlorofluorescein, LDADCF (3',6'-diacetyl-2',7'-dichlorofluorescein).

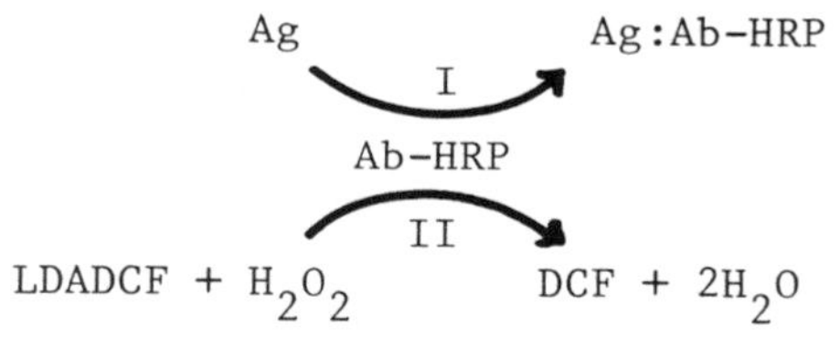

Fig. 2. The reaction scheme for an homogeneous fluorescence enzyme immunoassay with an antibody-labeled enzyme.

A dual-channel, dual-reagent reaction manifold makes up the core of the flow injection analysis system (Fig. 3.). This automatically performs the appropriate sequence of reactions before detection by a laser filter fluorimeter with a sheath flow cell (Kelly and Christian, 1981).

Various levels of the analyte, IgG, can be diagnostic of many disorders whether associated directly with the immune system or indirectly with various cancers or liver or kidney diseases (Brown et al., 1979). However, of greater importance is that macromolecules like IgG are usually analyzed by heterogeneous methods. Recent evidence indicates that is not necessarily so (Ngo and Lenhoff, 1981), giving added encouragement to the development of new homogeneous EIAs.

METHODS

Use phosphate-buffered saline solution (PBS; 0.02 M phosphate, 0.15 M sodium chloride, 1 % poly(ethylene glycol)(PEG-6000, Sigma P-2139), pH 7.4, boiled to prevent bacterial degradation of proteins) in making all solutions and dilutions. Make all final solutions fresh daily.

Optimization Studies

To study the binding equilibria of the antigen-antibody reaction and the activity of the conjugated enzyme, use a static spectrofluorimeter (Perkin-Elmer 650-10S), exciting with 488 nm light and monitoring the fluorescence at 525 nm. Monitor the fluorescence intensity changes for each of the following successive additions (made at 30 second intervals): (a) 2.8 ml of human IgG solution (25 ug/ml, Sigma I-4506), (b) various volumes of either HRP solution (35 IU/ml, Sigma P-8250) or anti-human IgG conjugated with HRP (1:20 dilution, Miles-Yeda, Ltd. 61-130)(Volume changes assumed to be negligible in the total final volume.), (c) 50 ul of activated LDADCF (10^{-5} M, synthesized according to the procedures of Brandt and Keston (1965)), and (d) 50 ul of hydrogen peroxide

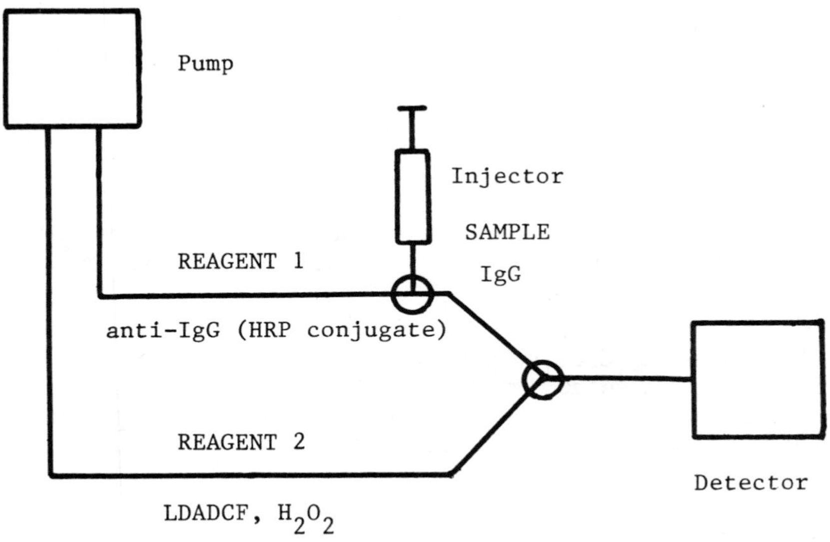

Fig. 3. Dual-reagent flow injection analysis system.

(10^{-5} M). Step (d) triggers the indicating reaction and is the intensity change of interest.

Analytical Procedures

Assemble the fluorimeter according to the specifications of Kelly and Christian (1981). Focus the argon ion laser (488 nm) on the sheath flow cell and allign the optics containing the 520 nm cut-off filter. This collects all the emitted radiation above 520 nm thus eliminating interferences and the scattered radiation.

Set-up the flow system as indicated in Fig. 3. using a 1.0 m length of 0.8 mm bore Teflon tubing between the 25 ul loop injector and the sheath cell. The same type of tubing should be used throughout the system to minimize excessive dispersion. The dye-peroxide stream (reagent 2) should enter half way to the detector. Use compressed air and reagent reservoirs as a means of pumping the system without pulsation. Two regulated air streams set at 0.36 and 0.32 atm (for reagent 1 and reagent 2 plus the sheath stream, respectively) drive the streams with the following velocities: reagent 1, 0.50 ml/min; reagent 2, 0.50 ml/min; sheath stream, 1.60 ml/min.

Make up the reagents and samples as follows: reagent 1, 1:250 dilution of the HRP conjugated anti-human IgG; reagent 2, 10^{-5} M activated LDADCF, 10^{-4} M hydrogen peroxide; sheath, distilled, deionized water; standards, appropriate concentrations of purified, crystalline human IgG; and serum samples, 1:700 dilution of samples (0.1 ul serum necessary per test) preanalyzed by radial immunodiffusion from Calbiochem, Tri-Partigen.

Allow 15 minutes for the system to warm-up and stabilize. Injections can be made every 60 seconds.

RESULTS

Optimization Studies

Initially the indicator reaction was characterized independently. This dye, LDADCF, was chosen because of its oxidation by a commonly used and versatile enzyme, HRP, and the properties of its fluorescent product, DCF. LDADCF is not fluorescent but DCF is highly fluorescent and utilizes excitation and emission wavelengths that avoid possible interferences from scattering or the sample matrix. A simplex technique was used to optimize the various solution conditions (Kelly and Christian, 1981). These were modified slightly in coupling this system to the immunochemical reaction. The poly(ethylene glycol) has been introduced as an agent that demonstrated its effectiveness in minimizing an autooxidation blank signal (Kelly and Christian, 1982) and accelerating the antigen-antibody binding (Buffone et al., 1975).

The instrumentation has also been previously evaluated (Kelly and Christian, 1981). The flow system is stable to within 2 %. The flow fluorimeter, when used with a similar dye, demonstrated a proportional response over four orders of magnitude with a minimum detectable concentration of 10^{-11} M.

With respect to the actual system of interest here, several specific features were observed in the response behavior. In the absence of IgG, the activities of free and conjugated enzymes were equal. When IgG was present, binding to the antibody was found to inhibit the activity of the conjugated enzyme. The extent of inhibition varied with the concentration ratio of antigen to antibody. No significant precipitation was observed yet complete binding was noted within 30 seconds. This makes possible not only the application as an immunoassay but the incorporation of flow injection analysis.

The inhibition served as a means of obtaining a "precipitin curve" (Fig. 4.) indicative of the binding stoichiometry. The precipitin curve was obtained by subtracting the signals of HRP-conjugated antibody bound to antigen from the signals for the free HRP alone. The maximum occurred at 0.56 IU/ml HRP, corresponding to a 1250-fold dilution of the antibody and an IgG concentration of 25 ug/ml or a 700-fold dilution of serum containing high IgG

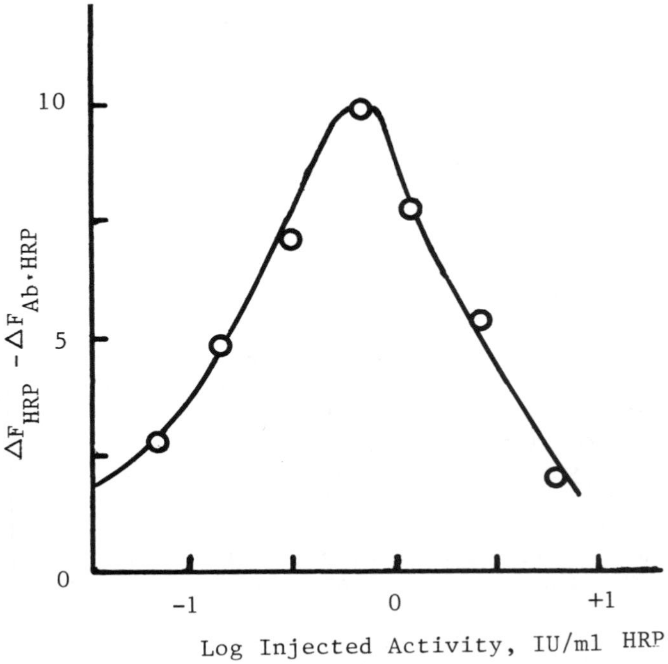

Fig. 4 Difference in the fluorescence intensity change with HRP activity between free HRP and HRP conjugated to bound antibody. Solution conditions were as follows: 25 ug/ml IgG, 10^{-5} M activated LDADCF, and 10^{-5} M hydrogen peroxide in PBS.

levels. These concentrations were used as a guide to the conditions for use with the flow system.

Immunoassay Performance

Since the conditions chosen for the flow system correspond to the maximal inhibition of HRP activity for high IgG levels, the entire normal range of serum IgG levels (8-14 mg/ml, or 11-20 ug/ml after dilution) should fall within the region of the precipitin curve in which the antibody is in excess, and allow for straightforward quantification. Pathologically elevated IgG levels would correspond to the region of the curve where the antigen is in excess and would give rise to signals that could be misinterpreted as corresponding to an antigen level much lower than the true value. To resolve this possible ambiguity the sample would have to be diluted further and re-analyzed. For the actual immunoassay of the flow system, a maximum response was observed at 31 ug/ml IgG

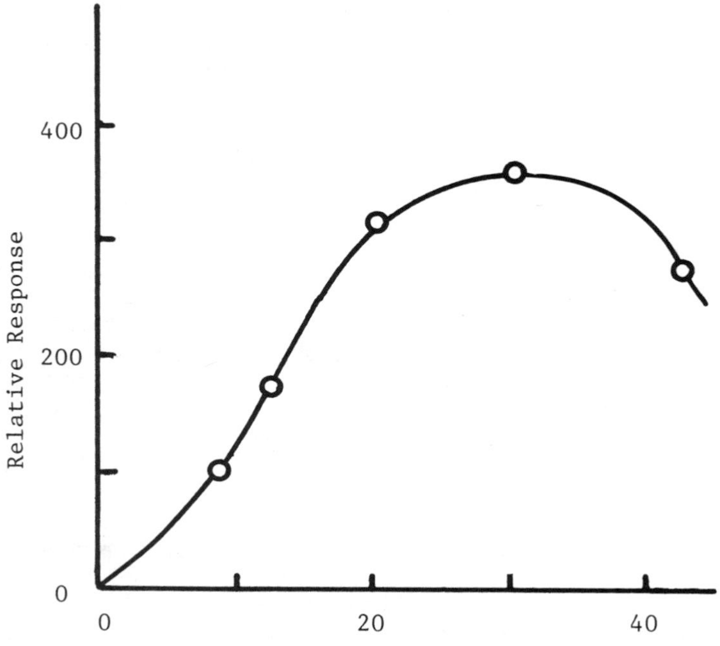

Fig. 5. Relationship of fluorescence response to concentration of IgG serum samples, i.e. a precipitin curve. 25 ul sample (prediluted 700-fold) volumes were injected in a reagent stream containing 1:250 HRP conjugated anti-human IgG in PBS. After reacting for 30 seconds the sample merged with a stream containing 10^{-5} M activated LDADCF and 10^{-4} M hydrogen peroxide in PBS. It reacted another 30 seconds before being detected in the flow cell.

(Fig. 5.). Calibration is possible over the range 2-30 ug/ml. The minimum detectable IgG concentration was 2 ug/ml (1.2×10^{-8} M or 50 ng in the 25 ul injected volume).

To determine the accuracy of the immunoassay system, recovery and comparison studies were performed. Twelve preanalyzed serum samples were analyzed in triplicate. The average serum IgG response was 96.7 % of the corresponding standard response. An average recovery of 103 % from serum spiked with IgG was observed.

The flow system exhibited a within-run precision of 9.8 % for determinations run at the maximum rate of 60 per hour.

No significant interferences (within common sample media) have been observed to adversely affect the indicating reaction (Kelly, 1981). Impurities in the antibody preparation could theoretically give rise to spurious results but none were seen in the analysis of real samples.

DISCUSSION

A few aspects of the immunoassay could bear some additional comments. The precision was adversely affected by a back-pressure developed during sample injection. This was perhaps compounded by the added viscosity from the presence of the poly(ethylene glycol). Either a by-pass stream for injection or a modified pumping system should result in significantly improved precision. A more sophistocated optical system would further enhance the sensitivity possible. The use of monoclonal antibodies will only serve to improve the assay's performance. The stability of the dye would benefit from a greater understanding of the catalyzed indicator reaction. Even without these considerations, the system is seen to provide an effective, new approach toward the development of homogeneous EIAs.

Relatively few homogeneous fluorescence EIAs have been developed. Burd and his associates (1977) developed the most widely used system of today. It is a substrate-labeled, competitive, rate method. The non-fluorescent umbelliferyl-β-galactoside is coupled to a drug of interest. This conjugate, substrate for the enzyme β-galactosidase, upon reaction gives a fluorescent product. Binding with an anti-drug antibody inactivates the substrate. Free drug in the sample competitively releases the conjugate for the enzyme reaction. The method has been used for the determination of several drugs (Burd et al., 1977 and 1978; Wong et al., 1979) and has also been used for proteins (Ngo et al., 1979 and 1981). The method displays detection limits of around 1 ug/ml using 1-2 ul sample volumes with good precision and accuracy. However, several manipulations and measurements are necessary causing the assay to take anywhere from a few up to twenty minutes to perform. Scientists at the Terumo Corporation of Japan (1983) has simplified this slightly by incorporating the method into a multilayer immobilized-filter test strip procedure.

Using a similar substrate-labeled EIA scheme, Burd (et al., 1977) tried a different combination of components. Umbelliferone (or fluorescein) is the fluorophore. Conjugating this to a biotin (or 2,4-dinitrophenyl derivatives) molecule with an ester linkage makes it non-fluorescent. Porcine esterase is the enzyme employed to cleave the bond. Binding with avidin (or anti-DNP antibody) inactivates the substrate. A number of complications have hindered its effectiveness.

Another enzyme system has recently been used in a strikingly different approach to the development of EIAs (Litman et al., 1980). Litman, Ullman, and their associates utilized the technique of enzyme channeling in an homogeneous immunoassay. It is where two enzymes (that catalyze consecutive reactions) are brought into close proximity by immunochemical reaction, thus enhancing the rate and specificity of the reaction. The antigen, IgG, is conjugated to one enzyme, hexokinase. The second enzyme, glucose-6-phosphate dehydrogenase, is immobilized along with the anti-human IgG antibody on sephadex beads. The analyte, IgG, competes with the first enzyme conjugate in binding to the beads. In the absence of IgG more complexes containing both enzymes will be formed thus accelerating the conversion of glucose, ATP, and NAD to gluconalactone-6-phosphate, ADP, and NADH. NADH is monitored fluorimetrically at 450 nm. The concept has significant potential demonstrated by the lowering of the detection limit to 1 ng/ml. Complexity, long analysis time, imprecision, and in this case interferences still plague the method however.

The demand for commercially available systems continues to grow. The potential of fluorescence EIAs has certainly been realized. New methods must demonstrate greater speed, versatility, and simplicity at reduced costs. Automation will be a key factor in making these improvements. There are many options with respect to enzymes and fluorophores used, means of automation, analytical schemes employed, that have yet to be investigated. The method presented here describes a new dye-enzyme combination for immunoassay, a non-competitive antibody-labeled system, one-point measurements, a 60 second analysis time, less than eight cents per run for reagents, and automation via flow injection analysis, in an attempt to meet these demands.

SUMMARY

An innovative homogeneous fluorescence immunoassay has been described here that incorporates specialized instrumentation, non-competitive, one-point detection, and automation via flow injection analysis. HRP (horseradish peroxidase), conjugated to anti-human IgG antibody, catalyzes the oxidation of LDADCF (leuco-diacetyl-dichlorofluorescein) by hydrogen peroxide to produce the highly fluorescent DCF (dichlorofluorescein). Binding of the antibody to its antigen, IgG, in the sample partially inactivates HRP thus causing a decrease in the fluorescence signal. Less than one microliter of serum is used in a determination that takes 60 seconds to perform and displays an average recovery of 103 % over the range of 2-30 ug/ml with a precision of 9.8 %.

ACKNOWLEDGEMENTS

The author wishes to acknowledge with appreciation the contributions of Dr. Gary Christian to the work described herein. His continual supervision, advise, and encouragement were instrumental in the successful outcome of this study.

REFERENCES

Brandt, R. and Keston, A.S. (1965). Synthesis of Diacetyldichlorofluorescin: A Stable Reagent for Fluorimetric Analysis. Anal. Biochem. 11: 6-9.

Brown, S.S., Mitchell, F.L. and Young, D.S., Eds. (1979). Chemical Diagnosis of Disease. Elsevier/North-Holland Biomedical Press, Amsterdam.

Buffone, G.F., Savory, J. and Hermans, J. (1975). Evaluation of Kinetic Light Scattering as an Approach to the Measurement of Specific Proteins with the Centrifugal Analyzer. II. Theoretical Considerations. Clin. Chem. 21: 1735-1746.

Burd, J.F., Carrico, R.J., Fetter, M.C., Buckler, R.T., Johnson, R.D., Boguslaski, R.C. and Christner, J.E. (1977). Specific Protein-Binding Reactions Monitored by Enzymatic Hydrolysis of Ligand-Fluorescent Dye Conjugates. Anal. Biochem. 77: 56-67.

Burd, J.F., Carrico, R.J., Kramer, H.M. and Denning, C.E. (1978). Homogeneous Substrate-Labeled Fluorescence Immunoassay for Determining Tobramycin Concentrations in Human Serum. In Enzyme Labeled Immunoassay of Hormone and Drugs. Walter de Gruyter & Co., Berlin, 387-403.

Burd, J.F., Wong, R.C., Feeney, J.E., Carrico, R.J. and Boguslaski R.C. (1977). Homogeneous Reactant-Labeled Fluorescence Immunoassay for Therapeutic Drugs Exemplified by Gentamicin Determination in Human Serum. Clin. Chem. 23: 1402-1408.

Kelly, T.A. (1981). Fluorimetric, Electrochemical, and Flow Injection Analysis of Enzyme and Immunochemical Systems. Doctoral Dissertation.

Kelly, T.A. and Christian, G.D. (1981). Fluorimeter for Flow Injection Analysis with Application to Oxidase Enzyme-Dependent Reactions. Anal. Chem. 53: 2110-2114.

Kelly, T.A. and Christian, G.D. (1982). Capillary Flow Injection Analysis for Enzyme Assay with Fluorimetric Detection. Anal. Chem. 54: 1444-1445.

Kelly, T.A. and Christian, G.D. (1982). Homogeneous Enzymatic Fluorescence Immunoassay of Serum IgG by Continuous Flow Injection Analysis. Talanta 29: 1109-1112.

Litman, D.J., Hanlon, T.M. and Ullman, E.F. (1980). Enzyme Channeling Immunoassay: A New Homogeneous Enzyme Immunoassay Technique. Anal. Biochem. 106: 223-229.

Monroe, D. (1984). Enzyme Immunoassay. Anal. Chem. 56: 921A-931A.

Ngo, T.T., Carrico, R.J. and Boguslaski, R.C. (1979). Homogeneous Fluorescence Immunoassay for Protein Using β-Galactosyl-Umbelliferone Label, paper presented at 2nd International Conference on Diagnostic Immunology, New England College, Henniker, New Hampshire.

Ngo, T.T., Carrico, R.J., Boguslaski, R.C., and Burd, J.F. (1981). Homogeneous Substrate-Labeled Fluorescence Immunoassay for IgG in Human Serum. J. Immunol. Methods 42: 93-103.

Ngo, T.T. and Lenhoff, H.M. (1981). Recent Advances in Homogeneous and Separation-Free Enzyme Immunoassays. Applied Biochem. & Biotechnol. 6: 53-64.

Ruzicka, J. and Hansen, E.H. (1981). Flow Injection Analysis. Wiley-Interscience, New York.

Soini, E. and Hemmila, I. (1979). Fluoroimmunoassay: Present Status and Key Problems. Clin. Chem. 25: 353-361.

Terumo Corp. Jpn. Kokai Tokkyo Koho. (September 7, 1983). Japan Patent No. 58,150,861. Immunoassay for the Determination of Drugs.

Wong, R.C., Burd, J.F., Carrico, R.J., Buckler, R.T., Thoma, J. and Boguslaski, R.C. (1979). Substrate-Labeled Fluorescence Immunoassay for Phenytoin in Human Serum. Clin. Chem. 25: 686-691.

SEPARATION-REQUIRED (HETEROGENOUS) ENZYME IMMUNOASSAY FOR HAPTENS AND ANTIGENS

Jim C. Standefer

Department of Pathology
University of New Mexico School of Medicine
Albuquerque, N. M. 87131

INTRODUCTION

Immunoassays that require separation of bound antigen from the unbound fraction have been developed for a large variety of haptens and antibodies (Avrameas,1969; Voller et al.,1976; Engvall and Carlsson,1976; O`Sullivan et al.,1979a). These immunoassay techniques may be divided into two catagories,i.e., sequential (noncompetitive) and competitive. The sequential, heterogenous immunoassay technique, otherwise known as the Enzyme Linked Immuno-Sorbent Assay or ELISA, has been most widely adopted.

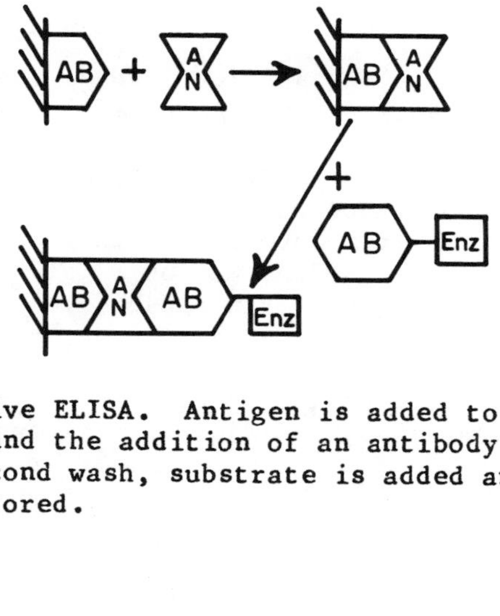

Fig.1. Noncompetitive ELISA. Antigen is added to bound antibody, followed by a wash and the addition of an antibody conjugated to an enzyme. After a second wash, substrate is added and formation of the product is monitored.

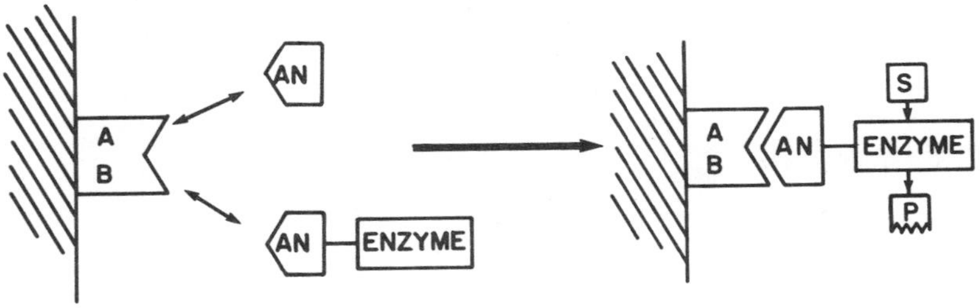

Fig.2. Competitive ELISA: Hapten and conjugated hapten are added to bound antibody. After a wash, substrate is added and appearance of product monitored.

The ELISA technique involves a two step procedure as illustrated in Fig.1. The antibody which is adsorbed or covalently linked to a solid phase reacts first with the antigen or other protein analyte. The residual from this reaction is washed from the reaction container and an enzyme-tagged antibody is added to complete the reaction. This antibody may be produced against a specific antigen or it may be more generally directed as a species specific antibody to an analyte protein. The residual from this second reaction is washed from the container before the substrate for the enzyme is added for the indicator reaction.

Competitive hetergenous assays involve a competition reaction in which the hapten or antigen (tagged and non-tagged) is mixed together with the antibody bound to the solid phase. After a short reaction period the unbound residual is washed from the container and a suitable substrate is added. This sequence is illustrated in Fig. 2. Although a simplier reaction sequence is required for competitive assays, they are generally less sensitive than the ELISA technique. Additionally, matrix effects may be more prominent due to the presence of sample interferents during the antigenic reaction.

Some of the most recent applications of the ELISA technique include methods for determining the presence of circulating antibodies to bacterial antigens (Elder et al.,1983; Hill et al., 1983; Sarafian et al.,1982), viral antigens (Sada et al.,1983; Isozaki et al.,1982; Franco et al.,1983) as well as circulating antibodies that arise from autoimmune diseases (Shimomiya et al.,1982; Ishaq and Ali,1983; Roman et al.,1984). A variety of other analytes including ferritin (Anaokar et al.,1979), galactosyltransferase (Verdon et al,1982), C5a antigen (Kunkel et al.,-1983), retinol-binding protein (Lucertini et al.,1984), anti-ds DNA (Myers et al,1984), prolactin (Sun et al.,1984), lutropin (Rathnam and Saxena,1984), and digoxin (Freytag et al., 1984) have been assayed successfully by this technique. The competitive technique has been applied with some success for assays of hCG (Yorde et al.,1976), gentamicin (Standefer and Saunders,1978), and digoxin (Fu et al., 1984). Regardless of whether the sequential or competitive assay is performed, certain common ingredients for a successful assay are required; these include the type and preparation of the solid phase, the type and quality of enzyme conjugate, the reaction sequence and the timing of the reaction steps.

The purpose of this chapter is to present some specific information regarding those common variables that must be controlled in order to insure a reliable and sensitive ELISA procedure. This will include selected procedures for binding antibody to various solid phases, some approaches for conjugating enzymes to antibodies or haptens, plus an examination of some of the procedural variables including conjugate concentration and incubation time.

METHODS

Conjugation of the Indicator Enzyme

Several methods have been proposed for covalently linking antibodies to indicator enzymes. One of the earliest of these methods used glutaraldehyde to link amino groups of the indicator enzyme with those of the conjugating antibody (Avramas,1969; Engvall and Pearlman,1971). While this conjugation method is relatively simple to perform, excessive intramolecular cross-linking limits both the enzymatic and immunologic activities. Other conjugation methods have been developed in order to promote covalent linking while preserving enzymatic and immunologic activities of the conjugate. A variety of maleimide derivitives, for example, have been used to link sulfhydryl groups of proteins. These maleimide derivatives are classified into two general catagories: Homobifunctional (single reactive group) and heterobifunctional (two or more reactive groups). A homobifunctional reagent , e.g., N,N′-o-phenylenedimaleimide, contains two reactive centers that each react with a sulfhydryl group of an amino acid in the protein (Kato et al.,1975). A heterobifunctional coupling reagent is one that has two reactive centers that each react with different functional groups of amino acids in the protein. For example, m-maleimidobenzoyl-N-hydroxysuccinimide ester has been used successfully to couple the sulfhydryl groups of IgG with the free amino groups of galactosidase (O′Sullivan et al.,1978). This process is illustrated in Fig. 3. The hydroxysuccinimide ester reacts rapidly under mild conditions (0.1 M phosphate, pH 7.0 at 30 C for 1 hour) with the available amino groups of the immunoglobulin. After chromatography on Sephadex G-25, the "activated" immunoglobulin is mixed with galactosidase treated with 0.01 M mercaptoethanol to reduce its

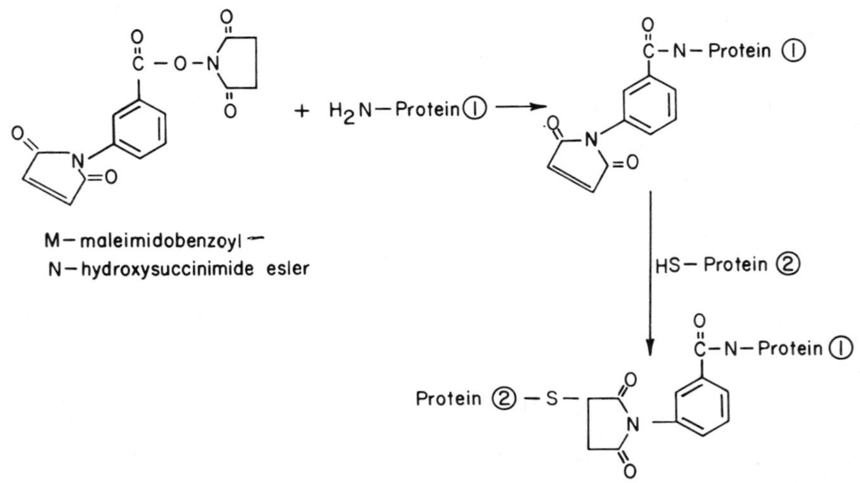

Fig.3. Enzyme:Antibody Conjugation: M-maleimidobenzoyl-N-hydroxy-succinimide is used to couple amino groups of IgG [Protein (1)] to sulfhydryl groups of an enzyme [Protein (2)].

sulfhydryl groups. The maleimide moiety reacts with the reduced sulfhydryl groups to complete the covalent link between the proteins.

Another approach to conjugation that requires an enzyme bound to carbohyrdrate residues is the periodate oxidation method illustrated in Fig. 4 and first described by Nakane and Kawaoi (1974). This method has been used to couple peroxidase, which is 10-15% carbo-hydrate by weight, to both antibodies (Anaokar et al.,1979) and haptens (Standefer and Saunders,1978). I describe below a slightly modified protocol for production of an IgG-ferritin conjugate.

Anti-ferritin rabbit IgG is coupled to horseradish peroxidase (300 U/mg) in a two-step process. The first step produces a "peroxidase aldehyde" as follows: Dissolve 5.0 mg of peroxidase in 1.0 mL of 0.3 mol/L sodium bicarbonate and gently mix with 0.1 mL of a 10 g/L solution of fluorodinitrobenzene in ethanol for 1 h at room temperature. Add 1.0 mL of a 80 mmol/L solution of sodium periodate in water and mix gently for 30 min at room temperature. Add 1.0 mL of a 0.16 mol/L solution of ethylene glycol and mix gently for 1 h at room temperature. Dialyze the mixture in carbonate buffer (10 mmol/L, pH 9.5) for one day at 4 C with several buffer changes. The "peroxidase aldehyde" is stable for at least one month at 4 C. In the second step of the coupling process, 5 mg of IgG is mixed with 5 mg of peroxidase aldehyde for 3 hours. The conjugate is passed through a Sepharose 6-B column as a dark brown band emerging in the void volume and is stable at -20 C in the presence of bovine serum albumin (20 g/L) for several months. Self-conjugation of the

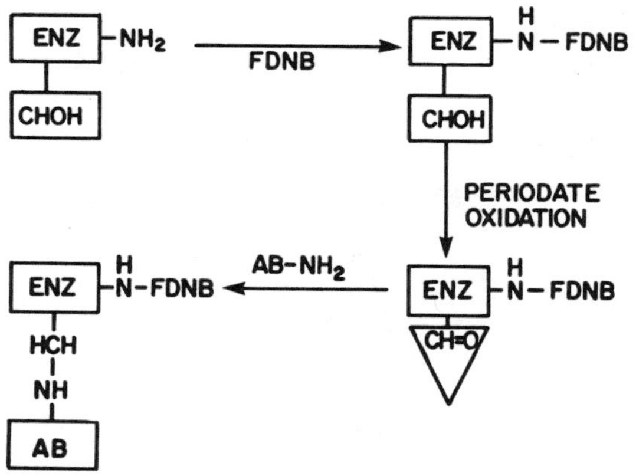

Fig.4. Enzyme:Antibody Conjugation: Oxidized carbohydrate residues (-C=O) of an enzyme are coupled to amine groups of IgG.

oxidized peroxidase can be reduced from 35% to 5% by adjusting the pH from 4.4 to 9.5 prior to the oxidation step (Wilson and Nakane, 1978).

Greater stability of the enzyme conjugate is one of the primary advantages of enzyme immunoassay techniques compared to radioimmunoassay. Maleimide derivitives are stable for 1-2 years when stored in 1% bovine serum albumin, 0.1% sodium azide and mercatoethanol (O`Sullivan,1979b). No significant loss of activity is observed for a peroxidase conjugate when it is stored for several months (Nakane and Kawaoi,1974).

Purification of the Antibody

Serum from animals injected with a hapten conjugate may contain a wide variety of antibodies in addition to extraneous proteins that may interfere with the binding of antibody to the solid phase. In order to obviate potential interferences, nonpurified anti-serum may be purified by repeated precipitation with 30% ammonium sulfate followed by dialysis (Anaokar et al,1979) or by ion-exchange chromatography (Rathman and Saxena,1984). While nonpurified serum may be used to coat the solid phase, protein adsorption to the solid phase is nonspecific and a more sensitive assay is possible when purified antibody is used in the reaction. Untreated serum from a rabbit that had been injected to raise antibodies to ferritin was compared to serum that had been purified by X3 precipitation with 30% ammonium sulfate. The data in Fig. 5 illustrate that when a nonpurified serum is used, the assay sensitivity is reduced to about one-third of that obtained when more purified antisera is used.

When excess antibody is applied to the solid phase there may be little change in the slope of the standard curve, even when the antibody solution is diluted several fold. Each of the two families of curves shown in Fig. 5 represent binding in an ELISA assay obtained when the antiserum was diluted over a ten fold range (10,000 to 100,000 fold) prior to coating the solid phase. A ten fold change in the antiserum concentration reduced the conjugate binding by less than 5% of that obtained with the more concentrated antiserum as evidenced by the change in absorbance values shown in Fig. 5. The optimal antibody dilution, i.e. the greatest dilution that will yield the required sensitivity and linearity through the desired range of concentrations, should be determined empirically. The antibody dilution that is optimal for a particular ELISA procedure may be determined by preparing a series of 10-fold dilutions of the antiserum and applying each of these dilutions to the solid phase. Use each series of antiserum dilutions to analyze a series of antigen standards whose concentrations are within the expected physiological range. Construct a series of standard curves from these data and select the curve that provides the best combination of sensitivity and linearity as judged by its slope, intercept and correlation coefficient.

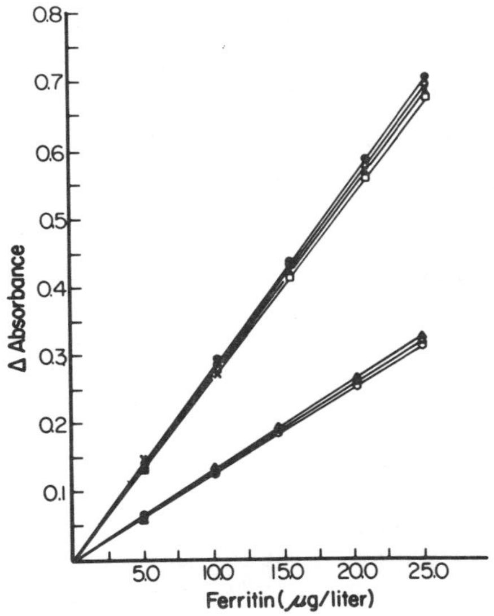

Fig. 5. Dose-response curves produced by coating solid phase with purified antibody (upper curves) or whole sera (lower curves). See text for details. Reprinted with permission from Clinical Chemistry.

Preparation of the Solid Phase

Antibody can be attached to the solid phase by "passive" adsorption or by "active" covalent linking. Passive adsorption of antibody is accomplished by soaking the solid phase in an antibody solution for a few hours so that the antibody has an opportunity to attach to the solid phase via hydrophobic bonds. After the initial soaking period, the excess antibody is washed off the solid phase and a dilute solution of albumin (1-3%) or gelatin (.05-0.1%) is applied in order to limit the nonspecific adsorption of conjugate. This process was first described by Catt and Tregear (1967). While the binding of proteins to solid phases such as polystyrene has been reported to be independent of pH and ionic strength of the buffer used in the coating process (Crosignani et al.,1970), an earlier report suggests that the pH and ionic strength of the coating buffer influence protein conformation and thereby influence protein binding to the surface (Oreskes and Singer,1961). Additionally, it is suggested that the increase in binding observed when proteins are at their isoelectric point is due to a reduction in electrostatic repulsion between closely packed proteins on the surface. It is clear that affinity of proteins for the solid phase is concentration dependant and the cumulative effect of pH and ionic strength on protein binding is more obvious at higher protein concentrations. At immunoglobulin concentrations below the saturation level of the polystyrene surface, increasing the pH from 4 to 9 increased protein binding 10%, while at protein concentrations high enough to saturate the solid phase surface the same pH change increased protein adsorption by 25%.

Adsorption of proteins to a polystyrene surface is directly related to contact time and temperature. Approximately two fold more protein is adsorbed at 37 C compared to 4 C. Furthermore, the adsorption is 80% complete within 3 hours of contact and appears to reach a maximum after 36 hours (Cantarero et al,1980). Treatment of the solid phase with a non-specific protein such as albumin or gelatin after the antibody has been applied fills the open binding sites that might contribute to non-specific binding of the conjugate. It is interesting to note that, at least in one instance, pretreatment of the solid phase with albumin prior to addition of the antibody appeared to work as well as post-treatment (Standefer and Saunders,1978).

The passive adsorption process can be made more efficient by pretreatment of either the polystyrene surface (Howell et al.,1981) or the antibody (Parsons,1981) prior to the adsorption step. Pretreatment of the polystyrene surface with a freshly diluted two percent solution of glutaraldehyde for two hours at room temperature appears to increase the antibody binding significantly. An alternative approach invloves pretreatment of the antibody with glutaraldehyde. Glutaraldehyde, 400 mL of a 0.02 M solution, is added to 2000 mL of a 1 to 2000 dilution of antiserum at pH 6.0 for 20 min at 37 C. This process will increase the binding of antibody to polystyrene during a one hour exposure at 37 C by 2 to 5 fold.

In order to minimize variations in the amount of antibody bound to the solid phase, covalent binding of antibody to the solid phase may be necessary. In this regard, antibodies have been attached covalently to latex (Quash et al.,1978), nylon (Hendry and Herrmann, 1980), as well as sepharose and cellulose (Bolton and Hunter, 1973). Generally, these procedures involve pretreatment of the solid phase to activate inert reaction sites and promote binding of the antibody. For example, free carboxyl and amino groups on nylon may be produced by incubation with 3.5 M HCl for 24 hours (Hendry and Hermann,1980) and cyanogen bromide is used to activate either cellulose or sepharose for covalent attachment of the antibody to the cyanogen bromide activated solid phase (Bolton and Hunter,1973).

Desorption of passively adsorbed antibody from the solid phase during the subsequent reaction steps may significantly reduce sensitivity and precision. It has been reported that 40% of passively adsorbed antibody may be lost during the subsequent reaction steps (Engvall et al.,1971). However, more recent studies indicate that the desorption process is time dependant, can be controlled, and the total amount of desorbed protein is insignificant (Howell et al.,1981; Cantarero et al.,1980). Polystyrene and glass surfaces were coated with IgG (80 ug/ml) and then soaked in an albumin solution (phosphate buffer containing 0.6% albumin and 0.1% Tween 20). Only a small portion of the adsorbed proteins (1.5% of a total of 32 ug bound) was desorbed with most of the protein loss occurring during the initial 48 hours (Howell et al,1981). If desorption of protein is a concern, pretreatment of the solid phase surface with glutaraldehyde (for polystyrene) or silane (for glass) will prevent most of the desorption.

RESULTS

Concentration of the Enzyme Conjugate

One measure of the sensitivity of immunoassays is the magnitude of the difference between the absorbance obtained in the initial reaction tube (the zero or no standard) and that obtained in the tube containing the next highest standard concentration (lowest standard in the series). One important factor which magnifies this difference is the concentration of the enzyme conjugate used in the assay. When the concentration of the conjugate is too high, excess non-specific binding of conjugate may occur. This may produce excessive background color development. In contrast, as the conjugate concentration is lowered, less substrate will be converted to product and the sensitivity of the assay may be limited. When three dilutions of a peroxidase:gentamicin conjugate were incubated with solid phase bound antibody, the subsequent substrate conversion was linear during a 10 minute reaction period only for the conjugate that had been diluted 500X. (Fig. 6). Nonlinear kinetics were observed when either 100X or 250X dilutions of the conjugate were used. Obviously, the requirement for high sensitivity must be balanced against the requirement to operate at zero order kinetics during the enzyme indicator reaction. As the concentration of conjugate is increased, more conjugate will be bound, and higher enzyme activity will be observed. Unfortunately, this higher enzyme activity may lead to substrate exhaustion during the reaction period as illustrated by the nonlinear increase in the absorbance after 10 minutes of reaction. This nonlinearity appears earlier during the enzyme reaction and is more pronounced at higher concentrations of enzyme conjugate.

Sensitivity

One aspect of ELISA assays that has been promoted as an advantage is their sensitivity. High sensitivity usually is attained by a) purification of the enzyme conjugate, b) increasing the incubation time, particularly during the enzyme reaction, and c) performing sequential reaction steps in which the antigen is equilibrated with the solid phase antibody followed by addition of a conjugated antibody.

When a conjugate is produced by covalently attaching an enzyme to an antibody, usually there is an excess of unreacted enzyme

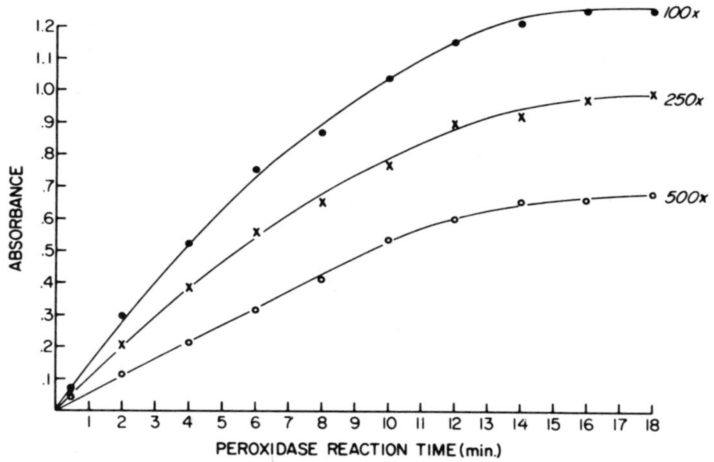

Fig.6. Effect of Conjugate Dilution on Peroxidase Reaction. Substrate conversion with dilutions of enzyme conjugate is linear at 500X dilution during 10 min reaction. See text for details. Reprinted with permission from Clinical Chemistry.

after the reaction is complete. This excess enzyme may reduce the sensitivity of the assay by contributing to nonspecific substrate conversion. Sensitivity may be improved dramatically by removing the excess enzyme. For example, the sensitivity of an ELISA assay for HBsAg was improved by 50 fold after the conjugate was purified by affinity chromatography and gel filtration (Porstmann et al., 1981).

Unfortunately, extended incubation times may not always result in increased sensitivity. It is important to determine whether longer reaction times actually lead to increased sensitivity or whether some loss of color may occur during extended periods, especially when elevated temperatures are used for the incubation. After an initial reaction with standards and a peroxidase:IgG conjugate, a substrate solution was incubated for various times at 25 C and at 37 C to determine whether some loss of color may occur during the enzymatic reaction (Fig.7). It is clear that some loss does occur after 30 minutes at 37 C compared to the continuing color development at 25 C.

An example of how the sequence of reagent addition may effect assay sensitivity is shown in Table 1. The absorbance values obtained from a competitive hCG assay (incubation time of 1 hour) are compared with those from a sequential assay in which the same reagents are incubated in two separate 1 hour incubations. Since absorbance values lower than .300 are associated with decreasing precision, one would expect less precision in assaying hCG levels lower than 100 mIU/ml when using the shorter, competitive assay.

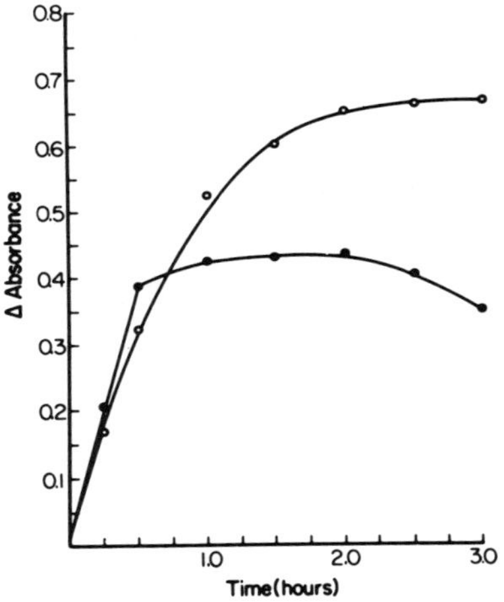

Fig.7. Effect of Temperature on Substrate Conversion. Peroxidase activity was measured at 25°C,(O-O), and at 37°C,(●-●). Reprinted with permission from Clinical Chemistry.

In this procedure the absorbance values required for accurate quantitation at low levels of hCG are not produced by a 1 hour competitive incubation. The sensitivity for this particular procedure has been improved several fold by increasing the amount of enzyme conjugate added, increasing the antigen-antibody reaction time, incubating at 37 C instead of at room temperature, increasing the reaction time of the enzyme reaction, and reducing the volume of the final stopping reagent (Mailliavin et al.,1984).

Circulating Immune Complexes

For ELISA assays that use a species specific second antibody for enzyme conjugation, the problem of circulating immune complexes may become important. An example is the assay for circulating antibodies to double stranded DNA in which the indicator reagent is anti-human IgG conjugated to an enzyme. The indicator conjugate is added to form a sandwich with the bound IgG (anti-DNA). If circulating immune complexes (which contain IgG) are present in the patient's serum during the initial reaction, they may also adsorb to the solid phase and react nonspecifically with the indicator IgG. This nonspecific reaction would lead to false positive results.

This phenomena is illustrated in Table 2 where data from two patients are compared. Serum from a patient with streptococcal glomerular nephritis, SGN, along with serum from a patient with serum lupus erythematosis,SLE, were assayed by an ELISA technique using a double stranded DNA bound to a solid phase and anti-human IgG conjugated to peroxidase as the indicator. Prior to analysis, a portion of the serum samples were ultracentrifuged at 17,000 X g for 10 minutes to remove immune complexes. The untreated sera of both patients show strong positive reactions in the sequential assay. However, when immuncomplexes are removed prior to assay, only the SLE serum continues to show a positive reaction.

Table 1. Comparison of Competitive and Sequential hCG Assays

	ABSORBANCE	
Sample (mIU/mL)	Competitive Assay (a)	Sequential Assay (b)
0	0.028	0.040
10	0.048	0.118
50	0.100	0.294

a) competitive single 1 hour incubation
b) noncompetitive sequential two 1 hour incubations

While the problem of false positive results due to circulating immune complexes must be considered when using an anti-IgG antibody in the conjugate, recent reports suggest that few false positives are observed in routine assays. In two recent studies (Myers et al.,1984; Haugen et al.,1983) a variety of patients were assayed for anti-DNA antibodies by ELISA techniques and only two false positive results were reported out of a total of approximately 80 patient samples that were potentially positive for SLE.

DISCUSSION

A major factor in the successful development of ELISA techniques is the adsorption of antibody to the solid phase in a reproducible manner that will limit nonspecific binding while supporting adequate sensitivity in the subsequent reactions. While this process may appear to be technically simple, several factors combine to cause nonreproducible results. First, all plastic surfaces are not uniform with respect to their hydrophobic regions. For example, a polystyrene surface may contain a variable amount of the releasing agent used during the extrusion process. This releasing agent may inhibit the binding of antibody by disrupting the hydrophobic forces involved in the binding so that a variable amount of protein (antibody) will be attached. Furthermore, the antibody that does bind may be so loosely attached that it may be lost during subsequent washing steps.

A second factor related to attaching an antibody to a solid

Table 2. Effect of Ultracentrifugation on Removal of Circulating Immune Complexes.

	Percent of Reaction	
	Blank	ds DNA
SGN Serum a)		
Non-treated	102	100
Ultracentrifuged	18	17
SLE Serum b)		
Non-treated	106	100
Ultracentrifuged	22	68

a) Streptococcal Glomerular Nephritis serum positive for circulating immune complexes
b) Serum Lupus Erythematosis serum positive for circulating immune complexes

phase is the concentration of the antibody in the solution that is placed in contact with the solid phase. It has been shown that adsorption of protein to polystyrene is linear with protein up to 100 nanograms per cm2 (Pesce et al.,1977; Cantarero et al.,1980; Howell et al.,1981). When protein is applied at higher amounts per surface area, the asdorption of protein to the surface is less efficient and more desorption occurs. This lowered affinity is attributed to the layering of proteins on the surface which leads to desorption of antibody during subsequent reaction steps (McClaren, 1981). Another explanation for the apparent decrease in antibody binding is a steric hindrance induced by interactions between closely packed antibody molecules. This steric hindrance leads to a nonlinear increase in binding to the solid phase at higher concentrations of antibody (Parsons,1981).

Since the steric effects of closely packed proteins may be important in solid phase adsorption, it is interesting to consider the effect of purifying the antisera before applying it to the solid phase. Since purification should remove a portion of the extraneous proteins from the antisera, relatively more of the specific antibodies should remain to bind to the solid phase. Two aspects of the purification process are critical in this regard. First, the binding of antibodies to polystyrene is not influenced by the presence of nonspecific proteins at concentrations lower than 1 ug/ml (Cantarero et al.,1980). Thus, it is important that the purification process remove extraneous protein to that extent. Second, most proteins bind to polystyrene with approximately the same affinity (Kenny and Dunsmoor,1983). This suggests that the type of protein present is not as important to the binding process as is the quantity of protein remaining after the purification process. However, as an exception to the rule, fibrinogen does bind to polystyrene with greater avidity than other proteins and may displace antibody from the polystyrene surface (Morrissey,1977). Thus, the purification process should at least remove most of the fibrinogen in order to prevent selective, albeit nonspecific, binding to the solid phase.

A third factor contributing to poor precision may be the deterioration of the link between the antibody and the solid phase during storage. Under controlled conditions, the adsorbed antibody remains avidly attached to the solid phase with only slow release of a small percent of the total bound protein. This release occurs primarily during the first 48 hours, but may continue at a slower rate for 30 days (Howell et al.,1981). In contrast to what is apparently a relatively slow release of protein from the solid phase surface, other factors may result in a greater and more variable release of the bound antibody of bound protein, including variations in humidity and elevated temperatures during shipping and storage of the reagents. One manifestation of this variable release of antibody is poor reproducibility of duplicates in the assay.

An important variable which directly affects assay sensitivity is the "specific activity" of the enzyme conjugate. If the specific activity is defined as the units of substrate converted per unit time per moles of enzyme present in the conjugate bound to the solid phase, some interesting calculations can be made. A reasonable objective for an enzyme immunoassay is to produce an absorbance of 0.100 in 10 minutes in a final reaction volume of 1 ml. This would require an absorbance change of 0.01 per minute. For an enzyme such as alkaline phosphatase using p-nitrophenyl phosphate as the substrate, this absorbance change would translate to the equivalent of approximately 0.5 mIU of enzyme activity. For most preparations of alkaline phosphatase 0.5 mIU would be associated with approximately 0.1 micrograms or 0.5 pmoles of enzyme. Thus, to produce the desired absorbance change of 0.1 per minute approximately 0.5 pmoles of enzyme conjugate would have to be bound to the antigen in the ELISA sandwich. Assuming approximately equal concentrations of antigen and conjugate, then 0.5 pmoles of antigen would be bound and this translates to some interesting detection limits if one assumes a sample volume of 0.1 ml. For example, 0.5 pmole of thyroxine in 0.1 ml is equivalent to 0.4 ug/dl which is well within the required detection limits for clinical situations. In contrast, 0.5 pmoles of digoxin per 0.1 ml of sample is equivalent to 3.8 ng/ml which is clearly higher than the detection limit required for clinical utility. It should be obvious that the apparent specific activity of the enzyme conjugate depends on several factors, including a) the molar ratio of enzyme to antibody in the conjugate, b) the purity of the conjugate, c) the specific activity of the enzyme used for conjugation, and d) the substrate chosen for the indicator reaction. Each of these parameters may be varied imperically in order to obtain the required sensitivity.

It is apparent that there is no conjugation method that will produce complete bonding of the antibody and enzyme while completely preserving both the enzyme activity and the antigenic activity. Each of the reagents which are used to form covalent bonds between the two proteins will also induce intramolecular bonding which will limit both the antigenic and enzymatic activities. However, there are some critical factors which may be controlled to limit intramolecular bonding and hence promote a successful conjugation procedure. The one-step glutaraldehyde procedure requires that both proteins have approximately the same reactivity with glutaraldehyde in order to produce a conjugate with acceptable reactivity (Ford et al.,1978). This equal reactivity is not the usual case and the choice of enzyme for conjugation can lead to dramatic differences in recovery of antigenic and enzymatic activities. For example, 21% of the antigenic activity survives when lactoperoxidase is coupled to IgG while only 1% of the original antigenic activity is retained when alkaline phosphatase is coupled to IgG. The former conjugate retained 50% of the original enzymatic activity, while the latter retained 70% (Engvall and Perlman,1971).

Regardless of the specific conjugation method, one must control the molar ratio of the antibody to enzyme so that optimum sensitivity can be obtained without sacrificing antigenic activity. Ordinarily, more than one or two molecules of enzyme per molecule of antibody will limit the antibody's antigenic activity. This equal molar ratio for the antibody:enzyme conjugate contrasts with the higher optimal ratio for hapten:enzyme conjugates. Maximal binding may be obtained only when several hapten molecules are attached to one enzyme molecule as reported for cortisol (Comoglio et al.,1976), triiodothyronine (O'Sullivan et al.,1978), and estradiol (Exely and Abuknesha,1978) assays. The exact optimal ratio of hapten:enzyme is empirically determined by examination of standard curves produced by using conjugates with a variety of hapten:enzyme ratios.

Within the limits of optimal hapten:enzyme ratios, analytical utility of the conjugate depends further on careful preparation of the conjugate. It is important to avoid blocking antigenic sites on the hapten and on the antibody. If antigenic sites on either molecule are lost, the quality of the ELISA assay will be diminished (Van Weeman and Shuurs,1976). Also, one should control the conjugation process to limit bridge binding. Bridge binding occurs when an antibody has been raised against the covalent link between the hapten and the carrier protein. If the hapten:enzyme conjugate has the same covalent link as that of the conjugate used to produce the antibody, then antibody binding during the ELISA assay will favor the conjugate to the exclusion of the free hapten. An insensitive assay may result. The bridge binding effect can be diminished by covalently linking the hapten to the enzyme at a site that is different than the one used to attach the hapten to the carrier protein for antibody production.

Another critical factor which must be considered in all immunoassays is the effect of sample matrix. Matrix effects may have a profound, direct influence on the analytical accuracy. One obvious example of matrix effects that occur in both ELISA and radiometric solid phase assays is the high-dose "hook" effect (Miles et al., 1974). The term is derived from the observation that at very high antigen concentrations, tagged antigen binding begins to decrease with increasing concentrations of non-tagged antigen. The "hook" effect has been attributed to changes in the affinity of the antibody when it is bound to the solid phase in a closly packed arrangement (Parsons,1981). When antibody molecules are closely packed on a solid surface, negative cooperativity may cause a decrease in antibody affinity which will lead to decreased ligand binding at high antigen concentrations. The effects of negative cooperativity usually can be overcome by using purified antibodies of high affinity for the solid phase coating process. Another explanation proposed for the high-dose "hook" effect at high antigen concentrations relates to the saturation of the primary antibody. After saturation of the primary, high affinity antibody, there would be increased competition for binding sites on antibodies that have lower affinity for the antigen. As more antigen is added an increase competition at the sites of lower affinity would promote an equilibrium in favor of an unbound conjugate-antibody-antigen complex. In this way, the added antigen would effectively strip away the bound conjugate and produce a standard curve that will decrease as more antigen is added. Neither hypothesis has been carefully evaluated. While the "hook" effect is not obvious in all ELISA assays (Ng et al.,1983), it has been demonstrated for a ferritin assay (Anido,1984). Thus, each ELISA assay should be evaluated for its linearity at very high antigen concentrations.

Most recent applications of the ELISA technique are relatively standard and have been directed toward identification of circulating antigens or antibodies resulting from infectious processes. However, there are some new applications that represent some unique adaptations of the ELISA methodology. One example is the affinity column mediated immunoassay technique (Freytag et al.,1984). In this technique, digoxin in a standard or patient sample reacts with an anti-digoxin antibody-galactosidase conjugate. This mixture is applied to an affinity column containing ouabain covalently linked to a solid support. As the sample-conjugate mixture passes through the column, conjugate that is not bound to drug will adhere to the ouabain and only the drug-conjugate complex passes through the column to mix and react with a substrate, ortho-nitrophenol-B-D--galactopyranoside. The formation of product, ortho-nitrophenol, is monitored at 405 nM and is proportional to the original concentration of digoxin in the sample or standard. This approach has been applied to other drugs, including theophylline and quinidine.

Another interesting adaptation of the ELISA technique is a "dipstick" technique in which the solid phase antibody is dipped into a series of reagents. In this relatively classic ELISA technique, antibody to hCG attached to a plastic dipstick is reacted with an hCG standard followed by a washing step and reaction with an enzyme conjugate. Following a second washing step, the dipstick is placed in a substrate solution for color development. The color development occurs on the surface of the plastic stick and its intensity is proportional to the hCG concentration in the sample. This unique use of a dipstick as the solid phase allows easy manipulation during the assay. Other versions of assays such as this which use monoclonal antibodies are being introduced to the clinical laboratory and represent a considerable advance with respect to sensitivity and specificity in nonisotopic assays.

Although the basic techniques for ELISA assays have been available for more than 15 years, practical applications for the clinical laboratory have only recently incorporated the necessary reliability and convenience to support routine use. These new applications maintain some of the obvious advantages of the ELISA technique including its sensitivity, its relative freedom from matrix effects and its relative simplicity while providing some of the other required features such as automation. Without doubt, many new applications will be available within the next few years and hopefully the sources of these applications will have established control of variables such as antibody specificity, preparation of the solid phase, conjugate stability and assay sensitivity so that these new methods may be easily adapted for routine use in the clinical laboratory.

REFERENCES

Anaokar, S., P.J. Garry, and J.C. Standefer (1979). Solid-phase enzyme immunoassay for serum ferritin. Clin. Chem. 25: 1426-1431.

Anido, G. (1984). Seven ferritin kits compared with respect to the "Hook" effect. Clin. Chem. 30: 500.

Avrameas, S. (1969). Enzyme immunoassays and related techniques: Development and limitations. Immunochemistry. 6: 43-52.

Bolton, A.E., and W.M. Hunter (1973). The use of antisera covalently coupled to agarose, cellulose, and sephadex in radioimmunoassay systems for proteins and haptens. Biochemic. et Biophysica. Acta. 329: 318-330.

Cantarero, L.A., J.E. Butler, and J.W. Osborne (1980). The adsorptive characteristics of proteins for polystyrene and their significance in solidphase immunoassays. Anal. Biochem. 105: 375-382

Catt, K.J., and G.W. Tregear (1967). Solid-phase radioimmunoassay in antibody-coated tubes. Science 158: 1570-1572.

Comoglio, S., and F. Celada (1976). An immunoenzymatic assay of cortisol using E. coli B-galactosidase as label. J. Immunol. Methods 10: 161-170.

Crosignani, P.G., R.M. Nakamura, D.N. Hovland, and D.R. Mishell, Jr. (1970). A method of solid phase radioimmunoassay utilizing polypropylene discs. J. Clin. Endocrinol. Metab. 30: 153-160.

Elder, E.M., A. Brown, J.S. Remington, J. Shonnard, and Y. Naot (1983). Microenzyme-linked immunosorbent assay for detection of immunoglobulin G and immunoglobulin M antibodies to Legionella pneumophila. J. Clin. Microbiol. 17: 112-121.

Engvall, E., and P. Perlmann (1971). Enzyme-linked immunosorbent assay (ELISA) quantitative assay of immunoglobulin G. Immunochemistry 8: 871-874.

Engvall, E., and H.E. Carlsson (1976). Enzyme-linked immunosorbent assay, ELISA. In First International Symposium on Immunoenzymatic Techniques INSERM Symposium No. 2 (Feldman, et al., Eds.). North-Holland Publishing Company, Amsterdam.

Exley, D., and R. Abuknesha (1977). A highly sensitive and specific enzyme-immunoassay method for oestradiol-17-B, FEBS Letters 91: 162-165.

Ford, D.J., R. Radin, and A.J. Pesce (1978). Characterization of glutaraldehyde coupled alkaline phophatase-antibody and lactoperoxidase-antibody conjugates. Immunochemistry 15: 237-243.

Franco, E.L., K.W. Walls, A.J. Sulzer, and J.C. Soto (1983). Diagnosis of acute acquired toxoplasmosis with the enzyme-labelled antigen reversed immunoassay for immunoglobulin M antibodies. J. Immunoassay 4: 373-393.

Freytag, J.W., J.C. Dickinson, and S.Y. Tseng (1984). A highly sensitive affinity-column-mediated immunometric assay, as exemplified by digoxin. Clin. Chem. 30: 417-420.

Fu, P.C., M. Vodian, M. Balga, V. Zie, and B. Haden (1983). Performance characteristics of a competitive solid phase enzyme immunoassay for serum Digoxin. Clin. Chem. 29: 1012.

Haugen, B.R., W.W. Ginsburg, and H. Markowitz (1983). An ELISA microplate assay for IgG and IgM antibody to ss- and ds-DNA. Clin. Chem. 29: 1212-1213.

Hendry, R.M., and J.E. Herrmann (1980). Immobilization of antibodies on nylon for use in enzyme-linked immunoassay. J. Immunol. Methods 35: 285-296.

Hill, H.R., and J.M. Matsen (1983). Enzyme-linked immunosorbent assay and radioimmunoassay in the serologic diagnosis in infectious diseases. J. Infect. Dis. 147: 258-263.

Howell, E.E., J. Nasser, and K.J. Schray (1981). Coated tube enzyme immunoassay: Factors affecting sensitivity and effects of reversible protein binding to polystyrene. J. Immunoassay 2: 205-225.

Ishaq, M. and A. Rashid (1983). Enzyme-linked immunosorbent assay for detection of antibodies to extractable nuclear antigens in systemic lupus erythematous, with nylon as solid phase. Clin. Chem. 29: 823-827.

Isozaki, M., H. Kuno-Sakai, and M. Kimura (1982). Simple enzyme-linked immunosorbent assay of mumps antibody for evaluation of immune status before and after vaccination. N. Engl. J. Med. 307: 1456.

Kato, K., Y. Hamguchi, H. Fukui, and E. Ishikayawa (1975). Enzyme-linked immunoassay; I. Novel method for synthesis of the Insulin -B-D-galactosidase conjugate and its applicability for insulin assay. J. Biochem. 78: 235-237.

Kenny, G.E., and C.L. Dunsmoor (1983). Principles, problems, and strategies in the use of antigenic mixtures for the enzyme-linked immunosorbent assay. J. Clin. Microbiol. 17: 655-665.

Kunkel, S.L., G.L. Manderino, W. Marasco, K. Kaercher, A. A. Hirata, and P.A. Ward (1983). A specific enzyme-linked immunosorbent assay (ELISA) for the determination of human C5a antigen. J. Immunol. Methods 62: 305-314.

Lucertini, S., P. Valcavi, A. Mutti, and I. Franchini (1984). Enzyme-linked immunosorbent assay of retinol-binding protein in serum and urine. Clin. Chem. 30: 149-151.

Mailliavin, A., A. Beaudonnet, J. Pichot, and M.C. Revenant (1984). Improvement of a Commercial EIA Kit for determination of serum B-Choriogonadotropin. Clin. Chem. 30: 597-598.

McLaren, M.L., J.E. Lillywhite, and C.S. Andrew (1981). Indirect enzyme linked immunosorbent assay (ELISA): Practical aspects of standardization and quality control. Med. Lab. Sciences 38: 245-251.

Miles, L.E.M., D.A. Lipschitz, C.P. Bieber, and J.D. Cook (1974). Measurement of serum ferritin by a 2-site immunoradiometric assay. Anal. Biochem. 61: 209-224.

Morrissey, B.W. (1977). The adsorption and conformation of plasma proteins: A physical approach. Ann. N.Y. Acad. Sci. 283: 50-64.

Myers, B., C. Manganaro, D. and Rippe (1984). A micro ELISA for the quantitative detection of antibodies to double stranded DNA. Clin. Chem. 30: 986-987.

Nakane, P.K., and A. Kawaoi (1974). Peroxidase-labeled antibody a new method of conjugation. J. Histochem. Cytochem. 22: 1084-1091.

Ng, R.H., B.A. Brown, and R. Valdes, Jr., (1983). Three commercial methods for serum ferritin compared and the high-dose "Hook Effect" eliminated. Clin. Chem. 29: 1109-1113.

Oreskes, I., and J.M. Singer (1961). The mechanism of particulate carrier reactions. I. Adsorption of human -globulin to polystyrene latex particles. J. Immunol. 86: 338-344.

O'Sullivan, M.J., E. Gnemmi, D. Morris, G. Chieregatti, M. Simmons, A.D. Simmonds, J.W. Bridges, and V. Marks (1978). A simple method for the preparation of enzyme-antibody conjugates. FEBS Lett. 95: 311-313.

O'Sullivan, M.J., J.W. Bridges, and V. Marks (1979a). Enzyme immunoassay: A review. Annals Clin. Chem. 16: 221-240.

O'Sullivan, M.J., E. Gnemmi, D. Morris, G. Chieregatti, A.D. Simmonds, M. Simmons, J.W. Bridges, and V. Marks (1979b). Comparison of two methods of preparing enzyme-antibody conjugates: Application of these conjugates for enzyme immunoassay. Anal. Biochem. 100: 100-108.

Parsons, G.H., Jr. (1981). Antibody-coated plastic tubes in radioimmunoassay. In Methods in Enzymology, Vol. 73 (Langone, J. and H. Van Vunakis, Eds.). Academic Press, New York.

Pesce, A.J., D.J. Ford, M. Gaizutis, and V.E. Pollak (1977). Binding of protein to polystyrene in solid-phase immunoassays. Biochimica. et Biophysica Acta. 492: 399-407.

Porstmann, T., B. Porstmann, H. Schmechta, E. Nugel, R. Seifert, and R. Grunow (1981). Effect of IgG-horseradish peroxidase conjugates purified on Con A-Sepharose upon sensitivity of enzyme immunoassay. Acta. Biol. Med. Germ. 40: 849-859.

Quash, G., A. Roch, A. Niveleaus, J. Grange, T. Keolouangkhot, and J. Huppert (1978). The preparation of latex particles with covalently bound polyamines, IgG and measles agglutinins and their use in visual agglutination tests. J. Immunol. Methods 22: 165-174.

Rathnam, P., and B.B. Saxena (1984). A "Sandwich" solid-phase enzyme immunoassay for lutropin in urine. Clin. Chem. 30: 665-671.

Roman, S.H., F. Korn, and T.F. Davies (1984). Enzyme-linked immunosorbent microassay and hemagglutination compared for detection of thyroglobulin and thyroid microsomal auto-antibodies. Clin. Chem. 30: 246-251.

Sada, E., G.M. Ruiz-Palacios, Lopez-Vidal, and Ponce de Leon (1983). Detection of mycobacterial antigens in cerebrospinal fluid of patients with tuberculous meningitis by enzyme-linked immunosorbent assay. Lancet 2: 651-652.

Sarafian, S.K., and H. Young (1982). Detection of gonococcal antigens by an indirect sandwich enzyme-linked immunosorbent assay. J. Med. Microbiol. 15: 541-550.

Shinomiya, Y., N. Kato, M. Imazawa, and K. Miyamoto, et al (1982). Enzyme immunoassay of the myelin basic protein. J. Neurochem. 39: 1291-1296.

Standefer, J.C., and G.C. Saunders (1978). Enzyme immunoassay for gentamicin. Clin. Chem. 24: 1903-1907.

Sun, M., M. Novotny, and R.C. Doss (1983). A nonisotopic immunoassay for human prolactin. Clin. Chem. 29: 1240.

Van Weeman, B.K., and A.H.W.M. Schuurs (1976). Sensitivity and specificity of hapten enzyme-immunoassays. In First International Symposium of Immunoenzymatic INSERM Symposium No. 2 (Feldman et al., Eds.). North-Holland Publishing Company, Amsterdam.

Verdon, B., T. Mandel, E.G. Berger, and H. Fey (1982). A solid-phase ELISA for human galactosyl-transferase. J. Immunol. Methods. 55: 27-33.

Voller, A., D.E. Bidwell, and A. Bartlett (1976). Enzyme immunoassays in diagnostic medicine; Theory and practice. Bull. World Health Organ. 53: 55-65.

Wilson, M.B. and P. Nakane (1978). Recent developments in the periodate method of conjugating horseradish peroxidase (HRPO) to antibodies. In Immunofluorescence and Related Staining Techniques (Knapp, et al, Eds.). North-Holland Biomedical Press, Amsterdam.

Yorde, D.E., E. Sasse, T. Wang, R. Hussa, and J. Garancis (1976). Conjugative enzyme linked immunoassay with the use of soluble enzyme/antibody immune complexes for labeling. 1. Measurement of human choriogonatropin. Clin. Chem. 22: 1372-1377.

ENZYME IMMUNOASSAY OF ANTIBODY

Seymour P. Halbert and Tsue-Ming Lin

Cordis Laboratories
2140 North Miami Avenue
Miami, Florida 33127

The detection of specific antibodies in body fluids has proven to be of considerable value in diagnostic medicine for a wide variety of diseases, including those of infectious as well as of auto-immune origins. A number of divergent methods have been developed for these purposes, such as complement fixation, passive hemagglutination, viral hemagglutination inhibition, immunofluorescence and latex agglutination procedures. All of these require serial dilution titration for quantitation, which is tedious and reduces precision, and all suffer from the fact that the readings are largely subjective. Weak reactions are often very difficult to distinguish from negative ones. The development of enzyme immunoassays, particularly of the "sandwich" ELISA type (Engvall and Perlmann, 1972), permitted quantitative estimations to be made with a single dilution of serum or plasma over a wide range of values. They also easily permit antibody activity in the various immunoglobulin classes to be distinguished. The results are objectively read in a photometer, and the intensity of the reading is directly correlated with the antibody level. By use of the appropriate antigen in the solid phase and the correct type of enzyme-labeled anti-immunoglobulin conjugate, numerous antibody tests can be designed using the same format.

Our laboratories have developed a series of such human antibody assays for diagnostic use in auto-immune disturbances and in certain infectious diseases, as shown in Table I. The present report will discuss some general principles involved in optimizing these determinations. In some cases, these principles will be exemplified by findings with certain of them.

Table 1. Diagnostic "Sandwich" Enzyme Immunoassay for Antibody Detection Developed at Cordis Laboratories

Disease Category	Test for Antibody to	Name of Assay	Disease	Reported Value	Ref.
Auto-immune	IgG (rheumatoid factor)	CORDIA RF	Rheumatoid arthritis	IU/ml*	Halbert et al., 1980
	DNA	CORDIA N	Systemic lupus erythematosus	IU/ml	Halbert et al., 1981
	Nucleoprotein	CORDIA NP	SLE & drug induced lupus	IU/ml	Halbert et al., 1981
	Thyroglobulin	CORDIA TG	Thyroiditis & other thyroid diseases	% of positive control	Halbert et al., 1983
	Immune complexes	CORDIA IC	Variety of diseases	μg eq./ml of agg. IgG	Lin et al., 1983
Infectious disease	Rubella virus (IgG)	CORDIA R	Rubella	IU/ml	Kleeman et al., 1983
	Toxoplasma gondii (Total Ig)	CORDIA T	Toxoplasmosis	IU/ml	Lin et al., 1980
	Toxoplasma gondii (IgM)	CORDIA T-M	Active, acute toxoplasmosis	% of positive control	Lin et al., 1984
	Cytomegalovirus (IgG)	CORDIA CMV	Cytomegalovirus infections	% of positive control	Kiefer et al., 1983
	Herpes simplex (IgG)	CORDIA HS	Herpes infections	% of positive control	Kiefer, 1984
	Entamoeba histolytica	CORDIA A	Amebic abscess & dysentery	% of positive control	Lin et al., 1981
	Paul-Bunnell heterophil antigen	CORDIA IM	Infectious mononucleosis	% of positive control	Halbert et al., 1982

* IU/ml: International units/ml, based on World Health Organization reference standards.

Solid phasing of antigen

The solid phase used in all of these tests is a small disc which allows covalent bonding of proteins and other antigen through isothiocyanate groups introduced into its surface. Once the antigen is chemically attached to the disc, it can be lyophilized and stored at 4°C in the presence of dessicant without appreciable deterioration for up to 4 years or more, in those cases where stability has been studied for this length of time. The use of covalent binding also offers the considerable advantages of uniformity of coating, and the prevention of antigen leaching from the solid phase. Both of these latter problems have been well documented with antigens physically adsorbed to carrier surfaces (e.g. Chessum and Denmark, 1978).

Purity of antigens

It is obvious that the antigen attached to the carrier must not be so crude that only a minute percentage of the solid phased material is immunologically reactive with the human antibodies encountered. These assays do not generally require very highly purified antigen preparations for quite satisfactory results, but the performance of the test usually improves with increased purification of the antigen used. It must also be kept in mind that in the case of infectious disease diagnosis, a large number of different antigen-antibody systems may be involved in the immune response after a particular infection. Some of them are undoubtedly more significant than others.

Purity of the enzyme labeled antibody

Here too, the antibody preparation need not be highly purified for satisfactory performance, but if the antibody level in a given antiserum is so low that the large bulk of the labeled immunoglobulin is non-reactive, the assay will not perform well. In our experience, immunospecifically purified antibody yields the best results, and were uniformly used in our assays. After the antisera are made specific for the immunoglobulin class being assayed, the globulin fractions of the antisera are passed through beds of the appropriate purified immunoglobulin antigen, and following thorough washing are harvested by dissociation at low pH.

The enzyme chosen for the label in our tests was calf intestinal alkaline phosphatase. This was used because it has a relatively high turnover number and both the substrate (p-nitrophenyl phosphate) as well as its end product (p-nitrophenol) are rather stable. In addition, this compound is not carcinogenic. A number of the substrates for the commonly used horseradish peroxidase label are known to cause malignancy. This is particularly true of the widely employed o-phenylenediamine-HCl, which caused cancer in three animal models studied at the National Cancer Institute (Weisburger et al, 1978). This latter substrate is also relatively unstable, and must be used rapidly after dissolving, preferably in the dark.

Assay procedure

In a typical assay, three incubation periods of 10 to 45 minutes each at 37°C are used, and the dilution of the test serum is usually 1:50 to 1:100. For example, in the CORDIA N assay, DNA coated discs are sequentially incubated 45 min. with 0.5 ml of 1:100 diluted test serum, washed, incubated 45 min. with enzyme labeled antibody, washed, and finally incubated 20 min. with 1 ml substrate. The reaction is stopped with 0.1 ml of 3 M NaOH, and the absorbance read at 405 nm.

Reporting test results

Since absorbance measurements with a given set of reagents may vary somewhat from test to test, day to day and laboratory to laboratory, depending on small differences in assay parameters such as incubation temperature, volumes delivered, substrate concentration, etc, it was felt that the assay results should be standardized to some other basis of reference. Since most antibodies are known to be extremely stable (Janeway et al, 1967), the procedures we have developed are standardized to reference serum preparations.

In those instances where WHO international reference standards were available, antibody levels in patient specimens are reported in IU/ml, as determined by standard curves set up in each run. The standard curves are constructed from the absorbance values found with the undiluted calibrated positive control serum and three serial dilutions of it. Fortunately, this is always a straight line relationship over a wide range when the IU/ml values are plotted on a log-log scale against the absorbance results seen. This is illustrated in a typical standard curve for the test measuring autoantibody to DNA, where two fold dilutions are used (Fig. 1). As can be seen, the correlation coefficient between absorbance and IU/ml is quite high, values of >0.995 usually being found. In addition to plotting the standard curves graphically, as shown in Fig. 1, they can be easily established on a log-log basis with inexpensive handheld calculators. The test sample values can then be rapidly determined with greater ease and precision.

In those assays for which international reference standards are not currently available, suitable reference primary standard positive sera were obtained, and numerous aliquots were stored at -70°C. In these cases, arbitrary units were not assigned to the positive control, and standard curves were not established in the assay. Instead, for simplicity, the test results were related to

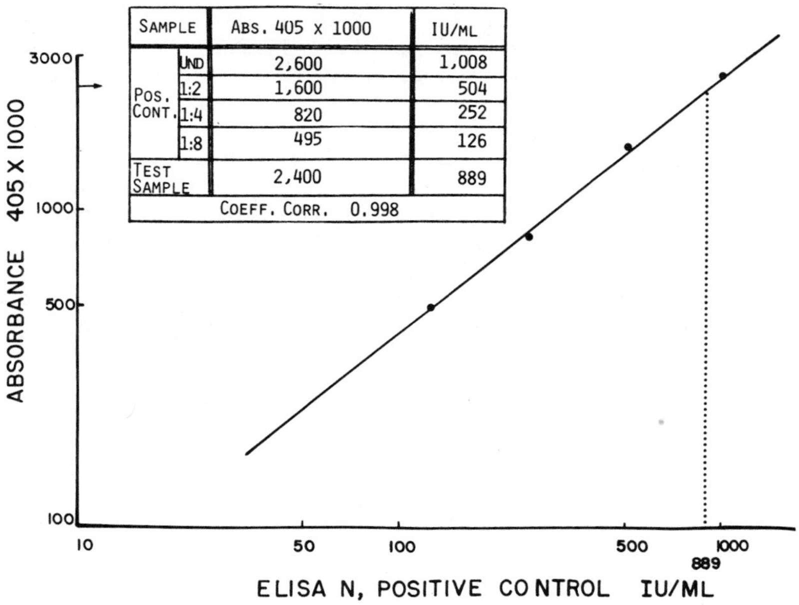

Fig. 1. Typical standard curve established in an assay for anti-DNA by CORDIA N. Coeff. corr., Coefficient of correlation of the absorbance and IU/ml values. From Halbert et al., 1981.

the kit standardized positive control absorbance, which was run in each assay. The test sample absorbance was divided by the positive control absorbance and multiplied by 100, to yield a % of positive control value. With every new lot of reagents, the kit positive control was adjusted to yield results as close as possible to the reference primary standard serum. The final assembled test set was also quality controlled against a panel of positive and negative sera, as was done with the tests employing the WHO standard.

For each procedure, it was necessary to establish the cut-off separating normal (or negative) values from abnormal (or positive) ones. For this purpose, a large number of normal sera negative for a given antibody by accepted procedures were tested with the standardized reagents in the ELISA test set. Usually, the mean of the results plus 2 or 3 standard deviations (SD) was taken as the level of antibody separating normal from abnormal. In some cases, however, the spread from normal to abnormal was sufficiently great so that the mean + 4 SD was used for the cut-off. This was true in the tests for infectious mononucleosis (Paul Bunnell heterophil antibody) and for thyroglobulin auto-antibodies. For certain assays, it was found helpful to include a borderline (equivocal) range of values, which could not be clearly classified as normal or abnormal.

In tests involving paired acute-convalescent sera for the diagnosis of the etiology of an infection in a given patient, it was sometimes not possible to readily obtain such clinical specimens because of the nature of the disease. However, it was possible to clearly establish the reproducibility of the assay procedure, which in most cases ranged between 5 and 10% for the coefficient of variation. Based on this, and adding a further factor to account for technical error, it was possible to determine a difference of CORDIA values which would represent a significant change in acute-convalescent sera when assayed simultaneously.

Incubation times

Perhaps because the reagents employed were all relatively highly purified, it was possible to use rather short incubation times for all of the assay procedures of Table 1. In most tests, incubations of 30 to 45 minutes were used for the first two incubations, with 20 to 30 minutes for the substrate incubation period. In some cases, as in the tests for infectious mononucleosis and for thyroglobulin antibodies, extremely short incubation periods could be used (three periods of 10 minutes each). In the case of thyroglobulin auto-antibodies, even with such short incubation times, absorbance values as high as 3.5 were seen with some strongly positive specimens, while the background absorbance with normal sera averaged about 0.02. These high levels of reactivity were seen even though the test sample dilution used in this procedure was 1:100.

Correlation with other assays

In each case, the quantitative results of the CORDIA procedures correlated extremely well with other assay methods for measuring the same antibodies. An example of this significant relationship is seen in Fig. 2, which reveals that the coefficient of correlation for rheumatoid factor determined by a latex titration assay and the CORDIA RF ELISA procedure was 0.94, $p < 0.001$. As can be seen, a number of sera (seven) were positive at low levels by ELISA, but negative by latex agglutination. All of these were confirmed to be true positives for rheumatoid factor by demonstrating that the reactions were blocked by pre-incubation of the specimens with heat aggregated human IgG, but they were unaffected by pre-treatment with native IgG. In the tests developed above, the correlation coefficients all were highly significant, but they tended to be somewhat lower when the reference test was more complex or had an inherently poorer reproducibility.

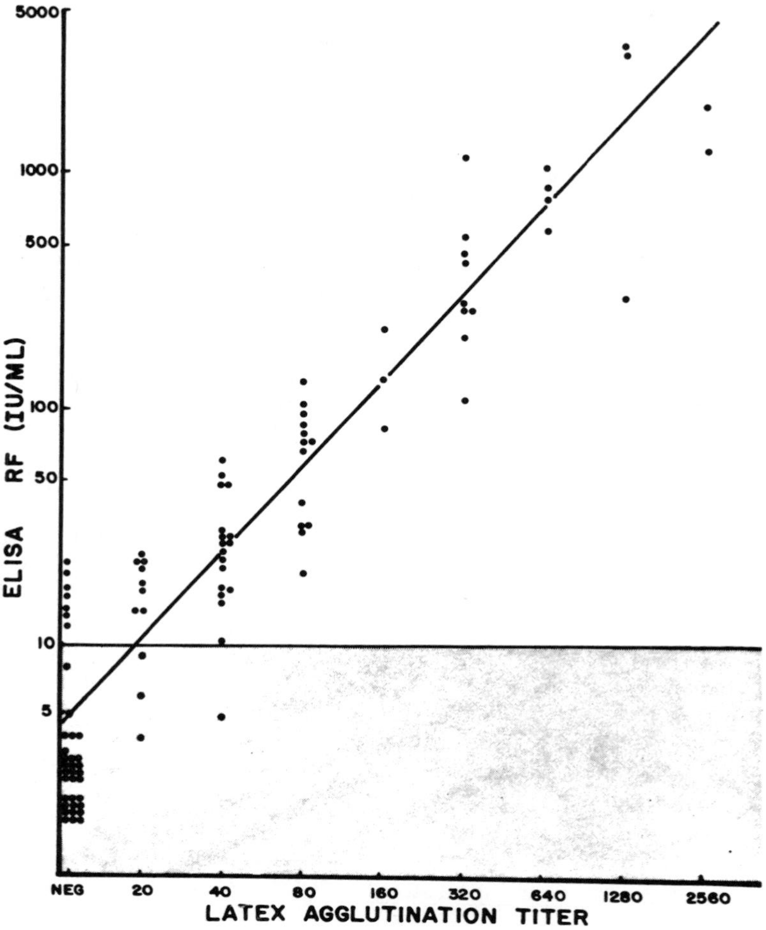

Fig. 2. Correlation of quantitative latex rheumatoid factor (RF)titers and ELISA RF values in 100 sera from patients and normal subjects. The coefficient of correlation (r) is 0.94. Shaded area, normal range. From Halbert et al., 1981.

Background reactivity

This has been reported to be a problem of varying severity by almost all investigators, probably because of the sensitivity of the reagents and the very high total levels of immunoglobulins in serum, particularly IgG. It was found that exposure of the antigen-coated solid phased discs to conjugate directly, or following incubation of the discs with specimen diluent alone resulted in almost no non-specific uptake of the enzyme labeled reagent. However, exposure of the disc to presumably antibody negative serum usually resulted in low levels of non-specific uptake of conjugate both on the walls of the vial as well as on the discs. This could be minimized by use of an appreciable test serum dilution (1:50 or 1:100), as well as by transfer of the disc after the incubation with the conjugate, so that the substrate reaction occurs in a clean vial. The non-specific uptake was also minimized with short incubation times, as in the case of the thyroglobulin auto-antibody assay. This problem was considerably less in the case of the detection of antibodies of the IgM class, as in infectious mononucleosis diagnosis, or rheumatoid factor detection.

Detection of class specific antibodies

In a number of test procedures developed, the total level of antibodies were determined by use of a mixture of enzyme labeled anti-IgG and anti-IgM which were not heavy chain specific. In some cases, it was readily possible to separately determine the levels of IgG and IgM antibodies by substituting the appropriate heavy chain specific antibody-enzyme conjugate into the assay. An example of this is shown in Table 2, where anti-DNA and anti-deoxyribonucleoprotein (NP) antibody levels of each class were determined in a number of systemic lupus erythematosus patients (Karsh et al, 1982). It may be seen that a large variety of patterns could be observed in the IgG and IgM reactions with these auto-antigens. Thus, the

Table 2. Determination of IgG and IgM Classes of Anti-DNA and Anti-deoxyribonucleoprotein (NP) in Systemic Lupus Erythematosus Patients.

Patient	Anti-DNA (IU/ml) IgG N < 89*	Anti-DNA (IU/ml) IgM N < 163	Anti-NP (IU/ml) IgG N < 24	Anti-NP (IU/ml) IgM N < 28
1	282	609	38	93
2	864	55	41	18
3	85	39	1,162	20
4	48	51	427	613
5	1,988	451	771	102
6	4,515	2,976	1,962	1,729
7	3,966	125	634	176
8	660	213	1,928	31

* N < , normal values are less than number shown.
From Karsh et al., 1982.

patient 1 showed moderately elevated levels of both nuclear antibodies, with the IgM responses predominating in each, while in patient 5, both antibodies were present, but primarily of the IgG class. Patient 3 only had elevated levels of anti-NP which were exclusively IgG, while patient 4 had only anti-NP in which the IgM class was greatest. Patients 5, 7 and 8 had high levels of both antibodies mostly of the IgG class, while patient 6 had extremely elevated values for both antibodies equally abnormal for IgG and IgM.

In the case of infectious mononucleosis, the predominant heterophil antibody response is of the IgM class, so this heavy chain specific enzyme labeled reagent is used in this ELISA test, together with discs coated with purified antigen isolated from bovine erythrocyte stroma. However, parallel tests performed with heavy chain specific anti-IgG conjugate revealed that infectious mononucleosis patients do show a variable but small portion of the heterophil antibody response in the IgG class, as seen in Table 3. It may be noted that the proportion of IgG over the IgM response

Table 3. IgM and IgG Infectious Mononucleosis Heterophil Antibody Reactions.

Test Sample	Sample No.	Absorbance with Anti-IgG conjugate	Absorbance with Anti-IgM conjugate	Absorbance IgG/IgM %
IM	1	0.145	1.740	8.3
	2	0.050	0.430	11.6
	3	0.025	0.935	2.7
	4	0.045	0.885	5.1
	5	0.055	1.685	3.3
	6	0.090	0.705	13.0
	7	0.230	0.690	33.3
	8	0.270	1.305	20.7
	9	0.285	0.650	43.8
	10	0.550	1.890	29.1
Normal	1	0.02	0.01	
	2	0.02	0.02	
	3	0.02	0.03	
Control IgG disks		1.435	0.000	
Control IgM disks		0.005	1.655	

From Halbert et al., 1982.

ranged from 2.7% to a high of 43.8%. Normal sera showed negligible background, and the conjugates were almost completely class specific as judged by control reactions with IgG and IgM coated discs.

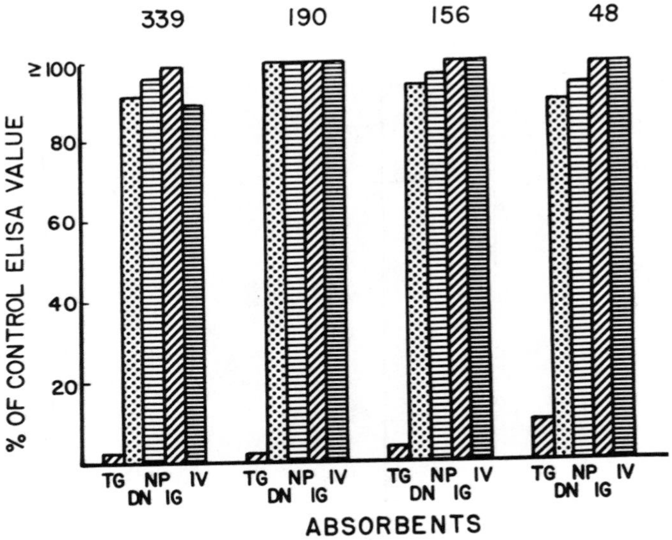

Fig. 3. Absorption experiments to confirm the specificity of the ELISA procedure for thyroglobulin autoantibodies. The numbers on top represent the unabsorbed control ELISA values for the four sera represented ($\geq$ 15 is abnormal). The absorbents were: TG, purified human thyroglobulin; DN, calf thymus DNA; NP, calf thymus deoxyribonucleoprotein; IG, heat aggregated human immunoglobulin G; IV, partially purified influenza virus, type A. From Halbert et al, 1983.

For the rheumatoid factor assay, the "antigen" is solid phased human IgG. Since the predominant RF is IgM, the enzyme labeled antibody is directed against IgM. The latter reagent obviously must be completely heavy chain specific to avoid direct reaction with the IgG on the disc. For the thyroglobulin (Tg) auto-antibody determination, non-heavy chain specific anti-IgG was used. This class predominates the response to Tg (Torrigiani et al, 1968), but other classes also occur, and would be detected to a considerable degree through the light chain antibody specificities.

Attempts to develop an IgM assay for rubella antibodies proved to be extremely difficult, however. Although this assay with solid phased rubella virus and heavy chain specific anti-IgM conjugate gave quite satisfactory results with convalescent phase sera from documented rubella patients, particularly after the IgG was largely removed by treatment with protein-A containing *Staphylococcus aureus*, an excessive number of apparent false positives were seen with presumed normal sera. The IgM fractions from the latter were not active in viral hemagglutination inhibition assays. Interestingly, these apparent false positive reactions for IgM rubella antibodies could be crisply blocked by pre-treatment with purified rubella virus antigen, but were quite unaffected by blocking with very high concentrations of uninfected tissue culture control extracts. A high proportion of false positive IgM rubella reactions were also seen in a panel of sera from documented infectious mononucleosis patients, in agreement with observations made by Morgan-Capner et al (1983).

Specificity of the ELISA reactions

In addition to the significant correlations seen between each of these ELISA assay results and values found with reference methods, the specificity of the ELISA reactions could be readily demonstrated by the classical immunological procedure of blocking or absorption. Sera were preincubated with a variety of antigens, and it was found in each of these procedures that only the antigen specific for the antibody being determined was capable of efficiently blocking the ELISA reactivity. This is illustrated in Fig. 3, in the case of the assay for detecting thyroglobulin autoantibodies. Only thyroglobulin blocked the reactions, even in the presence of high titers of these antibodies. DNA, deoxyribonucleoprotein, heat-aggregated IgG and influenza virus concentrates were without significant effect.

This blocking procedure was particularly valuable in documenting the correctness of results with those specimens which reacted positively by the ELISA procedure but were negative by the reference method. This is exemplified in Table 4, with sera from a number of patients with infectious mononucleosis (IM). The diagnosis in these patients was based on IgM anti-Epstein Barr virus capsid antigen findings, but their sera were negative by the horse erythrocyte

Table 4. Blocking Experiments with the CORDIA IM-positive, Horse Erythrocyte-negative Sera from Infectious Mononucleosis Patients.

Test sample	CORDIA IM value after absorption with: Diluent control	IM heterophil antigen	Bovine erythrocyte stroma	Aggregated human IgG
1	47*	10	4	45
2	38	7	9	33
3	21	4	2	22
4	52	22	5	55
5	20	8	7	23
6	23	7	8	27
7	26	10	7	32
8	21	2	3	21
9	31	2	3	23

* CORDIA IM values $\geq$ 20 = positive. From Halbert et al, 1982

agglutination test. It can be seen that the purified IM heterophil antigen and bovine erythrocyte stroma blocked the positive ELISA IM reactions dramatically, but heat aggregated human IgG was without effect. The latter, of course, blocks IgM rheumatoid factor very effectively, as pointed out above.

Such blocking tests were also very useful in confirming the validity of results obtained with some weakly reactive sera in the ELISA for rubella antibodies, which were not reactive by viral hemagglutination inhibition (HAI). Some of these findings are shown in Fig. 4, where it may be seen that the ELISA results were all clearly blocked by pre-incubation with purified rubella viral antigen, but were not affected by supernate from the uninfected culture (Kleeman et al, 1983). Even more convincing was the failure to block these reactions by pre-treatment with supernates of rubella virus harvests from which the virus had been removed by ultracentrifugation. This type of data has reinforced the speculations previously voiced that ELISA procedures can indeed detect low levels of rubella antibodies which are beyond the sensitivity of HAI (Kleeman et al, 1983; Serdula et al, 1984).

Diagnostic value of ELISA antibody determinations

The obvious purpose of these tests is to assist physicians in establishing the diagnosis of illness in a given patient, or to help in following the course of disease in a patient whose diagnosis is already documented. In some instances, the above tests have already been demonstrated to furnish very good correlations with

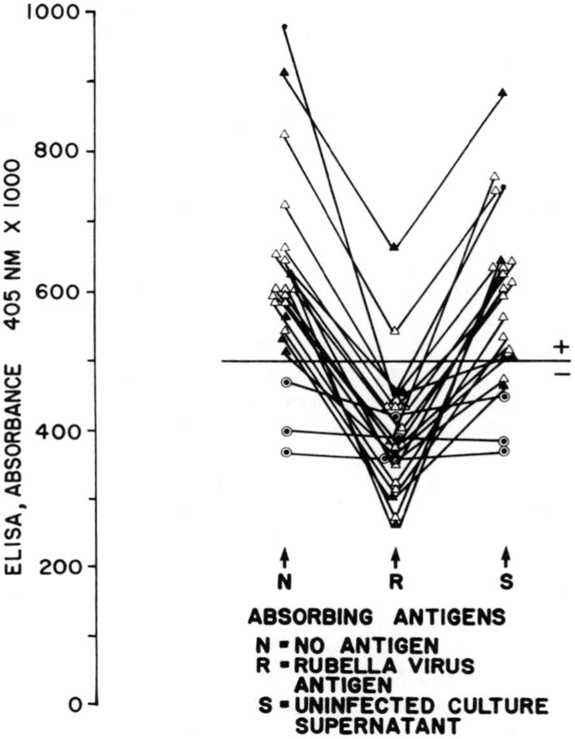

Fig. 4. Blocking of rubella antibodies in serum specimens positive by ELISA and HI (●), positive by ELISA and negative by HI with partial inhibition (△), positive by ELISA and negative by HI (▲), and negative by ELISA and HI (⊙). The specimens were preincubated before testing by ELISA with rubella antigen (R), uninfected BHK-21 culture supernatant (S), or no antigen (N). From Kleeman et al, 1983.

clinical course or with diagnosis. One of the clearest examples of this are studies by Healy and Walker (1982) in relation to the use of CORDIA A for the diagnosis of active Entamoeba histolytica infection. Of 55 patients with documented amebic liver abscess, 52 (95%) were positive by CORDIA A. Of 57 patients with stool positive clinically active intestinal amebiasis (amebic dysentery), 55 (96%) were positive by CORDIA A. On the other hand, only 1 of 25 (4%) of patients were positive who were asymtomatic but who revealed E. histolytica in their stools (carriers). A group of 83 control patients with other diseases showed only one (1%) of positive CORDIA A reactions. Many of the individuals in the latter control group had a variety of liver or gastro-intestinal illnesses, and the one positive reactor was a patient with etiologically undiagnosed polycystic liver disease. Similar findings relating CORDIA A test results to the diagnosis of amebic abscess were also made in a large study under field conditions in the Far East (Cross, J. H., personal communications, 1982).

In a study of CORDIA N (Estes, D., personal communication, 1983) a high percentage (91%) of 56 patients with active systemic lupus erythematosus (SLE) was positive for anti-DNA, many of the values being quite elevated. However, in patients whose disease was in remission (inactive SLE), only 10/31 (32%) were positive, mostly in the low abnormal range. In non-SLE patients with other connective tissue diseases including mixed connective tissue disease, progressive systemic sclerosis, and Sjogren's syndrome, only 7/50 (14%) were positive and all were in the low abnormal range. In agreement with more extensive earlier observations (Halbert et al., 1981; Karsh et al., 1982), only 1 of 35 normals (3%) showed a weak positive reaction.

In the follow-up of patients with documented SLE, it was found that the changes in CORDIA N and NP values correlated with Farr assay results during remissions and exacerbations (Karsh et al., 1982). Examples of these results are shown in Table 5, where the immunoglobulin classes of these antibodies were also analyzed. During the exacerbation of disease in patients 1 and 2, antibodies to both nuclear antigens increased sharply in value, and the increases were predominantly of the IgG class. In patient 3, exacerbation was associated with a sharp increase in the IgM antibodies to both DNA and NP, while the high IgG anti-DNA seen with both pairs was essentially unchanged. During the partial remission in #4, the decrease of both anti-DNA and anti-NP was primarily in the IgM class.

In the diagnosis of rubella infections by assay of paired sera (acute-convalescent), the CORDIA R procedure detected the required three fold or greater increase in IU/ml for the diagnosis to be established in 83/84 pairs tested, in comparison to 82/84 showing the required 4 fold change in titer by HAI.

Table 5. CORDIA N and CORDIA NP Values during Remissions and Exacerbations in Systemic Lupus Erythematosus

Patient	Interval between samples (days)	Farr Assay % binding N <25*	CORDIA N (IU/ml)		CORDIA NP (IU/ml)	
			IgG N < 89	IgM N < 163	IgG N < 24	IgM N < 28
1	120	15	167	48	112	10
		90	2,831	198	827	373
2	360	40	242	94	157	23
		97	3,541	658	1,529	203
3	240	22	1,462	132	41	199
		96	1,812	1,814	972	1,135
4	300	95	1,394	1,294	972	584
		64	1,900	341	743	130

* N < , normal values are less than the numbers shown.
From Karsh et al., 1982.

In summary, the ELISA principles offer a very satisfactory basis for the development of clinically useful diagnostic tests for antibody determinations. Meaningful quantitative, specific and objective results can be obtained by assay of a single dilution of serum, with ease and rapidity.

REFERENCES

Chessum, B. S. and J. R. Denmark (1978). Inconstant ELISA. Lancet, Jan., 161.

Engvall, E., and P. Perlmann (1972). Enzyme-linked immunosorbent assay, ELISA. III. Quantitation of specific antibodies by enzyme-linked anti-immunoglobulin in antigen-coated tubes. J. Immunol. 109: 129-135.

Halbert, S. P., J. Karsh, and M. Anken (1980). A quantitative enzyme immunoassay for IgM rheumatoid factor using human immunoglobulin G as substrate. Am. J. Clin. Path. 74: 776-784.

Halbert, S. P., J. Karsh, and M. Anken (1981). Studies on autoantibodies to deoxyribonucleic acid and deoxyribonucleoprotein with enzyme-immunoassay (ELISA). J. Lab. Clin. Med. 97: 97-111.

Halbert, S. P., M. Anken, W. Henle, and R. Golubjatnikov (1982). Detection of infectious mononucleosis heterophil antibody by a rapid, standardized enzyme-linked immunosorbent assay procedure. J. Clin. Microbiol. 15: 610-616.

Halbert, S. P., C. H. Bastomsky, and M. Anken (1983). A rapid standardized enzyme immunoassay for autoantibodies to thyroglobulin. Clin. Chim. Acta 127: 69-76.

Healy, G. R., and J. M. Walker (1982). Analysis of the sensitivity and specificity of the CORDIA A ELISA test for the diagnosis of amebiasis. 31st Annual Meeting of Am. Soc. Trop. Med. Hyg. Poster #86.

Janeway, C. A., F. S. Rosen, E. Merler, and C. A. Alper (1967). The Gamma Globulins. Little, Brown and Company, Boston 148 pp.

Karsh, J., S. P. Halbert, M. Anken, E. Klima and A. D. Steinberg (1982). Anti-DNA anti-deoxyribonucleoprotein and rheumatoid factor measured by ELISA in patients with systemic lupus erythematosus, Sjogren's syndrome and rheumatoid arthritis. Int. Arch. Allergy Appl. Immun. 68: 60-69.

Kiefer, D. J., D. A. Phelps, and S. P. Halbert (1983). Normalized enzyme-linked immunosorbent assay for determining immunoglobulin G antibodies to cytomegalovirus. J. Clin. Microbiol. 18: 33-39.

Kiefer, D. (1984). Personal communication.

Kleeman, K. T., D. J. Kiefer, and S. P. Halbert (1983). Rubella antibodies detected by several commercial immunoassays in hemagglutination inhibition-negative sera. J. Clin. Microbiol. 18: 1131-1137.

Lin, T. M., S. P. Halbert, and G. R. O'Connor (1980). Standardized quantitative enzyme-linked immunoassay for antibodies to Toxoplasma gondii. J. Clin. Microbiol. 11: 675-681.

Lin, T. M., S. P. Halbert, C. T. Chiu, and R. Zarco (1981). Simple standardized enzyme-linked immunosorbent assay for human antibodies to Entamoeba histolytica. J. Clin. Microbiol. 13: 646-651.

Lin, T. M., S. P. Halbert, R. Cort, and M. J. Blaschke (1983). An enzyme-linked immunoassay for circulating immune complexes using solid-phased goat C1q. J. Immunol. Methods 63: 187-205.

Lin, T. M., M. W. Chin-See, and S. P. Halbert (1984). To be published.

Morgan-Capner, P., R. S. Tedder, and J. E. Mace (1983). Rubella-specific IgM reactivity in sera from cases of infectious mononucleosis. J. Hyg. Camb. 90: 407-413.

Serdula, M. K., S. B. Halstead, N. H. Wiebenga, and K. L. Herrmann (1984). Serological response to rubella vaccination. J. Am. Med. Assoc. 251: 1974-1977.

Torrigiani, G., I. M. Roitt, and D. Doniach (1961). Quantitative distribution of thyroglobulin autoantibodies in different immunoglobulin classes. Clin. Exp. Immunol. 3: 621-630.

Weisburger, E. K., A. B. Russfield, F. Homburger, J. H. Weisburger, E. Boger, C. G. van Dongen, and K. C. Chu (1978). Testing of twenty-one environmental aromatic amines or derivatives for long-term toxicity or carcinogenicity. J. Environmental Path. Toxicol. 2: 325-356.

THE AMPLIFIED ENZYME-LINKED IMMUNOSORBENT ASSAY (a-ELISA)

J. E. Butler* and
J. H. Peterman

Department of Microbiology
The University of Iowa Medical School
Iowa City, Iowa 52242 USA

T. E. Koertge

Department of Periodontics
Medical College of Virginia
Virginia Commonwealth University
Richmond, Virginia 23298

INTRODUCTION

The current interest in immune regulation and in characterizing the immune response against pathogens and substances of interest to human and animal medicine, has prompted attempts to develop techniques capable of accurately and reliably measuring specific antibodies in serum and body fluids. Especially important is the ability to measure the response according to the isotype of the antibody. The enzyme-linked immunosorbent assay (ELISA) provides a sensitive and non-hazardous immunochemical basis for potentially making such determinations (Avrameas and Gilbert, 1972; Engvall et al. 1971). The amplified ELISA (a-ELISA) is a modification of the original ELISA which combines the principles of indirect detection with the use of a non-covalent enzyme-antibody detection system (Butler et al. 1978b). The latter avoids alteration of both enzyme and antibody, the complex in fact dissociating during the substrate reaction. The result is a system with improved sensitivity and broad flexibility. The a-ELISA serves as a useful means of mea-

*Research on the a-ELISA has been most recently supported by Grants 83-CRSR-2-2172 (USDA), HL22676 (NIH) and 5T32 DE 07007 (NIH).

suring the isotypic distribution of specific immune responses in many species.

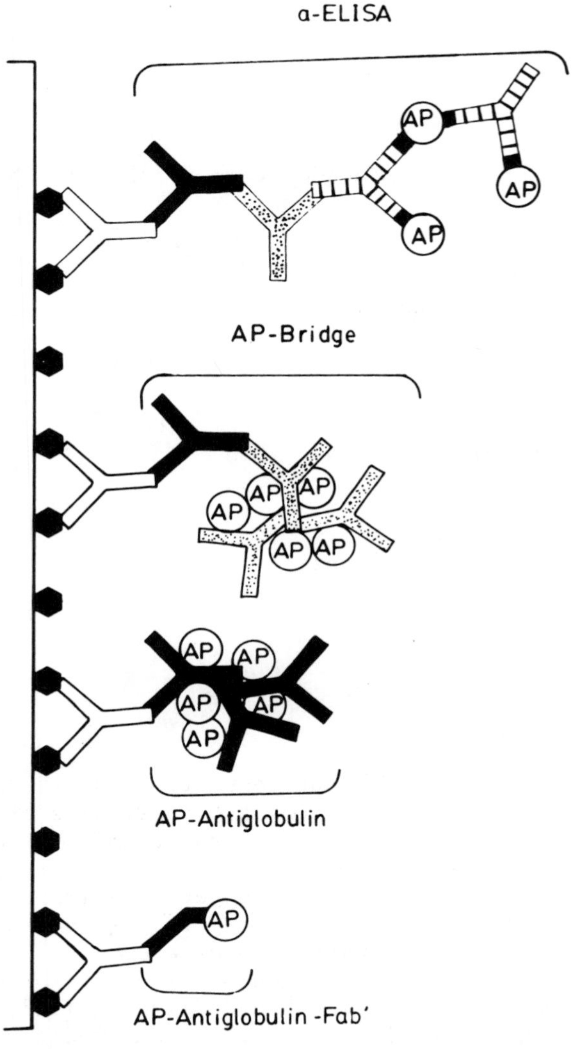

Fig. 1. Comparison of the reaction sequence of the a-ELISA and those used for other antigen-specific ELISAs. AP=alkaline phosphatase, hexagon=solid-phase antigen, open antibody=primary antibody; solid antibody= isotype-specific antiglobulin; dotted antibody=tertiary antiglobulin in one case conjugated with AP; hatched antibody=anti-AP. Conjugates depicted in middle two sequences are one-step glutaraldehyde conjugates.

The a-ELISA can also be adapted for measuring the distribution of antigens in biochemical separates by the version called the ELISA-based antigen distribution assay (EADA). Finally, the a-ELISA detection system can be used in indirect competitive immunoassays for the quantitative determination of small amounts of antigen.

The a-ELISA evolved as a solution to problems encountered when we applied the "original" ELISA to the measurement of antibodies according to subisotype (Sloan, 1975). First, we wished to circumvent the need to affinity-purify antiglobulins from weak, subisotype-specific sera in order to prepare useful subclass-specific antibody-enzyme conjugates. Second, we hoped to overcome the loss of both enzymic and immunological activity and improve the shelf-life of conjugates prepared using glutaraldehyde. We developed an indirect, antigen-specific ELISA using a soluble immune complex of antibody and enzyme which is "bridged" to the previously added subclass-specific antiglobulin through the use of a second, "bridging" antiglobulin. The reaction sequence of the a-ELISA is presented comparatively in Figure 1 together with a number of other configurations used for antigen-specific ELISAs.

The a-ELISA has been used as a model to study the stoichiometry of solid-phase assays for antibodies. These studies have provided immunochemical data upon which the selection of the proper type of ELISA can be based. We have investigated the characteristics of protein-plastic interaction, the role of antibody affinity and valence on detection, and most important, the relationship between the binding of antibodies to solid-phase antigen and their indirect detection using the a-ELISA. We have also been concerned with the absolute quantitation of antibodies using ELISA and with stream-lining data acquisition and analyses through the use of personal computers. This chapter will summarize our findings in these areas.

PREPARATION OF REAGENTS

A. Source of antiglobulins

1. Preparation of specific antiglobulins. Polyclonal rabbit antiglobulins specific for the IgA, IgG and IgM immunoglobulins (Igs) of rats, mice and humans have been prepared in our laboratory using both purified normal and myeloma Igs (or IgG-Fcs) as immunogens. Antibodies specific for bovine IgG1, IgG2, SIgA and IgM and swine IgG, IgA and IgM, have also been raised in the rabbits using purified polyclonal Igs of

the corresponding isotypes as immunogens. Antibodies specific for rabbit IgG and IgA have been prepared in guinea pigs and goats using normal IgG (or IgG-Fc) and SIgA isolated from serum and milk, respectively.

Rabbit or guinea pig antisera are rendered specific for their respective heavy chains by absorption on affinity columns to which the appropriate purified Ig or protein mixture has been covalently attached (Table I). All antisera are tested by immunodiffusion before affinity chromatography and the selection of the appropriate affinity column is determined beforehand by "in-well" absorption using the appropriate test antigens.

Table I. Immunosorbents used in the preparation of isotype-specific antiglobulins

Antiglobulin[a]	Affinity column(s) used for absorption[b]
IgG (mice, rats, swine, humans and guinea pigs)	$F(ab')_2$ of IgG of the same species
IgA (all species)	IgG; IgM-α_2 macroglobulin fraction of normal serum
IgM (mice, rats and humans)	IgG; myeloma IgA; α_2-macroglobulin
IgM (swine and cattle)	IgG/IgA fraction of colostrum and fetal serum (source of α_2-macroglobulin)
IgG1 (cattle)	IgG2
IgG2 (cattle)	IgG1

a/Animal name in parenthesis is the source of the antiglobulin.

b/Prepared by activation of Sepharose 4B with CNBr (Axen & Vretblad, 1971) as described by Cuatrecasus and Anfinsen (1971). Affinity columns usually contain 5-10 mg antigen/ml of gel; 2 ml of gel are usually adequate to adsorb 0.5 ml of serum when applied twice to such columns.

Bridging antiglobulins for use with rabbit anti-heavy chain reagents and rabbit anti-alkaline phosphatase (AP) are prepared in goats and only absorbed when "short-circuitry" problems are encountered (see C-1 below). The bridging antiglobulin for use with guinea pig antiglobulins is prepared in rabbits by immunization with guinea pig IgG.

2. Commercial reagents. Except for bridging antibodies which need not be highly isotype specific, we have found commercial reagents as supplied to be: (a) incompletely specific and (b) very low in antibody content; typically diluted. For these reasons, we prefer to prepare our own.

B. Preparation of the soluble immune complex (EIC)

1. Alkaline phosphatase (AP) and its antibody. Alkaline phosphatase (AP) is purchased from Sigma Chemical, St. Louis, MO either in the form of an ammonium sulfate suspension (bovine intestinal mucosa, type VII-5, lot P5521) or solubilized in sodium chloride (P6774). Antisera to AP are raised in rabbits or guinea pigs by sensitization using 1 mg of AP emulsified in complete Freund's adjuvant. Injections are given in the toe pads of rabbits or subcutaneously in guinea pigs. Booster immunizations are given i.v. (rabbits) or intradermally (guinea pigs).

Anti-AP precipitins are evaluated by immunodiffusion and those sera displaying similar serological characteristics are pooled together. Turbidimetric precipitin reactions followed by supernatant analyses by immunodiffusion, are used to establish equivalence.

2. Preparation and characteristics of the EIC. The EIC precipitate is prepared by incubating 1.0 ml of rabbit or guinea pig serum with a serologically equivalent quantity of AP (either lots P5521 or P6674) for 37°C, 1 hr and then overnight at 4°C. The precipitate is then collected by centrifugation and washed twice with cold PBS or borate buffered saline (BBS). The washed precipitate is then resuspended in a volume of 5 ml of PBS containing a nine-fold excess (w/w) of AP. The mixture is stirred overnight at 4°C and tested for activity (see C-3) before aliquoting. Aliquots may be stored at -20°C without further treatment or they may be treated with NaN_3 to 0.02% and diluted with an equal volume of glycerol (see B-3).

Figure 2a shows the sedimentation characteristics of enzymatically active EIC solubilized with a range of excess AP. These results show that changing the amount of excess AP used for solubilization does not alter the molecular

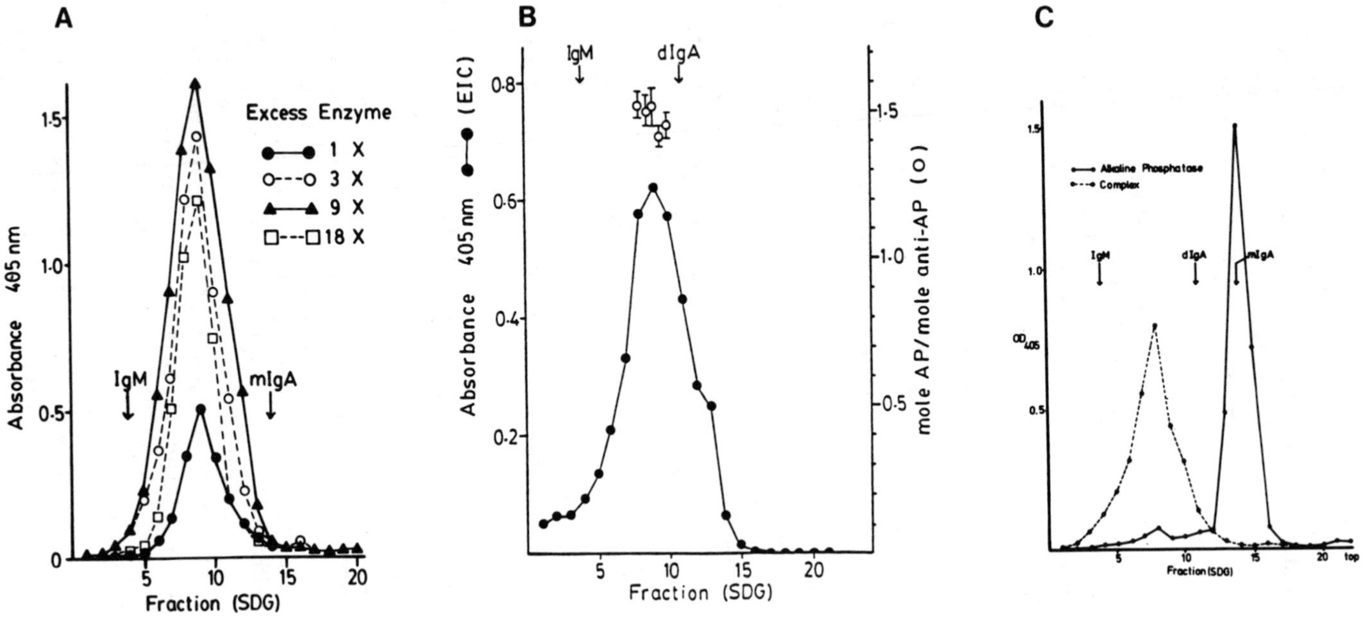

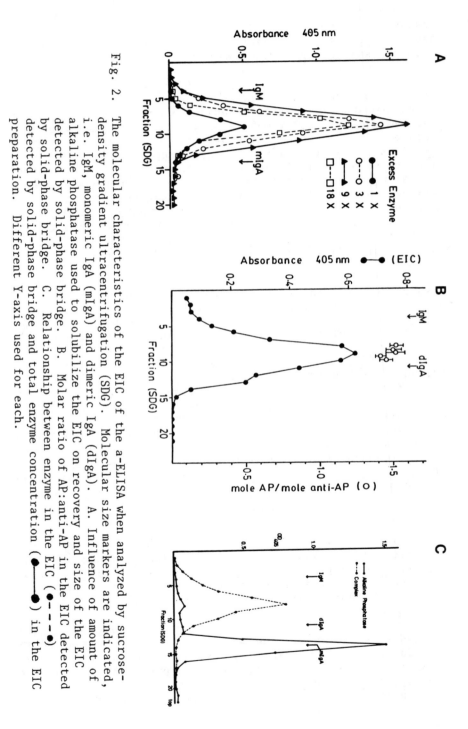

Fig. 2. The molecular characteristics of the EIC of the a-ELISA when analyzed by sucrose-density gradient ultracentrifugation (SDG). Molecular size markers are indicated, i.e. IgM, monomeric IgA (mIgA) and dimeric IgA (dIgA). A. Influence of amount of alkaline phosphatase used to solubilize the EIC on recovery and size of the EIC detected by solid-phase bridge. B. Molar ratio of AP:anti-AP in the EIC detected by solid-phase bridge. C. Relationship between enzyme in the EIC (●- - -●) detected by solid-phase bridge and total enzyme concentration (●——●) in the EIC preparation. Different Y-axis used for each.

size of the functional EIC but does influence the yield of EIC; the optimum being reached when a nine-fold excess is used.

Figure 2b shows the molecular composition of the EIC determined using ^{131}I-AP and ^{125}I-anti-AP as described by Koertge et al. (1985). The molar antigen:antibody ratio of the functional EIC is ca. 1.5:1 which considered together with its sedimentation behavior, i.e. equivalent to protein of ca. 650 Kd, suggest that the EIC is composed of 3 molecules of AP and 2 molecules of anti-AP (Fig. 1).

Figure 2c shows another characteristic of the EIC. In its final, utilitarian form, the EIC exist in a molecular environment of a large excess of free enzyme. Preliminary attempts to recover and re-use the excess enzyme appear to cause molecular alteration of the EIC so that such recovery attempts have been abandoned.

3. Storage of EIC. The EIC undergoes virtually no change in functional activity when stored in PBS at 4°C, -20°, or -70° over a one-year period (short-term) (Butler, 1981). The stability of the EIC during storage for five years (long-term) in different media has been evaluated both in terms of functional EIC activity (Table II) and in terms of molecular size (Koertge et al., 1985). These data show that EICs stored in 50% glycerol at -20° remained most stable although all EICs stored at -20° retained significant, functional activity while showing slight molecular alteration. Among the EICs stored long-term at 4°C, only those in 50% glycerol retained significant functional activity despite some minor molecular alteration (Koertge et al., 1985). These results are the basis for our recommendation that EICs be stored in 50% glycerol at -20° for long-term usage or in PBS or BBS at -20°C for up to one year. Although also stable at 4°C for short periods (Butler, 1981), -20° storage is preferable to avoid the nuisance of possible bacterial contamination.

C. Specificity testing

1. Short circuits in the reaction sequence. Multiple antibody ELISAs provide more opportunities for "short-circuit" cross-reactions than assays with fewer steps. The diagram in Fig. 3 illustrates the types of short circuits possible.

A common form of short-circuitry encountered is recognition of the carrier in a hapten-carrier antigen complex by the primary antibody (Type A). We eliminate this problem by using the albumin of the species of the primary antibody as the test hapten carrier; we have never observed homologous anti-albumin

Table II. Comparative activity of EICs stored for five years under different conditions.

Treatment of EIC	Dilution giving $OD^{405} = 0.5$	Relative Activity[c]
PBS, 4°C	1330	0.31
0.1% Tween-20, 4°C	ND[d]	ND
20% sucrose, 4°C	ND	ND
50% glycerol, 4°C	1900	0.44
0.1% Tween-20, -20°C	1050	0.24
20% sucrose, -20°C	2500	0.58
50% glycerol, -20°C	2300	0.54
New EIC in 50% glyercol	4300	1.00

c/Relative activity = $\frac{\text{dilution of stored EIC giving } OD^{405}=0.5}{\text{dilution of new EIC giving } OD^{405}=0.5}$

d/An OD^{405} of 0.5 was not reached during equal time.

antibodies. Short circuits between the antigen and the isotype-specific antiglobulin (Type B) or bridging antibody (Type C) require absorption of the antiglobulins to solve the problem. A type of short-circuiting often encountered is recognition of human IgG by goat anti-rabbit IgG used as the bridge. This requires absorption of the bridge on an affinity column of human IgG.

Short-circuits between the bridge and primary antibody (Type D) can be overcome by the same approach mentioned above for carrier cross-reactivity. Namely, the bridging antibody can be made in the same species as the primary. This is done in our a-ELISA for rabbit antibodies where we use guinea pig isotype-specific antiglobulins and anti-AP and a rabbit anti-guinea pig bridging antiglobulin.

Two advantages of the a-ELISA in dealing with short circuit problems are: (1) the EIC has only antibody activity to AP, not to some other species Igs as is the case with conventional conjugates and (2) isotype-specific antiglobulins can be tested for their specificity without first forming an enzyme-antibody conjugate with which to test them.

Short-circuitry is evaluated by adsorbing the potential antigen targets directly on the solid-phase and then adding the subsequent steps of the detection system. For example, to test whether the isotype-specific reagent recognizes the antigen directly (Type A short-circuit), the primary antibody is replaced by buffer. To determine whether the bridging antibody directly recognizes the primary antibody (Type D short-circuit), purified Ig of the same species and class as that of the primary antiserum, is adsorbed directly on the plastic and the reaction completed by adding the terminal reagents[e/].

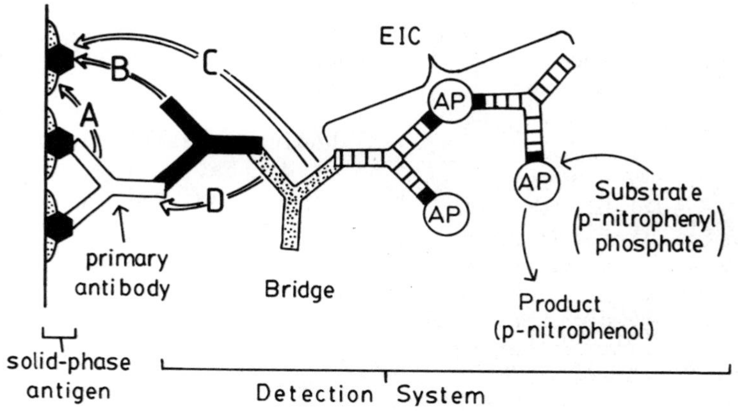

Fig. 3. Diagram of the a-ELISA reaction sequences illustrating various short-circuits. Hexagon depicts hapten and stippled region on each side depicts the carrier. The arrows indicate the four types (A-D) of circuit problems discussed in the text.

e/Terminal reagents of the a-ELISA include the bridging antiglobulin, the EIC and the substrate for AP.

2. Specificity testing of isotype-specific antiglobulins. The specificity of isotype-specific antiglobulins is a vital key to the accurate determination of antigen-specific immune responses according to class. These reagents are routinely tested using the a-ELISA system. Instead of antigen adsorbed on the solid-phase microtiter wells, we adsorb different amounts of purified Igs of the species of the primary antibody and test for cross-reactivity in the manner illustrated in Table III. Estimates of cross-reactivity can be made assuming that: (a) the amount of each Ig which adsorbs on the microtiter well is known and (b) each adsorbed Ig is known to be totally pure, i.e. free from Igs of other isotypes. Knowing the Igs to be pure and assuming that 60% of the added Ig binds to Immulon®[f/] 2 in the concentration range used[g/], the cross-reactivity of the reagents can be estimated (Butler et al. 1980a).

3. Titering of reagents in the a-ELISA. It is most convenient to titer the a-ELISA reagents using a modified version of the experimental layout that is used in Table III. Rather than broad range of concentrations, only a restricted number of different Ig concentrations are used in such preliminary tests. The amounts used should represent the minimum and maximum amounts of specific antibody that will be measured with the assay, e.g. 0.05 and 100 ng/well, as well as two intermediate values.

For titering the isotype specific antiglobulin, the bridge and EIC are used at arbitrary dilutions of 1:500 and 1:100, respectively (of equal or greater concentration than usual). Dilutions of the isotype-specific globulin can vary from 1:250 to 1:4000. Results are evaluated in terms of: (a) the slope of the titration plot, (b) the magnitude of O.D.405 obtained after a 1 hr incubation period and (c) economy of the reagent. The highest dilution providing good sensitivity (O.D.405 in 1 hr) and a long linear region with a slope between 0.9-1.0, is selected. For most isotype-specific reagents, this is usually 1:1000 or 1:2000.

The bridging antiglobulin is titered in a similar manner. The same layout as described above is used but the isotype-specific antiglobulin is used at a fixed concentration, e.g. 1:1000 and the bridging antiserum dilution is varied

---

f/Registered trademark of Dynatech, Alexandria, VA.

g/The actual percentage of bound Ig varies with the isotype (Cantarero et al., 1980). In general, polymeric Igs bind best (up to 80% of the added Ig) while IgGs bind less well.

Table III. Specificity testing of isotype-specific antiglobulin reagents using the a-ELISA

Added (ng)	SIgA h/	IgM	IgG1	IgG2
100	2.997 ± 0.065; 2.18†	3.123 ± 0.191; 6.11	3.073 ± 0.103; 3.34	2.949 ± 0.100; 3.40
30	1.389 ± 0.067; 4.79	1.485 ± 0.024; 1.62	1.470 ± 0.062; 4.22	1.404 ± 0.070; 4.99
10	0.647 ± 0.025; 3.86	0.863 ± 0.061; 7.07	0.785 ± 0.037; 4.71	0.794 ± 0.032; 4.03
3	0.224 ± 0.026; 11.61	0.331 ± 0.033;9.97	0.261 ± 0.065; 24.9	0.290 ± 0.045; 15.52
1	0.084 ± 0.005; 5.95	0.143 ± 0.017; 11.89	0.125 ± 0.012; 9.6	0.132 ± 0.024; 18.18
0.3	0.032 ± 0.005; 15.63	0.044 ± 0.006; 13.64	0.054 ± 0.008; 14.81	0.075 ± 0.007; 9.33
0.1	0.022 ± 0.005; 22.73	0.024 ± 0.006; 25.0	0.040 ± 0.003; 7.5	0.051 ± 0.006; 11.76
0.003	0.010 ± 0.002; 20.00	0.018 ± 0.005; 27.78	0.039 ± 0.007; 17.95	0.004 ± 0.005; 11.36
IgA----100†		0.152 ± 0.017	0.345 ± 0.076	0.102 ± 0.004
IgM----100†	0.008 ± 0.001		0.045 ± 0.05	0.065 ± 0.017
IgG1---100	0.036 ± 0.001	0.049 ± 0.005		0.084 ± 0.027
IgG2---100	0.064 ± 0.026	0.011 ± 0.005±0.086 ± 0.002		

h/Each column gives the O.D. 400nm obtained when various quantities (given on left) of a particular Ig (indicated at top of column) were tested, using an antiserum specific for that Ig; in this case tested using anti-IgA. †n = 3 for all data in this table. ‡ Last 4 rows indicate OD400nm values obtained when homologous antiglobulins, e.g. anti-IgA vs. IgA, were tested against 100 ng of the 3 heterologous Igs. Data are expressed as corrected OD400nm ± SD; coefficient of variation. All data are from Butler et al., 1980a.

from 1:500 to 1:4000. The same criteria are employed for selecting the proper dilution as when isotype-specific antiglobulins are being evaluated.

To titer the EIC, the optimal dilutions of the isotype-specific antiglobulin and bridge are used and the EIC is tested at dilutions from 1:500 to 1:8000. Again, the same criteria for selection of the optimal dilution as was described above, are employed. Typically our EICs are used at dilutions of either 1:1000 or 1:2000.

MEASUREMENT OF SPECIFIC ANTIBODIES IN MICROELISA SYSTEMS

A. Adsorption of antigen on microtiter wells

1. Binding characteristics of proteins and plastics. The microtiter a-ELISA depends on adsorption of proteins on plastic to create solid-phase antigen. Earlier we demonstrated that stable adsorption of protein to plastic is concentration dependent (Cantarero et al., 1980). We showed that there is a discrete range in which a linear relationship exists between the amount of protein added and the amount which binds, the so-called region of "independent binding". Studies in our laboratory using Immulon 1 and 2 reveal a similar relationship although the latter has a greater adsorption capacity for protein than Immulon 1. The addition of 100-200 ng/well of protein approaches the upper limits of the region of independent binding for Immunlon 2.

2. Influence of antigen concentration on non-specific binding. Components of the sample in the primary step and components of the detection system may bind directly to a polystyrene surface even in the presence of 0.05% Tween-20 (unpublished observations). Direct binding of sample components can result in the detection of non-antibody immunoglobulin, while the direct binding of the detection system can produce a background binding which distorts the titration curve. Both of these effects are fairly small and may be insignificant in an ELISA system of low sensitivity, however they can be very significant in ELISAs established for the detection of low levels of specific antibody.

The magnitude of the direct nonspecific binding of protein to the solid phase is reduced when relatively high levels of antigen are used. However, high levels of antigen on the plastic can lead to the shedding of significant antigen into the fluid phase during the assay, where it may act as a competitive inhibitor (Peterman, 1985). We have reduced the problem of direct binding through the use of "post-coating" (Fig. 4). With post-coating, the plate is first coated with the specific antigen for 3 hours at 37°C, the plate is emptied, and then the plate is coated with gelatin at 1 mg/ml in the coupling buffer (Butler et al., 1978b). Post-coating the plastic with a high concentration of gelatin substantially reduces subsequent

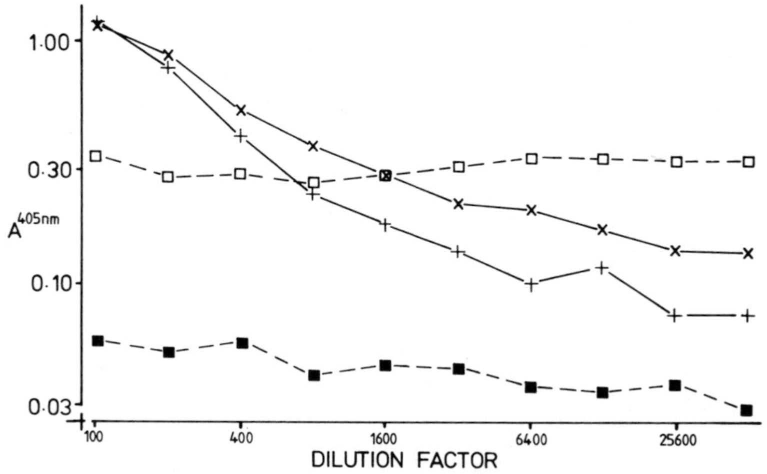

Fig. 4. Influence of a "post-coating" of microELISA plates (addition of 0.1%) gelatin in coupling buffer (see text) on the titration of IgA anti-fluorescein in PBS containing 0.05% Tween 20. X——X = titration on antigen-coated wells, +——+ = titration on antigen-coated wells after "post-coating", □- - -□ = titration on antigen-free wells, ■- - -■ = titration on antigen-free wells after "post-coating".

direct binding, while dissociation of gelatin into the fluid phase during the assay should not produce a problem. Post-coating with gelatin can reduce the sensitivity of some systems by blocking antigenic determinants (antigen-specific ELISAs) or antibody (sandwich assays), but the gains in specificity and linearity often make this procedure worthwhile.

B. Titration of the primary antibody

1. Preliminary titrations of primary serum or secretions. Quite often the investigator has some notion about the titer of specific antibodies in a serum. When this is not the case, we routinely prepare and test log dilution sequences as shown in Fig. 5a. We use both PBS containing 0.05% Tween-20 (PBS-T) as well as PBS-T supplemented with 0.1% gelatin or ovalbumin, as a diluent (depending on the system). This allows 20 unknown sera to be tested on a plate which also contains a doubling dilution sequence of a reference standard. When the highest concentration of reference standard reaches a pre-established maximum $O.D.^{405}$, the entire plate is either read automatically (see Section D-1) or the reactions stopped with 1 N NaOH. In interests of technician time (if dilutions are prepared manually), the same dilutions may be tested on parallel plates so that the antibody activity associated with different isotypes can be tested.

Typically one such preliminary titration with random samples selected from a large population study is adequate to allow the investigator to predict the correct initial dilution of test sera to be routinely tested for antibodies of a particular isotype.

2. Dilution sequences used in final titration of unknown samples. On the basis of preliminary titrations or prior experience, initial dilutions are made from the test sample in such a way as to eliminate dilution error; large dilutions are made through the use of intermediate dilutions and volumes are kept relatively large. We have found no significant differences between the accuracy of positive displacement or air displacement pipetting aids (Dierks, 1985).

Initial dilutions may be further diluted manually or transferred to microtiter plates for further dilution using automated diluters (e.g., Dynatech Autodilutor III; Flow Medimixer; Cetus Pro/Pette). With all autodiluters, the initial dilution must be made manually so that the amount of time saved may not be cost-efficient for some investigators. Figure 5b illustrates a typical microtiter plate layout as used in our laboratory. Columns 5 and 6 contain a one-half dilution sequence of a reference standard selected to span the linear region of the titration curve (section IV A, below) for the isotypes in question. The remainder of the plate is used to test four dilutions of a one-third dilution sequence of 20 unknown samples. We have found that no advantage is gained by testing dilutions in duplicate although it is advisable that less experienced personnel test each sample independently, or better,

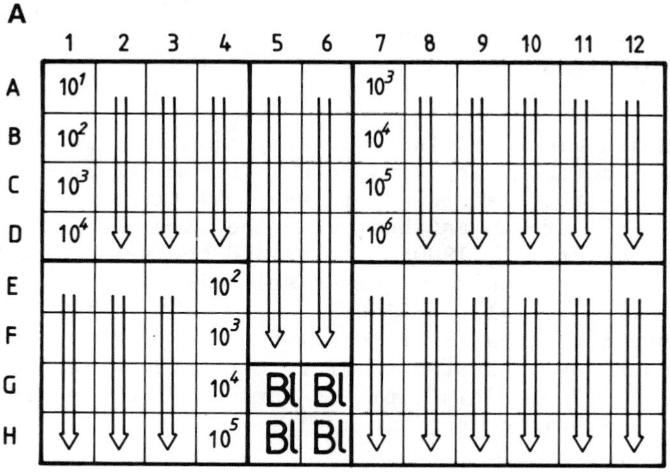

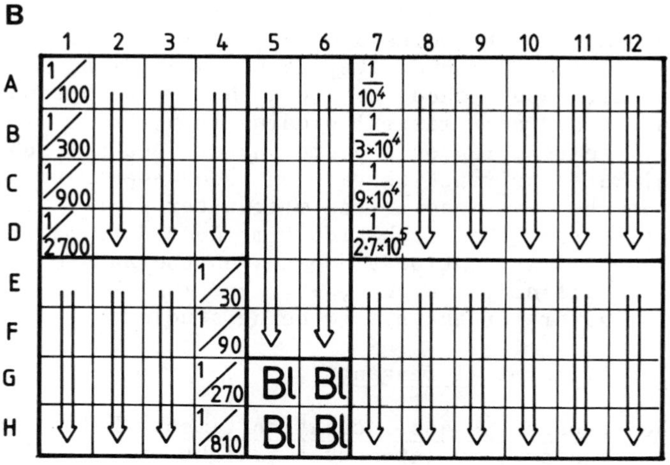

Fig. 5. Experimental design for antigen-specific microELISAs. Part A. Preliminary titration of test samples. A four log dilution sequence of each of 20 test samples and a one-half dilution sequence of a reference standard applied in duplicate [5(A-F) and 6(A-F)] are tested. Bℓ=control well receiving buffer instead of the primary antiserum. Arrows indicate progressive dilution of standard and test samples. Specific examples are given in 1(A-D), 4(E-H) and 7(A-D). Part B. Final 1/3 dilution sequence used for routine analyses of 20 test samples/microtiter plate. Actual dilutions are given for samples in 1(A-D), 4(E-H) and 7(A-D). Bℓ=control wells as in Part A.

to test each on two different occasions. We have found that the use of one-third dilution sequences of test samples provides adequate range and accuracy for estimating their antibody content.

3. Measurement of minor antibody classes. The measurement of minor antibody isotypes such as serum IgE in man or serum IgA in cattle, can result in atypical titration curves when high levels of IgG antibodies to the same antigen are present (Butler et al., 1980a; Butler and McGivern, 1985). Resolution of this problem requires selective removal of IgG by absorption on formalin killed *S. aureus*, chromatographic or ultracentrifugational fractionation of samples, or use of a reverse-phase assay system similar to that described by Zeiss et al. (1973). This topic is further discussed in the section on antigenic competition.

4. Incubation technique. In our laboratory, monomeric and dimeric antibodies of instrinsic Ka $\geq 10^7$ ℓ/m do not bind maximally to solid-phase antigen in 1 h at R.T. (Koertge, 1984). The dimeric antibodies approach maximal binding earlier and both are maximally bound after 24 hr incubation. In other studies (Butler, unpublished) we found that swine antibodies to fluorescein reached near maximal binding in 5½ hr at 37°C. Agitation of the reaction on a plate shaker does not substantially accelerate the reaction rate but does appear to improve uniformity of the reactions in the microtiter plate. These results suggest that the incubation time needed to maximize binding of the primary antibody depends on the nature of the antibody, i.e. affinity, valence, etc. and must be optimized by the investigator for the system being studied.

C. Addition of terminal reagents

1. Reaction steps and incubation times. Following incubation of the primary antibody overnight at 4°C, plates are washed (See C-2) and the isotype-specific antiglobulin added at a dilution of 1:500-1:2000. Reagents which are added uniformly to all wells are added using multichannel pipettes (Titertek; Flow Laboratories). Plates are then incubated 2 hrs with shaking, at R.T. Following the next wash, bridging antibody is added at dilutions ranging from 1:500 to 1:4000 (titering of reagents in the a-ELISA, p. 10), the plates incubated as in the previous step and the EIC added at a dilution of 1:1000 or 1:2000. We have found that sensitivity is increased if the EIC is incubated overnight at 4°C. This regime is also technically convenient allowing personnel to complete the assay the following morning.

The incubation times used may also be shorter if antibody titers are high and time schedules critical. The a-ELISA may be performed in a single day, provided antigen is adsorbed overnight on the day before, but with a corresponding loss of sensitivity.

Longer incubation times also allow reagents to be used at greater dilution so if commercial antisera must be purchased and time is of little consequence, longer incubation times are adviseable.

2. Plate washing methods and reagent loss. Washing the wells of microtiter plates is done using 0.15 M NaCl containing 0.05% Tween-20. Washing is accomplished using a combined dispenser-aspirator device developed by our laboratory in collaboration with H. Breier of the Institute for Animal Husbandry and Behavior, Mariensee, FRG (Fig. 6). Sixteen wells in two rows are simultaneously washed for 10 seconds each. The washer, equipped with a flow shut-off valve, is then manually transferred to the next two rows. An inline 8 μ Millipore filter is used to prevent particles from the gravity-fed reservoir from entering and clogging the dispenser nozzles. The 16 ℓ wash solution reservoir is periodically cleaned with chromic acid.

Washing of plates results in some desorption of antigen which in the region of "independent binding", accounts for but a few percent of the amount originally adsorbed (Cantarero et al., 1980). Greater amounts of antigen are desorbed if higher concentrations of antigen are used to coat the plate. When such desorption occurs during incubation of the primary antibody, antibody binding can be substantially effected through a mechanism of competitive inhibition (Peterman, 1985).

Washing also results in small losses of primary antibody and terminal reagents as shown in Figure 7. The percentage loss is small and it is unknown whether these labelled reactants are lost as individual entities or as complexes containing previously bound reactant, e.g. bridge lost as a complex with the isotype-specific antiglobulin.

D. The substrate-EIC reaction

1. Measurement of enzymatic reaction. After overnight incubation of the EIC with the previously bound reactants, the plate is again washed and a 1 mg/ml solution of p-nitrophenyl phosphate (substrate) is added. Substrate buffer, stored at 4°C, is allowed to come to room temperature before solubilization of substrate and addition to the plate. For the a-ELISA

run in our laboratory, the substrate reaction is stopped or measured after 30-120 minutes. Reactions may be stopped with 1 M NaOH or measured automatically when the well containing the highest concentration of the standard reaches a predetermined O.D.405 nm. Measurement of colorimetric enzyme reactions in microtiter plates is currently done using specialized plate readers, either manual (Dynatech MR 590; Biotek EL 309) or automated (Dynatech MR600; Biotek EL 310; Flow Multiscan) plate readers. Baseline control values are those obtained by measuring the O.D.405 nm of wells which had received reaction buffer (PBS-T or equivalent) instead of a dilution of the primary antiserum. To avoid the "edge effect" seen when using some microtiter plates, not all control wells are located at the edge of the plate (Fig. 5).

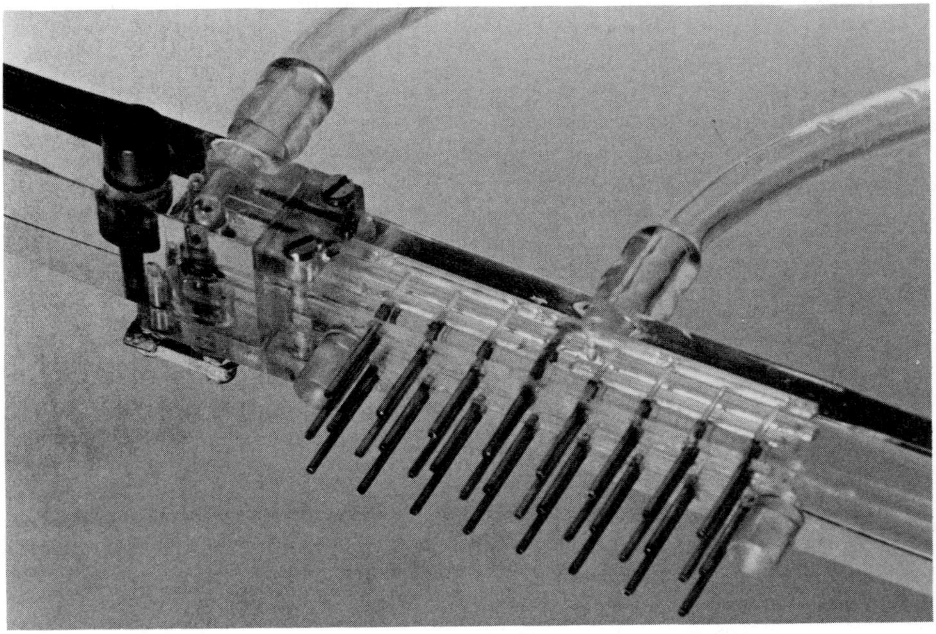

Fig. 6. Microtiter ELISA washer. Wash solution is fed by gravity to the 16 washer nozzles which discharge the solution near the bottom of the sixteen wells. "Used" wash solution is continuously removed by an aspirator through the shorter tubes adjacent to each nozzle. Designed by H. Breier, J. E. Butler and F. Klobasa, Institute for Animal Husbandry and Behavior, Mariensee, FRG.

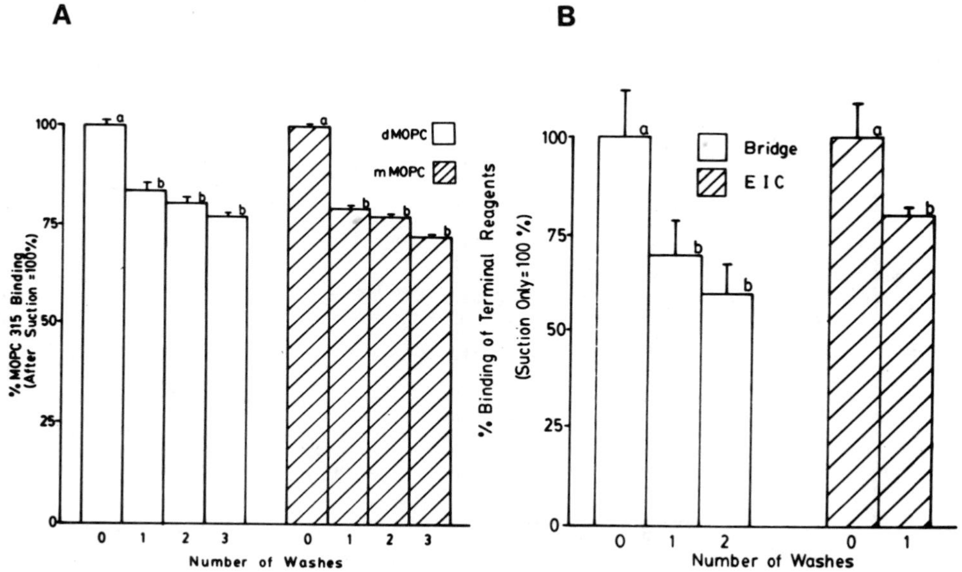

Fig. 7. Losses of primary antibody and reactants during performance of the a-ELISA. A. Losses of monomeric (mMOPC) and dimeric (dMOPC) MOPC 315 IgA anti-DNP from DNP-gelatin-coated microtiter wells. Amount bound after aspirating the unreacted antibody is designated 100%. B. Losses of the bridging antibody and EIC during performance of the a-ELISA. Legend on figure. One-hundred percent is the amount of reactant present after aspiration of the supernatant but prior to washing.

2. <u>Release of enzyme during substrate reaction</u>. The EIC is unique among enzyme-antibody conjugates used in ELISAs. The addition of substrate buffer at pH 9.8 results in gradual dissociation of the EIC liberating free enzyme as the substrate reaction proceeds (Figure 8a). Studies using double-labelled EIC (see Fig. 2b) reveal that the enzyme released is not part of soluble complexes, but free AP (Koertge et al., 1985; Koertge, 1984). The released AP was also shown to possess the same enzyme activity as that which remained bound (Fig. 8b). There is evidence however, that the AP of the EIC differs from stock AP (Koertge, 1984).

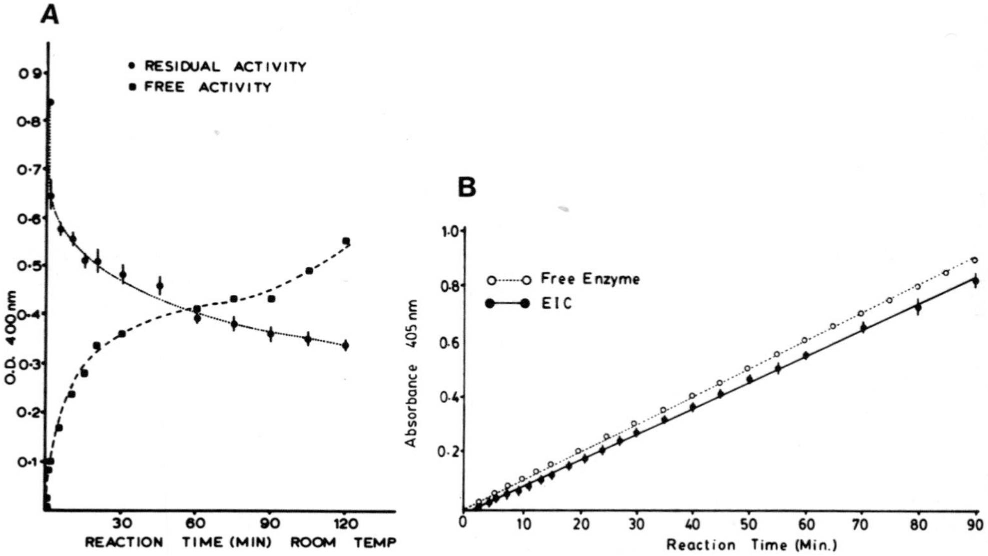

Fig. 8. The dissociation of alkaline phosphatase (AP) during the substrate reaction of the a-ELISA and the effect on its enzymic activity. A. Amount of residual and free AP present in the microtiter well during the first 120 min. after addition of the EIC. B. Enzymic activity of the free AP versus AP of residual EIC during first 90 min. after addition of the EIC.

SPECIAL CONSIDERATION IN THE MEASUREMENT OF SPECIFIC ANTIBODIES BY MICRO-ELISA

A. Relationship between primary antibody binding and detection by ELISA

1. Nature of the ELISA titration curve. The sigmoidal nature of the ELISA titration curve is familiar to all who use the assay for measurement of specific antibodies. While assumed to be the case, maximal binding of antibody in the linear portion of the sigmoidal curve has only recently been demonstrated (Koertge, 1984; Koertge & Butler, 1985). Figure 9 illustrates the relationship between the binding of primary antibody and the EIC of the a-ELISA over the linear region of the titration curve. These data also show a constant percentage of binding for the primary antibody in this region. The percentage of antibody which binds to solid-phase antigen, i.e.

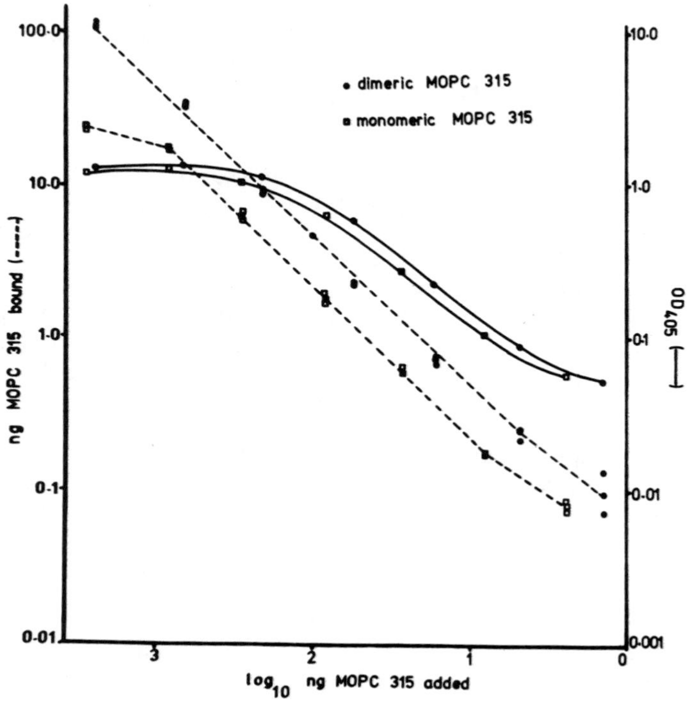

Fig. 9. Relationship between the binding of primary antibodies (monomeric or dimeric MOPC 315) to solid-phase DNP-gelatin (dotted lines) and their detection using the a-ELISA (solid lines).

hapten on a carrier, is equal to the percentage of antibody that can bind to the same hapten in a highly antigen-excess, totally independent system (DNP-*E. coli* suspensions) (Koertge, 1984; Koertge & Butler, 1985). Hence, for the model system studied, the linear region does represent the region where maximal, constant antibody binding occurs. Figure 9 also illustrates that while primary antibody binds in a linear fashion over a nearly 3-log range, its indirect detection shows a different pattern. Hence, the sigmoidality of ELISA titrations cannot be explained by the binding of primary Ab to solid-phase antigen in the micro-ELISA system.

2. <u>Influence of conjugate size</u>. Sigmoidality in the upper portion of the ELISA titration plot, i.e. the region of high input of primary antibody, suggests: (a) progressive steric hindrance of the detection system in the region of more concentrated antibody or (b) selection for binding of conjugates with different antibody-enzyme ratios or with enzymatic activity different than those which binds in regions of low antibody input. The second possibility seems unlikely because of the homogeneity of the EIC of the a-ELISA (Fig 2). Therefore we tested the influence of antibody-enzyme conjugates of varying degrees of complexity and hence of potentially different degrees of steric hindrance, on the sigmoidality of the ELISA titration. The systems used are those illustrated in Figure 1. Size was determined by sucrose density gradient ultracentrifugation similar in principle to the method used to generate the data of Figure 2. Such tests confirmed that the glutaraldehyde conjugates behaved as large aggregates while the Fab' conjugate was homogeneous with a mol. wt ca. 140 Kd. When these various detection systems were used to measure the anti-DNP activity against DNP-gelatin bound in microtiter wells, the results shown in Figure 10 were obtained. These results are consistent with the conclusion that sigmoidality in the region of high antibody input is the result of steric hindrance; the conjugate composed of 1 Fab' and 1 AP gave a titration which most closely approximated the binding of primary antibody.

The data of Figure 10 also show that the titration obtained with the a-ELISA, i.e. the EIC, does not produce the "hook" effect seen at high antibody input. The latter may result from inhibition produced by free antibody contaminating the glutaraldehyde conjugates. The constant plateau region of the a-ELISA is believed to result from the fact that reactants are added piecemeal, i.e. "built up in situ" on the primary antibody, so that no free antibody is available to produce inhibition (Koertge, 1984; Koertge & Butler, 1985).

3. <u>Antigenic competition</u>. The problem of antigenic competition, especially in the measurement of minor antibody classes such as IgE, is theoretically illustrated in Table IV. In this example, several assumptions are made: (a) the percentage of specific antibody except IgE is considered to be 10%; IgE specific antibody is 25% of total IgE, (b) 100 ng/well of the antigen can be stabily adsorbed to Immulon 2 wells and (c) the concentrations of the various Igs in human serum fall within the normal range. The example in Table IV shows that under conditions favorable to binding 100% of the antibody in serum, i.e. 2-log excess of antigen (see section IV B), an ELISA assay for IgE would have to be sensitive to

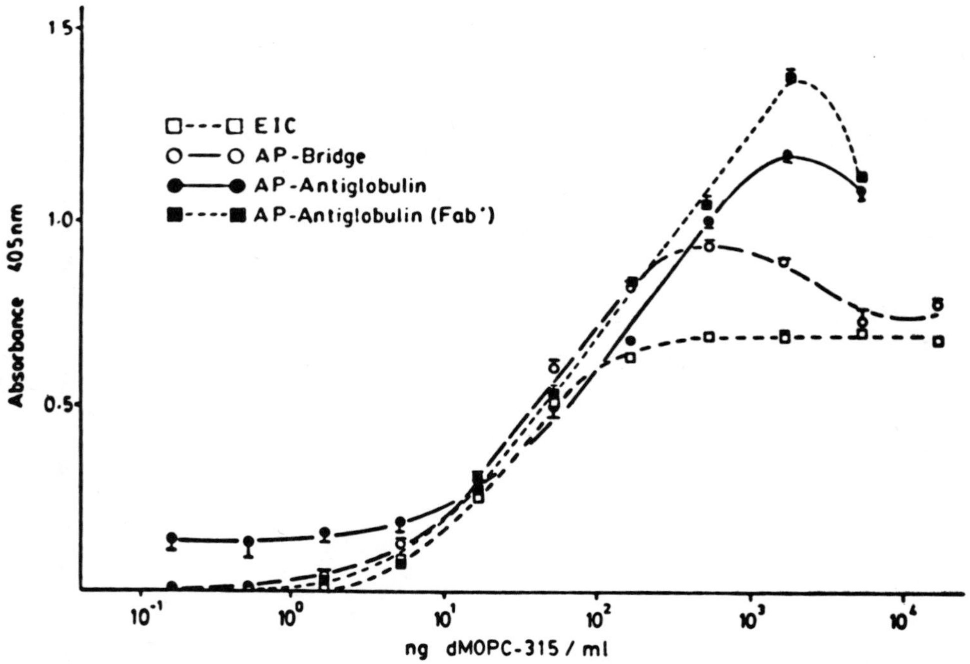

Fig. 10. The influence of the size of the detection system used for antigen-specific ELISA. The legend is on the figure and corresponds to the detection systems depicted in Figure 1. Reaction times were adjusted so that the magnitude of the O.D. 405nm in the linear regions of the various curves were similar.

Table IV. Antigenic Competition in Micro-ELISA: Quantitative Theory

Antigen or antibody added	μg	μM bound[a]	ng Ab bound[a]	Molar Ag:Ab[b]
Antigen added[c]	200	2×10^{-3}		
Specific Ab in human serum				
IgG (1:25)	48.4	3×10^{-4}	48,400	6.7
IgG (1:500)	2.42	1.5×10^{-5}	2,420	133
IgA (1:25)	10	1×10^{-5}	10,000	20
IgA (1:500)	0.5	3.12×10^{-6}	500	641
IgM (1:25)	3.72	2.33×10^{-5}	3,720	85.8
IgM (1:500)	0.18	1.86×10^{-7}	186	1×10^{4}
IgE (1:25)	5×10^{-3}	2.8×10^{-8}	5	7.1×10^{4}
IgE (1:500)	2.5×10^{-4}	1.39×10^{-9}	0.25	1.44×10^{6}
TOTAL		1:25 = 3.3×10^{-4}		
		1:500 = 1.84×10^{-5}		

Ratio Total Ag:Total Ab in 1:25 dil. = 6
Ratio Total Ag:Total Ab in 1:500 dil. = 108

a/Assumes 50% of added antigen and 100% of added antibody binds.
b/Molar ratio of antigen (Ag): bound antibody (Ab) assuming no competition and 100% binding of the latter.
c/ Molecular weight = 50 Kd.

0.25 ng/well or 1.25 ng/ml. At a serum dilution in which 5 ng/well would be present, only a 6-fold excess of antigen would be present. Hence, most antibodies would not be of sufficient affinity to totally bind to the solid-phase antigen under such conditions. Hence, affinity selection would occur and if IgE antibodies were of lower affinity, than a substantial portion of the IgG antibodies, they would go undetected in such an assay. Inspite of their concentration and affinity, multivalent IgM and dimeric IgA might be favored in such situations because of their higher avidity. Examples of the former have been observed in our laboratory for bovine IgM versus IgG antibodies (Butler et al., 1980) and for dimeric MOPC-315 versus monomeric MOPC-315 (Koertge, 1984).

Solutions to the above problem include: (a) increasing the concentration of solid-phase antigen, (b) eliminating the competing antibodies e.g. IgG, that are present in high concentrations or (c) using a reverse assay which first binds IgE and then assays for its antigen-binding activity. The last approach has been successfully employed by Zeiss et al. (1973) while we have employed the second. The first approach is likely to be least effective because as described in IV-A above, steric hindrance might well inhibit detection long before antigen saturation becomes a problem.

B. The role of antibody affinity and avidity

1. Theoretical aspects. The polyclonal nature of the immune response means that each serum or secretion to be tested contains a mixture of antibodies of different isotypes and affinities. As the surface area of plastic in a microtiter well is limited and the adsorption capacity of plastic is limited, situations are certain to occur in which the ratio of antigen to antibody is favorable for the binding of certain antibodies but not for others. The early work on the ELISA by Engvall and Perlman (1972) suggested that the ELISA could be used as an affinity-independent assay for specific antibody, however an increasing number of authors are suggesting that the ELISA is significantly influenced by antibody affinity (Butler et al., 1978; Lehtonen and Eerola, 1982; Peterfy et al., 1983; Nimmo et al., 1984). If affinity does not play a role then conditions should exist where all antibodies can be detected using ELISA technology. These hypotheses may be examined theoretically utilizing the Mass Law.

The Mass Law may be rearranged (Equation 1) such that when antibody bound is graphed arithmetically against antibody added, a plot with a slope dependent on both the affinity of the antibody and the free antigen concentration is obtained.

Equation 1 $$[Ab] = \frac{K[Ag]}{1+K[Ag]} * [Abt] + 0$$

A graphical examination of the logarithmic transformation of the simple Mass Law (Equation 2) yields a plot in which the Y-axis intercept is dependent on both antibody affinity and antigen concentration.

Equation 2 $$\log [Ab] = \log \frac{K[Ag]}{1+K[Ag]} + \log[Abt]$$

Hence the apparent influence of affinity on ELISA titrations depends on the manner in which the data are plotted. Unfortunately, titrations are often plotted semi-logarithmically which, according to discussions presented here, may not be the appropriate method, and may only complicate the interpretation of the effects of affinity. However, whenever the product of the free antibody concentration and the antibody affinity is 100 or higher, the value of $K[Ag] \div 1 + K[Ag]$ approximates unity. Thus a system will be independent of affinity above a particular value, depending on the concentration of available antigen.

The influence of affinity may also be complicated by antibody heterogeneity. The Mass Law may be modified to include the influence of heterogeneity (Lew, A. M., 1984), as expressed in the following equation:

$$[AbAg] = \frac{K[Ag]^a}{1 + K[Ag]^a} X [Ab_t]$$

This equation reveals that the influence of affinity on antibody binding diminishes as antibody heterogeneity increases. Thus data concerning the effect of antibody affinity on the ELISA obtained with the use of monoclonal antibodies (Peterfy et al., 1983; Nimmo et al., 1984; Sips, 1948) may not apply to heterogeneous samples. However the Sips modification has limited validity (Bruni et al., 1976) and may not be totally applicable to ELISA data.

The practical influence of affinity can be an underestimation of low affinity antibody and the non-detection of very low affinity antibodies (Butler, 1981; Butler et al., 1978). Although one author has suggested that this effect is sufficient to prevent the use of ELISA for quantitative determinations (Sips, 1948), the data available are currently insufficient to warrent its abolition as a means of estimating specific antibody concentrations.

2. Evidence for the influence of affinity on titrations. Various investigators have demonstrated effects that were ascribed to affinity when using ELISA to measure specific antibody (Butler et al., 1978; Lehtonen & Eerola, 1982; Peterfy

et al., 1983; Nimmo et al., 1984). The preliminary data shown in Figure 11a and 11b, demonstrate the influence of affinity when data are expressed on arithmetic and log-log titration plots respectively. These data support the predicted effects based on

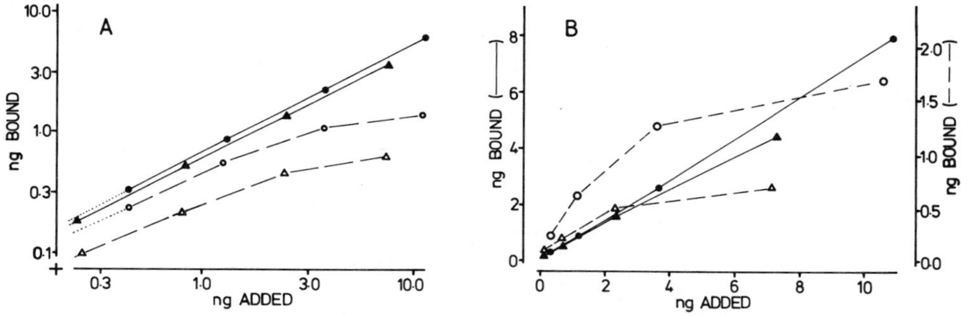

Fig. 11. The relationship between antibody binding plots, antibody affinity, and the amount of antigen used to coat microtiter wells. With either a logarithmic plot (A) or a linear plot (B), wells coated with 100ng of antigen (———) do not substantially distinguished between antibody of high affinity ([●, ○] with Ka of 2 x 10^{10} 1/M) and antibody of moderate affinity ([▲, △] with Ka of 3 x 10^{7} 1/M). A logarithmic plot of antibody binding to wells coated with only 10ng of antigen (— — — —) shows that the system does distinguish between high and moderate affinity antibodies in terms of the Y-intercepts of their respective titration plots. Linear plots of antibody binding to well coated with 10ng of antigen show that such a system distinguishes between high and moderate affinity antibodies in terms of the slopes of their linear regions.

rearrangement of the Mass Law (Equations 1 and 2). Unfortuantely normal immune resopnses are polyclonal, and the influence of affinity demonstrated with homogeneous antibodies (Fig. 11) remains to be determined for mixed populations.

3. Influence of polyvalency. Studies on bovine serum antibodies to human albumin suggested that the selective detection of IgM at low serum dilution, might be explained by polyvalent-dependent avidity (Butler et al., 1980a). To test this hypothesis we purified monomeric and dimeric MOPC-315, monoclonal antibodies with the same intrinsic Ka, and compared their ability to bind DNP-gelatin adsorbed to microtiter wells. When compared separately, we found no differences in their ability to bind DNP or to be detected by the a-ELISA after 24 hr incubation. We found, however, that more than a 10-fold higher concentrations of DNP-glycine (free hapten) was requred to inhibit the binding of dMOPC-315 compared to mMOPC-315 (Koertge and Butler, 1985). Furthermore, the rate at which mMOPC-315 became stably bound to DNP was markedly slower than that for dMOPC-315 (Koertge, 1984). The higher aviditiy and rate of stable binding of dMOPC-315 might result in selection for these antibodies in situations where antigen is limiting but not in situations where sufficient antigen is available.

C. Data acquisition and analyses

1. Analytical hardware. A major advantage of the micro-ELISA system is its suitability for automated analyses. Three manufacturers currently market automatic microtiter plate readers (Dynatech MR600; Flow Multiscan or Biotek El 310). Such machines are capable of reading and recording the optical densities of all 96 wells of a standard microtiter plate in ca. 60 seconds. Each of them can be readily interfaced with computers; computers may be used as controllers and/or for data acquisition and eventual analyses. Both Apple and IBM-compatible systems are popular. Instruments like the IBM-PCjr are reasonably-priced, compact and with sufficient capabilities for control, data acquisition and analyses.

2. Computer programs for data analyses. Various hardware manufacturers such as those which manufacture plate readers, offer simplified programs for data analyses. In most cases these programs fail to consider ELISA immunochemistry, contain no inbuilt means for comparing the slope of the titration of standard and test samples, and are too often designed only to treat simple end-point determinations. To overcome these disadvantages, we have designed a data analysis program based on our current understanding of ELISA immunochemistry.

The validity of "endpoint" or "single point" determinations, in which the absorbance values obtained from single dilutions of samples are compared with a standard curve obtained from a standard antibody preparation, requires that the titration curves of each sample parallel that of the standard. As antibody titration curves may have different shapes dependent on the affinity of the antibody being measured (see section IV B), an absorbance value obtained from the linear region of the titration curve of an unknown may not be within the linear region of the titration curve for the standard. Furthermore, the titration curve of a sample and that of the standard may have linear regions with different slopes. Therefore investigators who apply linear regression analysis to compare single dilution O.D. values of samples with a standard, can be introducing a large error. Therefore it is our belief that valid comparisons of samples with a standard titration curve must include values from a range of sample dilutions.

The examination of ELISA data to determine its validity requires graphic and regression analysis of the titrations of many samples and standards. To aid our analyses we have developed a computer program (for the IBM-PC or PCjr) which enables data from samples and a standard tested on a single plate (Fig. 5b), to be rapidly analyzed. The program performs the following functions:

1. Obtains final absorbance values via an interface with the ELISA plate reader.
2. Rearranges the data into a format facilitating analysis and generates the appropriate dilution values using either standard formats (e.g. those shown in figure 5b) or information relevant to an individual plate provided by the operator.
3. Graphs the data from the standard and allows the operator to select the linear region of the standard curve.
4. Performs linear regression analysis of the linear region of the standard and obtains the equation of the standard curve for use in sample analysis.
5. Utilizes the equation of the standard curve to examine data from the samples; the dilution of the standard which would have given the same final absorbance as a particular dilution of a sample is first calculated. The program then uses the actual dilution of the sample to calculate a value in units/ml for the undiluted sample.
6. Compares the values obtained for a range of dilutions of each sample, excluding any values obtained from absorbance values outside of the range of the linear region of the standard curve.

7. Allows the operator to select those consecutive dilutions of each sample which produce similar units/ml values (which must therefore be from within a region of the sample titration curve which has the same slope as the linear region of the standard curve), and uses the mean of these values as the concentration of the undiluted sample in units/ml.

8. Allows the operator to examine selected data by linear regression analysis to ensure that the apparent linear region is actually linear and parallel to that of the standard.

The above protocol maximizes the validity of data obtained using the ELISA, however it cannot allow for errors in estimating the concentration of the standard, and it assumes that the amount of antibody bound in the ELISA is both proportional to its concentration and independent of its affinity.

D. Absolute quantitation of specific antibodies

1. Review of current methods. At least four different methods have been proposed and/or described for the measurement of antibodies in absolute quantities. Each of those so far described possesses either theoretical and/or practical problems or uncertainties. The *reference standard method* uses a reference serum, the antibody content known by independent criteria, to comparatively quantify unknowns. To be valid, the affinity of sample and *reference standard* must be within the range of affinities, in which binding is affinity-independent (section IV B). The magnitude of error made by such an assumption is evidenced by previous studies (Butler et al., 1978). When multivalent antigens are used, reference and samples must also have the same specificity. The so-called "*direct-ELISA method*"*/, which uses purified Igs adsorbed on plastic for standards for the primary antibody, also depends on a dangerous assumption. Namely, that the adsorbed Ig is recognized by the isotype-specific anti-Ig in the same manner as when primary antibodies of the same isotype are bound on antigen. Recent data from our laboratory indicates that this assumption is invalid; antibody bound to antigen is more easily recognized by anti-Ig than Ig of the same isotype adsorbed on plastic (Dierks, 1985). A third method of quantitation depends on inhibition of the isotype-specific antiglobulin by purified Ig of the same isotype, i.e. *antiglobulin-inhibition method*. This method is technically complicated by the fact that: (a) soluble purified Ig competes differently with anti-

*/Laboratory jargon for assay based on using purified Ig of the same species and isotype as the primary antibody being measured, as standards adsorbed directly on plastic. This is better referred to as the "*Ig standard*" method.

globulin than antibodies of the same isotype bound to antigen and (b) soluble complexes are formed during inhibition which are probelmatic. To overcome these objections so as to make them "relative effects", a reference standard is required which then become subject to the criticisms of the *reference standard method* per sec.

Finally, quantitation by *reaction stoichiometry* has been proposed and tested. While theoretically sound, the failure to obtain a constant relationship between primary antibody bound and the bound enzyme in the a-ELISA (Fig. 9) raises doubts about the eventual value of this method The concept might still be workable if simple conjugates of uniform size and long-term stability could be prepared which because of their lower steric hindrance, might show a constant relationship between the primary antibody and the enzyme of the detection system (Fig. 10).

A major question which to this point has not been satisfactorily resolved, but which underlies all methods of quantitation, is whether the amount of antibody bound in ELISA represents all or nearly all of the antibody in the sample measured. The theoretical considerations in IV-B1 predict that the percentage of antibody bound in the linear region is affinity dependent so that the total antibody content of sera containing only low affinity antibodies, might be seriously underestimated. Our studies on a homogeneous anti-DNP with an intrinsic K_A of $>10^7$ did show maximal binding in the linear region after 24 hr although not after 1 hr (Koertge & Butler, 1985). Maximal binding was defined as the amount of MOPC-315 which could bind to a nearly infinite excess of DNP-*E. coli*. Hence, anti-hapten antibodies with $K_A > 10^7$ appear to bind maximally to their solid-phase hapten provided sufficient incubation time is used. Corresponding data on low affinity antibodies, those directed to protein antigens and the influence of population heterogeneity on this parameter, are not available. Therefore, provided quantitative methods could be developed for measuring the amount of bound antibody, data are still required to show that the amount of bound antibody is representative of the total antibody of that isotype in the sample being measured.

OTHER APPLICATIONS OF THE A-ELISA

A. The ELISA based antigen distribution assay (EADA)

1. Theoretical considerations. The EADA assay (Butler, 1981) is derived from studies of protein-plastic interaction which demonstrated that a "region of independent binding" could

be identified (Cantarero et al., 1980). In this region there is a linear relationship between the amount of antigen added and the amount bound so that the actual percentage of the added antigen which binds is constant. The value of this constant binding is characteristic for each protein and plastic. The binding percentages for Sarstedt polystyrene vary from 20-80% but for many proteins, the value is 40-50%. In the microtiter system, the region of independence for Immulon 2 extends up to ca. 100 ng/well for most proteins when 200 µl of sample are incubated in the well. The region of independent binding represents a region wherein competition for binding sites does not occur when a mixture of antigens is added; all proteins present in solution will be represented on the surface of the plastic after incubation and removal of unbound protein (Cantarero et al., 1980). The antigens present can then be identified using the detection system of the a-ELISA and a species-compatible antigen-specific antibody. The assay is an impractical quantitative measure of the amount of the antigen in the original sample because: (a) the binding percentage for a given protein antigen in a mixture of protein is usually unknown and (b) adsorption is subject to day-to-day variation of up to 10-20%.

2. Applications of the EADA. The EADA is ideal as a sensitive assay to measure the distribution of an antigen in biochemical separates, e.g. sucrose-density gradient fractions, low and high performance chromatographic separations and antigen eluted from polyacrylamide gels (Butler, 1981, 1985; Butler et al., 1980a,b; Pringnitz et al., 1985b; Pringnitz & Butler, 1985). The assay is valid when one wishes to compare the relative antigen activity among different fractions. This can be done by normalizing the enzymic activity to total protein content, i.e. OD405/µg of sample protein. Like its parent assay, the EADA readily detects <0.5 ng of bound antigen with a good signal:noise ratio. The system should be ideal for protein HPLC where concentrations are low but known through the use of highly calibrated detectors. The EADA is also a useful means of evaluating the specificity of an antisera which will eventually be used in the ELISA. The data shown in Table III are technically speaking, EADA data.

3. Potential problems. We have encountered three problems with the EADA of which the user should be aware. First, when a fraction is >90% composed of a single antigen, e.g., IgG1, the EADA activity of the major protein appears to be underestimated. Dilution of these samples to correct this problem often causes problems for the detection of the minor antigens in the same samples and normalization of data among widely different dilutions can be difficult. Second, if

detergents have been used to extract protein antigen as in studies on biological membranes, the detergent must be removed beforehand to allow binding to plastic to occur. Subsequent fractionation of detergent containing membrane extract on sucrose-density-gradients tends to effect such removal (Pringnitz & Butler, 1985; Pringnitz et al., 1985b). Finally, the day-to-day variation in the percentage of protein which binds and in microtiter systems, the large variation which we have observed among plates from different manufacturers and among wells on the same plate, can be disturbing. ELISAs used to measure antigen-specific antibodies in which the antigen added is in excess and not directly measured, do not reveal the type of interplate and interwell variations observed when using the EADA because variation in protein-plastic interactions, i.e. the quantitiy of antiglobulin adsorbed, is but a minor factor.

B. Competitive enzyme-linked immunoassay (CELIA)

1. Configuration of indirect competitive assay. The CELIA is an indirect competitive assay for the quantitation of antigen described by Yorde et al. (1976). Indirect competitive assays are defined as those in which the excess antibody remaining after addition of the unknown reference antigen is measured (Yorde, 1976) as opposed to measuring the amount of antigen directly (Engvall et al., 1971). Indirect assays are useful when: (a) antigens are univalent, e.g. haptenic, and typical sandwich assays are not applicable, (b) the detection system is so large or heterogeneous, e.g., an enzyme, that when attached to the competitive antigen in a direct competitive assay, free and enzyme-labelled antigen do not equally or consistently compete.

2. Application of the CELIA. The CELIA was originally described for the measurement of human choriogonadotropin but we have successfully used it for measuring β_2-microglobulin, secretory component and the membrane protein BAMP (Pringnitz et al., 1985a, b).

The CELIA, while fully functional, is admittedly cumbersome because of the many steps involved in the a-ELISA. It is still valuable in laboratories where the a-ELISA is used, as an initial test to determine whether an indirect ELISA is a feasible assay for measurment of the antigen in question because no special preparation of reagents are necessary for its application. Use of an enzyme conjugate containing antibody to the antigen to be quantitated, is preferable for routine work.

SUMMARY

The a-ELISA has proven to be a sensitive and reliable method for the measurement of antigen-specific antibodies. The stability of the soluble immune complex, its ease of preparation, lack of enzyme denaturation and adaptability to various species with only the change of a single reagent, has made it a popular system in our laboratory for measurement of specific antibodies. Its adaptability in the study of antigen distribution, i.e. EADA, in indirect competitive immunoassays for antigens and its more recent application to the study of ELISA immunochemistry and stoichiometry, add to its reputation as a practical "work horse" immunochemical system. In essence, the a-ELISA is but a type of detection system. The hard questions such as the influence of affinity and avidity or the development of methods for absolute quantitation of antibodies, await future immunochemical studies.

REFERENCES

Axen, R. and P. Vretblad. Chemical fixation of proteins to water-insoluble carriers. in: Protides of the Biological Fluids. Vol. 18. H. Peeters, ed., Pergamon Press, Oxford (1971).

Avrameas, S. A. and B. Guilbert. Enzyme-immunoassay for the measurement of antigens using peroxidase conjugates. Biochimie 54:837 (1972).

Bruni, C., A. Germani, G. Koch and R. Strom. Derivation of antibody distribution from experimental binding data. J. Theor. Biol. 61:143 (1976).

Butler, J. E. The amplified ELISA: Principles and applications for the comparative quantitation of class and subclass antibodies and the distribution of antibodies and antigens in biochemical separates. Methods in Enzymology, Vol. 73, Immunochemical Methods. H. J. Vunakis and J. J. Lagone, ed. (1981).

Butler, J. E., T. L. Feldbush, P. L. McGivern and N. Stewart. The enzyme-linked immunosorbent assay (ELISA): A measure of antibody concentration or affinity? Immunochemistry 15:131 (1978a).

Butler, J. E., P. L. McGivern and P. Swanson. Amplification of the enzyme-linked immunosorbent assay (ELISA) in the detection of class-specific antibodies. J. Immunol. Methods 20:365 (1978b).

Butler, J. E., and P. L. McGivern. The measurement of IgE and other classes of antibodies in the sera and bronchial lavage fluids of patients to *Alternaria*, house dust and ragweed. (1985, Submitted).

Butler, J. E., P. L. McGivern, L. A. Cantarero, and L. Peterson. Application of the amplified enzyem-linked immunosorbent assay: Comparative quantitation of bovine serum IgG1, IgG2, IgA and IgM antibodies. Am. J. Vet. Res. 41:1479 (1980a).

Butler, J. E., L. Peterson and P. L. McGivern. A reliable method for the preparation of bovine secretory immunoglobulin A (SIgA) which circumvents problems posed by IgG1 dimers in colostrum. Mol. Immunol. 17:757 (1980b).

Cantarero, L., J. E. Butler and J. W. Osbourne. The binding characteristics of six proteins to polystyrene and its implications to solid-phase immunoassays using polystyrene tubes. Analy. Biochem. 105:375 (1980).

Cuatrecasas, P., and C. B. Anfinsen. Affinity chromatography. Annual Rev. Biochem. 40:259 (1971).

Dierks, S. E. Absolute quantitation of IgG and IgA antiprotein and anti-hapten antibodies using a microELISA system. M.S. Thesis, University of Iowa (1985).

Engvall, E., K. Jonsson and P. Perlmann. Enzyme-linked immunosorbent assay. II. Quantitative assay of protein antigen, immunoglobulin G, by means of enzyme labelled antigen and antibody coated tubes. Biochim. Biophys. Acta. 251:472 (1971).

Engvall, E., and P. Perlmann. Enzyme-linked immunosorbent assay, ELISA. III. Quantitation of specific antibodies by enzyme-labelled anti-immunoglobulin in antigen-coated tubes. J. Immunol. 109:129 (1972).

Koertge, T. E. A study of the quantitative capability of the enzyme-linked immunosorbent assay (ELISA) with regard to the use of specific antibodies to study the transport of monomeric and polymeric antibodies. Ph.D. Thesis, University of Iowa (1984).

Koertge, T. E., and J. E. Butler. The relationship between the binding of primary antibody to solid-phase antigen in microtiter plates and its detection by the ELISA. J. Immunol. Methods (1985, submitted).

Koertge, T. E., J. E. Butler and S. E. Dierks. Evaluation of the soluble immune complex (EIC) of the amplified enzyme-linked immunosorbent assay (a-ELISA) and the use of this assay for quantitation by reaction stoichiometry (1985, submitted).

Lehtonen, O.-P., and E. Eerola. The effect of different antibody affinities on ELISA absorbance and titer. J. Immunol. Methods 54:233 (1982).

Lew, A. M. The effect of epitope density and antibody affinity on the ELISA as analyzed by monoclonal antibodies. J. Immunol. Methods 72:171 (1984).

Nimmo, G. R., A. M. Lew, C. M. Stanley and M. W. Steward. Influence of antibody affinity on the performance of different antibody assays. J. Immunol. 72:177 (1984).

Peterfy, F., P. Kunsela and O. Makela. Affinity requirements for antibody assays mapped by monoclonal antibodies. J. Immunol. 130:1809 (1983).

Peterman, J. H. The influence of antibody affinity, valency and heterogeneity on binding behavior of antifluorescein antibodies in solid-phase micro-ELISA systems. Ph.D. Thesis, University of Iowa (1985).

Pringnitz, D. J., and J. E. Butler. Bovine-associated mucoprotein (BAMP). III. De novo synthesis by non-mammary tissues. J. Dairy Sci. (1985, in press).

Pringnitz, D. J., J. E. Butler and A. J. Guidry. In vivo proleolytic activity of the mammary gland. Contribution to the origin of secretory component, β_2-microglobulin and bovine-associated mucoprotein (BAMP) in cows milk. Vet. Immunol. Immunopath. (1985a, in press).

Pringnitz, D. J., J. E. Butler and A. J. Guidry. Quantitation of bovine β_2-microglobulin: Occurrence in body fluids, on milk fat globules and origin in milk. Mol. Immunol. (1985b, in press).

Sips, R. On the structure of the catalyst surface. J. Chem. Phys. 16:490 (1948).

Sloan, G. J. Enzyme-linked immunosorbent assay (ELISA): Application to the quantitation by subclass of bovine antibodies against *Staphylococcus aureus*. M.S. Thesis, University of Iowa (1975).

Yorde, D. E., E. A. Sasse, T. Y. Wang, R. O. Hussa and J. C. Garancis. Competitive enzyme-linked immunoassay with use of a soluble enzyme/antibody immune complex for labeling. I. Measurement of human choriogonadotropin. Clin. Chem. 22:1372 (1976).

Zeiss, C. R., J. J. Pruzansky, R. Patterson and M. Roberts. A solid phase radioimmunoassay for the quantitation of human reaginic antibody against ragweed antigen E. J. Immunol. 110:414 (1973).

AFFINITY COLUMN MEDIATED IMMUNOENZYMOMETRIC ASSAYS

J. William Freytag

E. I. du Pont de Nemours & Co., Inc.
Biomedical Products Department
Glasgow Research Laboratory
Wilmington, DE 19898

INTRODUCTION

Immunometric assays, by definition, are immunoassays configured in such a way that labeled-antibodies are utilized as the quantifying entity. This is in contrast to competitive radioimmunoassays or enzyme-immunoassays in which labeled-antigens are utilized in competition with sample antigen to bind to a limited number of antibody sites. The label applied to the antibody can take the form of a radioisotope, enzyme, fluorophore, luminescent tag, or almost any tag ultimately capable of generating a signal, even through some complex coupling mechanism. Two fundamentally different immunometric assay configurations have been described: the one-site immunometric assay and the two site (sandwich) immunometric assay.

The one-site immunometric assay dates back to almost the days of the first radioimmunoassay, but actually was not highly implemented until 1971, with the description of the first enzyme-linked immunosorbent assay (ELISA) by two independent groups, Schuurs & Van Weeman (1971) and Engvall & Perlmann (1971). The one-site immunometric assay, including the many iterations to follow that employ second anti-immunoglobulin antibodies, protein-A, or avidin/biotin as signal amplifiers (Porstmann et al., 1982; Holbeck et al., 1983; Yolken et al., 1983), utilize the same principle assay configuration whereby the antigen to be assayed inhibits competitively the binding of labeled-antibody to antigen that has been immobilized on a solid support. This assay format is suitable for antigens of all sizes (haptens and large proteins) and has found extensive utility in research and diagnostic medicine. It, nonetheless, suffers of several shortcomings. One in particular is the sensitivity limit imposed on the assay because of its competitive configuration; the amount of labeled antibody used in the assay cannot exist in large excess over the molar amount of antigen to be measured, otherwise significant modulation cannot be achieved.

The two-site sandwich assay was first described in rudimentary form in 1970 by Habermann, and then in a fully implemented form in two reports by Maiolini & Masseyeff and Ling in 1975. In the assay described by Maiolini and Masseyeff, the antigen measured (alpha-fetoprotein) was first captured by exposing the sample to an excess of specific antibody which had been insolubilized on a solid support. Following an incubation period and a brief wash step, a molar excess (with regard to the antigen) of enzyme-labeled antibody was added to the antigen that had been captured by the solid support. Following another wash step to remove excess enzyme-labeled antibody, the amount of enzyme-labeled antibody, that was in turn captured on the support through the antigen sandwich, was quantified.

This particular assay format has evolved through many variations to include different sequences of reagent addition as well as the implementation of two (Porstmann et al., 1983) or even three (Nomura et al., 1983) monoclonal antibodies that are known to recognize different epitopic sites on the antigen, allowing for simultaneous addition of reagents, thus minimizing the number of wash steps. All in all, the two-site sandwich immunometric assay has proven to be the most sensitive and robust immunoassay system available — capable of quantifying as little as one attomole [10^{-18} mole] of antigen (Ishikawa et al., 1982). One severe disadvantage of this assay format, however, is the fact that the assay requires that the antigen must be large enough to support two simultaneous epitopic binding sites — a requirement that precludes many important clinical analytes. The non-competitive, excess-reagent configuration, on the other hand, is highly desirable for high sensitivity and versatility.

PRINCIPLE OF AFFINITY COLUMN MEDIATED IMMUNOMETRIC ASSAY

A new immunometric assay configuration described recently (Freytag et al., 1984), called Affinity Column Mediated Immunometric Assay [ACMIA], combines the advantages of the non-competitive nature of the sandwich assay and the requirement of only one binding site of the one-site immunometric assay, to form a very sensitive and versatile immunoassay system. In this assay (also described in Figure 1), excess labeled (enzyme) monovalent antibody is incubated with a sample containing the antigen that is to be measured. Even when the antigen concentration is extremely low, addition of a large excess of monovalent antibody forces the binding reaction to completion such that all analyte is bound rapidly. The excess labeled monovalent antibody, which does not acquire an antigen in

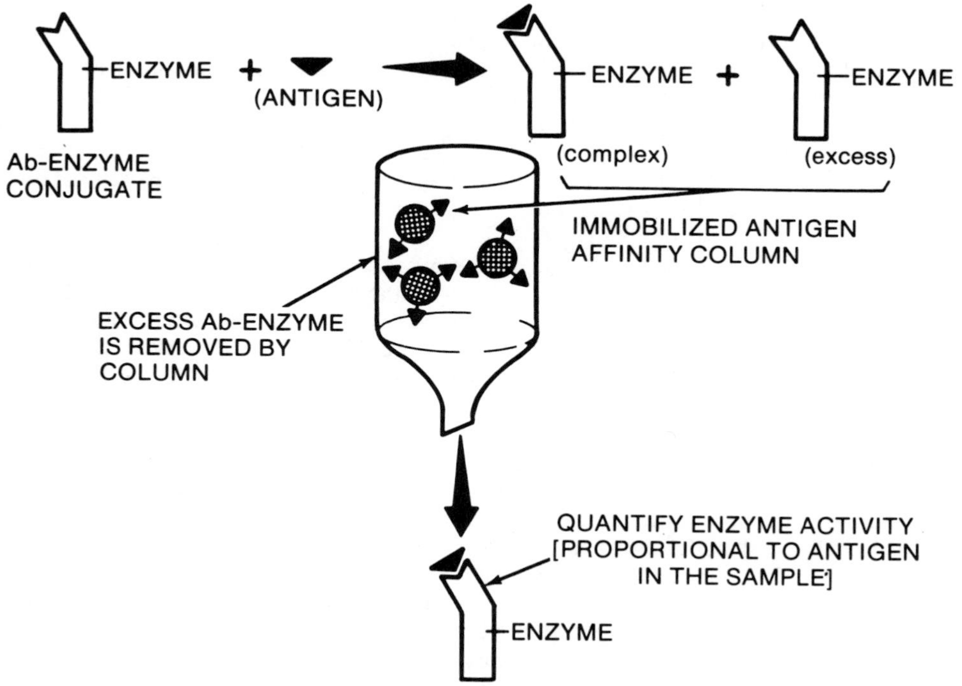

Fig. 1. Schematic description of assay mechanism of affinity-column-mediated immunoenzymometric assay (ACMIA).

its binding site is removed by passing the mixture through a porous affinity column containing immobilized antigen (or its analog) present in a large molar excess. Only the labeled monovalent antibody that possesses an antigen in its binding site elutes from the column in the unbound fraction. The label (enzyme activity) in this fraction is then quantified and is directly proportional to the original concentration of antigen in the sample. This assay, therefore, is limited in sensitivity only by its detection system and not the antibody affinity constant for the antigen. The use of an enzyme as a label, therefore, is highly advantageous in that extended time and can be utilized to achieve adequate signal when necessary. Furthermore, the ACMIA principle can be used for both small and large antigens alike.

PROCEDURES

It should be emphasized that the use of native divalent antibodies or enzymes conjugated with multiple copies of monovalent antibodies will limit the assay sensitivity to a competitive mode, because an antibody containing an antigen in one of its two binding sites cannot be differentiated from an antibody with no antigens — both will be retained by the affinity column. For many analytes, however, the sensitivity requirement is such that non-monomeric/monovalent antibody-enzyme conjugates will suffice, particularly in the case where high affinity antibodies are available. An example of this is the development of a sensitive (lower limit of detection is 0.2 nanomolar) and highly precise (C.V.'s of less than 3%) automated assay for digoxin using the ACMIA approach (Leflar et al., 1984). For this assay a non-monomeric $F(ab')_2$-β-galactosidase conjugate was employed.

One the other hand, when the sensitivity requirement exceeds that which is achievable with a given antibody due to the limitations of the antibody equilibrium binding constant [which rarely exceeds 10^{10} L/mol, and more often is between 10^9-10^7 L/mol] (Pecht, 1982), a monomeric/monovalent antibody-enzyme conjugate is required. In these instances, routine conjugation chemistries such as the one-step or two-step glutaraldehyde procedures described by Avrameas and others (Avrameas, 1969; Avrameas & Ternynck, 1971) are not effective since large aggregated complexes are created. The more site-specific procedures developed by Ishikawa and others (Ishikawa et al., 1983; Ishikawa et al., 1981) which utilize homo- and hetero-bifunctional crosslinking reagents better approximate the need, yet are still inadequate because of the inability to purify monomeric conjugates away from large quantities of uncoupled enzyme and antibody. Most recently, however, Ishikawa and co-workers (Imagawa et al., 1984; Hashida et al., 1984) described two novel but simple procedures for preparing and purifying monomeric Fab'-β-galactosidase or Fab'-horseradish peroxidase conjugates. In these procedures, IgG (generally affinity purified) is first converted to the $F(ab')_2$-fragments by limited digestion with pepsin and then labeled with a simple haptenic group like 2,4-dinitrophenol or fluorescein. Only a limited number, 3-4 of the haptenic groups is required. The hapten-labeled $F(ab')_2$-fragments are then split into the monovalent Fab'-SH fragments by selective reductive cleavage on the inter-heavy chain disulfides. Subsequent coupling of the Fab'-SH fragments to β-galactosidase is achieved by a two-step procedure using the homobifunctional thiol crosslinking reagent o-phenylenedimaleimide. In the case of horseradish peroxidase, the heterobifunctional crosslinking reagent, N-succinimidyl 4-(N-maleimidomethyl)cyclohexane-1-carboxylate, was used. The coupling step is performed under conditions where the enzyme is present in large molar excess to ensure only one Fab' molecule is crosslinked to a single enzyme molecule. The final monomeric Fab'-enzyme conjugate is then purified away from the excess enzyme by affinity chromatography using anti-dinitrophenol of anti-fluorescein columns. Elution of the conjugate from the antibody columns can be achieved under non-denaturing conditions using free hapten (2,4-dinitrophenol or fluorescein) which do not interfere with the immunochemical reactivity of the Fab'-enzyme. The use of monomeric/monovalent conjugates (Fab'-β-galactosidase) prepared in this manner in an

ACMIA assay set-up for human IgG permitted the quantification of 0.1 fmol of IgG.

In principle, virtually any enzyme can be used in the preparation of an appropriate antibody-enzyme conjugate for use in an ACMIA assay; β-galactosidase, however, offers many distinct advantages. For one, it possesses a very high enzymic turnover number (>450,000 min^{-1}) for a simple colorimetric substrate system (o-nitrophenyl-β-D-galactopyranoside; molar extinction coefficient of o-nitrophenol is ca. 3100 L/mol/cm). It is readily available in highly pure form from several commercial vendors (we found Boehringer Mannheim's product to be superior in quality). β-Galactosidase possesses multiple (12-20) free sulfhydryl groups which are available for site-specific coupling without concomitant inactivation of the enzyme. Finally, since the enzyme activity of the unbound fraction is quantified in the ACMIA format, enzyme activities and/or enzyme inhibitors endogeneous to the sample specimen have the potential of influencing the results. For this reason, when human serum is the test specimen, the popular enzymes, alkaline phosphatase and horseradish peroxidase, are much less desirable than is β-galactosidase which does not seem to be expressed or influenced by human serum.

Although, under the strictest of definitions, ACMIA is a heterogeneous immunoassay, the separation step is so facile that the assay is perceived to be as simple and rapid to perform as most homogeneous assays available today — particularly with the aid of instrumentation. After combination of the sample with the enzyme-labeled monovalent antibody, requiring only a simple pipetting step, the time required for the incubation, being a simple reversible bimolecular reaction is purely diffusion limited and therefore, is very much a function of the concentration of antigen and antibody present. Table 1 presents the kinetics, derived by calculations, of antibody-antigen association as a function of their respective concentrations. The only assumption made in these calculations is a bimolecular forward rate constant of $1X10^7$ (L/mol/s), a number which is consistent with that which has been measured for most small molecule-antibody interactions (Pecht, 1982). From this table it can be understood that even when the analyte concentration falls to picomolar concentrations, if a large excess of antibody-enzyme conjugate is used (nanomolar), sufficient binding occurs such that an assay can still be completed in less than 5-10 minutes. Figure 2 substantiates this theoretical understanding with actual data measured for the binding of digoxin by a monomeric Fab'-β-galactosidase conjugate. The concentration of digoxin in this experiment was 0.2 nanomolar and the concentration of Fab'-β-galactosidase was approximately 10 nanomolar. As can be seen, virtually complete binding has occurred before the first time point (5 minutes) was taken.

TABLE 1
KINETICS OF
ANTIGEN-ANTIBODY ASSOCIATION

CONCN, mol/L		
ANTIBODY	ANTIGEN	$t_{90\%}$, s[a]
1×10^{-8}	1×10^{-9}	25
1×10^{-9}	1×10^{-10}	245
1×10^{-8}	1×10^{-10}	23
1×10^{-10}	1×10^{-11}	2454
1×10^{-9}	1×10^{-11}	232
1×10^{-8}	1×10^{-11}	23
1×10^{-10}	1×10^{-12}	2311
1×10^{-9}	1×10^{-12}	230
1×10^{-8}	1×10^{-12}	22

[a] *Assumes a bimolecular forward rate constant of* $1 \times 10^7\ M^{-1}s^{-1}$

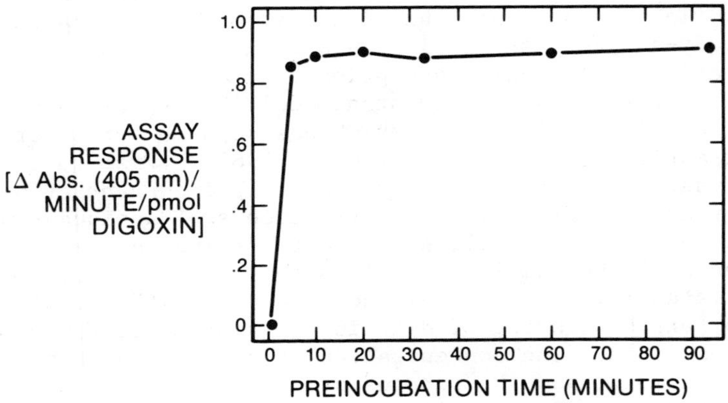

Fig. 2. Kinetics of binding of digoxin (0.2 nanomolar) to Fab'-β-galactosidase conjugate (∿ 10 nanomolar).

Once the antibody-enzyme conjugate has bound the free antigen in the test sample, the separation step serves to only remove the excess antibody-enzyme conjugate which has no antigen in its binding site. The time required to achieve this, again, is very much dependent on the relative concentrations of the respective binding partners. If the immobilized antigen is presented in vast excess, this separation is virtually instantaneous. It is important that time taken to achieve the separation step be an interval that is less than the time in which significant dissociation of the antigen-antibody complex will have occurred. This, of course, is a direct function of the antibody affinity constant; the half-time of dissociation for an antibody-antigen complex with an equilibrium binding constant of $1X10^{-10}$ L/mol is about 10 minutes.

There are numerous chemistry procedures reported in the literature for immobilization of antigens of all kinds to insoluble supports of all kinds (Yarmush & Colton, 1983; Weetall, 1973). The important parameters of the affinity column component in the ACMIA assay are no different than that of any other solid support employed in solid-phase immunoassays. Solid matrices with high surface area, such as that provided by most column chromatography resins (agaroses, dextrans, porous and nonporous silicas, polyacrylamides, etc.) are generally all acceptable. We have found success with: Sephadex G-10, G-25, G-50; Sepharose 4B, 6B, Bio-Gel biobeads; Affi-Gel-10; controlled pore glass; Eupergit®; Zipax®; finely ground nylon 66; and many others. Since the unbound liquid fraction is analyzed in the ACMIA assay, it is not as important as it is for the two-site sandwich and ELISA assays, that the solid support not exhibit non-specific adsorption, although this is a desired property. The chemistry of attachment requires nothing unusual: cyanogen bromide activation or periodate oxidation/borohydride reduction of Schiff bases of the polysaccharide supports; carbodiimide coupling to amino or carboxy terminating supports; N-hydroxy-succinimidyl activation, etc., are all suitable chemistries. For tests that will be manufactured for commercialization, however, great care must be taken to ensure extreme stability in the coupling chemistry. Our best success, in which we have demonstrated no measurable leakage of the antigen from the resin throughout one year storage at 4°C, has been achieved with the periodate oxidation/borohydride reduction of Sephadex G-10. The use of protein or synthetic organic spacers between the antigen and the solid support sometimes proves advantageous — this generally must be determined on a case by case study.

RESULTS AND DISCUSSION

The column configuration or geometry can also assume almost any form. Imagawa et al. (1984) describe the use of 0.02 mL of wet resin (human IgG-Sepharose 4B) packed into a Gilson C-200 Blue pipet tip, whereas Freytag et al. (1984) describe the use of a plastic tube (0.5 x 8 cm) fitted with rubber stoppers at both ends to contain Sephadex G-10/human serum albumin/ouabain resin. Actually, a simple batch procedure will suffice if accurate timing of the interaction of the sample/antibody-enzyme conjugate mixture with the affinity support can be achieved. We have experienced, however, the best performance of this assay when a good control of flow rate of the sample mixture is maintained in a chromatographic format.

The data presented in Figure 3 indicates the effect of flow rate on the assay performance. This data was collected for an assay for digoxin using a monomeric Fab'-β-galactosidase conjugate and Sephadex G-10/human serum albumin/ouabain affinity resin packed into a column having dimensions of 0.5 x 2 cm. [The use of ouabain instead of digoxin on the resin will be addressed below]. The flow rate of sample elution through the column was controlled by a syringe pump possessing variable speed control. Within certain boundaries, the flow rate seems to influence only the "background" enzyme activity and not the assay slope sensitivity. [The background enzyme activity is defined as the enzyme activity observed at zero antigen concentration. In theory, this value should be zero, but because of the presence of small amounts of either free enzyme or non-immunoreactive Fab'-β-galactosidase, this background activity is frequently observed to be some finite portion of the total enzyme activity. It is recommended that the background activity be kept to less than 10% of the total enzyme activity.]

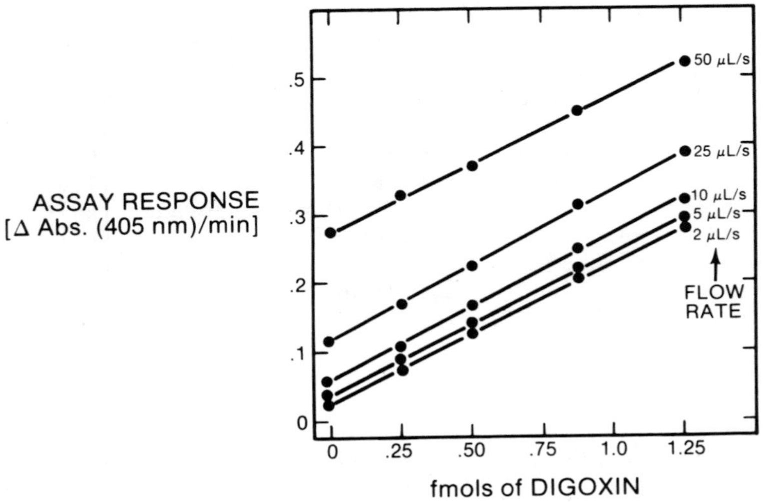

Fig. 3. Elution flow rate dependency of assay performance.

The use of more antigen affinity resin to effect the separation in the assay has no influence on the background activity observed, demonstrating that the resin is in sufficient excess. In fact, we have shown that one of these small columns can be used with over 60 consecutive test samples to yield identical results without any washing or regeneration inbetween. Higher background activities at higher flow rates can be partially offset by larger column capacities.

The affinity resin used in the column-mediated immunometric assay can be prepared by immobilizing on a solid support the actual antigen, a derivative of the antigen, or an analog. One might expect that the actual antigen would provide the best solid phase for performing the separation step. This, however, was not the result observed for the digoxin system, a system that we have extensively characterized and utilized as a model antigen for several years now. Ouabain is a structural analog of digoxin that possesses slightly higher water solubility and a four-log lower affinity for the digoxin antibody. Ouabain and digoxin were covalently coupled to Sephadex G-10 (Pharmacia Fine Chemicals) and Affi-Gel-10 (Bio Rad Laboratories) via the organic spacer triethylenetetraamine or the protein spacer, human serum albumin. The coupling chemistries were first optimized for maximum capacity and stability. The resins were then evaluated for their performance with regard to slope sensitivity, background, and precision, in a digoxin assay using several antibody-enzyme conjugates, the monovalent Fab'-β-galactosidase conjugate in particular. In all cases, the ouabain resin outperformed its counterpart.

We are not sure why the ouabain resins consistently outperformed the digoxin resins in almost all aspects. From theoretical considerations of column-mediated immunometric assays, one would predict that an affinity column prepared with the same molecule as the antigen (or hapten) to be determined would perform better than a column prepared with an analog, particularly an analog for which the binding constant for the antibody is >10,000-fold less than that of the antigen. This prediction is based on the assumption that once the labeled antibody had bound to the free antigen from the test sample, the column would serve only to remove the unbound fraction. Given a flow rate of 34 μL/s used in these experiments, the residence time of the antibody-antigen complex on the column is only 17 seconds, which is relatively short compared with an antigen-antibody first-order dissociation rate constant of ca. 2×10^{-3} s^{-1}. [The affinity constant of this antibody for digoxin was measured by equilibrium binding analysis to be 5×10^{9} L/mol.] We are currently investigating whether the same observation holds true for other antigens. If so, this may prove to be extremely beneficial for the development of diagnostic tests for antigens that are rare or difficult to purify; particularly the rare protein antigens like polypeptide hormones and cancer/tumor

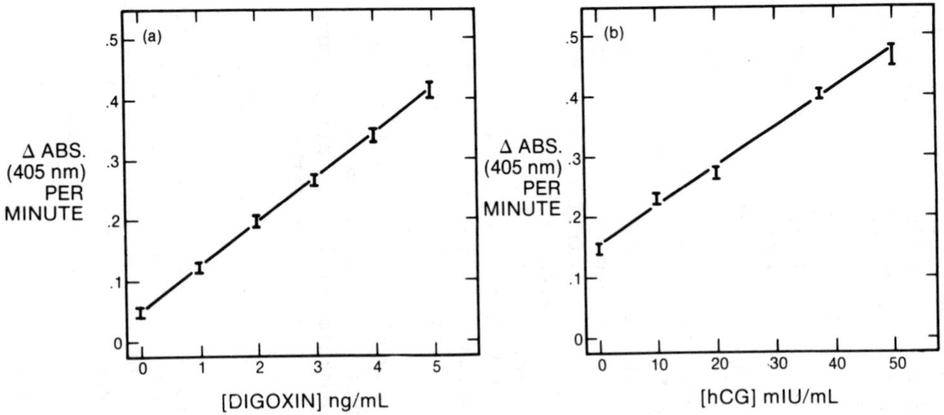

Fig. 4a & b. Dose-response curves for digoxin and human chorionic gonadotropin.

markers. In such cases, instead of immobilizing the purified (or partially purified) antigen, a synthetic peptide fragment, prepared by automated peptide synthesizers or by molecular cloning techniques, could provide inexpensive 'analogs' of the antigen for the preparation of the affinity column.

We have developed affinity-column mediated immunometric assays for several antigens (digoxin, vitamin B_{12}, human chorionic gonadotropin [hCG], IgG, ferritin, and several others) which vary in molecular weight from 700 to >500,000 daltons. These analytes vary in concentration by several log units as well. Assay curves for digoxin and hCG are shown in Figures 4a & b.

For digoxin, sample (20 μL) was incubated with monomeric Fab'-β-galactosidase conjugate (20 μL) for ten minutes. This mixture was eluted through a 320 μL column of ouabain-HSA-Sephadex at a flow rate of 5 μL/s. An aliquot of the effluent was quantified for β-galactosidase activity using o-nitrophenyl-galactopyranoside as a substrate. Enzyme activity was expressed as a change in absorbance at 405 nm per min. For hCG, a 25 μL sample was incubated with a 25 μL aliquot of Fab'-β-galactosidase for 15 minutes prior to being eluted through a 320 μL affinity column hCG immobilized onto controlled pore glass (CDI-CPG) Glycophase from Pierce Chemical Co.). Again, as in the digoxin assay, the β-galactosidase activity in the unbound fraction was quantified in a kinetic assay.

In both assays, the dose response curves are linear and sensitive to antigen levels that are clinically useful. Furthermore, the time taken to complete these assays is substantially less than any other assay configuration that is reported in the literature to date.

In conclusion, the affinity-column mediated immunoenzymometric assay represents another approach to the development of sensitive immunological tests for low level analytes. The assay sensitivity limit is not governed by the antibody affinity constant and has been shown to be capable of quantifying as little as 0.1 fmol of analyte. The ACMIA assay is also extremely fast requiring only a brief incubation step followed by a single facile separation step, both of which are readily adapted to automation. Antigens of all sizes, mono- or poly-epitopic in nature, are all applicable to the ACMIA approach. The synthesis of two key reagents is required for ultimate performance; a monomeric/monovalent antibody-enzyme conjugate and a stable antigen column. The synthesis of the former has been simplified by a procedure we reported very recently.

REFERENCES

Avrameas, S. (1969). Coupling of enzymes to proteins with glutaraldehyde. Use of the conjugate for the detection of antigens and antibodies. Immunochemistry, 6: 43-52.

Avrameas, S. and T. Ternynck (1971). Peroxidase labeled antibody and Fab conjugates with enhanced intracellular penetration. Immunochemistry, 8: 1175-1179.

Engvall, E. and P. Perlmann (1971). Enzyme-linked immunosorbent assay (ELISA). Quantitative assay of immunoglobulin G. Immunochemistry, 8: 871-874.

Freytag, J.W., J.C. Dickinson, and S.Y. Tseng (1984). A high sensitivity affinity-column-mediated immunometric assay, as exemplified by digoxin. Clinical Chemistry, 30: 417-420.

Habermann, V.E. (1970). Ein neus prinzip zur quantitativen bestimmung hochmolekularer antigene. Z. klin. Chem. u. Klin. Biochem., 8: 51-55.

Hashida, S., M. Imagawa, E. Ishikawa, and J.W. Freytag (1985). A simple method for the conjugation of affinity-purified Fab' to horseradish peroxidase and β-galactosidase. Molecular Immunology (in press).

Holbeck, S.L. and G.T. Nepon (1983). Enhanced detection of immunoglobulin binding by a modified ELISA. J. Immunol. Methods, 50: 47-52.

Imagawa, M., S. Hashida, E. Ishikawa, and J.W. Freytag (1984). Preparation of a monomeric 2,4-dinitrophenyl Fab'-β-galactosidase conjugate for immunoenzymometric assays. J. Biochem., 96: 1727-1735.

Ishikawa, E., T. Kawai, and K. Miyai (1981). Enzyme Immunoassays. Igaku-Shoin, Tokyo. 280 pp.

Ishikawa, E., M. Imagawa, S. Yoshitake, Y. Niitsu, I. Urushizaki, M. Inada, H. Imura, R. Kanazawa, S. Tachibana, N. Nakazawa, and H. Ogawa (1982). Major factors limiting sensitivity of sandwich enzyme immunoassay for ferritin, immunoglobulin E, and thyroid-stimulating hormone. Ann. Clin. Biochem., 19: 379-384.

Ishikawa, E., M. Imagawa, S. Hashida, S. Yoshitake, Y. Hamaguchi, and T. Ueno (1983). Enzyme labeling of antibodies and their fragments for enzyme immunoassay immunohistochemical staining. J. Immunoassay, 4: 109-327.

Leflar, C.C., J.W. Freytag, L.M. Powell, J.C. Strahan, J.J. Wadsley, C.A. Tyler, and W.K. Miller (1984). An automated affinity-column-mediated enzyme-linked immunometric assay for digoxin on the Du Pont aca® discrete clinical analyzer. Clinical Chemistry, 30: 1809-1811.

Ling, M.C. (Feb. 18, 1974). United States Patent No. 3,867,517. Direct Radioimmunoassay for antigens and their antibodies.

Maiolini, R., and R. Masseyeff (1975). A sandwich method of enzymoimmunoassay I. Application to rat and human alpha-fetoprotein. J. Immunol. Methods, 58: 293-300.

Nomura, M., M. Imai, and K. Tasahashi (1983). Three-site radioimmunoassay with monoclonals for a sensitive determination of human alpha-fetoprotein. J. Immunol. Methods, 58: 293-300.

Pecht, I. (1982) Dynamic Aspects of Antibody Function. In The Antigen (M. Sela, Eds.). Academic Press, New York, NY. 33-42.

Porstmann, B., T. Porstmann, R. Seifert, H. Meisel, and E. Nugel (1982). Comparison of direct and indirect two-site binding enzyme immunoassays. Clinica Chimica Acta, 122: 1-9.

Van Weeman, B.K. and A.H.W.N. Schuurs (1971). Immunoassays using antigen-enzyme conjugates. FEBS Letters, 15: 323-236.

Weetall, H.H. (1973). Affinity Chromatography. Sep. and Purif. Meth., 2: 199-229.

Yarmush, M.L. and C.K. Colton (1983). Affinity Chromatography. In Comprehensive Biotechnology. Pergamon Press.

Yolken, R.H., F.J. Leister, L.S. Whitcomb, and M. Santosham (1983). Enzyme immunoassays for the detection of bacterial antigens utilizing biotin-labeled antibody and peroxidase biotin-avidin complex. J. Immunol. Methods, 56: 319-327.

STERIC HINDRANCE ENZYME IMMUNOASSAY (SHEIA)

Albert Castro[1] and Nobuo Monji[2]

Department of Pathology[1], University of Miami School of Medicine, Miami, FL 33101; Genetic Systems Corporation[2], 3005 First Avenue, Seattle, WA 98121

INTRODUCTION

We have developed a new enzyme immunoassay for the quantitation of small antigens. The assay is called SHEIA (Steric Hindrance Enzyme Immunoassay) since this assay involves the interference of complex formation between the antigen (or hapten) coupled enzyme and the enzyme inhibitor attached to agarose by antibody to the antigen (or hapten), probably due to steric hindrance (Monji and Castro, 1979). The assay principle is shown in Figure 1.

In EIA as well as in RIA, separation of bound and unbound ligand is a critical step (Ratcliff, 1976; Odell et al., 1975; Collins et

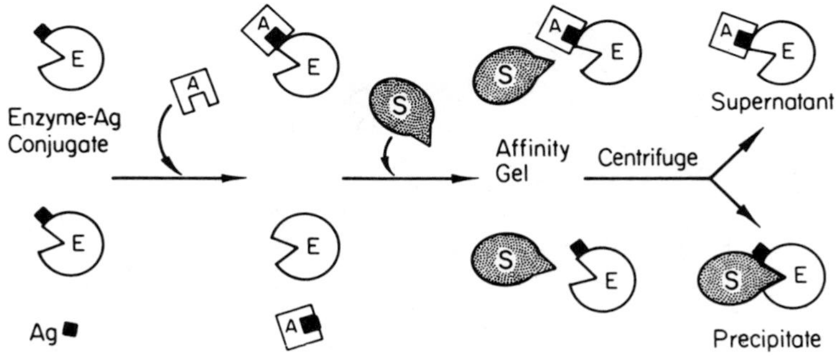

Fig. 1. Assay outline for SHEIA (Steric Hindrance Enzyme Immunoassay).

al, 1975). The most widely used separation method in RIA for small antigens is an adsorption technique. This system precipitates antibody unbound tracer using adsorbent materials, such as dextran coated charcoal, talc, resins, and has advantages in its simplicity and reproducibility. EIA so far developed rely mainly on either second antibody precipitation or solid phase adsorption, except homogeneous enzyme immunoassay (Rubenstein et al., 1972). The double antibody method which involves precipitation of antibody bound enzyme-antigen conjugate has been so far the most sensitive, reliable and reproducible method. This assay system, however, often requires a long incubation time as well as frequent washings of the precipitates and involves complex reaction kinetics. To avoid such disadvantages, we have developed a novel separation technique using enzyme affinity gels. This system precipitates only unbound enzyme-antigen (or hapten) conjugates. The SHEIA system developed for choriomammotropin (CMT), and thyroxine (T_4) are described in this chapter.

METHODS

SHEIA for CMT

Antiserum: CMT antibody was produced in rabbits by the method of Castro et al., (1976). The affinity constant for this antibody was $3.1 \times 10^9\ M^{-1}$.

Preparation of CMT-ß-galactosidase (GAL):

To couple GAL to CMT, m-maleimidobenzoyl-N-hydroxysuccinimide ester (MBS, Pierce Chemical Co.) is used as a bifunctional reagent (Fig. 2). MBS is first conjugated to CMT as follows: Mix 800 µg of MBS in 0.16 ml of THF with 15.4 mg of CMT in 2 ml of sodium phosphate buffer (50 mM, pH 7.0). Incubate the mixture at room temperature for 30 min and stop the reaction by adding 1 ml of citrate

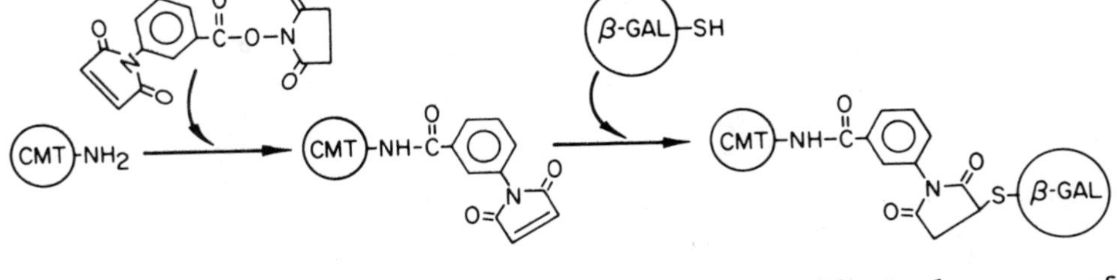

Fig. 2. Conjugation of choriomammatropin to sulfhydryl groups of ß-galactosidase through m-maleimidobenzoyl linkage.

buffer (1.0 M, pH 5.0). Collect the resulting precipitate by centrifugation at 2,000 x g for 15 min, and wash twice with 10 ml of citrate buffer (10 mM, pH 5.3). In the second step, the derivatized CMT is conjugated to GAL through thioether linkage: Redissolve the precipitate of CMT derivative in 1.0 ml of the pH 7.0 sodium phosphate buffer and add that solution to 5 mg of GAL dissolved in 1.0 ml of the same buffer. Mix the solution thoroughly and incubate at room temperature for 2 h. To remove the unreacted CMT, chromatograph the mixture on Sephadex G-75 column (1 x 40 cm) using the pH 7.0 sodium phosphate buffer as an eluate.

To test the relative amount of GAL that is conjugated to CMT, mix the conjugate with an excess of rabbit anti-CMT antibody. Then, add goat anti-rabbit γ-globulin in excess. After washing off the unbound enzyme conjugate, examine the enzyme activity in the precipitate. It should contain about 80% of the enzyme activity initially present. A control with normal rabbit serum in place of anti-CMT antibody gives no detectable enzyme activity in the precipitate.

Preparation of the affinity purified CMT-GAL conjugate

CMT-GAL conjugate isolated in the previous section is further purified by affinity chromatography as follows: To AG-GA (Agarose-6-aminocaproyl-ß-D-galactosylamine, pharmacia P-L-Biochemicals) column (1 x 5 cm) equilibrated with 0.05 M phosphate buffer, pH 7.0, apply CMT-GAL conjugate dissolved in the same buffer. Wash the column extensively with the same phosphate buffer. When no more enzyme activity is detected in the eluate, elute the column with borate buffer, pH 10.0, 0.1 M. Collect the fractions containing the major enzyme activity and then dialyze for 24 h in one liter of sodium phosphate buffer. After dialysis, add 0.05% sodium azide and 2 µl/ml of 1M $MgCl_2$ and store at 4°C. Measure the enzyme activity by the method of Dray et al., (1975). To determine the relative amount of enzyme that is conjugated to CMT, mix the conjugate with an excess of rabbit anti-CMT antibody. Add anti-rabbit γ-globulin in excess. The resulting precipitate contains more than 90% of the enzyme activity initially present. A control with normal rabbit serum in place of the CMT antibody gives no detectable enzyme activity in the precipitate.

Assay Procedure

The procedure is outlined in Figure 1. To a glass tube (16 x 10 mm), add 150 µl of phosphate buffer, followed by addition of 5 µl of the standard containing either 0, 2, 4, 6 or 10 mg/l of CMT in 5% BSA solution. Add 100 µl each of CMT-GAL conjugate of appropriate dilution and 0.5% BSA in phosphate buffer successively. After mixing, add 100 µl of 1:1,000 dilution of rabbit CMT antiserum and incubate for one hour at 4°C. Add 100µl of phosphate buffer containing 5 µl of the swollen AG-GA gel and place the tube in a shaker and oscilate gently (100 rpm) for 60 min. Remove the tube from the shaker and centrifuge for 10 min at 2,000 x g. Take 250 µl of the supernate and assay for enzyme activity.

SHEIA for T_4

Antiserum and T_4-GAL conjugate were prepared by the methods described on page 299 of this book. AG-GA affinity purified T_4-GAL was prepared by the same procedure as that described for CMT-GAL conjugate in this text. The assay procedure was essentially same as that for CMT SHEIA.

RESULTS

To examine the extent of conjugation, CMT-GAL conjugation optimization and assay performance, the following studies were carried out.

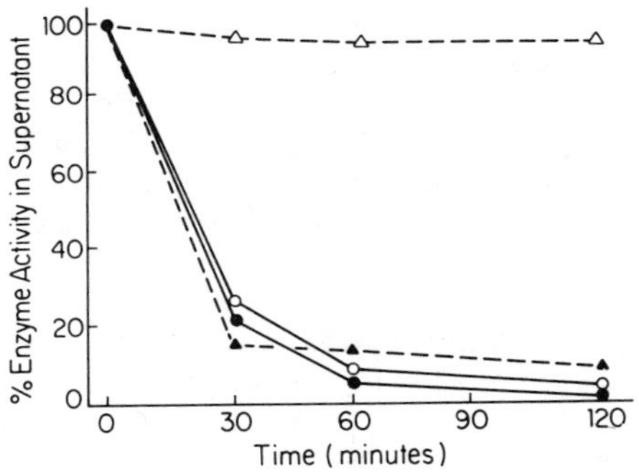

Fig. 3. Effect of incubation time on CMT-ENZ and AG-GA complex formation. CMT was conjugated to ß-galactosidase according to the method described in the text. CMT:ENZ ratio in each conjugate was not directly determined; CMT:ENZ ratios shown here are the expected value calculated from the results obtained by Kitagawa and Aikawa (1976). CMT-ENZ conjugates used in here were purified by affinity chromatography prior to the experiment. ●, CMT-ENZ (2:1) alone; O, CMT-ENZ (2:1) incubated one hour with CMT antibody prior to the separation step; ▲, CMT-ENZ (8:1) alone; △ , CMT-ENZ (8:1) incubated one hour with CMT antibody prior to the separation step.

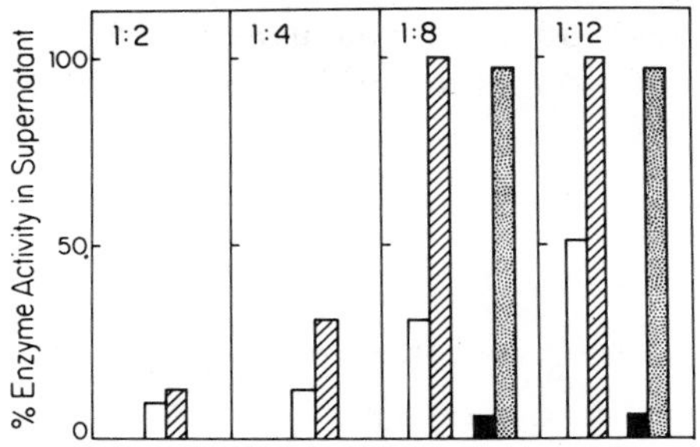

Fig. 4. Effect of CMT:ENZ molar ratio on CMT-ENZ and AG-GA complex formation. ▭ , CMT-ENZ conjugate before purification by affinity chromatography; ■ , CMT-ENZ conjugate purified by affinity chromatography; ▨ , CMT-ENZ conjugate before purification by affinity chromatography was incubated one hour with CMT antibody prior to the separation step; ▒ , CMT-ENZ conjugate purified by affinity chromatography was incubated one hour with CMT antibody prior to the separation step.

In order to precipitate the enzyme from an assay solution, AG-GA was added to the solution and incubated in a shaker (100 rpm) for 30, 60, 90 and 120 minutes. The solid phase was then precipitated by centrifugation and enzyme activity in the supernate was examined. The results showed that only 5% of the enzyme remained in the supernate, suggesting about 95% of the added enzyme being bound to approximately 5 µl (wet volume) of AG-GA with 60 min or more of incubation (Fig. 3).

When CMT and GAL conjugation reaction was carried out at various molar ratios and when no purification of the conjugate by affinity column was done, enzyme activity remained in the supernate was found to depend on the molar ratio employed for CMT-GAL conjugation. With the expected CMT and GAL ratio (this term is described in the legend of Figure 3) in the conjugate of 8:1 and 12:1, about 30% and 50% respectively, of the total enzyme activity remained in the supernate, while the reduction of the expected molar ratio to 4:1 or less resulted in about 10% of the enzyme activity remaining in the supernate (Fig. 4). This reduction of the CMT-GAL binding to AG-GA possibly suggested a steric hindrance by the over conjugated CMT. When these conjuates were purified in AG-GA affinity column, however, the enzyme activity remaining in the supernate was reduced to about 5%, suggesting the removal of over-conjugated CMT-GAL (Fig. 4). Binding of these purified CMT-GAL with anti-CMT antibody resulted in com-

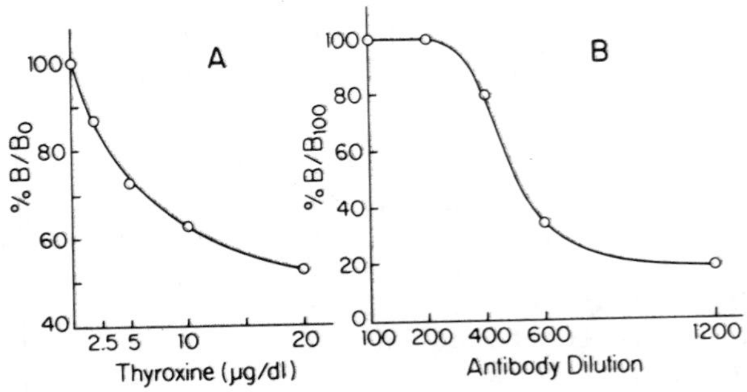

Fig. 5. Standard curve for T_4 (A) and T_4 antibody titration curve (B).

plete inhibition of complex formation between CMT-GAL and AG-GA. Such complete inhibition of CMT-GAL and AG-GA binding was not seen by the CMT-GAL conjugates which have the expected molar ratio of 4:1 or less (Fig. 4). These results suggested the requirement of optimal molar ratio between CMT and GAL in the CMT-GAL conjugate in order to conduct an effective SHEIA. Both reduction of the background value by purification of CMT-GAL cnojugate in AG-GA column and inhibition of CMT-GAL and AG-GA binding by anti-CMT antibody permitted the successful performance of CMT SHEIA. The assay showed a maximum sensitivity of 4 ng/tube. In a typical standard curve, the assay system was adjusted to cover the range of 0-10 mg/l with maximum sensitivity between 0-4 mg/l using 5 µl sample size.

As for the SHEIA for T_4, assay optimization as well as procedure were essentially same as those for CMT SHEIA. Antibody titration curve for obtaining optimal antibody concentration and the standard curve for T_4 suggested that SHEIA principle could be applied for small haptens (Fig. 5).

DISCUSSION

One of the best inhibitors for ß-galactosidase isolated from <u>E. coli</u> is ß-D-galactosylamine (k_i = 0.225 mM) (Lai and Axelrod, 1973). Agarose attached 6-aminocaproyl derivative of galactosylamine has been used widely in affinity chromatography for the purification of ß-galactosidase from various sources (Hapaz et al., 1974). Steers and Cuatrecasas (1974) have shown that ß-galactosidase would not bind to an inhibitor unless a long arm (at least 6 carbon chain) was used to increase the distance between the ß-D-galactosylamine and the agarose back bone, indicating rather strict physicochemical requirement for their bindings. We utilized the hindrance of this rigid steric requirement by antibody to develop the enzyme immunoassay. Although we employed the batch method to separate antibody bound and unbound CMT-GAL, it may be quite possible to apply this SHEIA method using affinity column in an automated system on a continuous flow system. With respect to the use of other affinity gels, it was found more recently that agarose attached 6-aminocaproyl or 4-aminobutyl derivative of p-aminophenyl ß-D-galactopyranoside was superior affinity gel to precipitate the ß-galactosidase enzyme (requiring only 15 min shaking time with rspect to 60 min; unpublished data). We are also investigating the possible use of coenzyme attached affinity gel for SHEIA.

REFERENCES

Castro, A., I. Priesto, C. Wunsch, G. Ertingshausen and H. Malkus (1976). Automated radioimmunoassay of choriomammotropin (human placeatal lactogen). Clin. Chem. 22: 1655-1658.

Collins, W. P., G. J. R. Barnard and J. F. Hennan (1975). Factors affecting the choice of separation techniques, in Steriod Immunoassay, Proceeding of the Fifth Tenivus Workshop (E. H. D. Cameron, S. G. Hillier and K. Griffiths, eds.). Alpha Omega, Wales, UK. p. 223-225.

Dray, F., J. E. Andrieu and F. Renaud (1975). Enzyme immunoassay of progesterone at the picogram level using ß-galactosidase as label. Biochem. Biophys. Acta. 403: 131-138.

Harpaz, N., H. M. Flowers and N. Sharon (1974). Purification of coffee bean ß-galactosidase by affinity chromatography. Biochem. Biophys. Acta. 341: 213-221.

Kitagawa, T. and T. Aikawa (1976). Enzyme coupled immunoassay for insulin using a novel coupling reagent. J. Biochem. (Toyko) 79:233-236.

Lai, H. Y. L. and B. Axelrod (1973). 1-Aminoglycosides, a new class of specific inhibitors of glycosides. Biochem. Biophys. Res. Commun. 54: 463-468.

Monji, N. and A. Castro (1979). Steric Hindrance Enzyme Immunoassay (SHEIA); a novel method in enzyme immunoassay. Res. Commun. Chem. Path. Pharm. 26: 187-196.

Odell, W. D., C. Silver and P. K. Grover (1975). Competitive protein binding assays methods of spearation of bound from free, in Steroid Immunoassay, Proceeding of the Fifth Tenovus Workshop (E. H. D. Cameron, S. G. Hillier and K. Griffiths, eds.). Alpha Omega, Wales, UK. p. 207-222.

Ratchliff, J. G. (1976). Radioimmunoassay and saturation analysis; separation techniques in saturation analysis. Brit. Med. Bull. 30: 32-37.

Rubenstein, K. E., R. S. Schneider and E. F. Ullman (1972). "Homogeneous" enzyme immunoassay. A new immunochemical technique. Biochem Biophys. Res. Comm. 47: 846-851.

Steers, E. and P. Cuatrecasas (1974). ß-Galactosidase. Methods in Enzymology 34: 350-358.

MALEIMIDE DERIVATIVE OF HAPTEN FOR ENZYME COUPLING IN EIA

Nobuo Monji[1] and Albert Castro[2]

Genetic Systems Corporation[1], 3005 First Avenue
Seattle, WA 98121; Department of Pathology[2]
University of Miami School of Medicine, Miami, FL 33101

INTRODUCTION

For labeling of antibodies or macromolecular antigens with various enzymes, glutaraldehyde has been most widely used. In the case of ß-galactosidase, maleimide derivatives have been used for coupling the enzyme to proteins, such as immunoglobulins and their fragments (Ishikawa, 1983). Use of these coupling agents showed no appreciable reduction of enzyme activity due to the involvement of sulfhydryl groups of the enzyme in the coupling process. Coupling of haptens to the enzyme, on the other hand, is often identical to the preparation of hapten-protein conjugates for immunization. Such coupling involves either amino or carboxyl groups of the enzyme, resulting in either low efficiency of coupling (Dray et al., 1975) or reduction of enzyme activity (Comoglio and Celada, 1976). In order to obivate such disadvantages, we utilized m-maleimidobenzoyl derivative of a hapten in order to couple it to sulfhydryl groups of the enzyme; a high efficiency of binding to the enzyme without appreciable reduction in enzyme activity were found. There are several advantages for using ß-galactosidase as an enzyme of choice: 1) It can be obtained in a highly purified form; 2) it has a high catalytic number; 3) conjugation to sulfhydryl groups of the enzyme does not reduce enzyme activity; 4) the enzyme and its conjugate are stable up to one year when stored at 4°C; and 5) human or animal ß-galactosidase activity is not detectable in the pH range used in immunoassay. Preparation and the use of these m-maleimidobenzoyl derivatives are described here for cortisol, thyroxine and digoxin. All of the assays described here showed sensitivities comparable to those of RIA and were highly specific to the compound to be analyzed (Monji et al., 1978; 1980 A; 1980 B).

METHODS

Thyroxine EIA

Synthesis of m-Maleimidobenzoyl Derivative of L-Thyroxine Methyl Ester. Meta-maleimidobenzoic acid (MBA) was first prepared as follows: Synthesize m-carboxymaleanilic acid from m-aminobenzoic acid and maleic anhydride by the method of Parola (1934), then cyclize with acetic anhydride to give MBA by the method of Searle (1948). Prepare L-thyroxine methyl ester next by the method of Ashley and Harington (1929). Dissolve MBA (200 μg) in 3 ml of thionyl chloride ($SOCl_2$) and reflux for 30 min. Evaporate excess $SOCl_2$ under diminished pressure. Keep pale yellow powder of m-maleimidobenzoyl chloride (MBC) overnight in a calcium desiccator. Then, dissolve the dried MBC in 10 ml of tetrahydrofuran (THF) and add dropwise to a stirred THF solution containing L-thyroxine methyl ester (400 μg) and a slurry of sodium carbonate (400 μg). Reflux the reaction mixture for 30 min. Examine the reaction by thin layer chromatography (TLC) using ethyl acetate as a eluting solvent. Filter the reaction mixture and remove the solvent under diminished pressure to yield crude pale-yellow product. Purify m-maleimidobenzoyl L-thyroxine methyl ester (MBTM) by silica gel column chromatography (1.5 cm x 30 cm) using chloroform as eluting solvent. The isolated white powder of MBTM gives a single spot on TLC, Rf = 0.56, using ethyl acetate as a solvent. Confirm the presence of maleimide group in the isolated product by IR and by its ability to react with cysteine using the method of Grassetti and Murray (1967). Melting points 137-141°C.

Conjugation of ß-Galactosidase (GAL) to MBTM. Add 50 μl of a solution of MBTM in THF (0.2 mg/ml, 10 nmoles) to 1.5 ml of 0.05 M phosphate buffer (pH 7.0) containing GAL (0.5 mg, 0.93 nmole, Boehringer Mannheim Biochemicals). Incubate the mixture for 2 h at room temperature. Following overnight dialysis in the same phosphate buffer, chromatograph the mixture on a Sephadex G-25 column (1.5 x 40 cm). Use the fractions of eluate containing the peak of enzyme activity for the T_4 assay. Assay GAL activity by the method of Dray et al., using o-nitrophenyl ß-D-galactoside as substrate.

Antiserum. Antiserum to T_4 was produced in rabbits by injection of T_4 conjugated to bovine serum albumin prepared by the method of Gharib et al., (1971). Cross reactivity with triiodothyronine was minimal. Goat anti-rabbit immunoglobulin antibody was obtained from Calbiochem, La Jolla, California.

EIA Procedure. Add to a glass tube (16 x 100 mm) 100 μl of a 100-fold dilution of the peak fraction of MBTM-enzyme conjugate eluted from Sephadex G-25 column followed by addition of 500 μl of the phosphate buffer (pH 7.0, 0.05M) containing 0.5 mM 8-anilinolnaphthalenesulfonic acid-sodium salt. Add 50 μl standards containing either 0, 2, 4, 8, 12 or 20 µg/100 ml of L-thyroxine in serum followed by 100 µl of a 400-fold dilution of rabbit anti-thyroxine antiserum. After incubation for 45 min at room temperature, add 100 µl of 20-fold dilution of normal rabbit serum followed by addition of 100 µl of goat anti-rabbit IgG. Incubate the tube for 15 min in ice water after vortex mixing and centrifuge for 10 min at 3,000 rpm. Wash the pellet two times with the phosphate buffer and resuspend in 0.5 ml phosphate buffer-BSA (pH 7.5, 0.05 M, 0.1% BSA). Assay the enzyme activity using o-nitrophenyl β-D-galactoside as a substrate. Incubation time for enzyme activity is 45 min. Measure the amount of o-nitrophenol produced at the end of incubation time by a spectrophotometer at 420 nm wavelength.

Cortisol EIA

Synthesis of Cortisol-21-m-Maleimidobenzoate (CT-MB). Add 100 mg of MBC dissolved in 2 ml THF dropwise to 148 mg cortisol dissolved in 2 ml THF. Reflux the mixture for 1 h, while monitoring the synthesized product by TLC using ethyl acetate as an eluting solvent (Rf=0.67). Isolate the product by silica gel column chromatography (1.5 x 30 cm) using chloroform-ethyl acetate (3:1) as a solvent system. The isolated product gives a single fluorescent spot under UV light on TLC after sulfuric acid spray (10% concentration of sulfuric acid in ethanol), followed by heating. Confirm the esterification at position 21 by its immunoreactivity to cortisol antiserum and the presence of a maleimide group on the compound by its ability to react with cysteine using the method of Grassetti and Murray. The structure of CT-MB is shown in Fig. 1. The melting points 134-136°C.

Preparation of m-Maleimidobenzoyl Derivative of Cortisol-21-Hemisuccinate (CHS) Through p-Phenylenediamine Linkage (CHS-MB). Dissolve CHS (100 mg) in 4 ml THF and cool to -10°C, followed by the addition of 20 µl tributylamine. Add 20 µl of isobutyl chloroformate and, after mixing, incubate for 30 min. Add the activated CHS dropwise to a solution of p-phenylendiamine (400 mg) in THF and incubate for 60 min. Monitor production of the p-phenylenediamine derivative of CHS on TLC (Rf = 0.22) using ethyl acetate-chloroform (90:10) as a solvent system. Isolate the product by silica gel column chromatography (1.5 x 40 cm) using the same solvent system. The isolated product gives a single spot on TLC: it is ninhydrin positive and fluoresces under UV light after sulfuric acid spray, followed by heating. Redissolve the isolated gummy product in THF and add dropwise to a solution of MBC(20 mg) in THF containing a slurry of sodium carbonate (200 mg). Reflux the solution for 30 min. Monitor the synthesis of the final product by TLC (Rf = 0.26) using ethyl acetate-chloroform (3:1) as a solvent system. Isolate the product by column chromatography using the same solvent system. The isolated product gives a single spot on TLC after sulfuric acid spray, followed by heating. Confirm the presence of the maleimide group in the isolated product by its reactivity to cysteine using the method of Grassetti and Murray. The structure of CHS-MB is shown in Fig. 1. The melting points 161-163°C.

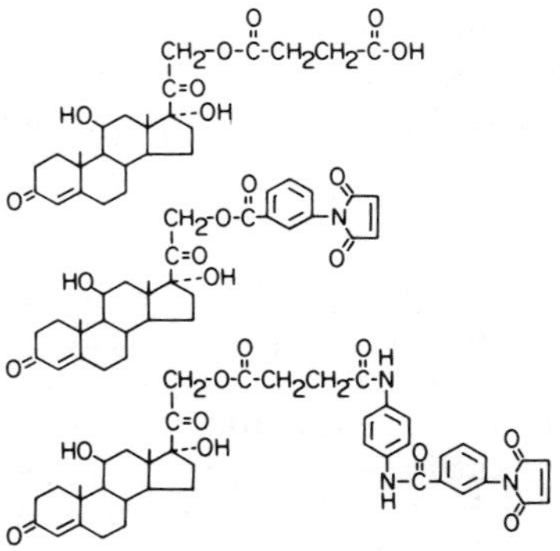

Fig. 1. Structures of cortisol-21-hemisuccinate (CHS), Cortisol-21-m-maleimidobenzoate (CT-MB), and m-maleimidobenzoic acid conjugated with cortisol-21-hemisuccinate through a p-phenylenediamine linkage (CHS-MB).

Preparation of CHS-GAL Conjugate (CHS-GAL). Conjugate CHS to GAL by a mixed acid anhydride method using the procedure of Comoglio and Celada (1976).

Preparation of CHS-MB or CT-MB-GAL Conjugate (CHS-MB-GAL and CT-MB-GAL). Conjugation of CHS-MB or CT-MB with GAL in 0.05 M phosphate buffer (pH 7.0; buffer A) as well as isolation of the enzyme hapten conjugate by Sephadex G-25 column chromatography are carried out by the same procedure reported in T_4-enzyme conjugation and isolation.

Production of Cortisol Antiserum. Antiserum to CHS-BSA was raised in rabbits following the procedure of Ruder et al, (1972). This antiserum was highly specific to cortisol when examined either by RIA or by EIA.

Cortisol Standard. Dilute a stock solution of 100 µg/ml cortisol in ethanol with 0.05 M phosphate buffer-0.1% BSA (pH 7.3; buffer

B), and store in a deep freezer 10-ml volumes of the standard solution (0-50 μg/dl). These standard solutions are stable for at least 6 months.

EIA Procedures. Mix volumes of 20 μl of standards or plasma samples with 30 μl buffer A in glass tubes (16 x 100 mm). After thorough mixing, heat the solutions in a boiling water bath for 10 min. Cool the mixtures and add 0.4 ml buffer B followed by 0.1 ml of the enzyme conjugate solution. After mixing, add 0.1 ml of a 1:500 dilution of cortisol antiserum (final dilution, 1:3,000) and proceed the incubation for 60 min at room temperature. Then add 100 μl of a 1:40 dilution of normal rabbit serum followed by 100 μl goat anti-rabbit immunoglobulin G. After mixing on a vortex mixer, incubate the tubes for 2 h at 4°C and then centrifuge for 10 min at 2,000 x g. Wash the pellet twice with buffer B. Assay the enzyme activity in the pellet according to the procedure of Dray et al., using o-nitrophenyl-ß-D-galactoside as substrate. Incubation time for enzyme activity is 60 min. Measure the amount of o-nitrophenol produced at the end of the incubation at a 420 nm wavelength.

Digoxin EIA

Preparation of Digoxigenin-3-O-Succinate. Synthesis of digoxigenin-3-O-succinate was carried out by the procedure of Oliver et al (1968). Dissolve both 0.7 g of succinic anhydride and 2.0 g of digoxigenin in 12 ml of pyridine. Protect the solution from light and allow to react at room temperature for 90 days. Pour the solution into 75 ml of cold 2 N H_2SO_4. Isolate the solid product by filtration and wash with cold water. Redissolve the isolated product in 150 ml of chloroform-methanol (2:1). Wash the chloroform-methanol solution once with 25 ml of 1N H_2SO_4 and three times with water. Add 25 ml of methanol after each washing. Dry the organic phase over anhydrous sodium sulfate and on a rotary evaporator. Dissolve the residue in 15 ml of hot ehtanol, and then add hot water to turbidity. Allow the solution to cool to room temperature and then leave it at 4°C for 48 h. Isolate the resultant crystals by filtration and wash three times with cold ethanol water (3:2). The final product is a white powder with melting points of 190-197°C. Confirm esterification at position 3 by its immunoreactivity to digoxin antiserum.

Synthesis of p-(Digoxigenin-3-O-Succinamido)Aniline. Dissolve digoxigenin-3-O-succinate (50 mg) in 2 ml of THF, and add 25 ml of tributylamine. Then cool the solution to -10°C. Add 25 ml of isobutylchloroformate and incubate at -10°C for 1 h. Filter the precipitate at -5°C and add the supernate dropwise to the cooled (-5°C), stirred solution of THF (2 ml) containing 200 mg of p-phenylenediamine. Incubate the solution further for another 2 h at room temperature. Monitor synthesis of the product (Rf = 0.28) by thin layer chromatography using ethyl acetate as solvent. A spot for the product gives ninhydrin positive and fluoresces under UV light. Isolate the product by column chromatography using ethyl acetate as eluting solvent. Melting points 172–178°C.

Synthesis of m-Maleimidobenzoyl Derivative of p-(Digoxigenin-3-O-Succinamido) Aniline (DSA-MB). Dissolve p-(digoxigenin-3-O-succinamido)aniline (50 mg) in THF (5 ml), and add to that solution 50 mg of Na_2CO_3. Dissolve MBC (30 mg) prepared by the procedure described previously in 2 ml of THF and add dropwise to the solution of p-(digoxigenin-3-O-succinamido)alinine. Reflux the solution for 30 min and monitor on TLC (Rf = 0.35) using ethyl acetate as eluting solvent. The product gives ninhydrin negative and fluoresces under UV light. Isolate the product by silica gel column using chloroform-ethyl acetate (3:7) as solvent system. Melting points 250–256°C. Confirm the presence of maleimide group in the product by its ability to react with cysteine using the method of Grassetti and Murray (1967).

Production of Anti-Digoxin Antibody. Digoxin-bovine serum albumin (BSA) conjugate was prepared by the procedure of Smith et al., (1970).

Preparation of Enzyme-Hapten Conjugate. Conjugation of DSA-MB to GAL is done by the procedure similar to that described in T_4 EIA section. Add a solution of DSA-MB in THF (0.2 mg/ml, 10 nmoles) to 1.5 ml of 0.05 M phosphate buffer (pH 7.0) containing GAL (0.5 mg, 0.93 nmole). Incubate the mixture for 2 h at room temperature. Following overnight dialysis in the same phosphate buffer, chromatograph the mixture on a Sephadex G-25 column (1.5 x 40 cm). Use the fractions of eluate containing the peak of enzyme activity for the digoxin assay.

Digoxin Standard. Serially dilute a stock solution of 1 mg/ml of digoxin dissolved in dimethyl sulfoxide with phosphate buffer - BSA (pH 7.3, 0.05 M, 0.1% BSA), and store 10 ml volumes of the standard solutions (0–5 ng/ml) at -20°C.

EIA Procedure. Add 50 µl of a standard or plasma sample to a glass tube (12 x 75 mm) containing an appropriate dilution of enzyme conjugate in 0.4 ml phosphate buffer-BSA. After mixing, add 100 µl of goat anti-digoxin antiserum and incubate overnight. Then add solid phase - rabbit anti-goat immunoglobulin antibody (0.4 ml, Immunobeads from BIO-RAD Laboratories) and, after mixing, incubate for 2 h. Precipitate the solid phase by centrifugation at 2,000 x g for 10 minutes. Wash the precipitated solid phase three times with 1 ml phosphate buffer-BSA. Assay the enzyme activity in the precipitate according to the procedure described previously. Incubation time for enzyme activity is 120 min. Determine the amount of o-nitrophenol produced at the end of incubation at 420 nm wavelength.

RESULTS AND DISCUSSION

By double antibody precipitation method in excess of anti-T_4 antibody in T_4 EIA development, more that 97% of the enzyme was found to be conjugated with MBTM when MBTM and GAL conjugation was carried out at molar ratios of over 5 to 1. The number of moles of MBTM conjugated per enzyme was not directly determined. Since GAL possesses about 10-12 sulfhydryl groups readily reacting to p-chloromercuribenzoic acid (Wallenfels et al., 1964) and iodoacetic acid, (Jorwall et al., 1978), the maximum number of MBTM attached per enzyme is about 12. The practical limit of solubility of MBTM in the conjugation solvent is 10 μg/ml, restricting the molar ratio of MBTM to enzyme. The maleimide group of MBTM was found to be labile; about 50% was destroyed in 3 hours when 1 μg of MBTM was dissolved in 1 ml phosphate buffer (pH 7.0, 0.05M). The conjugation reaction was, therefore, carried out immediately after dissolution of MBTM in the buffer. Enzyme activities examined before and after the conjugation step did not show any difference, suggesting full retention of the enzyme functional groups. With the final antiserum dilution of 2,400-fold, a reproducible T_4 enzyme immunoassay was successfully demonstrated (Fig. 2). Incubation times used in the assay were 45 and 15 minutes, respectively, for competitive binding of T_4 and MBTM-enzyme conjugate and double antibody precipitation steps. The highest sensitivity in the assay was observed at 0-10 μg/100 ml range.

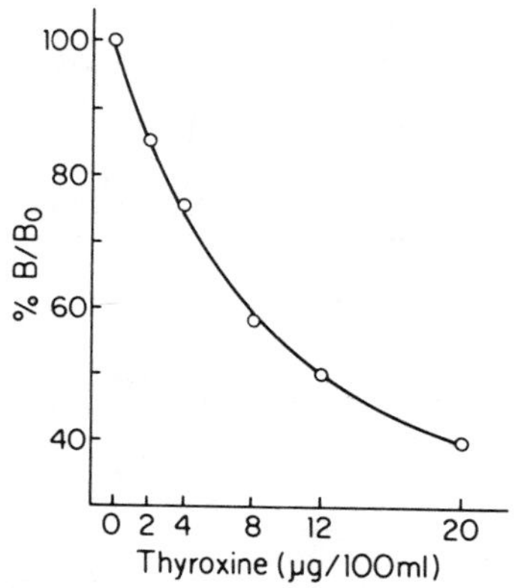

Fig. 2. Standard curve for thyroxine enzyme immunoassay.

For hapten-enzyme conjugation in cortisol EIA development, CHS-MB or CT-MB was synthesized and conjugated to sulfhydryl groups of GAL. We compared our method with the conventional conjugation method, i.e. CHS being conjugated to amino groups of GAL using a mixed acid anhydride method. The latter method resulted in only 40% of the enzyme being labeled, as examined by the double antibody precipitation method in excess of cortisol antibody, and 20% reduction of enzyme acitivity, whereas our conjugation method showed full retention of enzyme acitivity and over 95% of the enzyme being labeled with hapten derivatives (Table 1). The number of moles of CHS-MB or CT-MB conjugated per enzyme was not determined. However, as in the case of T_4-enzyme conjugate, about 12 molecules are thought to be attached to one molecule of GAL. With respect to immunoreactivity of the enzyme-hapten conjugates and to assay specificity and when the immunoreactivities of these conjugates to the produced cortisol antiserum were examined, both CHS-GAL and CHS-MB-GAL conjugates showed excellent immunoreactivity but poor displacement with 5 ng of added cortisol. CT-MB-GAL conjugate, on the other hand, showed not only a high immunoreactivity to the antibody, but also displaced well with the added cortisol (Fig 3). Modification around the bridge of cortisol derivative may have resulted in increased sensitivity to cortisol. Cross-reactivity, as examined by the method of Abraham (1969), to other steroids in the assay was 150% for cortisol-21-acetate, 10% for cortisone, 7% for corticosterone, and less than 0.1% for 11-deoxycortisol, 17-hydroxyprogesterone, prednisone, and dexamethasone. These immunochemical results strongly indicated that CT-MB was esterified at position 21 by the m-maleimidobenzoyl group. A typical standard curve for EIA as well as RIA of cortisol is shown in Fig. 4. The measurable range was 1-50 μg/dl using a 20-μl sample size. A miximum sensitivity of 1 μg/dl was measured

Table 1. Efficiency in enzyme-hapten coupling and effects of hapten -enzyme conjugation on enzyme Activity.

Hapten Derivative	Enzyme Coupling Efficiency (%)*	Enzyme Activity Recovered (%)	Functional Group of Enzyme Involved in Coupling Process
CHS	40	80	Amino group
CHS-MB	96	100	Sulfhydryl group
CT-MB	97	100	Sulfhydryl group

*Antibody bound enzyme was measured after second antibody precipitation in the presence of excess anti-cortisol antibody.

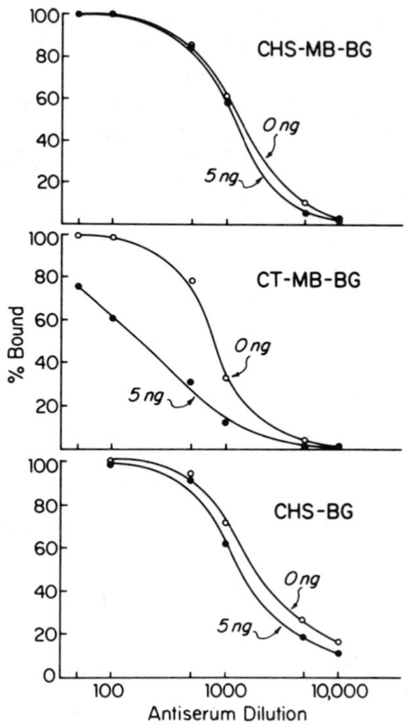

Fig. 3. Cortisol antiserum dilution curves using CT-MB-GAL, CHS-GAL and CHS-MB-GAL as labels either with (5 ng) or without (0 ng) the addition of cortisol.

at B/B_o = 90%. When serial dilutions of plasma samples (n=3) containing various concentrations of cortisol were made, and their cortisol levels were determined by EIA (Fig. 5), a linear relation was seen between the cortisol concentration and the dilution in each sample. When the cortisol levels in 24 plasma samples were determined by EIA for correlation with RIA, a good correlation was seen between the values determined by the two methods ($Y = 0.93X + 0.42$, $r = 0.99$, $n = 24$).

When the maleimide derivative of digoxigenin-3-O-succinate was conjugated to β-galactosidase in digoxin EIA development, about 97% of the enzyme was labeled with the hapten derivative. No reduction in enzyme activity was observed after the conjugation reaction. The results obtained in digoxin EIA were consistent with those obtained for other haptens. Antiserum was diluted 5,000-folds and one tenth ml of that was used for the routine assay. Displacement with the unlabeled digoxin in EIA showed the assay range of 0-5 ng/ml with a maximum assay sensitivity of 0.6 ng/ml with a coefficient of variation of about 3% (Fig.6). Solid phase - second antibody precipitation method we used thus gave reliable and reproducible results. The results on cross reactivity studies showed that cross reactions of 142% and 3.8% were obtained, respectively, for digoxigenin and digotoxin, suggesting that the EIA is specific for digoxin and that the digoxigenin was succinylated at the posi-

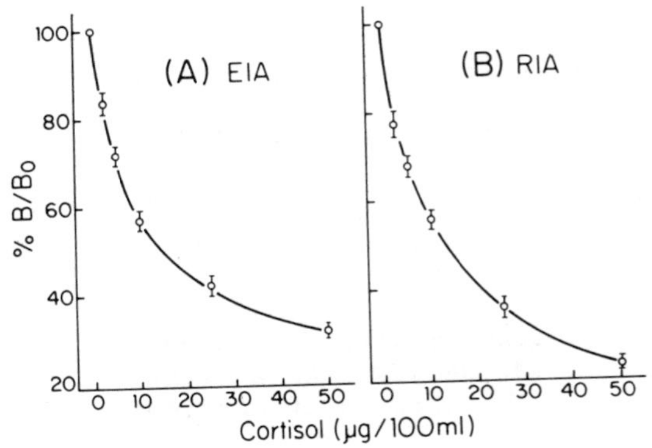

Fig. 4. Standard curves for cortisol EIA (A) and RIA (B). Each point represents the mean ±SD from six replicate determinations.

tion 3 during the succinylation reaction. In previous sugar containing hapten enzyme conjugation procedures, ε-amino group of lysine of enzymes have been covalently linked to a terminal sugar moleity through Schiff's base formation, followed by reduction with $NaBH_4$ (Lauer and Erlanger, 1974). This procedure is rather tedious and difficult to replicate. Our method of hapten conjugation to ß-galactosidase is simple and can easily be replicated because of the involvement of the sulfhydryl groups of the enzyme in hapten conjugation reaction.

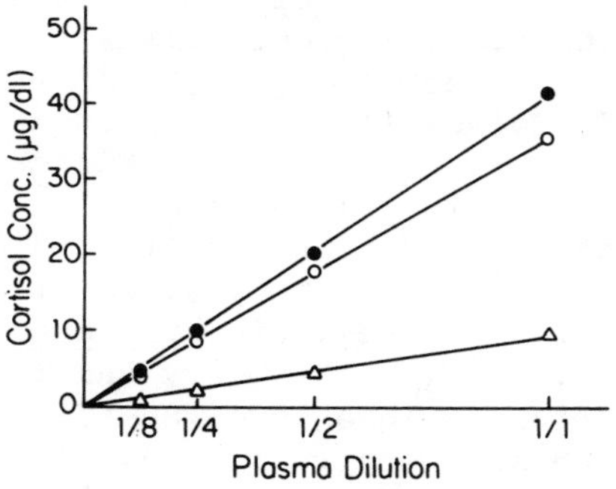

Fig. 5. Cortisol concentrations in several dilutions of plasma samples.

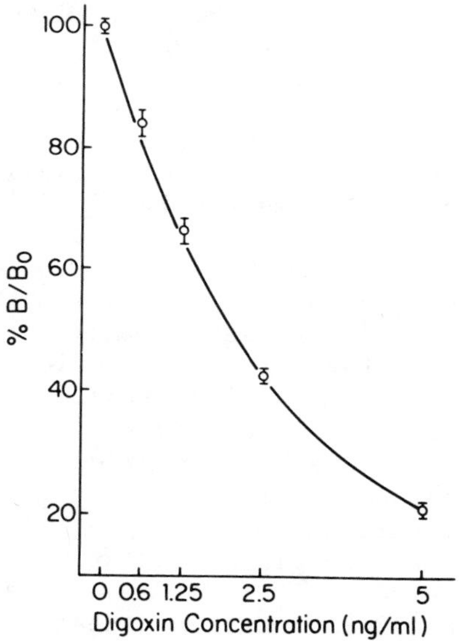

Fig. 6. Standard curve for digoxin in digoxin enzyme immunoassay.

REFERENCES

Abraham, G. E., (1969). Solid phase radioimmunoassay of estradiol -17 beta. J. Clin. Eudocrinol. Metab. 29: 866-870.

Ashley, H. N. and C. R. Harington, CLXXIX (1929). Some derivatives of thyroxine. Biochem. J. 22, 1436-1439.

Comoglio, S. and F. Celada (1976). An immuno-enzymatic assay of cortisol using E. coli ß-galactosidase as label. J. Immunol. Methods 10: 161-170.

Dray, F., J. E. Andrieu and F. Renaud (1975). Enzyme immunoassay of progesterone at the picogram level using ß-galastosidase as label. Biochem. Biophys Acta. 403: 131-138.

Gharib, H., R. J. Ryan, W. E. Mayberry and T. Hackert (1971). Radioimmunoassay for triiodothyronine (T_3): Affinity and specificity of the antibody to T_3. J. Clin. Endocr. 33: 509-518.

Grassetti, D. K. and J. F. Murray, Jr. (1967). Determination of sulfhydryl groups with 2,2'- or 4,4'-dithiodipyridine. Arch. Biochem. Biophys. 119: 41-49.

Ishikawa, E. (1983). Enzyme-labeling of antibodies. J. of Immunoassay 3: 209-327.

Jorwall, H., A. V. Fowler and I. Zabin (1978). Probe of ß-galactosidase - Structure with iodoacetate; Differential reactivity of thiol groups in wild type and mutant forms of ß-galactosidase. Biochemistry 17: 5160-5168.

Lauer, R. C. and Erlanger, B. F. (1974). An enzyme-immunoassay of antibody specific for adenosine using ß-galactosidase. Immunochemistry 11:533-540.

Monji, N., H. Ali and A. Castro (1980 A). Quantification of digoxin by enzyme immunoassay: synthesis of a maleimide derivative of digoxigenin succinate for enzyme coupling. Experientia 36: 1141-1142.

Monji, N., N. O. Gomez, H. Kawashima, H. Ali and A. Castro (1980). Practical enzyme immunoassay for plasma cortisol using ß-galactosidase as enzyme label. J. Clin. Endocr. Metab. 50: 355-359.

Monji, N., H. Malkus and A. Castro (1978). Maleimide derivative of hapten for coupling to enzyme: a new method in enzyme immunoassay, Biochem. Biophys. Res. Commun. 85: 671-677.

Oliver, G. C. Jr., B. M. Parker, D. L Brasfield and C. W. Parker (1968). The measurement of digitoxin in human serum by radioimmunoassay. J. Clin. Invest. 47: 1035-1042.

La Parola, G. (1934). Azione dell'anidride maleica sulle lasi aldeido cumminiche. Gazz. chim. Ital. 64: 919-925.

Ruder, H. J., L. G. Robert and M. B. Lippett (1972). A radioimmunoassay for cortisol in plasma and urine. J. Clin. Endocrinol. Metab. 35: 219-224.

Searle, N. E. (July 6, 1948). United States Patent No. 2,444,536. Synthesis of N-aryl-maleimides.

Smith, T. W., V. P. Butler, Jr. and E. Haber (1970). Characterization of antibodies of high affinity and specificity for the digitalis glycoside digoxin. Biochemistry 9: 331-337.

Wallenfels, K., B. Muller-Hill, D. Dabich, C. Streffer and R. Weil (1964). Untersuchungen an milchzuckerspaltenden Enzymen, XVI Zahl und Reactivitat der SH-Gruppen der ß-galaktosidase aus E. coli. Biochem. Z. 340: 41-47.

ENZYME IMMUNOASSAYS USING TAGGED ENZYME-LIGAND CONJUGATES

T.T. Ngo and H.M. Lenhoff

Department of Developmental and Cell Biology
University of California, Irvine
Irvine, CA 92717

INTRODUCTION

Background

We have developed an approach to separation enzyme immunoassays that uses a tag molecule linked to an enzyme-linked conjugate. This approach differs from most separation enzyme immunoassays in which the separation process usually depends upon the interaction between the ligand (or antigen) and antibody, one of which is immobilized on a solid material (Schuurs and van Weemen, 1977; Pal, 1978; Engvall and Pesce, 1978; Borrebaeck and Mattiasson, 1979; Ngo and Lenhoff, 1982).

In our assay, the enzyme label is linked both to the ligand and also to "tag" molecules; those tag molecules can bind tightly to an insolubilized receptor for that tag. We call this assay AMETIA, i.e. it is an "antibody masking the tag" immunoassay. In the model system we describe, DNP-lysine is the ligand, β-galactosidase is the enzyme, biotin is the tag, and avidin immobilized to Sepharose is the insoluble receptor.

Principles of the AMETIA Assay

(i) The ligand (L) and the tagged enzyme-ligand conjugate (L-E-T) compete in a homogeneous solution for antibody (Ab) [Scheme 1, reactions (a) and (b)]. (ii) The insolubilized receptor (R-I) is added to the solution and mixed. After a brief period, the suspension centrifugated to separate the soluble tagged enzyme-ligand conjugate complexed with the antibody (i.e. Ab:L-E-T) from the tagged enzyme-ligand conjugate

SCHEME 1.

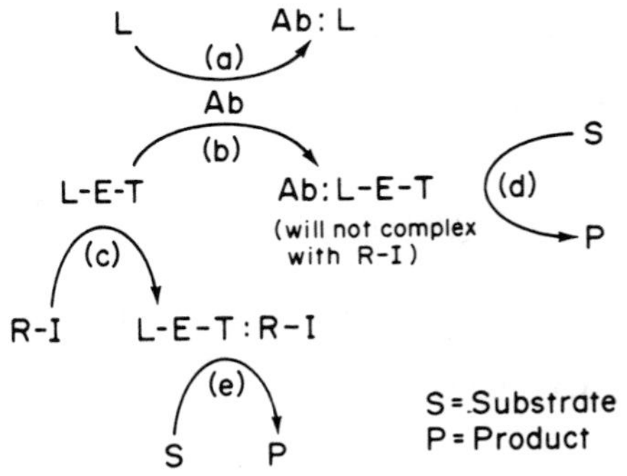

made insoluble by complexing with the insoluble receptors (i.e. L-E-T:R-I) by reaction (c). The tag of Ab:L-E-T is not able to complex with R-I because of the presence of Ab. (iii) The enzyme activity of either the supernatant fraction Ab:L-E-T or the insoluble fraction L-E-T:R-I is measured [reaction (d) or (e) respectively].

In AMETIA fixed concentrations of L-E-T, Ab and R-I are used. The amount of free L-E-T available to complex with R-I depends upon the amount of the analyte L because both L and L-E-T compete for free Ab [reactions (a) and (b)]. Thus, the lower the concentration of L, the more Ab is available to complex L-E-T [reaction (b)] so that less L-E-T is available to complex with R-I by reaction (c). When L-E-T complexes with Ab [reaction (b)] to form Ab:L-E-T, the T of L-E-T cannot bind to R-I, presumably because the Ab in Ab:L-E-T physically hinders the interaction of T of L-E-T with R-I.

Thus, in AMETIA low concentrations of the analyte leads to a high enzyme activity in the supernatant phase [reaction (d)] and a low on in the insoluble fraction [reaction (e)]. On the other hand, a higher concentration of L in sample to be tested will tie up more Ab [reaction (a)] thereby freeing L-E-T so that it will complex with R-I (reaction (c)] to give a lower enzyme activity in the supernatant and a higher one in the insoluble fraction.

MATERIALS AND METHODS

Chemicals and their sources are: 2,4-dinitrophenyl-ε-lysine, o-nitrophenyl-β-D-galactopyranoside and avidin (Sigma Chemical Co.); *E. coli* β-galactosidase (Boehringer Manneheim Biochemicals); AH-Sepharose (Pharmacia Fine Chemicals); rabbit anti-DNP serum (Miles Labs., Inc.); and m-maleimidobenzoyl-N-hydroxy-succinimide ester (Pierce Chemical Co.).

Immobilization of Avidin on Aminohexyl-Sepharose (AH-Sepharose)

Immobilize avidin on AH-Sepharose using glutaraldehyde as a crosslinker. Suspend washed AH-Sepharose (10 ml in 400 ml 5% glutaraldehyde in 0.2 M $NaHCO_3$, pH 8.5. Stir the suspension for 1 hour at room temperature (RT); then wash the gel successively with 0.5 ℓ 0.5 M NaCl and 0.5 ℓ 0.1 M $NaHCO_3$ pH 8.5. Suspend glutaraldehyde activated gel in 5 ml avidin (50 mg in 5 ml 0.1 M $NaHCO_3$, pH 8.5); stir the suspension at RT for 1 hour and at 4°C for 20 hours. Wash the avidin gel conjugate sequentially with 1 ℓ 0.5 M NaCl in 0.1 M sodium phosphate, pH 8.0 and 1 ℓ 0.1 M sodium phosphate, pH 8.0. This method should yield 13 mg of bound avidin per 10 ml packed gel.

Synthesis of m-Maleimidobenzoyl-DNP-Lysine

Add 10 μmoles m-maleimidobenzoyl-N-hydroxysuccinimide ester, 11 μmoles Na_2CO_3 to 2 ml mixture of tetrahydrofuran and dimethylformamide (1:1 mixture) and stir at RT. The reaction should be completed after 24 hours.

Labeling of β-Galactosidase with Biotin

Dialyze β-Galactosidase against 0.1 M sodium phosphate, pH 8.0. To the dialyzed enzyme (2.8 nmoles in 2 ml 0.1 M phosphate, pH 8.0) add 281 nmoles N-hydroxysuccinimide-biotin in 0.1 ml dimethylsulfoxide. Stir the solution at RT for 3 hours and at 4°C for 5 hours, and then dialyze at 4°C.

Labeling of Biotinyl-β-Galactosidase with m-Maleimidobenzyol-DNP-Lysine

To 1 ml biotinyl enzyme (1.5 nmoles) add 0.2 ml m-maleimidobenzoyl-DNP-lysine (1 μmole). Stir the solution at RT for 2 hours. Separate the excess unreacted m-maleimidobenzoyl-DNP-lysine from the labeled enzyme on a column (2 x 45 cm) of Sephadex G-50 C. Use 0.1 M sodium phosphate, pH 7.2, as the eluant.

Enzyme activity

Assay for the enzyme activity in the supernatant or the insoluble fractions using o-nitrophenyl-β-galactopyranoside as a substrate (Ngo, Narinesingh, and Laidler, 1976).

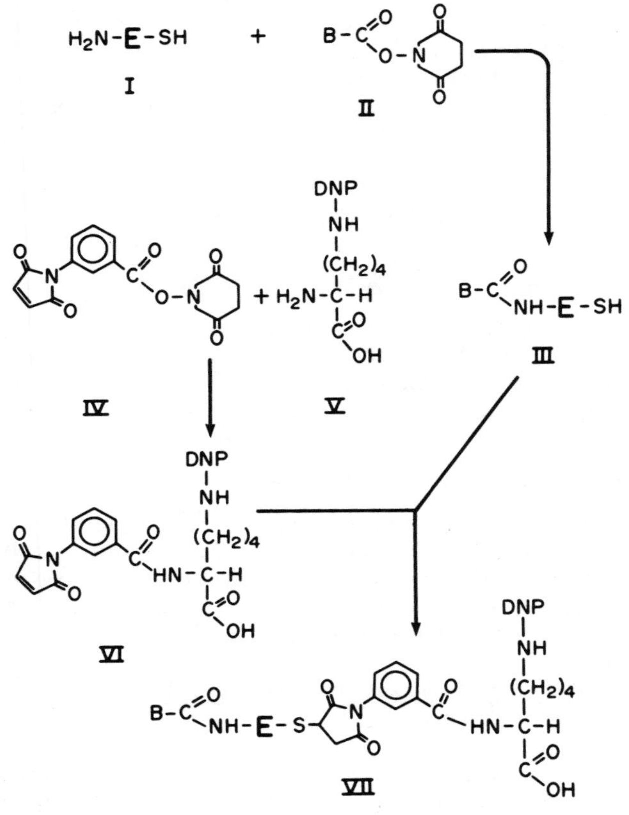

Fig. 1. Chemical processes for preparing β-galactosidase linked to both DNP-lysine and biotin.
I = β-galactosidase; II = N-hydroxysuccinimido-biotin; III = biotinyl-β-galactosidase; IV = m-maleimidobenzoyl-N-hydroxysuccinimide ester; V = DNP-lysine; VI = m-maleimidobenzoyl-DNP-lysine; VII = β-galactosidase labeled with biotin and DNP-lysine.

TYPICAL RESULTS

Preparing model L-E-T, i.e. β galactosidase (E) linked to biotin (T) and DNP-lysine (L)

The procedure outlined in Figure 1 is for preparing β-galactosidase (E) linked to biotin (tag) and to DNP-lysine (ligand). The enzyme (I), which possesses both free amino and thio groups, reacts with activated biotin (B), i.e. N-hydroxysuccinimidobiotin (II). The resultant biotinyl enzyme (III) reacts with m-maleimidobenzoyl-DNP-lysine (VI) obtained by reacting m-maleidobenzoyl-N-hydroxysuccinimide ester (IV) with DNP-lysine (V). The final product is the tagged enzyme-ligand conjugate (VII) consisting of β-galactosidase linked to DNP-lysine (VI) through m-maleimidobenzoate and to biotin tags. In our hands, each molecule of enzyme prepared this way was linked to 21 biotin tags and 37 DNP-lysines.

Reaction of Soluble L-E-T with Insoluble R-I

Addition of increasing amounts of avidin-gel, i.e. the insolubilized receptor (R-I), to a fixed concentration of biotin tagged enzyme-ligand conjugate (L-E-T) gave a decrease in enzyme activity in the supernatant fraction and a concommitant increase in activity in the insoluble avidin-gel fraction (Fig. 2). Hence, reaction (c) of the scheme takes place.

Inhibition of Enzyme Activity of L-E-T When the T is Complexed with Ab

When fixed amounts of avidin-gel (R-I) were added to solution containing fixed amounts of tagged enzyme-ligand conjugate (L-E-T) and increasing amounts of anti-DNP serum (Ab), the enzyme activity in the supernatant increased and that of the insoluble fraction decreased (Fig. 3). Hence, these results demonstrate that (1) antibodies bind to the DNP-residue (L) of the tagged enzyme-ligand conjugate (L-E-T), and (2) the biotin tag molecules (T) of the resultant antibody conjugate complex (Ab:L-E-T) was not able to bind to the insoluble receptor avidin-gel (R-I), presumably because the antibody masked the biotinyl residues on the enzyme.

Standard Curve: Competition of L and L-E-T for Ab

The DNP-lysine residues (L), of the conjugate (L-E-T) competed successfully for the antibodies (Ab) with free analyte DNP-lysine (L). When increasing amounts of DNP-lysine (L) were added to fixed amounts of conjugate (L-E-T), of antibody (Ab) and of avidin-gel (R-I), the enzyme activity in the supernatant

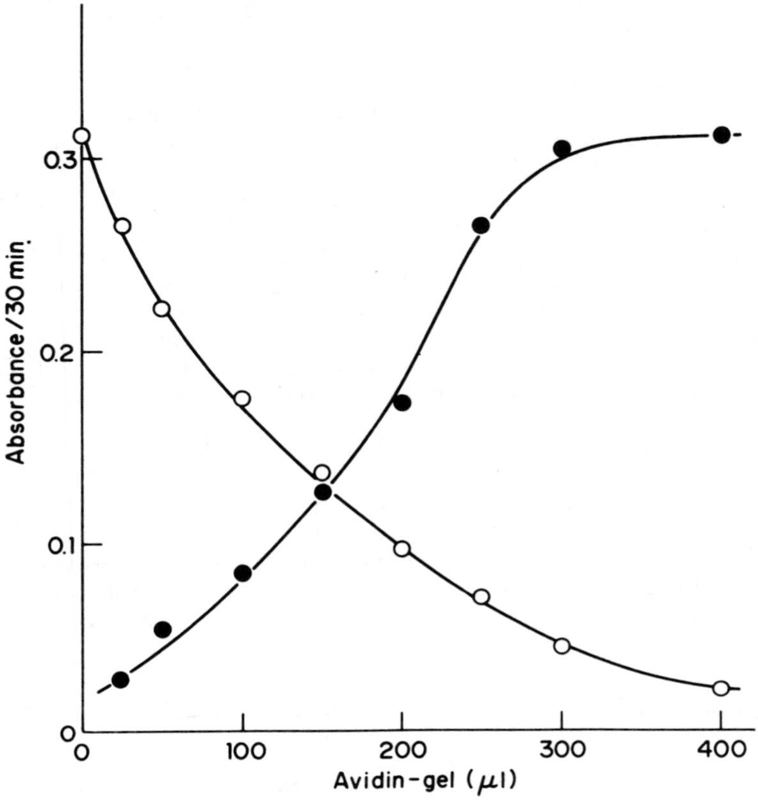

Fig. 2. Binding of β-galactosidase linked to DNP-lysine and biotin to sepharose bound avidin. Varying amounts of 10% avidin-gel suspension were added to 200 μℓ β-galactosidase linked to DNP-lysine and biotin (1.36 nM). The suspensions were adjusted to 600 μℓ with 0.5% gelatin in 0.1 M sodium phosphate, pH 7.2 and incubated at 25°C for 30 min. The suspensions were centrifuged for 5 min using an Eppendorf centrifuge. The supernatants (500 μℓ) were assayed for β-galactosidase activity using o-nitrophenyl-β-galactopyranoside as the substrate (open circles). The pellets after centrifugation were washed by suspending them in 1 ml 0.5% gelatin in 0.1 M sodium phosphate, pH 7.2 followed by centrifugation for 5 min. The washing was repeated three times. The final pellets were assayed for enzyme activity by suspending them in 3 ml substrate solution (closed circles). All assays were incubated at 25°C for 30 min.

(L-E-T) decreased while simultaneously that in the insoluble fraction (L-E-T:Ab) increased (Fig. 4). Presumably by having DNP-lysine (L) compete with ligands of tagged enzyme-ligand conjugates (L-E-T) for antibodies (Ab), there are more of the conjugate uncomplexed (L-E-T) with antibody and free to bind to insolubilized receptor giving L-E-T:R-I.

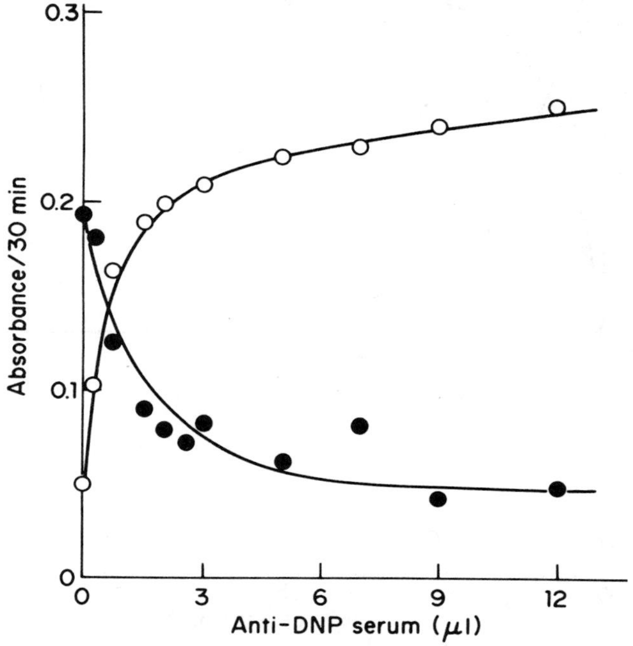

Fig. 3. Antibody to DNP-lysine prevents the β-galactosidase linked to DNP-lysine and biotin from binding to avidin-gel. Varying amounts of anti-DNP serum were added to 200 μℓ β-galactosidase linked to DNP-lysine and biotin (1.36 nM). The solutions were adjusted to 600 μℓ. Constant amounts of 10% avidin-gel suspension (400 λ) were added. The mixtures were incubated at 25°C for 30 min and were centrifuged for 5 min. The supernatants (800 μℓ) were assayed for enzyme activities (open circles). The pellets were washed three times and assayed for activities (closed circles) as described in legend of Figure 2.

Thus, the concentration range of 1-25 μM of the analyte, DNP-lysine, can be measured by this assay (Fig. 4).

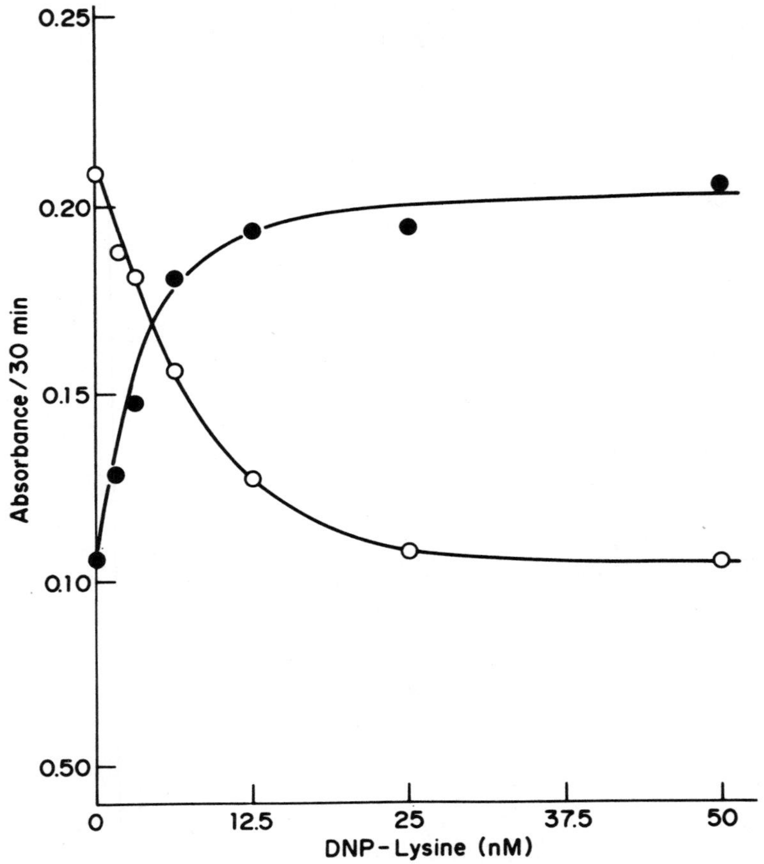

Fig. 4. Standard curve for measuring DNP-lysine by AMETIA. Solutions of 200 μℓ containing various amounts of DNP-lysine were added to 200 μℓ solutions containing a fixed concentration of β-galactosidase linked to DNP-lysine and biotin (1.36 nM). To these solutions were added 0.5 μℓ anti-DNP serum and 500 μℓ 10% avidin-gel suspension. The mixtures were incubated at 25°C for 30 min and centrifuged for 5 min. The supernatants (800 μℓ) were assayed for enzyme activity (open circles). The pellets were washed three times and assayed for enzyme activity (closed circles) as described in legend of Figure 2.

Interfering substance

If biotin, which is a natural consituent of biological systems, exists in higher than normal concentrations in the serum of the patient being tested, then those high levels of biotin would interfere with the immunoassay that we describe. In such cases, the high level of serum biotin can interfere with the binding of L-E-T to R-I (if the tag, T, is biotin) by competing with the T of L-E-T for the insolubilized avidin. One case in which highly elevated levels of biotin can exist is in the serum of patients who are being treated with biotin and who have a deficiency of holocarboxylase (Roth et al., 1981).

DISCUSSION

The data presented in Figures 2, 3, and 4 for our model system supports the validity of the principles of AMETIA: that is, of a separation-type enzyme immunoassay in which the antibody binds with the ligand of a ligand-enzyme-tag conjugate (L-E-T) such that the conjugated tag cannot complex with the insoluble receptor (R-I). As a result, only the free conjugate (L-E-T) binds to insoluble receptor (R-I) such that the insoluble complex can be separated and analyzed free of the supernatant.

Novelty of AMETIA

All heterogeneous enzyme labeled immunoassays thus far described (Borrebaeck and Mattiasson, 1979; Engvall and Pesce, 1978; Pal, 1978; Schuurs and Van Weemen, 1977) (a) the complexing of antibody with either antigen or ligand, and (b) either immobilized antigen or ligand, or immobilized antibody as the means for separating the unbound from the antibody bound fractions. In AMETIA, however, reactions between either ligand or antigen with antibody are not the basis of the separation step.

In AMETIA, the antibody serves two unique functions as regard to EIA: It serves to bind either the analyte or an analyte derivative (L) of L-E-T as well as to mask the tag molecules (T) of L-E-T. The unique aspect of the separation step in AMETIA is the binding of the enzyme conjugate (L-E-T) that is not complexed with antibody to the insolubilized receptor (R-I) to form the insoluble L-E-T:R-I. The portion of L-E-T complex with antibody (i.e. Ab:L-E-T) cannot bind to R-I, and hence, remains in the supernatant.

In all heterogeneous EIA's described (Borrebaeck and Mattiasson, 1979; Engvall and Pesce, 1978; Engvall and Pulmann,

1971; Pal, 1978; Schall et al., 1978; van Weemen and Schuurs, 1971, 1972, 1974; van Hell et al., 1979), the enzyme-ligand conjugates bound to the insoluble fractions are inevitably those complexed with antibodies. In AMETIA the opposite is true, i.e. the conjugates bound to the insoluble fraction are always uncomplexed free conjugates.

Another feature distinguishing AMETIA from other competitive heterogeneous EIA's is that at low analyte concentration the percentage of enzyme-ligand conjugate bound to the insoluble phase increases with increasing concentrations of the analyte. Therefore the standard curve in AMETIA has positive slopes (Fig. 4, closed circles) rather than negative ones as is the case for other separation EIA's (Schall et al., 1978; van Hell et al., 1974).

Molecules other than biotin can be used as tags provided there exists receptors for them having a high affinity. The biotin-avidin system is ideal because of their extraordinary affinity to each other (Green, 1975). One can tag the enzyme with avidin rather than with biotin and use immobilized biotin as a molecular hook to fish out the enzyme-ligand conjugate (L-E-T) in which avidin is the tag.

Applications and Future Potential

In AMETIA, the insolubilized receptor is a "universal" separating reagent because it can be used to develop EIA's for different ligands, as long as the enzyme is linked to the same tag. The centrifugation step in AMETIA can be eliminated by using a receptor coated tube, a column of insolubilized receptor, or receptor coated beads.

The assay can be converted to one using a fluorescent label instead of an enzyme label. One way in which this could be done is to make a conjugate in which instead of the enzyme, there is a carrier on which the ligand, the tag, and the fluorescent label is attached.

Limitations

A disadvantage for the widespread use of AMETIA as currently designed rests on the relative cost of avidin and the requirement of a separation step.

SUMMARY

We describe an approach to separation EIA that uses a tag molecule linked to an enzyme-linked conjugate. The insoluble phase is an insolubilized receptor to that tag. The antibody to the ligand, in addition to complexing either the free ligand or the one covalently linked to the tagged enzyme, also serves to mask the tag on the tagged enzyme-ligand conjugate so that it can no longer bind to the insolubilized receptor. Accordingly, the proportion of enzyme conjugate associated with the insoluble fraction is proportional to the amount of analyte ligand being assayed. This separation system based on the "antibody masking the tag" is called AMETIA.

For our model system, we use DNP-lysine as the ligand, β-galactosidase as the enzyme, biotin as the tag, and avidin immobilized to Sepharose as the insoluble receptor.

REFERENCES

Borrebaeck, C. and B. Mattiasson (1979). Recent developments in heterogeneous enzyme immunoassay. J. Solid Phase Biochem. 4: 57-67.

Engvall, E. and P. Perlmann (1971). Enzyme-linked immunosorbent assay (ELISA): quantitative assay of immunoglobulin G. Immunochem. 8: 871-874.

Engvall, E. and A.J. Pesce (1978). Quantitative enzyme immunoassay. Scand. J. Immunol. 8: Suppl. 7.

Green, N.M. (1975). Avidin. Adv. Prot. Chem. 29: 85-133.

Ngo, T.T. and H.M. Lenhoff (1981). New approach to heterogeneous enzyme immunoassays using tagged enzyme-ligand conjugates. Biochem. Biophys. Res. Comm. 99: 496-503.

Ngo, T.T. and H.M. Lenhoff. (1982). Enzymes as versatile labels and signal amplifiers for monitoring immunochemical reactions. Mol. Cell. Biochem. 43: 3-12.

Pal, S.B. (Ed.) (1978). Enzyme Labeled Immunoassay of Hormones and Drugs. de Gruyter, Berlin, 475 pp.

Roth, K.S., L. Allan, W. Yang, J.W. Foreman and K. Dakshinamurti (1981). Serum and urinary biotin levels during treatment of holocarboxylase synthetase deficiency. Clin. Chim. Acta 109: 337-340.

Schall, F.R. Jr., A.S. Fraser, H.W. Hansen, E.W. Kern, and J.W. Tenoso (1978). A sensitive manual enzyme immunoassay for thyroxine. Clin. Chem. 24: 1801-1804.

van Hell, H., J.A.M. Brands and A.H.W.M. Schuurs (1979). Enzyme-immunoassay of human placental lactogen. Clin. Chim. Acta 91: 309-316.

van Weemen, B.K. and A.H.W.M. Schuurs (1971). Emmunoassay using antigen-enzyme conjugates. FEBS Letters 15: 232-235.

van Weemen, B.K. and A.H.W.M. Schuurs (1972). Immunoassay using hapten-enzyme conjugates. FEBS Letters 24: 77-81.

van Weemen, B.K. and A.H.W.M. Schuurs (1974). Immunoassay using antibody-enzyme conjugates. FEBS Letters 43: 215-218.

FLUORIMETRIC MEASUREMENTS IN ENZYME IMMUNOASSAYS

Kristin H. Milby

Monsanto Company
800 N. Lindbergh Blvd.
St. Louis, MO 63167

INTRODUCTION

The sensitivity of enzyme immunoassays (EIAs) depends on the specificity and avidity of the immunochemical reagents, the specific activity of the enzyme label, and the detection limit of the enzyme product. When the antibody and enzyme systems have been optimized, the ability to detect the enzyme product controls assay sensitivity. When applicable, one of the simplest and most sensitive quantitative methods is fluorimetry. Colorimetry is limited to the parts per billion (10^{-3}μg/mL) level whereas fluorimetry achieves parts per trillion (10^{-6}μg/mL) and, with the use of laser excitation, even sub-parts per trillion detection limits (Bradley and Zare, 1976, and Imasaka and Zare, 1979). This advantage is because fluorimetry measures a signal increase above relatively low background luminescence while colorimetry measures a small decrease due to absorbance from a large signal of transmitted light. Using fluorimetric microscopy, Rotman (1961) was able to detect single molecules of the enzyme β-galactosidase. Under more practical circumstances, the detection limits of fluorimetry are frequently limited by background fluorescence such as that arising from serum or other biological samples. For this reason, fluorimetric EIA methods have become practical only for heterogeneous systems, preferably when a solid phase bound antibody is used to facilitate washing.

Fluorimetric enzyme assays have been in use since the early sixties (Guilbault, 1968). For some enzymes, it is possible to take advantage of the fluorescent properties of the reduced pyridine nucleotide coenzymes, NADH and NADPH. For other systems, synthetic substrates have been found which form highly fluorescent enzyme products. As shown by examples below, the detection limit of an enzyme by use of a fluorigenic substrate can be orders of magnitude lower than if a colorigenic substrate is used. This sensitivity advantage in measuring enzyme activity can lead directly to lower detection limits for analytes by EIA. It is also possible to take advantage of this sensitivity enhancement by using smaller sample sizes, more dilute reagents, and/or shorter incubation times. As discussed below, by utilizing laser excitation for fluorimetry with EIA systems, there is an even greater potential to minimize detection limits, sample and reagent volumes, and analysis times (Lidofsky, Hinsberg and Zare, 1981; Hinsberg, Milby, Lidofsky and Zare, 1981; Hinsberg, Milby and Zare, 1981).

SUBSTRATES

When choosing a fluorigenic substrate, there are several important considerations. The substrate should be readily available in a pure and stable form. It should be nonfluorescent under the detection conditions to be used and should have no significant nonenzymatic degradation to fluorescent products. Of course, it must act as a very good substrate for the enzyme label. The resultant enzyme product must be stable and highly fluorescent. Detection is facilitated if there is a large Stokes shift, that is, a significant wavelength difference between the product's excitation and emission maxima. It is also preferable to have a long excitation wavelength, as there will be lower background fluorescence at lower excitation energies.

The three enzymes that have been used as labels for EIA with fluorimetric detection are β-galactosidase (EC 3.2.1.23), alkaline phosphatase (EC 3.1.3.1), and horseradish peroxidase (EC 1.11.1.7). The most popular fluorigenic substrates for each and the fluorescent properties of the products are given in Table 1 and the structures are given in Figure 1.

In choosing between alkaline phosphatase and β-galactosidase as a label, Ishikawa and Kato (1978) and Neurath and Strick (1981) chose the latter although the β-galactosidase activity for 4-methylumbelliferyl-β-D-galactoside (4MUG) is lower than the alkaline phosphatase activity for 4-methylumbelliferyl phosphate (4MUP). Both groups found 4MUP to have a significantly higher fluorescence background level than 4MUG. This is probably due to nonenzymatic hydrolysis of 4MUP (Neurath and Strick, 1981) and could be lessened by further purification and desiccator storage.

Kato, Hamaguchi, Fukui and Ishikawa (1975a) used both fluorescein-di-(β-D-galactopyranoside) and 4MUG as substrates for a β-galactosidose label. They had similar results with both

Table 1. Enzyme-substrate Systems Used for Fluorimetric EIA

Enzyme	Substrate	Product	Excitation Max. (nm)	Emission Max. (nm)	Relative Signal[a]
β-galactosidase	4-methylumbelliferyl-β-D-galactoside	4-methylumbelliferone	360	450	10[b]
alkaline phosphatase	4-methylumbelliferyl phosphate	4-methylumbelliferone	360	450	10[b]
horseradish peroxidase	p-hydroxyphenylacetic acid	dimer[c]	317	414	0.03[d]

[a] (Fluorescence coefficient of product) ÷ (that of quinine sulfate) measured on the same instrument.
[b] Yolken and Stopa, 1979.
[c] See Figure 1.
[d] Guilbault, Brignac and Juneau, 1968.

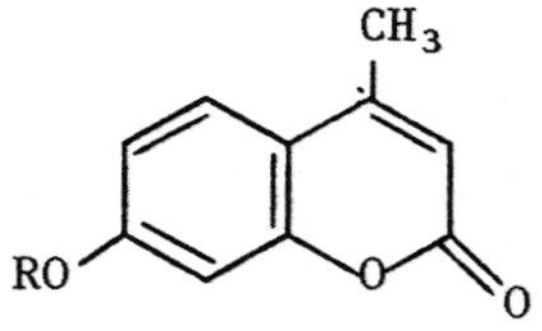

4-methylumbelliferone R = H

4-methylumbelliferyl-β-D-galactoside R =

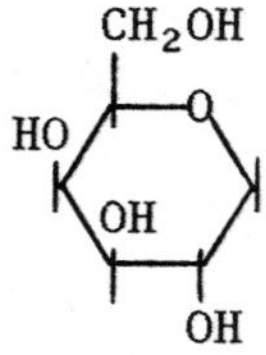

4-methylumbelliferyl-phosphate R =

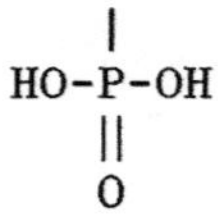

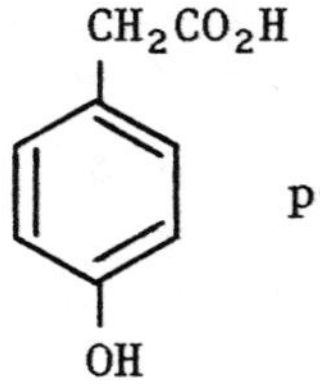

p-hydroxyphenylacetic acid

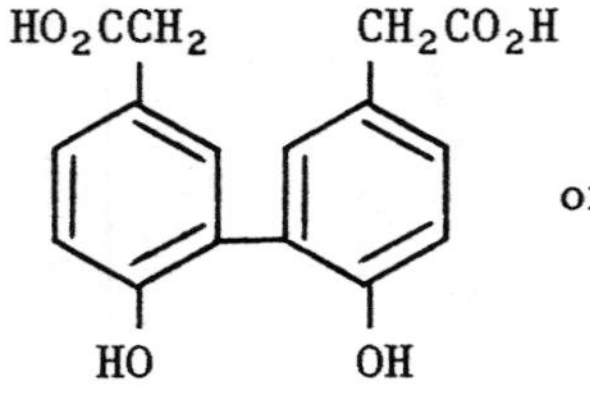

oxidized dimer

Figure 1. Structures of Substrates and Products

substrates but used 4MUG for the continuation of that work (Ishikawa and Kato, 1978).

Yolken and Stopa (1979) tested three fluorigenic and one colorigenic substrates for an alkaline phosphatase label. The best was 4MUP which could detect 5×10^{-19} moles of enzyme with 100 minutes of incubation versus 5×10^{-17} moles detected by colorimetry using p-nitrophenyl phosphate as a substrate. Flavone 3-diphosphate and 3-0-methyl fluorescein phosphate resulted in less sensitivity than 4MUP but more than the colorigenic substrate.

Many fluorigenic substrates for horseradish peroxidase have been tested, especially by Guilbault, Bignac and Juneau (1968). Those used for fluorimetric EIA include tyramine, homovanillic acid, p-hydroxyphenylacetic acid (HPA), and p-hydroxyphenylpro-prionic acid. HPA has become the most common choice because of cost, sensitivity and stability advantages. Hydrogen peroxide is always included as the second substrate.

INSTRUMENTATION

Most researchers using fluorimetric EIA have employed conventional fluorimeters which have required manually transferring samples to cuvettes and measuring their signals individually. Simply adding a flowcell and autosampler or sipper to a conventional fluorimeter would significantly decrease the labor involved. There are also instruments on the market such as the Syva "Advance", Baker Instruments "Encore", and Abbott Labs' "TDx" which automatically sample, measure and calculate signals for fluorescence immunoassays. With appropriate filter sets and software, these should be equally suitable for fluorimetric EIA applications.

Because it is emitted in all directions, there is more versatility in designing detectors for fluorescence than for absorbance. For example, reflectance fluorimetry has been used with thin layer chromatography (Butler and Poole, 1983). A similar detector design could be used for fluorescence measurements directly from the wells of microtiter plates. Forghani, Dennis, and Schmidt (1980) have designed a light box for the visual reading of fluorescence directly from microtiter plates.

With laser excitation, fluorescence can be measured from sub-microliter sample volumes. Such a detection system could be combined with a miniaturized EIA system (Leabeck and Creme, 1980 and 1981) to quantitate antigens using minimal volumes of sample and reagents.

APPLICATIONS

A listing of fluorimetric EIA applications is given in Table 2 which illustrates the variety of analytes and immunoassay methodologies employed. Space prohibits inclusion of many details, but detection limits are tabulated to demonstrate that the levels obtained are equal to or better than those achieved by radioimmunoassays. The reports specifically comparing fluorimetric and colorimetric EIA applications will be highlighted here.

Neurath and Strick (1981) reported a three orders of magnitude increase in sensitivity for the detection of β-galactosidase by fluorescence versus absorbance. Theoretically the detection limit for an antigen by EIA using a β-galactosidase label would be a thousandfold lower by utilizing fluorimetry rather than colorimetry. In reality, they found that although their enzyme-linked fluorescence immunoassay (ELFA) was about two orders of magnitude more sensitive than RIA methods for the detection of human hepatitis B surface antigen, they could not reach the level predicted on the basis of their detection of the enzyme. Their sensitivity was limited by fluorescence signals obtained by antigen-negative samples. Such a problem may be due to nonspecific binding by the antibodies used that could only be solved by more specific immunochemistry. Or the problem could be background signal due to fluorescent contaminants. In that case, purer reagents should be used and/or the final signal should be measured from the enzyme product after chromatographic separation (Hinsberg, Milby and Zare, 1981). Automated high performance liquid chromatography (HPLC) with a fluorescence detector is as fast and simple as a scintillation counter.

Despite the many reports of success in using β-galactosidase as an enzyme label (see Table 2), Kato, Umeda, et al (1979) report on interference from serum when using this label for an insulin EIA. By using a solid-phase sandwich method and a gelatin-containing buffer, they were able to overcome the interference. This same problem was encountered by Lidofsky (1980), also in an insulin EIA using β-galactosidase. In this case, the solutions recommended by Kato, Umeda, et al (1979) did not overcome the interference. No other similar problems with β-galactosidase as a label have been reported. Fluorimetric EIAs for insulin using horseradish peroxidase as a label have been successfully carried out in serum samples (Lidofsky, Hinsberg and Zare, 1981; Hinsberg, Milby and Zare, 1981; Hinsberg, Milby, Lidofsky and Zare, 1981; Matsouka, Maeda and Tsuji, 1979).

Two studies have been reported which directly compare fluorimetric and colorimetric EIAs by using the same procedures and reagents except for the substrates and detection systems.

Table 2. Applications of Fluorimetric Enzyme Immunoassays

Analyte	Immunoassay Method[a]	Detection Limit	References
A. Enzyme label = β-galactosidase			
ferritin	S-Ab-Ag-AbE	0.25 μg/L	Konijn, et.al. (1982) and Watanabe, et.al. (1979)
ferritin	Same	0.03 fmoles	Ishikawa and Kato (1978)
hepatitis B surface antigen	Same	5-10 pg	Neurath and Strick (1981)
hepatitis B surface antigen	Same	0.03-0.05 fmoles	Ishikawa and Kato (1978)
human IgG	Same	5 fmoles	Kato, Hamaguchi, et.al. (1975)
human IgG	Same	3 fmoles	Kato, Hamaguchi, et.al. (1976)
human IgG	Same	0.3 fmoles	Kato, Fukui, et.al. (1976)
α fetoprotein	Same	2 fmoles	Ishikawa and Kato (1978)
ornithine δ-aminotransferase	Same	30 attomoles	Ishikawa and Kato (1978)
chorionic gonadotropin	Ab'-Ab-AgE vs. Ab'-Ab-Ag	0.4 mIU/mL (10× more sens. than RIA)	Kikutani, et.al. (1978)

(continued)

Table 2. Applications of Fluorimetric Enzyme Immunoassays (cont.)

Analyte	Immunoassay Method[a]	Detection Limit	References
gentamicin	<u>Ag</u>-AbE	<5 μg/mL	Leabeck and Creme (1981)
insulin	Ab'-Ab-AgE vs. Ab'-Ab-<u>Ag</u>	20 pg	Kitigawa and Aikawa (1976)
insulin	S-Ab-AgE vs. S-Ab-<u>Ag</u>	0.5 μU	Kato and Hamaguchi (1975)
insulin	S-Ab-<u>Ag</u>-AbE	20 attomoles	Ishikawa and Kato (1978)
insulin	Same	5 mIU/L	Kato, Umeda, et.al. (1979)
cAMP	S-Ab'-Ab-AgE vs. S-Ab'-Ab-<u>Ag</u>	5 fmoles	Yamamoto, and Tsuji (1980)
cCMP	Same	0.5 fmoles	Yamamoto, Takai, et.al. (1982)
clenbuterol	Same	0.5 pg	Yamamoto and Iwata (1982)
befunolol and its metabolite	Same	10 pg/mL and 75 pg/mL	Sato and Yamamoto (1983)
B. Enzyme label = alkaline phosphatase			
human rotovirus	S-Ab-<u>Ag</u>-Ab-Ab'E	10^{-5} dilution of standard vs. 10^{-3} by colorimetry	Yolken and Stopa (1979)

Analyte	Immunoassay Method[a]	Detection Limit	References
polyribose phosphate	Same	10 pg/mL vs. 640 pg/mL by colorimetry	Yolken and Leister (1982)
anti-DNA antibodies	S-Ag-Ab-Ab-E	16× more sens. than by colorimetry	Ali and Ali (1983)
cytomegalo-virus antibodies	Same	Not reported	Forghani, et.al. (1980)
C. Enzyme label = horseradish peroxidase			
testosterone in plasma and saliva	S-Ab-AgE vs. S-Ab-Ag	500 fg or <30 pM	Turkes, et.al. (1979) and Turkes, et.al. (1980)
oestradiol	Same	50 pg	Numazawa, et.al. (1977)
thyroxine	Same	1.7 μg/dL	Arakawa, et.al. (1982)
17α-hydroxy-progesterone	S-Ab'-Ab-AgE S-Ab'-Ab-Ag	1.0 pg	Arakawa, et.al. (1983)
neocarzino-statin	Same	0.5 ng	Tsuji, et.al. (1978)

(continued)

Table 2. Applications of Fluorimetric Enzyme Immunoassays (cont.)

Analyte	Immunoassay Method[a]	Detection Limit	References
thyroid stimulating hormone	Same	0.06 μU	Tsuji, et.al. (1978), Kato, Naruse, et.al. (1979), and Kato, Ishji, et.al. (1980)
insulin	Same	2.5 μU/mL	Tsuji, et.al. (1978), and Matsuoka, et.al. (1979)
insulin	S-Ab-<u>Ag</u>-AbE	1.1 μU/mL or 7.9 pM	Lidofsky, et.al. (1981), Hinsberg, Milby, Lidofsky and Zare (1981) and Hinsberg, Milby and Zare (1981)

[a] Immunoassay methods are indicated by shorthand notation of the reactions involved. Ab=antibody; Ab'=second antibody; Ag=antigen; E=enzyme label; S=solid phase. The analyte is indicated by underscoring.

Yolken and Stopa (1979) found the fluorimetric method to be one hundred times more sensitive for the detection of human rotovirus in stool specimens than either RIA or the colorimetric EIA methods. Ali and Ali (1983) report a sixteen-fold sensitivity enhancement by fluorimetric versus colorimetric EIA for anti-DNA antibodies as a marker for systemic lupus erythematosus. In both these studies, the advantage realized by utilizing fluorimetry made a significant improvement in the screening of clinical samples.

An interesting comparison of fluorigenic and colorigenic substrates for EIA was made by Yolken and Leister (1982). The analyte was Haemophilus influenzae polyribose phosphate. A solid-phase sandwich EIA was used with alkaline phosphatase as the label. Using a 10-minute substrate incubation time, the fluorimetric method was 64-times more sensitive than the colorimetric method. As the incubation time was increased, however, the sensitivity of the colorimetric method improved while that of the fluorimetric method did not. By 240 minutes the sensitivities of both methods were equal. The authors interpret these results to mean that the assay is limited in sensitivity by the immunochemistry which no improvement in enzyme product detection can overcome. Even so, the use of fluorimetry offers a significant time advantage.

LASER FLUORESCENCE ENZYME IMMUNOASSAY

The above applications demonstrate the value of more sensitive enzyme product detection for EIAs. To further extend this advantage, it is possible to combine fluorimetry with laser excitation. There are three major advantages to using a laser as a fluorescence excitation source in immunoassays. First, the high intensity of laser radiation leads directly to a more intense fluorescence signal. This is true for both the signal of interest and the background signal, so extra precautions must be taken (examples below) to keep background signals low. Second, laser radiation is monochromatic. Therefore, potentially interfering Rayleigh and Raman bands are narrow and relatively easy to filter filter out from the signal. Third, because a laser beam is collimated, it is easily focused into tiny volumes. Flowcells, such as the one used in this work (see Figure 2) can be small and simple.

Imasaka and Zare (1979) demonstrated the above advantages of laser fluorimetry versus conventional fluorimetry or colorimetry in the detection of enzyme products. Lidofsky, Imasaka and Zare (1979) used laser excitation for a fluorescence immunoassay. The combination of laser fluorimetry and EIA (LF-EIA) was developed using insulin as a model antigen (Lidofsky, Hinsberg and Zare,

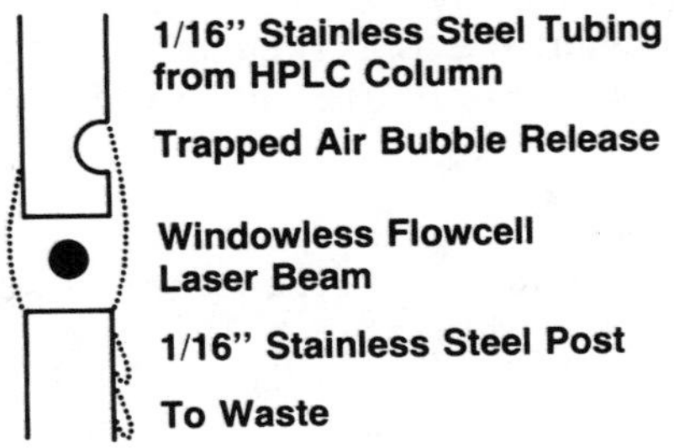

Figure 2. Close-up View of Flowing Droplet Laser Fluoresence Detector

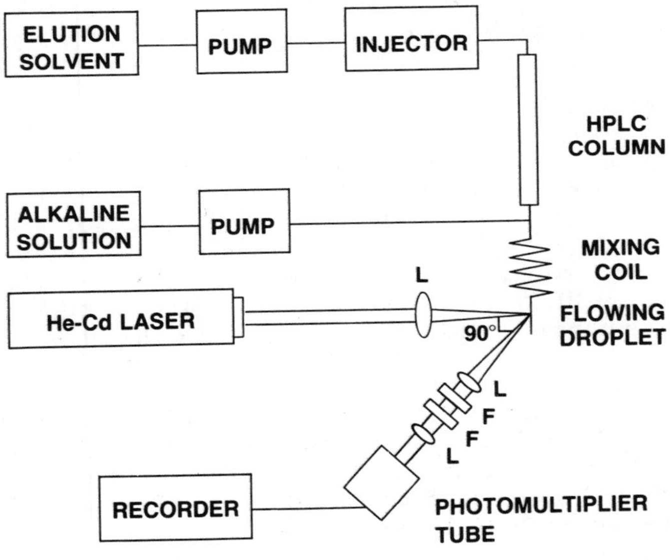

Figure 3. Schematic Diagram of HPLC with Laser Fluorimetric Detector. L=Lens, F=Filter.

1981; Hinsberg, Milby and Zare, 1981; and Hinsberg, Milby, Lidofsky and Zare, 1981) and will be discussed here in detail.

To avoid potential problems from background fluorescence, a variety of precautions were taken. A solid-phase sandwich method was chosen because the serum, which has a high level of fluorescence, is easily washed away before substrate incubation. Doubly distilled water was used throughout. The substrate p-hydroxyphenylacetic acid (HPA) was purified before use by preparative reverse phase chromatography. Finally, the enzyme product was separated by the HPLC prior to laser fluorescence detection. Because the resultant assay was not limited by background signal, the HPLC step could be replaced by flow injection analysis or a cuvette system. However, the HPLC method offers further sensitivity if other limitations are overcome and it could be readily automated.

The immunoassay reagents were purchased in kit form ("Insulotec Mochida EIA kit", Mochida Pharmaceutical Co., Tokyo, Japan). This included solid-phase bound antibody and horseradish peroxidase-labeled antibody. The colorigenic substrate, 5-aminosalicylic acid, was replaced by the fluorigenic HPA. The kit protocol was followed except that all three incubation times were shortened, and simple precautions to reduce nonspecific binding were taken (Hinsberg, Milby and Zare, 1981).

The instrumentation used is shown schematically in Figure 3. An aliquot of the final incubation solution was injected onto a reverse phase HPLC column. To optimize the fluorescence signal, the column effluent is made alkaline before it passes through the flowing droplet detector (shown in detail in Figure 2). Exciting radiation is from a helium-cadmium ion laser operating as a continuous source at 325 nm. Fluorescence from the flowing droplet is imaged onto the photomultiplier tube after passing through liquid filters to isolate the spectral region between 410 and 490 nm. The output of the photomultiplier tube is displayed directly on a stripchart recorder. It took four minutes to run the chromatograms from each sample and peak heights were used as the measure of fluoresence signal. Using this detection system, it was possible to detect 10^{-17} moles of horseradish peroxidase (Lidofsky, 1980) which is two orders of magnitude lower than earlier reports (Guilbault, Brignac and Zimmer, 1968).

With the LF-EIA it was possible to use a total of 45 minutes for the three incubation times and achieve a detection limit of 1.1 μU/mL insulin (7.9 pM). The colorimetric kit method requires at least three hours of incubation and has a detection limit of 5μU/mL insulin. For the LF-EIA, sensitivity is limited by nonspecific binding in insulin-free samples. Detection of the enzyme product has a signal-to-noise ratio of over 100 so any improvements in immunochemical specificity may be directly translated into improved sensitivity, reduction in sample size and/or further decrease in incubation times.

CONCLUSION

Utilizing fluorimetry with enzyme immunoassays is a simple modification of existing methods. There is no doubt that its use would be more widespread if convenient commercial instruments were available.

Improving enzyme product detection cannot compensate for all the limiting factors in an enzyme immunoassay such as low avidity or nonspecific binding. However, as this review has shown, often the change from colorimetric to fluorimetric detection results in enhanced sensitivity and more flexibility in incubation times and reagent amounts.

REFERENCES

Ali, A., and Ali R. (1983), Enzyme-linked Immunosorbent Assay for Anti-DNA Antibodies Using Fluorogenic and Colorigenic Substrates, J. Immunol. Meth., 56:341-346.

Arakawa, H., Maeda, M., Tsuji, A., Ishii, S., Naruse, H., and Kleinhammer, G. (1982), Fluorophotometric Enzyme Immunoassay of Thyroxine in Dried Blood Samples on Filter Paper, Bunseki Kagaku, 31:E55-E61.

Arakawa, H., Maeda, M., Tsuji, A., Naruse, H., Suzuki, E., and Kambegawa, A. (1983), Fluorescence Enzyme Immunoassay of 17α-hydroxyprogesterone in Dried Blood Samples on Filter Paper and Its Application to Mass Screening for Congenital Adrenal Hyperplasia, Chem. Pharm. Bull., 31:2724-2731.

Bradley, A. B., and Zare, R. N. (1976), Laser Fluorimetry. Sub-part-per-trillion Detection of Solutes, J. Am. Chem. Soc., 98:620-621.

Butler, H. T., and Poole, C. F. (1983), Optimization of A Scanning Densitometer for Fluorescence Detection in High Performance Thin-layer Chromatography, J. High Res. Chrom. Chrom. Comm., 6:77-81.

Forghani, B., Dennis, J., and Schmidt, N. J. (1980), Visual Reading Reading of Enzyme Immunofluorescence Assays for Human Cytomegalovirus Antibodies, J. Clin. Microbiology, 12:704-708.

Guilbault, G. G. (1968), Use of Enzymes in Analytical Chemistry, Anal. Chem., 40:459R-471R.

Guilbault, G. G., Brignac, P. J., and Juneau, M. (1968), New Substrates for the Fluorometric Determination of Oxidative Enzymes, Anal. Chem., 40:1256-1263.

Guilbault, G. G., Brignac, P., and Zimmer, M. (1968), Homovanillic Acid as a Fluorometric Substrate for Oxidative Enzymes, Anal. Chem., 40:190-196.

Hinsberg, W. D., Milby, K. H., Lidofsky, S. D., and Zare, R. N. (1981), Application of Laser Fluorimetry to Enzyme-linked Immunoassay, Soc. Photo-optical Instrumentation Engineers, 286:132-138.

Hinsberg, W. D., Milby, K. H., and Zare, R. N. (1981), Determination of Insulin in Serum By Enzyme Immunoassay with Fluorimetric Detection, Anal. Chem., 53:1509-1512.

Imasaka, T., and Zare, R. N. (1979), Enzyme Amplification Laser Fluorimetry, Anal. Chem., 51:2082-2085.

Ishikawa, E., and Kato, K. (1978), Ultrasensitive Enzyme Immunoassay, Scand. J. Immunol., 8 Suppl. 7:43-55.

Kato, K., Fukui, H., Hamaguchi, Y. and Ishikawa, E. (1976), Enzyme-linked Immunoassay: Conjugation of the FAB' Fragment of Rabbit IgG with β-D-galactosidase from E. Coli and Its Use and Its Use for Immunoassay, J. Immunology, 116:1554-1560.

Kato, K., Hamaguchi, Y., Fukui, H. and Ishikawa, E. (1976), Enzyme-linked Immunoassay: Conjugation of Rabbit Anti-(Human Immunoglobulin G) Antibody with β-D-galactosidase from Escherichia Coli and Its Use for Human Immunoglobulin G Assay, Eur. J. Biochem., 62:285-292.

Kato, K., Hamaguchi, Y., Fukui, H. and Ishikawa, E. (1975a), Enzyme-linked Immunoassay: Novel Method for Synthesis of the Insulin-β-D-galactosidase Conjugate and its Applicability for Insulin Assay, J. Biochem., 78:235-237.

Kato, K., Hamaguchi, Y., Fukui, H., and Ishikawa (1975b), Enzyme-linked Immunoassay: A Simple Method for Synthesis of the Rabbit Antibody-β-D-galactosidase Complex and Its General Applicability, J. Biochem., 78:423-425.

Kato, N., Ishii, S., Naruse, H., Irie, M. and Tsuji, A. (1980), Fluorophotometric Enzyme Immunoassay of Thyroid-Stimulating Hormone Using Peroxidase as Label, J. Pharmacobio-Dyn, 3:S-28.

Kato, N., Naruse, H., Irie, M. and Tsuji, A. (1979), Fluorophotometric Enzyme Immunoassay of Thyroid-Stimulating Hormone, Analyt. Biochem., 96:419-425.

Kato, K., Umeda, Y., Suzuki, F., Hayashi, D. and Kosaka, A. (1979), Evaluation of a Solid-Phase Enzyme Immunoassay for Insulin in Human Serum, Clin. Chem., 25:1306-1308.

Kikutani, M., Ishiguro, M., Kitagawa, T., Imamura, S. and Miura, S. (1978), Enzyme Immunoassay of Human Chorionic Gonadotropin Employing β-galactosidase as Label, J. Clin. Endocrinol. and Metab., 47:980-984.

Kitagawa, T. and Aikawa, T. (1976), Enzyme Coupled Immunoassay of Insulin Using a Novel Coupling Reagent, J. Biochem., 79:233-236.

Konijn, A. M., Levy, R., Link, G. and Hershko, C. (1982), A Rapid and Sensitive ELISA for Serum Ferritin Employing a Fluorogenic Substrate, J. Immunological Methods, 54:297-307.

Leaback, D. H. and Creme, S. (1980), A New Experimental Approach to Fluorometric Enzyme Assays Employing Disposable Micro-Reaction Chambers, Analyt. Biochem., 106:314-321.

Leaback, D. H. and Creme, S. (1981), Extremely Economical Micro-ERMA Procedures for Performing 'Sequential' Fluorogenic Enzyme Assays and Fluorogenic Enzyme Immunoassays on Human Serum, Biochem. Soc. Trans., 9:580.

Lidofsky, S. D. (1980), Laser Fluorescence Immunoassay, Ph.D. Dissertation, Columbia University.

Lidofsky, S. D., Hinsberg, W. D. and Zare, R. N. (1981), Enzyme-linked Sandwich Immunoassay for Insulin Using Laser Fluorimetric Detection, Proc. Natl. Acad. Sci. USA, 78:1901-1905.

Lidofsky, S. D., Imasaka, T. and Zare, R. N. (1979), Laser Fluorescence Immunoassay of Insulin, Anal. Chem., 51:67-69.

Matsuoka, K., Maeda, M. and Tsuji, A. (1979), Fluorescence Enzyme Immunoassay for Insulin Using Peroxidase-Tyramine-Hydrogen Peroxide, Chem. Pharm. Bull., 27:2345-2350.

Neurath, A. R. and Strick, N. (1981), Enzyme-linked Fluorescence Immunoassays Using β-galactosidase and Antibodies Covalently Bound to Polystyrene Plates, J. Virological Methods, 3:155-165.

Numazawa, M., Haryu, A., Kurosaka, K. and Nambara, T. (1977), Picogram Order Enzyme Immunoassay of Oestradiol, FEBS Letts., 79:396-398.

Rotman, B. (1961), Measurement of Activity of Single Molecules of β-D-galactosidase, Proc. Natl. Acad. Sci. USA, 47:1981-1991.

Sato, S. and Yamamoto, I. (1983), Enzyme Immunoassays for β-adrenoreceptor Blocking Agent, Befunolol and Its Main Metabolite, M1, J. Immunoassay, 4:351-371.

Tsuji, A., Maeda, M. Arakawa, H., Matsuoka, K., Kato, N., Naruse, N. and Irie, M. (1978), Enzyme Immunoassay of Hormones and Drugs by Using Fluorescence and Chemilumenescence Reaction, Enzyme Labelled Immunoassay of Hormones and Drugs (Pal, S.B., Ed.), Walter de Gruyter, Berlin and New York: 327-339.

Türkes, A. O., Türkes, A., Joyce, B. G. and Riad-Fahmy, D. (1980), A Sensitive Enzyme Immunoassay with a Fluorimetric End-Point for the Determination of Testosterone in Female Plasma and Saliva, Steroids, 35:89-101.

Türkes, A. O., Türkes, A., Read, G. F. and Fahmy, D. R. (1979), A Sensitive Fluorometric Enzyme Immunoassay for Testosterone in Plasma and Saliva, J. Endocrinol., 83:31P.

Watanabe, N., Niitsu, Y., Ohtsuka, S., Koseki, J., Kohgo, Y., Urushizaki, I., Kato, K. and Ishikawa, E. (1979), Enzyme Immunoassay for Human Ferritin, Clin. Chem., 25:80-82.

Yamamoto, I. and Iwata, K. (1982), Enzyme Immunoassay for Clenbuterol, an β_2-Adrenergic Stimulant, J. Immunoassay, 3:155-171.

Yamamoto, I., Takai, T., and Tsuji, J. (1982), Enzyme Immunoassay for Cytidine 3',5'-Cyclic Monophosphate (Cyclic CMP). Immunopharmacology, 4:331-340.

Yamamoto, I., and Tsuji, J. (1981), Enzyme Immunoassay of Cyclic Adenosine 3',5'-Monophosphate (AMP) Using β-D-galactosidase as Label, Immunopharmacology, 3:53-59.

Yolken, R. H. and Leister, F. J. (1982), Comparison of Fluorescent and Colorigenic Substrates for Enzyme Immunoassays, J. Clin. Microbiology, 15:757-760.

Yolken, R. H. and Stopa, P. J. (1979), Enzyme-linked Fluorescence Assay: Ultrasensitive Solid-Phase Assay for Detection of Human Rotovirus, J. Clin. Microbiology, 10:317-321.

RADIAL PARTITION ENZYME IMMUNOASSAY

Joseph L. Giegel

American Dade Company, American Hospital Supply Corp.
1851 Delaware Parkway
Miami, FL 33152

INTRODUCTION

Radial Partition Immunoassay permits highly sensitive measurements of both low and high molecular weight ligands. The entire immunoassay is conducted on a small section of filter paper and assays can be done in less than 10 minutes. A variety of labels can be used with the technology including enzymes, fluorophores, chromophores or radioisotopes. We have achieved a high degree of sensitivity using enzyme labels and fluorescent readout of enzyme activity. The general principles of the radial partition enzyme immunoassay system are illustrated in Fig. 1.

Antibody to the analyte to be measured is first immobilized on filter paper and the paper is inserted into a plastic tab holder for transport. After drying these tabs are quite stable. Three different types of assays can be conducted as shown in Fig. 1. For the sequential assay, sample containing the analyte is added to the dry tabs. This is allowed to react for approximately 2 minutes. An excess of enzyme labeled analyte is then added to saturate any remaining antibody sites on the tab. Following a 3 minute incubation period, substrate for the enzyme is added to the tab in such a way as to effect a radial chromatographic wash of excess, unbound labeled analyte. This step is common to all three types of assay and is an essential feature of the technology.

The center portion of the tab containing the antibody bound fraction is monitored by front surface fluorescence to determine the rate of enzyme activity. This rate is then converted to concentration units by reading from a suitable calibration curve.

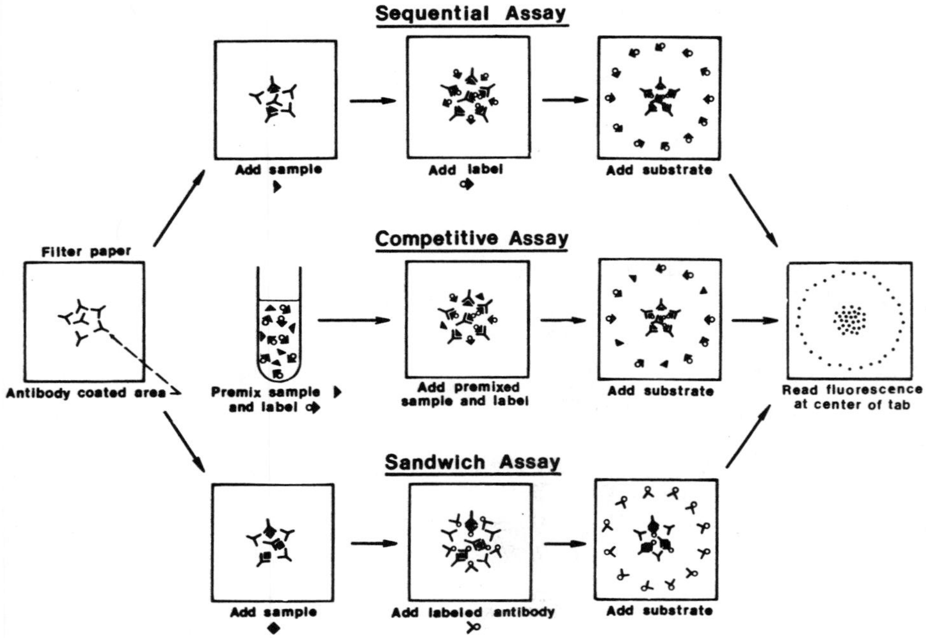

Fig. 1. Outline of Radial Partition Enzyme Immunoassay

The central area of the filter paper is coated with antibody and dried. Three different types of assays can be conducted as indicated.

Sequential Assay: Sample is added to the filter paper. An excess of enzyme labeled antigen is then added.

Competitive Assay: The sample is premixed with enzyme labeled antigen. An aliquot of this premix is added to the antibody coated tab.

Sandwich Assay: Sample is added to the filter paper. An excess of enzyme labeled antibody is then added.

Following the above steps, substrate for the enzyme is added to the center of the paper to wash excess label to the edge of the paper. The rate of change of fluorescence is monitored in the center of the tab.

Sequential assays are useful for the direct determination of low concentration analytes in serum such as digoxin.

The second type of assay is conducted in a competitive mode. In this case sample and enzyme label are premixed prior to addition to the antibody tab. The substrate is added as in the sequential assay and the enzyme rate in the center of the tab is measured. Competitive assays are useful for high concentration analytes such as most therapeutic drugs. In addition the competitive mode can be used for assays such as thyroxine in which it is necessary to displace thyroxine from its binding proteins prior to measurement.

A third type of assay is an immunometric or "sandwich" assay in which an antibody is labeled with an enzyme. These assays are performed in a similar manner to the sequential assays. Sandwich assays are useful for large, multivalent antigens such as ferritin, TSH or hepatitis virus. Because of the high ratio of surface area to volume of the glass fiber filter paper large amounts of antibody can be coated on the paper to accelerate the rate of reactions for these assays.

An automated system for performance of radial partition immunoassay has been described by Giegel(1982). This instrument handles sample addition and the substrate wash and reads each filter paper tab by front surface fluorescence.

This article will review the performance of three procedures representing each type of assay described above. The procedures are the measurement of digoxin in serum (sequential) the measurement of thyroxine in serum (competitive) and the measurement of ferritin in serum (sandwich). Each of these assays can be conducted in less than 10 minutes.

MATERIALS AND METHODS

Instrumentation

The steps outlined in Fig. 1 are performed on an automated instrument available from the American Dade Division of American Hospital Supply Corporation. The instrument is outlined schematically in Fig. 2. An appropriate number of antibody coated tabs are loaded into the tab load station and the corresponding number of samples are loaded into the sample carousel. If pre-dilution of sample and conjugate are required as for the competitive assays, this is done off-line and transferred to a sample cup. If needed, the enzyme labeled conjugate is placed in its proper position at the conjugate pump and the substrate for the enzyme is placed at the substrate pump.

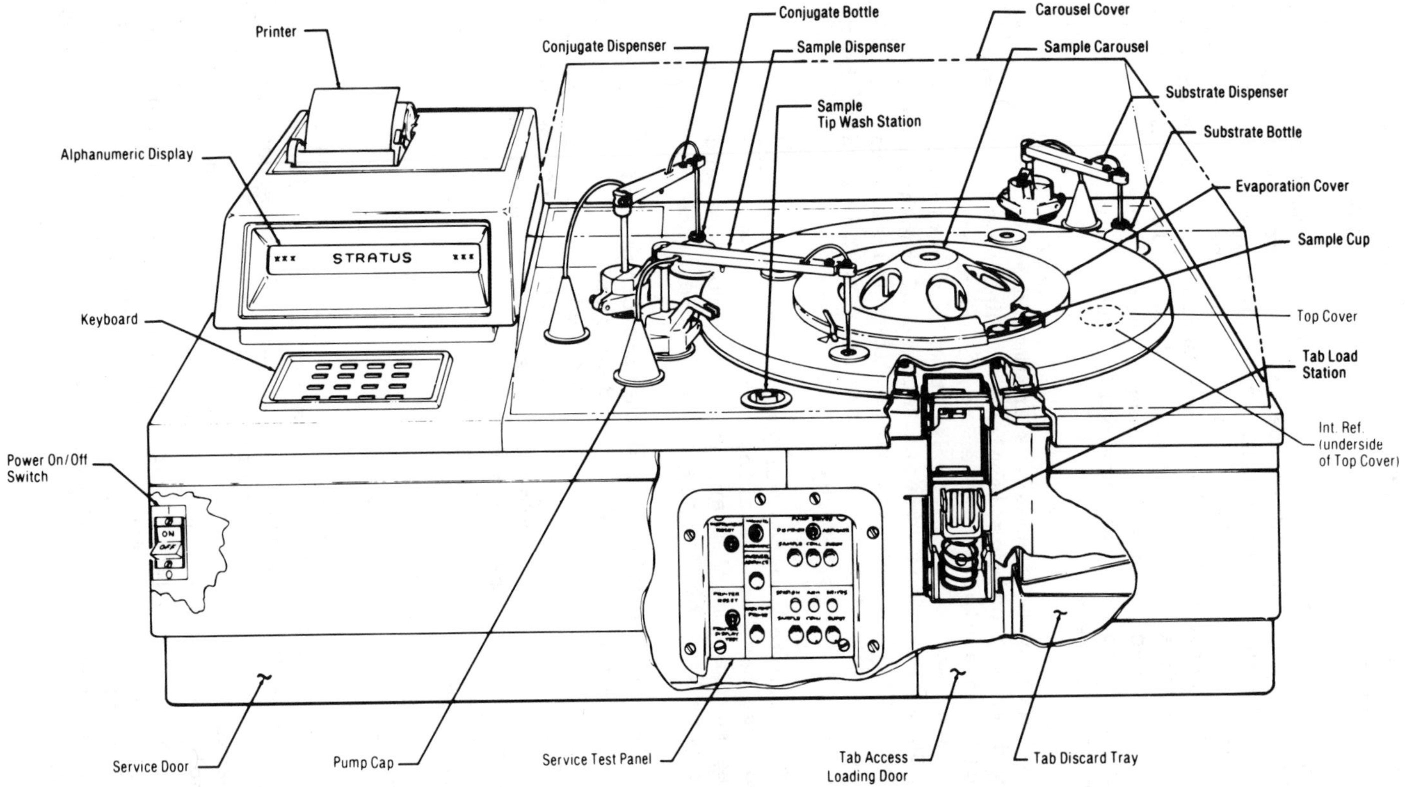

Fig. 2. Schematic outline of Stratus instrument.

The appropriate test number is selected from the keyboard and the microprocessor will automatically select the proper parameters for each test.

The first step in the procedure is the addition of sample or prediluted sample to a tab which is indexed under the sample station. Indexing generally occurs at the rate of one tab per minute. Eight indexing steps are required to bring the tab to the read station, thus the total time to complete an assay is approximately eight minutes. Following the addition of sample, the tab is indexed to the conjugate station and conjugate is added for either the sequential or sandwich assays. Following this step, the tab is indexed to the substrate station where the substrate wash occurs. This is the key feature of the technology. The addition of substrate to the center of the tab causes anything which is not bound to the antibody to be washed to the edge of the tab. At the same time the wash is occuring, the enzyme cleaves the substrate producing fluorescence. The rate of enzyme activity is then read at the read station by front surface fluorescence.

A schematic outline of this read station is shown in Fig. 3. A small section of the center of the tab is read continuously and the rate of change of fluorescence is calculated over an approximatey 30 second time period. Since the free antigen is outside of the field of view of the instrument, its contribution to the rate is not detected. The aperature is usually smaller in diameter than the antibody coated area, therefore only a portion of the bound antigen is read for each tab. The enzyme rate observed for each tab is proportional to the concentration of analyte being measured. An appropriate standard curve is prepared for each test using a six level series of calibrators and patient results are calculated using a modification of the four parameter logistic fit of Rodbard (1978).

PREPARATION OF REAGENTS

<u>Preparation of Antibody Coated Tabs</u>

Rabbit antibody against digoxin was obtained from Antibodies, Inc. Rabbit antibody against thyroxine was obtained from Cambridge Medical Diagnostics, Inc. Antibody to ferritin was produced by injection of New Zealand White rabbits with human liver ferritin.

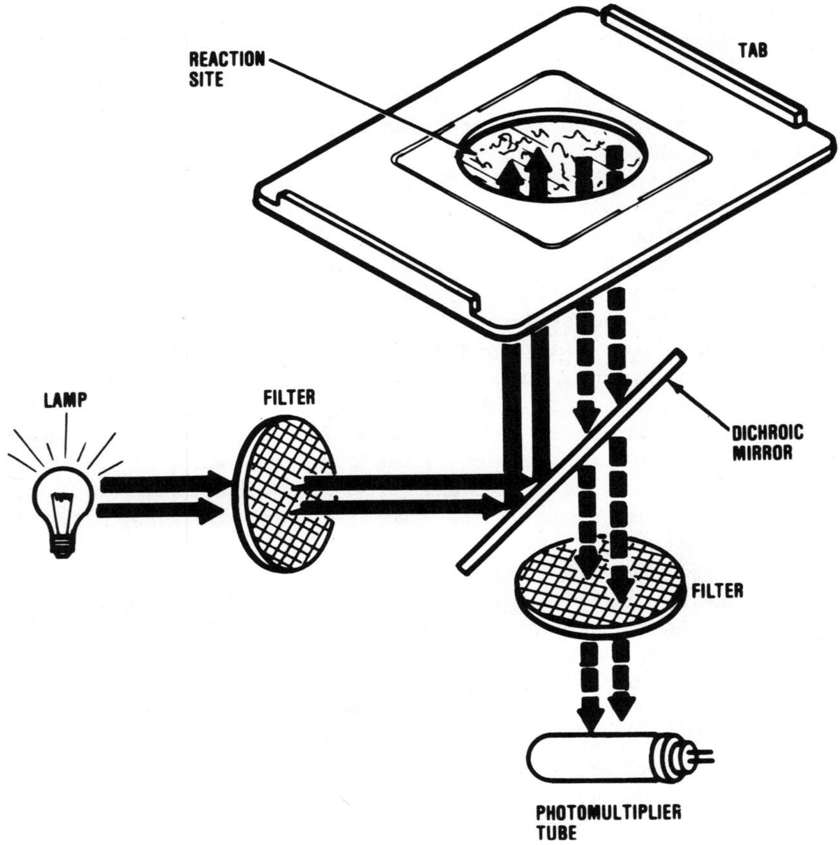

Fig. 3. Schematic outline of front-surface fluorometric read station.

The ferritin was isolated by modification of the procedure described by Granick (1942) and was further purified by chromatography on sepharose 6B. Rabbits were injected with 100 ug each of purified ferritin mixed with equal volumes of Freund's adjuvant and after eleven weeks rabbits were boosted with a second 100 ug dose of antigen. After an additional eight weeks the rabbits were bled and the antisera from five rabbits was pooled and used for coating antibody. Goat antirabbit antibody was obtained from Antibodies, Inc. Glass fiber filter paper, grade GF/F was obtained from Whatman, Inc. and cut into one inch squares. These squares are then placed into a plastic tab holder to conduct an assay. Although some antibody may be coated onto the glass fiber filter paper directly, we have found that premixing the primary antibody with a second antibody and allowing a soluble complex to form results in a substantially improved level of binding to the filter paper. An example of the optimization of antibody concentration for a ferritin assay is shown in Fig. 4. In this figure three different concentrations of primary antibody, (rabbit antiferritin) are mixed with different dilutions of goat antirabbit antiserum in the percentages indicated. The immune complex is allowed to form for approximately 24 hours. Any particulate material is filtered out using a membrane filter and the remaining soluble complex is spotted onto filter paper tabs which are then dried. The tabs are stored dry at 2-8C and have been found to be stable for one year under these conditions. Similar optimizations are conducted for digoxin and thyroxine antibodies.

Preparation of enzyme labeled conjugates

Digoxin was conjugated to *E.coli* alkaline phosphatase by a modification of the procedure of Smith (1970) which involves periodate oxidation of digoxin and coupling to alkaline phosphatase via a Schiff base formation. The product is purified by gel filtration. Thyroxine is also conjugated to *E.coli* alkaline phosphatase by first thiolating alkaline phosphatase and then coupling this with an amine derivitive of thyroxine using a heterobifunctional coupling agent.

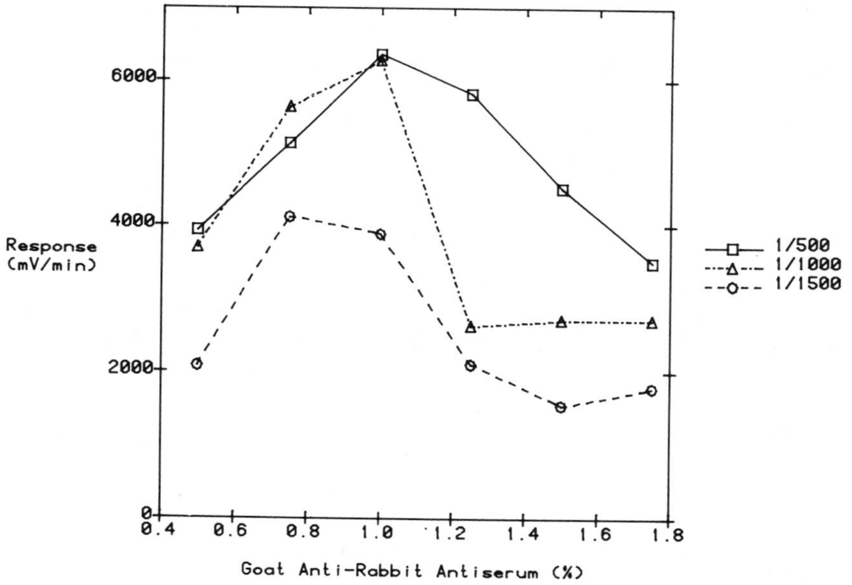

Fig. 4. Optimization of antiferritin antibody coating of filter paper tabs.

For the ferritin conjugate, B-galactosidase was coupled to an affinity purified Fab' fragment of rabbit antiferritin by the method of Ishikawa (1983). For this procedure an IgG cut of rabbit antiferritin is prepared using QAE sephadex chromatography and the IgG cut is digested with pepsin to produce the $F(ab')_2$ fraction which is then isolated by ACA 44 Gel This fraction is purified by affinity chromatography using human spleen ferritin coupled to agarose beads. This affinity purified fraction is then reduced with 2-mercaptoethylamine to produce the Fab' monomer. Maleimide groups are then introduced with N-N'-o-phenylenedimaleimide. The Fab' maleimide intermediate is isolated by gel chromatography on G25 Sephadex and coupled to B-galactosidase. The Fab' enzyme conjugate is isolated by chromatography on Sepharose 6B. The final conjugate is collected and diluted into a buffer consisting of 30 mM sodium phosphate, pH 7.5, 150 mM sodium chloride, 1% bovine albumin and 1% normal rabbit serum with 0.1% sodium azide as a preservative. This conjugate has been found to be stable for approximately one year when stored at 2-8C.

Preparation of substrate wash solution

For the alkaline phosphatase coupled conjugates the substrate wash solution consists of 370 mM 4-methylumbelliferyl-phosphate and 1mM magnesium chloride, 1M Tris adjusted to 9.0. For B-galactosidase the substrate is 300 mM 4-methylumbelliferyl B-D-galactopyranoside which is prepared by dissolving 200 mg in 40 ml of dimethylformamide and added to 2 liters of phosphate buffer (50 mM sodium phosphate pH 7.4, 50 mM sodium chloride 0.1% Tween and 0.1% sodium azide). Both these substrate wash solutions have been found to be stable for more than one year when stored at 2-8C.

Preparation of Calibrators

Calibrators for digoxin were prepared from normal human serum collected from donors who were not receiving digoxin. For thyroxine, serum was stripped by charcoal to remove endogenous T4 and calibrators were prepared from T4 free serum. Calibrators for ferritin were prepared by removing endogenous ferritin from human serum by affinity chromatography. A purified preparation of human liver ferritin was then added to the stripped serum to prepare the calibrators.

Measurement Procedures

The measurement of digoxin is done directly on serum. No pretreatment is necessary. The calibration curve routinely used was from 0-4 ng/dl. The ferritin procedure is conducted in a similar fashion, directly on serum without prior treatment. The calibrator range for ferritin was from 0-600 ng/ml. For thyroxine it is necessary to predilute the sample into a buffer containing 8-anilino-1-napthalene sulfonic acid (ANS) which diplaces thyroxine from its binding proteins. For this procedure, 50 ul of serum is combined with 20 ul of conjugate and 200 ul of the ANS diluent prior to addition to the Stratus instrument. The calibrator range for thyroxine is from 0-25 ug/dl.

RESULTS AND DISCUSSION

Sequential Assays - Digoxin

The measured output for each tab on the instrument consists of a rate measurement of enzyme activity which is usually expressed as millivolts per minute. A typical series of rate curves for a set of digoxin calibrators is shown in Fig. 5. The highest rates are observed at the lowest concentrations and as the concentration increases the rate decreases. These rates are used for the calculation of the calibration curve. The signal to noise ratio of these rates is extremely good and linearity of the enzyme activity is observed over a several minute time period.

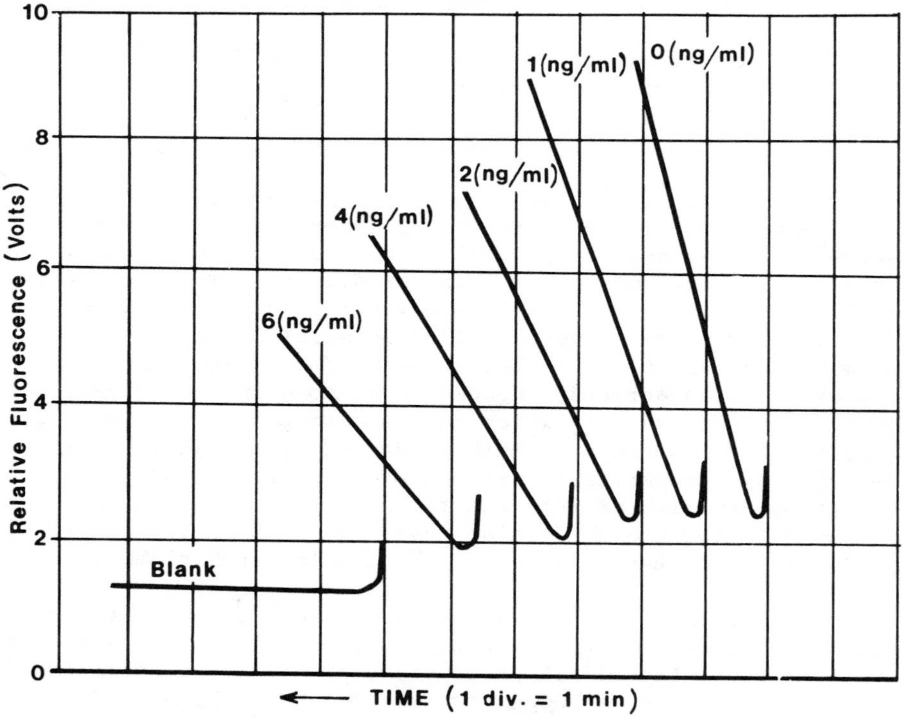

Fig. 5. Enzyme rates at various concentrations of digoxin calibrator.

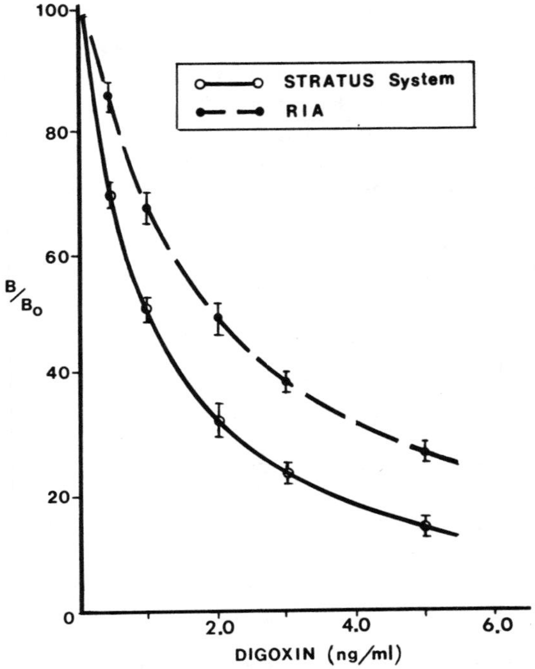

Fig. 6. Digoxin calibration curves for the Stratus[R] system and for an RIA procedure.

Typically the rate is measured over a 30 second period for each tab. Of particular interest in Fig. 5 is the low blank rate observed with the system. The blank consists of an antibody tab, coated with an antibody other than for digoxin. Thus there is no specific binding to the tab. The extremely low blank rate indicates that there is very little non-specific binding in the system. The glass fiber filter paper is fairly inert and after being coated with antibody, becomes even more inert. Occasional preparations of conjugate which have become polymerized will give higher blank rates, however methods are available which routinely give conjugate preparations without polymer formation. These low blank rates are also of interest for sandwich assays in which the low concentration standards also have low rates and sensitivity may be limited by the non-specific binding of such blanks.

In Fig. 6 the rates shown in Fig. 5 are plotted in a more conventional manner with B/Bo plotted vs concentration. Also shown is a conventional RIA procedure using a double antibody separation technique with the same antibody and the same calibrators as for Fig. 5. This gave very similiar curves.

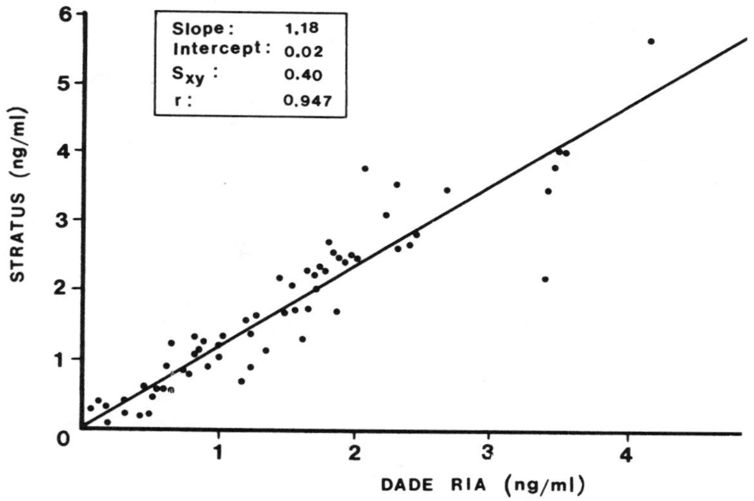

Fig. 7. Correlation between the Stratus[R] digoxin method and an RIA procedure for patients' samples.

In Fig. 7 the correlation between the Stratus method and an RIA procedure on patient samples is shown and agreement between the two methods is quite good. Additional comparisons with other methods have been conducted with the Stratus system for digoxin and have shown good correlation with a variety of techniques.

To study the timing of the system, the kinetics of binding of both digoxin and enzyme labeled digoxin to the tabs was analyzed. In Fig. 8 the binding curve for a sample containing 2 ng/ml of digoxin is illustrated. In this study conjugate was added at the intervals indicated after the 2 ng/ml sample was spotted. At 90 seconds the digoxin was approximately 90% bound to the tab indicating that this first reaction is quite rapid. Fig. 8 also illustrates the binding of conjugate for samples containing no digoxin and for a sample containing 2 ng/ml of digoxin. In this case the conjugate is added 2 minutes after the sample is added. The reaction is essentially 90% complete in less than 4 minutes. The molecular size of the conjugate is much larger than the free drug, and could account for the difference in binding rates.

Since the instrument provides good control of timing, it is not necessary for the reaction to go to completion to provide a useful measurement. For digoxin we have routinely used 2 minutes for the first step and 3 minutes for the second step of this reaction. Even shorter time intervals could provide useful assays and we have recently obtained useful digoxin curves with less than 4 minutes total assay time.

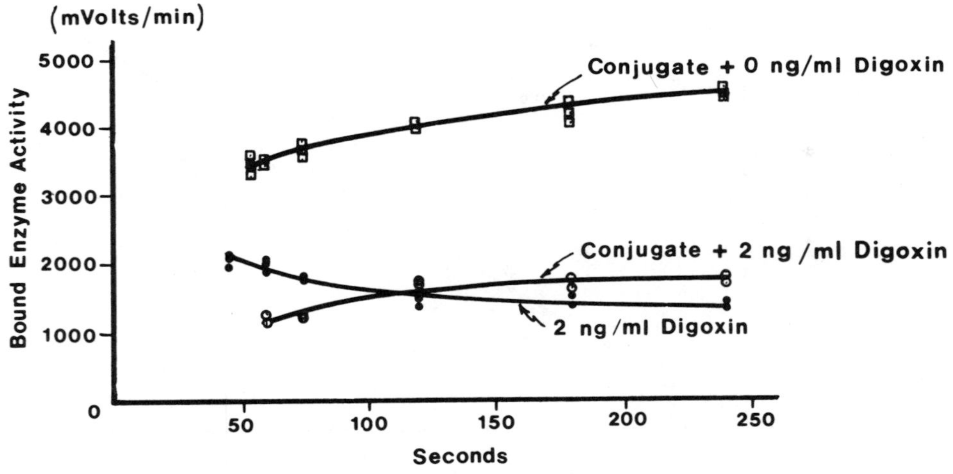

Fig. 8. Rate of binding of digoxin and enzyme-labeled digoxin to antibody tabs.

Competitive Assays - Thyroxine

The Stratus instrument has been designed to achieve long term stability by using a stable fluorescent reference material and adjusting the gain at the beginning of each run for any drift which has occured between runs. Calibration may be stable for periods in excess of two weeks. Table I shows an example of clinical data obtained over a 20 day period for thyroxine. Forty-eight samples of a stable control material at three different levels were analyzed in triplicate on each day. This represents 16 runs. Over this time period no drift in the system was observed and precision was very good.

Table 1. Thyroxine Calibration Curve Stability (period: 20 days)

	Level I	Level II	Level III
n	48	48	48
Mean (ug/dl)	3.815	7.684	14.672
S.D.	0.254	0.371	0.773
CV Total	6.66%	4.83%	5.27%

Table 2. Assay Precision for Thyroxine

	Level I	Level II	Level III
n	82	82	82
Mean (ug/dl)	3.640	7.050	15.212
S.D.	0.221	0.266	0.607
CV Within	4.18%	2.85%	3.69%
CV Between	4.48%	2.53%	1.54%
CV Total	6.06%	3.77%	3.99%

A further example of the assay precision is shown for thyroxine in Table 2 with 82 samples for the same controls. The precision data is shown for within run and between run. The between run is similar to the within run, indicating very low drift in the system over an extended time frame.

To further analyze the differences between methods, an extensive study was conducted with thyroxine. Table 3 gives a summary of four RIA methods and one method using fluorescence polarization technology. All five methods gave similar slopes against the radial partition immunoassay procedure and correlation was good both for studies conducted in-house as well as in outside laboratories.

TABLE 3. Method Comparison for Thyroxine

		n	Intercept	Slope	r	Sy.x
NML/RIA	(In House)	85	0.618	0.93	0.89	0.92
NML/RIA	(Site 1)	124	0.847	0.882	0.92	1.34
Abbott/TDx*	(Site 1)	111	1.38	0.862	0.94	1.23
Clinical Assays/RIA	(Site 2)	113	1.04	0.888	0.94	1.39
NML/RIA	(Site 3)	110	0.389	0.930	0.93	1.03

An example of a recovery study for thyroxine is shown in Table 4 in which two samples of serum were spiked with either 5, 10 or 15 ug/dl of thyroxine and then analyzed on the Stratus instrument. The percent recovery is quite good. Additional studies with other analytes have been conducted and similar recoveries were observed.

---

*TM: Abbott Laboratories, Chicago, IL 60064

TABLE 4. Thyroxine Recovery Study

Sample #	Thyroxine Expected (ug/dl)	Thyroxine Recovered (ug/dl)	% Recovery
1	5.0	4.83	96.6
1	10.0	10.43	104.3
1	15.0	14.49	96.6
2	5.0	4.90	98.0
2	10.0	9.77	97.0
2	15.0	14.97	99.8

The effect of various types of plasma on the system has been examined and data for thyroxine is shown in Table 5. Four samples were collected, either as serum or as plasma. No significant difference was observed between either serum or four different types of plasma for thyroxine. In addition, other drug assays, including digoxin, have been examined and either serum or plasma could be used for all assays analyzed thus far. This could provide a clinical advantage since it may be possible to do testing faster on plasma than waiting for blood to clot to produce serum.

TABLE 5. Serum / Plasma Equivalence for Thyroxine (Means from four normal individuals, ug/dl)

		PLASMA				OVERALL	
SAMPLE	SERUM	EDTA	HEPARIN	OXALATE	CITRATE	MEAN	CV%
1	7.28	7.11	7.43	6.96	6.57	7.07	4.2
2	6.44	6.41	6.63	6.08	5.86	6.28	4.4
3	6.67	6.00	6.53	5.79	7.02	6.40	7.0
4	7.31	6.73	7.28	6.59	6.38	6.86	5.4

Sandwich Assays - Ferritin

The sandwich assays are performed in a similar manner to sequential assays. Sample is added to sample cups and placed in the instrument. At the conjugate position, enzyme labeled anti-ferritin is added and substrate for the enzyme is added at the substrate position. Fig. 9 shows a typical calibration curve for ferritin up to 600 ng/ml and the effect of different incubation times on the calibration curve. For this experiment the tabs are indexed at either 50, 60 or 70 second intervals. With the 50 second index time the incubation period for the first reaction with antiferritin is 100 seconds and the second incubation with enzyme labeled antiferritin is 150 seconds. Total assay time is less than 7 minutes with a 50 second cycle time, approximately 8 minutes with a 60 second cycle time and approximately 9 minutes with a 70 second cycle time. The major impact of increased incubation time is at the high end of the curve. Very little increase in low end sensitivity is observed in going from 50 seconds to 70 seconds. Since sensitivity was adequate at 60 seconds this time interval was selected for further investigation.

A second factor which can influence sensitivity of the assay is the concentration of the enzyme labeled antiferritin. Fig. 10 shows the effect of doubling the concentration of conjugate on the calibration curve. These assays were conducted at a 60 second

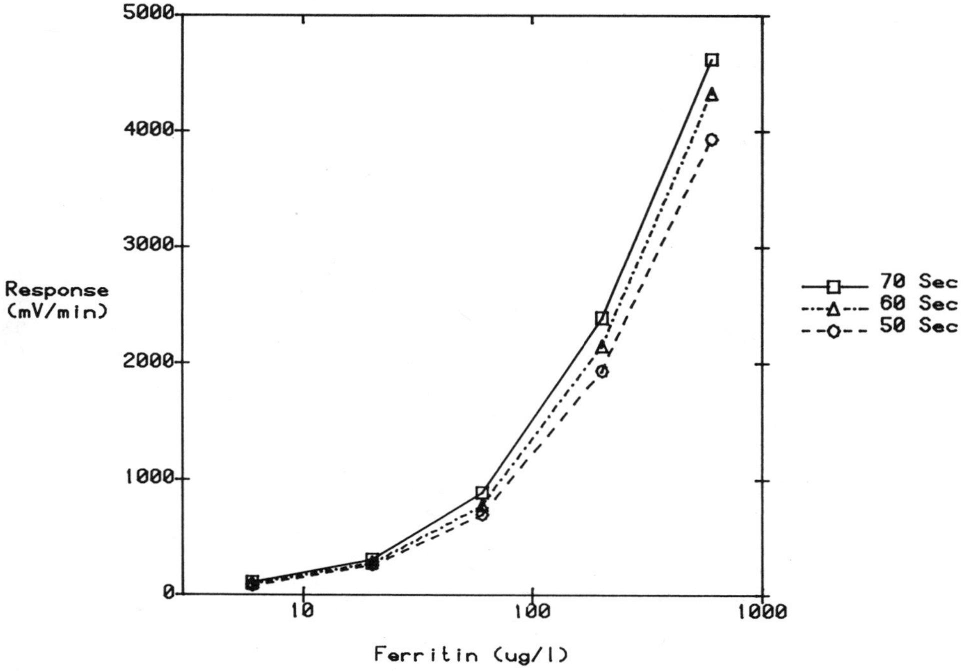

Fig. 9. Calibration curve for ferritin assay at different incubation times.

index interval. The major effect of increasing conjugate concentration is an improved signal at the high end of the curve. Some increase in background is observed with increasing concentration of conjugate which limits the improvement in low end sensitivity. The conjugate concentration is optimized to achieve the best low end sensitivity while still giving a useful dynamic range.

A more detailed analysis of the calibration curve and sensitivity is shown in Table 6 which gives the results of the calibration curve repeated seven times over an eight day period. Each calibrator was run in triplicate and the results were pooled to produce the mean rate for 21 observations. The SD of each rate was calculated and sensitivity was determined according to the method of Rodbard (1978). Based on this calculation a sensitivity of 0.8 ng/ml was achieved. This sensitivity is very adequate for clinical use and permits precise evaluation of low ferritin concentrations. Table 6 also illustrates the excellent stability achieved in the overall system. No trends or drifts in the data were observed over this eight day period suggesting that both the instrument and reagents are quite stable.

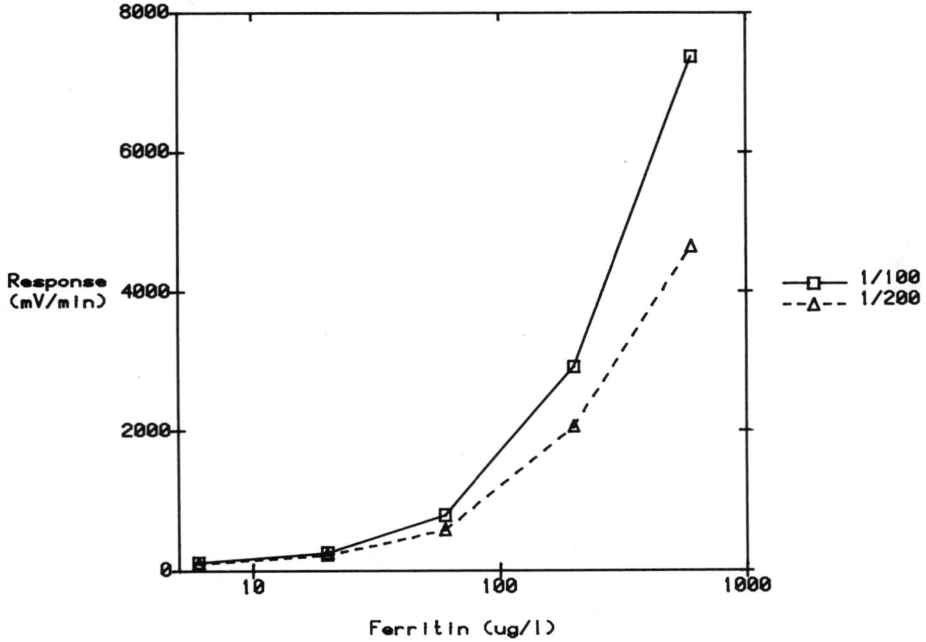

Fig. 10. Ferritin calibration curves with two different concentrations of enzyme labeled antibody.

Table 6. Stability of ferritin calibration curve over an 8 day period.

For each calibration run, triplicate samples of each level of calibrator were analyzed. This procedure was repeated 7 times over an 8 day period.

Number of observations	Concentration of calibrator (ug/L)	Mean rate millivolt/minute	SD of Rates
21	0	28.7	6.7
21	6	129	7.5
21	20	407	7.1
21	60	1184	31
21	200	3227	83
21	600	6117	135

Sandwich assays appear to have improved precision over both sequential and competitive assays in routine use. Table 7 gives an example of precision obtained on three controls over an eight day period. Each control was analyzed in quadruplicate for each run. Total CV's were approximately 3%. The improved precision of sandwich assays is probably related to the fact that each tab is coated with an excess of antibody and slight variations in the coating process do not cause a variation in the final result. Both the sequential and competitive assays require an exact amount of antibody on each tab. Any variability in antibody coating will result in some degradation of precision.

TABLE 7. Precision for ferritin over an eight day period.

Three controls were analyzed over eight days. For each run quadruplicate samples were run for each level of control. A total of seven runs were conducted with results calculated from a single calibration curve.

Number of Observations	Mean Value (ug/L)	S.D.	Total CV(%)	Within Assay CV(%)	Between Assay CV(%)
28	41.1	1.2	3.0	2.6	1.4
28	127	3.6	2.9	2.1	2.0
28	193	6.1	3.3	1.7	2.8

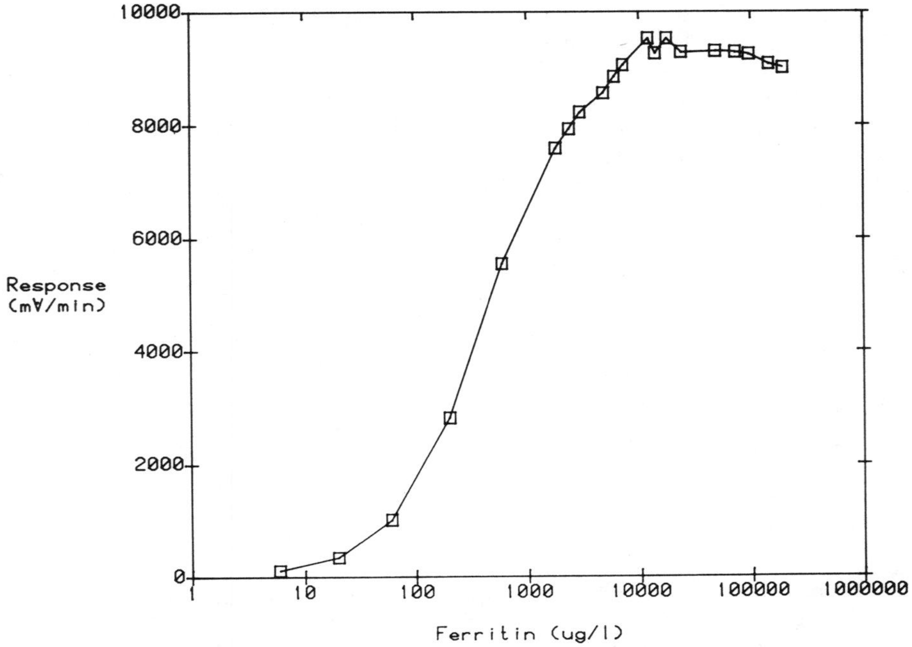

Fig. 11. Extended calibration curve for ferritin and the absence of a high dose hook effect.

To evaluate the high dose hook effect described by Miles, et al. (1974). Samples containing purified human liver ferritin were analyzed up to a concentration of 192,000 ng/ml. These data are shown in Fig. 11. The assay reaches a plateau at approximately 10,000 ng/ml, however no hook effect is evident even at 192,000 ng/ml. Ng, et al. (1983) has reported values on clinical samples as high as 47,000 ng/ml, therefore the absence of a hook effect is an important assay parameter. A useful assay could be obtained up to 1,000 ng/ml and still achieve good low end sensitivity. This is one of the largest dynamic ranges reported for ferritin.

The hook effect can be attributed to two factors, one of which is competition with excess antigen from the sample for the enzyme labeled second antibody. This is minimized or non-existent in radial partition immunoassay because the addition of the enzyme labeled antibody serves to wash out any antigen which has not been bound to the antibody in the first step. The second factor which could cause a hook effect is loss of antigen bound in the first reaction during the second reaction. Since the second reaction is generally less than three minutes this loss of antigen appears to be minimal.

DISCUSSION

Radial partition immunoassay has proven to be a very versatile technology for enzyme immunoassay combining rapid assays with high sensitivity. The technology is adaptable to assays for both high and low molecular weight ligands. The speed of the reaction can be attributed to several factors. The filter paper has a very high surface area to volume ratio. We have calculated that the filter paper has approximately 2,000 times greater surface area to volume ratio than a coated tube. In addition diffusional distances are very small between antibody and ligand. The average distance between fibers in the paper is 1 micron. Therefore molecules are in fairly intimate contact with antibody on the solid phase throughout the reaction.

In addition, for sequential and sandwich assays, no dilution of the sample is required for the first step of the reaction. The wash step is extremely efficient and serves to remove unbound material very completely. This achieves a low blank rate and when coupled with the low non-specific binding observed with antibody coated filter paper, sensitivity is very good. The use of fluorometric readout for the enzyme activity further improves sensitivity.

Additional assays currently under investigation in our laboratory include the detection of hepatitis surface antigen in serum, the measurement of B-HCG, TSH and CK-MB. Additional applications being considered are detection of cancer antigens as well as serological and microbiological applications.

REFERENCES

A. Giegel, J. L., Brotherton, M. M., and Cronin P. et al. 1982. Radial partition immunoassay. Clin. Chem. 28:1894-1898.

B. Granick, S., (1942), Ferritin I physical and chemical properties of horse spleen ferritin. J. Biol. Chem. 14:451-461.

C. Ishikawa, E., Imagawa, M., and Hashida, S. et al. 1983. Enzyme labeling of antibodies and their fragments for enzyme immunoassay. J. Immunoassay 4:209-327.

D. Miles, L.E.M., Lipschitz, D. A., Bieber, C. P., and Cook,J.D., 1974, Measurement of serum ferritin by a 2-site immunometric assay. Anal. Biochem. 61:209-224.

E. Ng, R. H. Brown, B. A., and Valdes, R. Jr., 1983. Three commercial methods for serum ferritin compared and the high dose "hook effect"eliminated, Clin. Chem. 25: 326-329.

F. Rodbard, N., 1978, Statistical estimation of the minimal detectable concentration ("sensitivity") for radioligand assay. Anal. Biochem. 90:1-12.

G. Smith, T. W., Butler, V. P., and Haber, E., 1970,Characterization of antibodies of high affinity and specificity for the digitalis glycoside digoxin. Biochem. 2:331.

DEVELOPMENT OF IMMUNOCHEMICAL ENZYME ASSAYS FOR CARDIAC ISOENZYMES

Richard Wicks and Magdalena Usategui-Gomez

Hoffmann-La Roche Inc.
Department of Diagnostic Research
340 Kingsland Street
Nutley, NJ 07110

INTRODUCTION

Background: Lactate Dehydrogenase

The term isoenzyme is applied to proteins that catalyze the same chemical reaction, yet are structurally different. This heterogeneity of enzymes has been known for several decades (Vessell, 1958; Cahn, 1962; Markert, 1963; Nisselbaum, 1963). In the last few years, isoenzyme determinations have become very important in the clinical laboratory because they have proven to be powerful diagnostic tools since the various tissues and organs of man exhibit characteristic isoenzyme profiles. When an infusion of particular isoenzymes into the circulation occurs due to damage to a specific tissue or organ, an isoenzyme determination can often pinpoint the site of injury. After a myocardial infarction (MI), isoenzymes of lactate dehydrogenase (LD) and creatine kinase (CK) are released from the myocardial tissue into the circulation. We have utilized the different immunochemical properties of CK and LD isoenzymes to develop simple, rapid and precise immunoassays for determining the specific cardiac isoenzymes of both enzymes.

Lactate dehydrogenase (EC 1.1.1.27) appears in human serum as five isoenzymes which are tetramers composed of two types of monomer subunits. The isoenzymes comprise all five combinations of the two monomers, H (for heart-derived) and M (for skeletal muscle-derived), as represented by the formulas H_4, H_3M, H_2M_2, HM_3, and M_4. It is common practice in the clinical laboratory to refer to the isoenzymes according to their electrophoretic mobilities, the fastest anodally migrating isoenzyme being termed "LD-1" (H_4) and the slowest "LD-5" (M_4). The synthesis of each kind of subunit is believed to be under the control of a separate gene. It has been suggested by Markert (1963) that both subunits assort randomly with one another during the formation of the LD molecule. Therefore, the characteristic distribution of isoenzymes so important in clinical diagnosis is determined by the relative abundance of each subunit in a particular tissue.

The reactions of LD isoenzymes with antisera prepared by injecting highly purified human or rhesus monkey LD-1 or LD-5 into goats or rabbits have been thoroughly studied (Markert, 1963; Nisselbaum, 1963; Burd, 1973a and b). Antibodies produced against LD-1 do not cross react with LD-5, and vice versa. Nevertheless, as would be expected from their hybrid tetrameric structure, LD-2, LD-3, and LD-4 react with antibodies to either LD-1 or LD-5.

Early Immunochemical Methods for LD-1

Burd, et al. (1973a) prepared antibodies against highly purified rhesus monkey muscle M_4 lactate dehydrogenase, and found that the antiserum specifically inhibited the enzymatic activity of purified human lactate dehydrogenase isoenzymes in proportion to their M monomer content, indicating the existence of four active sites per tetramer. They developed a very simple test based on the inhibition of all M subunits in the serum of patients. By obtaining an accurate and precise measurement of the percent H subunits in patient's serum, the authors were able to clearly differentiate those patients having suffered a myocardial infarction. Excellent correlation with electrophoretic techniques was demonstrated. Boll, et al. (1974) carried out precipitation studies with the LD isoenzymes of human origin and antisera against the human isoenzymes H_4 and M_4. These authors found that anti-M_4 precipitated all multiple forms consisting of at least one M subunit, while the multiple forms consisting of at least one H subunit could be precipitated by anti-H_4 in the presence of polyethylene glycol. Therefore, they were able to determine in extracts of human tissue H_4, M_4 and the total hybrid enzyme content. This type of approach was not readily applicable to the routine determination of isoenzymes.

Development of Immunochemical LD-1 Assay

Purification of LD-5. The purification of M_4 from human liver has been described in great detail (Burd, et al., 1973a). Solubilization of the lactate dehydrogenase was achieved by grinding the material followed by high speed homogenization in a Waring blender. The cell debris was removed and the homogenate ready for a series of chromatographic and salt fractionation steps as indicated in Table I. The enzyme preparation obtained was isoenzymically homogeneous and in a high degree of purity.

Table I. Purification of LD-5 Isoenzyme from Human Liver Tissue

Step		Protein (mg)	Spec. Act. (units/mg)	Purification (fold)
1.	Homogenate	125,000	521	--
2.	Calcium phosphate	32,700	1,560	3.0
3.	$(NH_4)_2SO_4$	9,300	3,420	6.6
4.	CM-Sephadex	1,190	23,700	45.5
5.	DEAE-Sephadex	81.9	233,000	446
6.	Sephadex G-100	62.5	240,000	460

Preparation of First and Second Antibodies. New Zealand white male rabbits were given initial injections of human M_4 in complete Freund's adjuvant equally divided among the four foot pads (total of 1-2 mg/rabbit). Boosters were given intravenously at 3 week intervals using 1 mg on both the 21st and 23rd day after the previous booster. Blood was collected 10 days after the boost. Goats were immunized using the same schedule with levels of the enzyme antigen 5 times that of the rabbits.

Donkey anti-goat gamma-globulin was conjugated to polyvinylidene fluoride floccules by a method described by Newman, et al. (1980). The polymer antibody conjugate concentration was 10% in 20 mM Tris pH 7.5, 1 g azide, and 5 g of bovine serum albumin per liter. The binding capacity was approximately 30 ug of goat IgG per milliliter. The antibody to goat IgG was prepared in donkeys by immunization with chromatographically pure IgG in complete Freund's adjuvant. Precipitation analysis of the pooled serum showed 6 mg of precipitable antibody protein per milliliter.

Properties of Antisera. Large differences in performance have been observed in antiserum obtained from different animals and/or bleedings. Mixtures of inhibitory and precipitating antibodies were elicited in all animals, however the strength of the antiserum as determined by enzyme inhibition varied dramatically from animal to animal. An example of an antisera with a low inhibition titer can be seen in Figure 1A. After the addition of 20 ul of antiserum, 40% of M_4 activity still remained. However, 10 ul of the same antiserum contained enough precipitating antibodies to remove from solution most of the M_4 isoenzyme. In contrast, Figure 1B shows the results obtained with a very potent inhibitory antiserum which required less than 5 ul to completely inhibit and also precipitate all of the isoenzyme present. Therefore, in producing antisera the investigator has to clearly analyze each bleeding in order to carefully single out those sera with the right inhibitory or precipitating properties as the case may be. Good quality control of antisera is essential for consistency of performance.

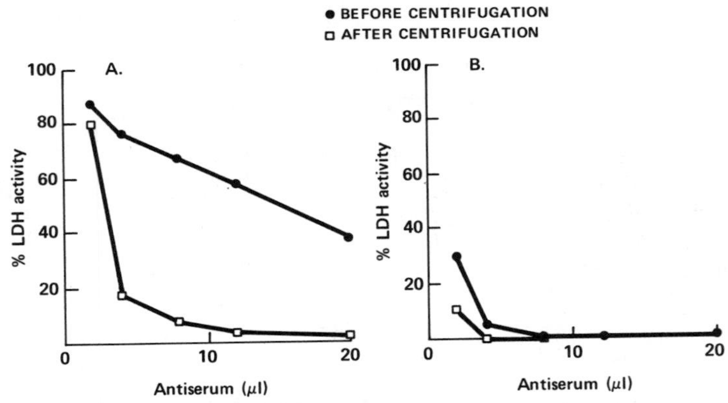

Fig. 1. Comparison of rabbit antisera with low (A) and high (B) inhibitory titers. Varying levels of antisera were incubated overnight with human M_4. The enzyme activity was measured before (●———●) and after o———o centrifugation at 2500 x g for 30 minutes.

The extent of inhibition of the purified human isoenzymes by M_4 antibodies was carefully studied. Also, the effect of the antiserum on the activities of the five human isoenzymes for the conversion of both lactate to pyruvate and pyruvate to lactate was examined. As shown in Figure 2, the degree of inhibition of the isoenzymes was proportional to their M subunit content, regardless of the enzyme activity assay employed.

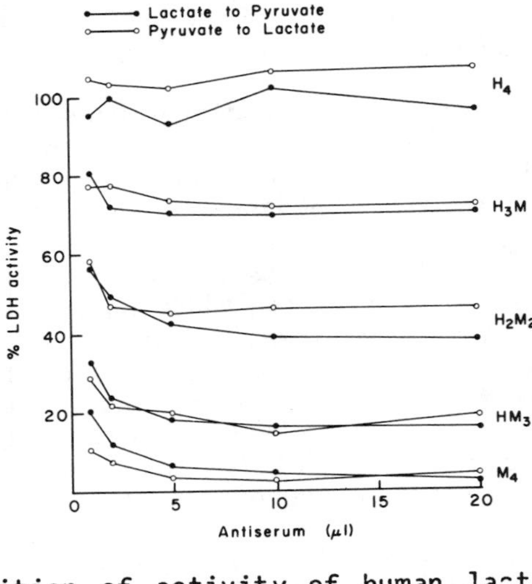

Fig. 2. Inhibition of activity of human lactate dehydrogenase isoenzymes by goat anti-monkey M_4 antiserum. Enzyme activity was measured for the conversion of lactate to pyruvate (●———●) and pyruvate to lactate (o———o).

Assay Procedure. Goat anti-LD-5 serum (50 ul) diluted in tris (hydroxymethyl) methylamine buffer is added to 200 ul of patient's serum, gently mixed and incubated for 5 minutes at ambient temperature. Polymer-second-antibody conjugate (200 ul) is then added, gently mixed, and again incubated for 5 minutes at ambient temperature. After centrifugation at 1000 x g for 5 minutes to remove the insoluble complexes of all isoenzymes containing M subunits, the supernatant fluid is tested for LD-1 activity employing a suitable reagent system.

It is essential to add enough goat anti-LD-5 serum to insure complete precipitation of purified LD-5 isoenzyme up to a concentration of 500 U/L. The effect of various volumes of goat anti-LD-5

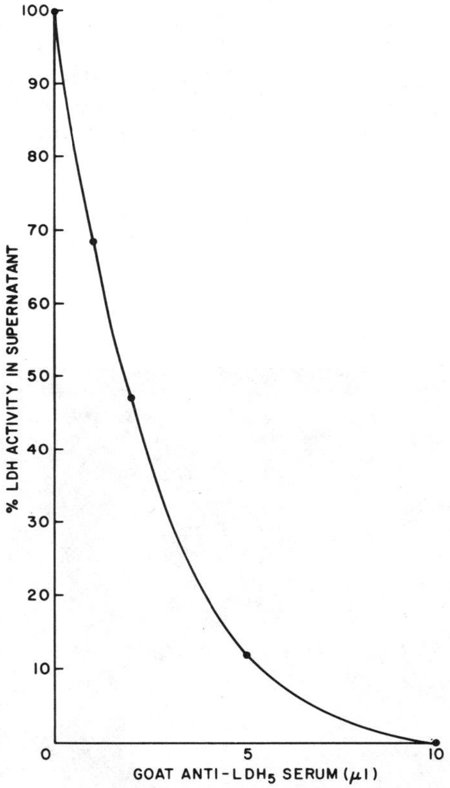

Fig. 3. Lactate dehydrogenase LD-5 precipitation as a function of volume of antiserum. Various amounts of goat anti-LD-5 serum (diluted 20-fold) were incubated with purified human LD-5 for 5 minutes. After adding insoluble donkey anti-goat gamma-globulin and centrifuging to remove insoluble antigen-antibody complexes, the supernatants were assayed for enzyme activity.

serum on precipitation of purified LD-5 isoenzyme can be seen in Figure 3. When enough antiserum is added to bind the available isoenzyme, no activity is found in the supernatant after centrifugation and removal of the LD-5 antibody complexes. The effectiveness of the separation of LD-1 from the four other isoenzymes is demonstrated in Figure 4. After treatment of a positive MI serum and a normal serum with anti-LD-5 followed by centrifugation, the electropherogram of the supernatants clearly shows the removal of all four isoenzymes containing the M subunit.

Electrophoresis has been the classical procedure for the determination of LD isoenzymes since the early 1950's. Therefore, correlation studies were performed comparing LD-1 activities obtained by both technologies. As shown in Figure 5, good correlation was obtained between the immunochemical and electrophoretic procedures. The deviation of the regression line slope from unity can be explained by the fact that LD-1 and LD-5 have different substrate affinities and it is not feasible to use a gradation of lactate concentrations in the staining solutions for a single electropherogram. The value of this slope will depend on the reagents used for determining LD-1 activity and for developing the electropherogram.

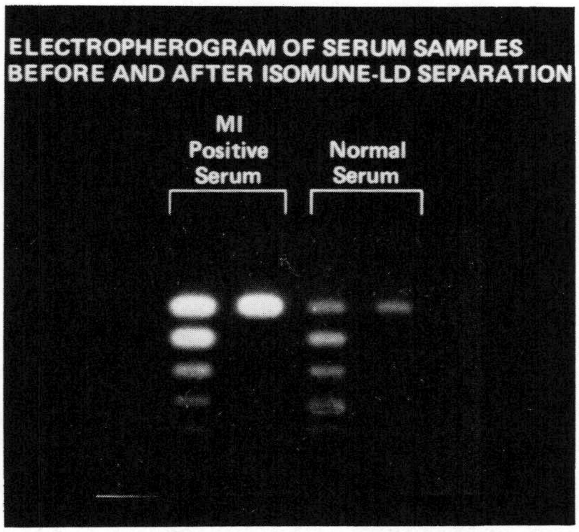

Fig. 4. LD electropherogram of serum from a patient with and without a myocardial infarct, before and after treatment with first and second antibodies.

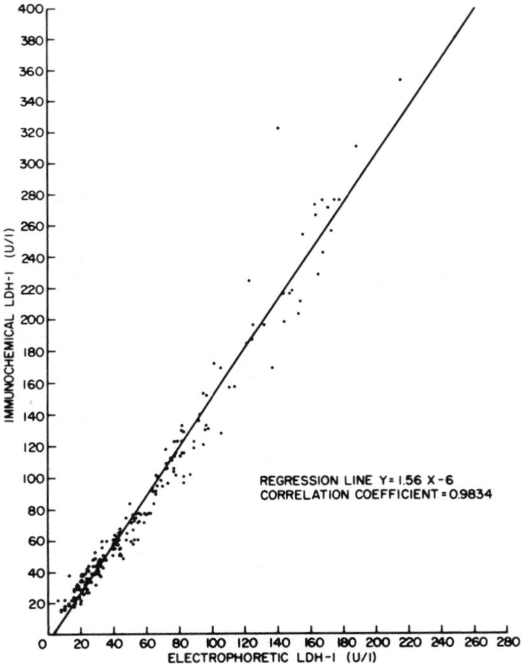

Fig. 5. Correlation between results of the immunochemical and electrophoretic determinations of LD-1.

Clinical Studies

A number of investigators have published their results on the use of this immunochemical LD-1 assay (Isomune-LD^R) as an aid in the diagnosis of myocardial infarction. Without exception, their conclusions have confirmed that this immunochemical determination of LD-1 is a precise, accurate, and rapid assay which can provide excellent sensitivity and specificity for the diagnosis of myocardial infarction. Figure 6 illustrates the overall clinical behavior of the LD-1 assay as first reported by Usategui-Gomez, et al. (1979). Selectivity appears to be maximum on day 2, although many patients exhibited a diagnostically useful LD-1 value on day one and even at the time of admission. Bruns, et al. (1981) carefully studied LD-1 levels in 65 patients admitted to the coronary care unit. Blood was sampled on admission, at four hour intervals during the first twenty-four hours of hospitalization and at eight to twelve hour intervals during the second hospital day. Thereafter, samples were drawn once every twenty-four hours. By twelve hours after admission, 90% of LD-1 results were positive; all were positive by twenty-four hours after admission. In this study, the authors found that the ratio of LD-1 to total LD was a better discriminator for myocardial infarction than was the absolute value of LD-1. The LD-1 to total LD ratio remained increased for several days after infarction, whereas almost 40% of the CK-MB results had reverted to normal by forty-eight hours after admission.

Fogh-Anderson, et al. (1982) used LD-1 determinations on 113 consecutively admitted patients with suspected myocardial infarction. Blood samples were drawn on admission and in the morning of the 1, 2, 3, 5, and 7th days. Selectivity in this group of patients appeared to be maximum on days 2 and 3 with sensitivities of 97% for both days and specificities of 95 and 98% respectively. Even at day 5 the sensitivity was still 97% and the specificity 90%. These investigators did not find the ratio of LD-1

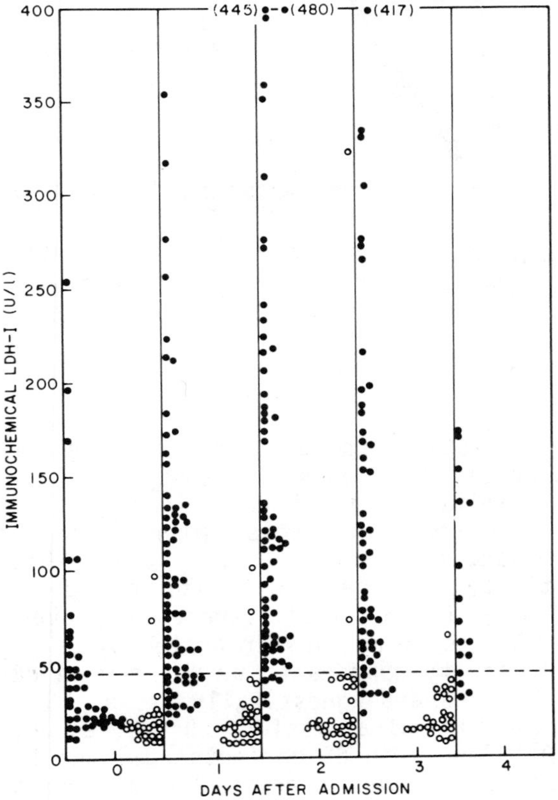

Fig. 6 Serum LD-1 activities on the day of admission and four days thereafter in all non-myocardial infarct (o) and myocardial infarct (●) patients studied. The dashed line represents the decision level (43 IU/L) used as cutoff.

to total LD to be useful. Based on their results, they recommended a future analysis program for patients with suspected myocardial infarction consisting of creatine kinase (CK-MB) and LD-1. McCoy, et al. (1983) obtained multiple serum samples from 85 patients on admission and at 12, 24, and 48 hours. The sensitivities of LD-1 on admission and at 12, 24, and 48 hours was 43%, 84%, 95%, and 100% respectively, and the specificity at these same time intervals was 95%, 100%, 97%, and 97% respectively.

As indicated above, a key question currently being debated is whether LD-1 or the LD-1 to LD total ratio is the better discriminator to distinguish between an MI and a non-MI patient. Gerhardt (1983) pointed out that the ratio might have a higher diagnostic sensitivity in minor infarctions in patients with low-normal total LD activity. However, they indicated how total LD activity is increased in a variety of non-MI conditions, including heart failure with liver stasis, thus decreasing the ratio and causing false negatives. Their own results indicated LD-1 levels to be more reliable than the LD-1 to total LD ratio.

Ali et al. (1981) published in their study of 100 consecutive patients hospitalized for clinical suspicion of myocardial infarction an interesting analysis of the time required to perform LD-1 as compared to electrophoresis. The time required to perform assays on 100 sera was 320 minutes by Isomune LD and 548 minutes by electrophoresis. Therefore, there was a saving of three hours and 48 minutes for the LD-1 assay. The time study for electrophoresis included the time required for preparing LD electrophoresis graphs for mounting on the laboratory reports.

Foo et al. (1981) examined several methods proposed by the Association of Clinical Biochemists for their ability to differentiate between serum samples from patients with cardiac or liver diseases. The techniques investigated were: immunochemical determination (Isomune-LD), electrophoresis, α – hydroxybutyrate dehydrogenase, urea stable LD, and heat stable LD. The number correctly classified was greatest for LD-1 by immunoprecipitation, closely followed by heat stable LD and by LD-1 using electrophoresis. α – hydroxybutyrate and urea stable LD did not show statistically significant improvements over total LD measurements.

Because there is more LD-1 than LD-2 isoenzyme in the myocardium, a sensitive assay to detect the presence of increased LD-1 in serum should signal the existence of myocardial necrosis before the time when LD-1 exceeds LD-2 in the peripheral blood. As Usategui-Gomez et al. (1979) have shown, the LD-1 assay becomes of diagnostic value earlier in 20 to 25% of the patients when compared to the criteria of positive LD-1/LD-2 ratio by electrophoresis.

In conclusion, LD isoenzyme determinations have not been applied extensively in the routine diagnosis of acute myocardial infarction in spite of their proven diagnostic value. This was probably due, in part, to the limitations of electrophoresis, which is a tedious and time-consuming technique, and to the lack of specificity of assays that depend on gradations of physical and chemical properties of isoenzymes. The immunochemical LD-1 system is rapidly becoming the method of choice in the clinical laboratory. This system offers the advantages of simplicity, high sensitivity and specificity, avoidance of any additional set of enzyme reagents or instrumentation, and ease of incorporation into the routine of any clinical laboratory that performs LD assays.

Background: Creatine Kinase

The CK-MB isoenzyme of creatine kinase (ATP: creatine N-phosphotransferase EC 2.7.3.2) is widely accepted as the most sensitive indicator of myocardial injury (Galen, 1975, and 1978; Lott, 1980). The clinical utility of this isoenzyme has led to a multitude of assay techniques for its measurement. Due to the specificity of antigen-antibody reactions, immunochemical reactions of CK isoenzymes have been thoroughly investigated as diagnostic tools for the measurement of CK-MB activity.

In 1964, electrophoretic techniques revealed the existence of three isoenzymes of creatine kinase with differing mobilities on agar gels (Burgher 1964). These isoenzymes were found to be dimers composed of two different subunit types: MM (muscle derived), BB (brain derived), and the hybrid MB isoenzyme. The isoenzymes are also designated CK-1 (BB), CK-2 (MB), and CK-3 (MM), according to their electrophoretic migration. CK-BB is the fastest anodally migrating isoenzyme, CK-MM the least anodally migrating, and the hybrid CK-MB isoenzyme has an intermediate migration. The existence of three separate CK isoenzymes prompted various investigators to purify the individual isoenzymes and to produce and characterize antisera to them (Neumeier et al. 1976; Jockers Wretou et al. 1975; Zweig et al. 1978; Morin et al. 1974; Keutel et al. 1972; Roberts et al. 1977). It soon became apparent that the two types of subunits were immunologically distinct. Antisera produced to the CK-MM isoenzyme do not react with CK-BB and vise versa. The hybrid CK-MB isoenzyme predictably reacts with both BB and MM antibodies. These observations led us to develop a specific and rapid immunochemical assay for CK-MB activity in human serum.

Development of a Simple and Rapid Immunochemical CK-MB Assay (Isomune[R] - CK)

This immunochemical method for CK-MB utilizes both the inhibitory and precipitating properties of goat antiserum to the CK-MM isoenzyme (Wicks et al. 1982). In this assay, an excess of inhibitory M-subunit antibodies is added to a serum specimen in order to completely block the enzymic activity of M-subunits from CK-MM and CK-MB. In a second tube, the same antibodies are used to precipitate all CK-MM and CK-MB in the patient's serum. The difference in activity between these two tubes gives a quantitative measurement of the B-subunit activity of the hybrid CK-MB isoenzyme. The following is a detailed description of the development of this assay.

Preparation of reagents. The CK-MM was purified from human skeletal muscle as described previously (Usategui-Gomez et al. 1981). Antibodies were produced in goats by weekly subcutaneous injections of 2 mg of purified CK-MM emulsified in complete Freund's adjuvant. For this immunochemical method, the antibodies were titrated against a purified preparation of CK-MM diluted in human serum (figure 7).

Essentially 100% of the M-subunit activity is inhibited by the antibodies after a 10 to 20 minute incubation at room temperature. Considerable variation in antibody titer and inhibitory properties were found between the goats immunized. Approximately one goat out of five demonstrated sufficiently high titers and inhibition of CK-MM to be useful in this assay.

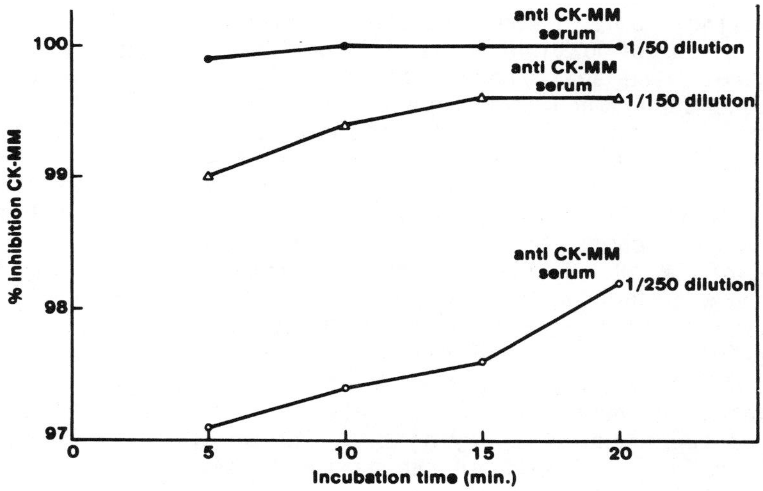

Fig. 7. Kinetics of the immunoinhibition reaction: A control serum containing purified human CK-MM was used to test various dilutions of anti-CK-MM serum in the immunochemical CK-MB assay. Incubation time of the immunoinhibition tube (tube 1) was also varied to examine the kinetics of the reaction.

Antibodies to goat IgG were raised in donkeys by weekly subcutaneous injections of 2.5 mg of purified goat IgG. The donkey anti-goat IgG serum was fractionated by ammonium sulfate precipitation, and coupled to polyvinylidene fluoride floccules by hydrophobic absorption (Newman et al. 1980). The binding capacity of this polymer-antibody conjugate was 400 ug goat IgG per milliliter of a 20% (w/v) suspension. This high capacity solid phase second antibody suspension was used to quantitatively bind the large excess of CK-MM antibodies required for the immunoprecipitation step.

The goat antibodies to CK-MM were also titrated for the specific precipitation step. A control serum containing 80% purified CK-MM and 20% purified CK-MB activity in human serum (total CK activity 1300 U/L at 37°C) was incubated for five minutes at ambient temperature with various dilutions of the antiserum. An excess of the solid phase antibody was added and again incubated for five minutes. After centrifugation, the supernates were tested in electrophoresis (after a five fold concentration) to check for complete precipitation of CK-MM and CK-MB. Antiserum dilutions of 50 to 200 fold were generally necessary to fulfill the inhibitory and precipitating requirements of the assay.

Immunochemical CK-MB procedure. The diluted anti-CK-MM serum (250 uL) is added to 200 uL of patient's serum and incubated for a minimum of twenty minutes at ambient temperature. While this tube is incubating, 50 uL of the same diluted antiserum is incubated with another 200 uL of patient's serum for five minutes at ambient temperature. The solid phase anti-goat IgG (200 uL) is added to this second tube, incubated five minutes, and then centrifuged at 1000g for five minutes to remove the insoluble complexes of CK-MM and CK-MB. The residual activity in this "blank" tube and in the first tube are then measured using an appropriate CK-substrate reagent. In the first tube, the residual activity not inhibited by the antibodies comes from four possible sources: the B-subunit activity of CK-MB, B-subunit activity from CK-BB or atypical isoenzymes with B-subunits, mitochondrial CK, and adenylate kinase. The residual activity in the supernate of the blank tube will be essentially the same as the first tube for all of these sources, except for the B-subunit activity of CK-MB isoenzyme which is precipitated in the blank tube. Subtracting the residual activity in the two tubes and multiplying by a factor of two will give the original activity of CK-MB in the patient's serum.

Characteristics of immunochemical CK-MB assay. In order to assess the effect of CK-BB on this immunochemical assay, we assayed CK-MB positive samples before and after adding purified CK-BB to the sera. Within the limits of error of the assay, CK-BB had no apparent effect on the CK-MB results (figure 8 and table II). Precision of the assay at 100 IU/L and 11.2 IU/L CK-

MB was 1.3% and 2.4%, respectively, for intra-assay variation, and 4.5% and 6.3% for inter-assay variation. Linearity was excellent for a positive control serum diluted serially in saline, giving a regression line of $y = 1.002\,x - 0.6$ (x = theoretical CK-MB, Y = actual CK-MB).

Table II. Influence of CK-BB on the Immunochemical Assay for CK-MB

	CK-MB, U/L	
Sample	Before adding BB	After adding BB
1	108.0	104.8
2	77.0	78.8
3	63.2	66.0
4	42.0	40.0
5	45.6	47.8
6	49.0	48.8
7	63.2	65.6
8	113.4	100.2

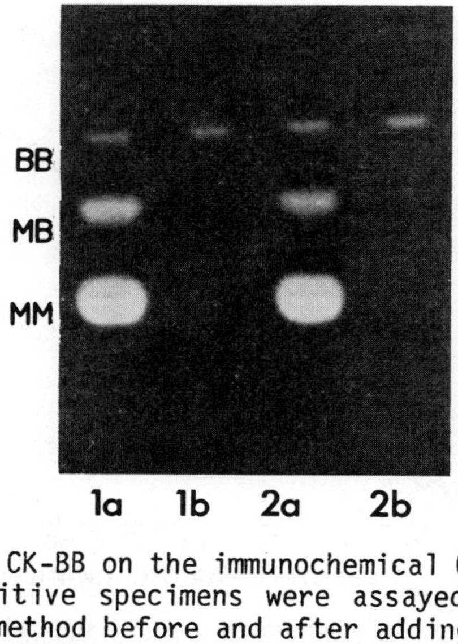

Fig. 8. Effect of CK-BB on the immunochemical CK-MB assay: Eight CK-MB positive specimens were assayed with the immunochemical method before and after adding 30 IU/L of CK-BB. The electropherogram shows sample 1 and 2 (from table II) before and after precipitation with CK-MM antiserum and second antibody suspension.

Our original clinical studies with this immunochemical assay demonstrated essentially equivalent diagnostic efficiency to an electrophoretic method for CK-MB detection. Forty patients who did not have myocardial infarcts (by clinical and biochemical criteria) provided a cutoff value for CK-MB activity in this study. Serial samples from these patients gave a mean and SD of 2.4 and 2.1 U/L respectively. An upper limit of normal of 8.7 U/L (mean plus 3 SD) was used to calculate diagnostic sensitivity in this study.

Using this cutoff value, 43 out of 43 patients with myocardial infarcts demonstrated a positive CK-MB value on at least one serial specimen. Electrophoresis on agarose gels also had a sensitivity of 100% in this limited study. In the 40 patients without myocardial infarct, the specificity of the immunochemical method was 95%. One patient had a peak CK-MB value of 10.6 U/L with a diagnosis of myocardial ischemia and a positive CK-MB band on electrophoresis. A second patient also had a trace band of CK-MB by electrophoresis, with 9.8 U/L CK-MB by the immunochemical method. Extensive clinical studies (Ali et al. 1982; Bruns et al. 1983; Seckinger et al. 1983) have confirmed the high diagnostic efficiency of this immunochemical method.

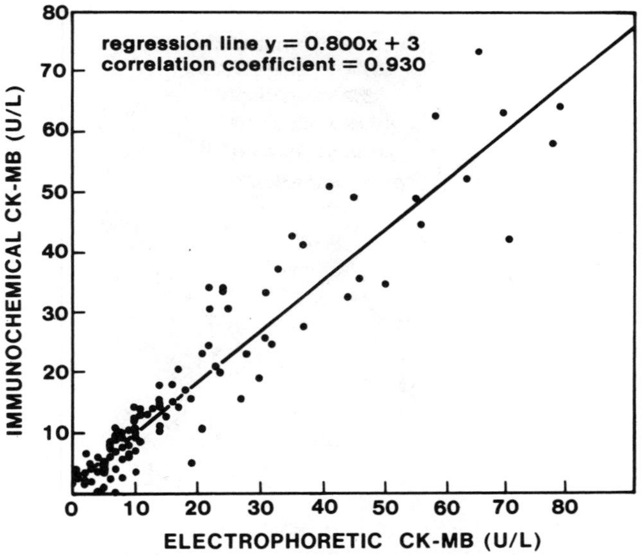

Fig. 9. Correlation of immunochemical and electrophoretic CK-MB values: Results on CK-MB positive specimens. Quantitation by electrophoresis was done by scanning densitometry.

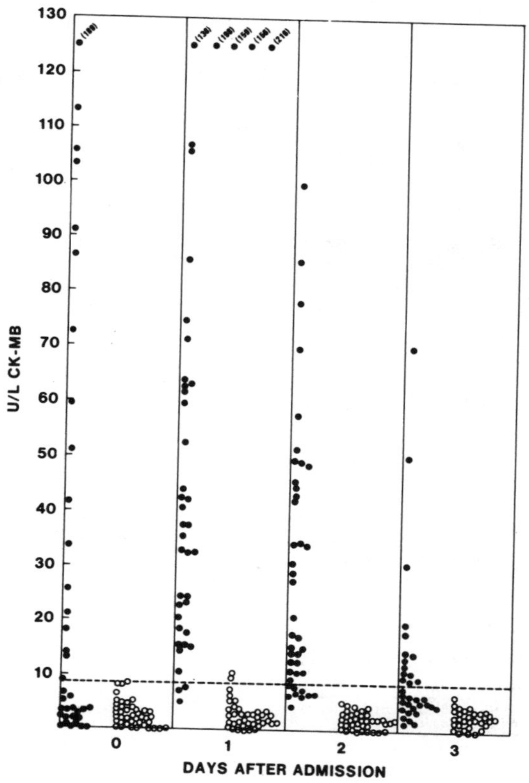

Fig. 10. Time course of immunochemical CK-MB: (•) patients diagnosed as having MI, and (o) patients with some other diagnosis. The dotted line represents the cutoff value of 8.7 IU/L.

<u>Atypical Isoenzymes.</u> During the course of the clinical studies on the immunochemical CK-MB method, we observed two patients who demonstrated the presence of atypically migrating CK bands on electrophoresis. These bands migrated between CK-MM and CK-MB on electrophoresis, and had elevated activity in both tubes of the immunochemical assay. Subtraction of the activity in the two tubes, and multiplying by two gave CK-MB activities below the cut-off value, confirming the specificity of this method with atypical CK samples (figures 11 - 13). Other authors have confirmed the specificity of this immunochemical assay with samples containing CK-BB and atypical isoenzymes (Seckinger, 1983 and Pesce, 1983). We have since investigated many clinical specimens for the presence of atypical CK isoenzymes, and have done binding studies with both CK-MM and CK-BB antibodies on these specimens. In agreement with the considerable body of literature on this subject, we have found two major types of atypical CK.

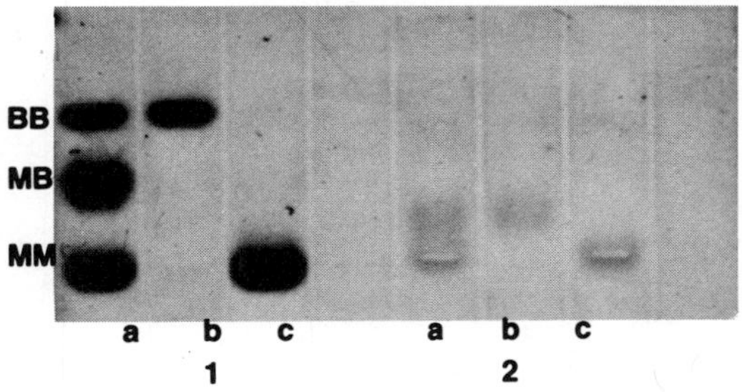

a = sample treated with buffer
b = sample treated with CK-MM antibody and second antibody suspension
c = sample treated with CK-BB antibody and second antibody suspension

	TOTAL CK	TUBE 1	TUBE 2	CK-MB
1 (control)	523	171.2	98.8	144.8
2	55	38.4	39.8	0

Fig. 11. Testing of atypical CK specimens in the immunochemical CK-MB assay.

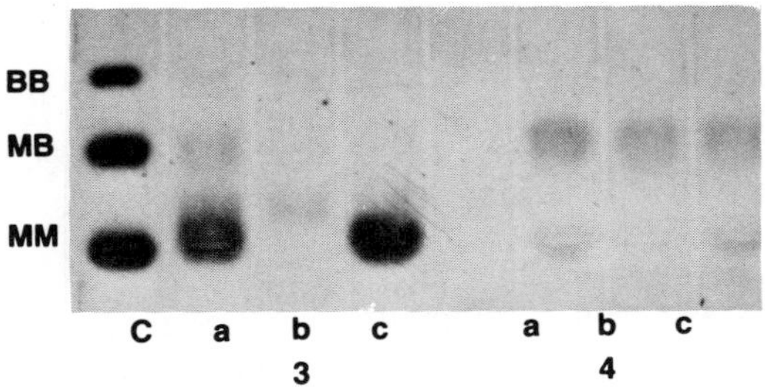

a = sample treated with buffer
b = sample treated with CK-MM antibody and second antibody suspension
c = sample treated with CK-BB antibody and second antibody suspension

	TOTAL CK	TUBE 1	TUBE 2	CK-MB
3	345	50.1	44.8	10.6
4	33	23.2	22.9	0.6

Fig. 12. Testing of atypical CK specimens in the immunochemical CK-MB assay.

The atypical CK most frequently observed is one which migrates between CK-MM and CK-MB on electrophoresis. This atypical CK isoenzyme is commonly referred to as type I, and has been shown to be a complex of CK-BB with immunoglobulin G in the patient's serum. A number of experimental approaches have been used to confirm this. By gel filtration chromatography, this isoenzyme

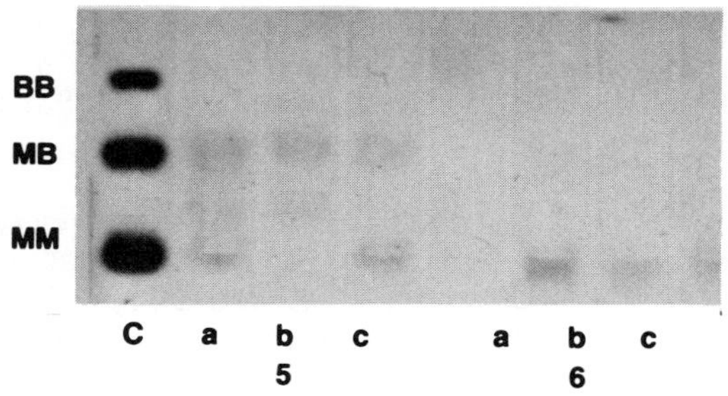

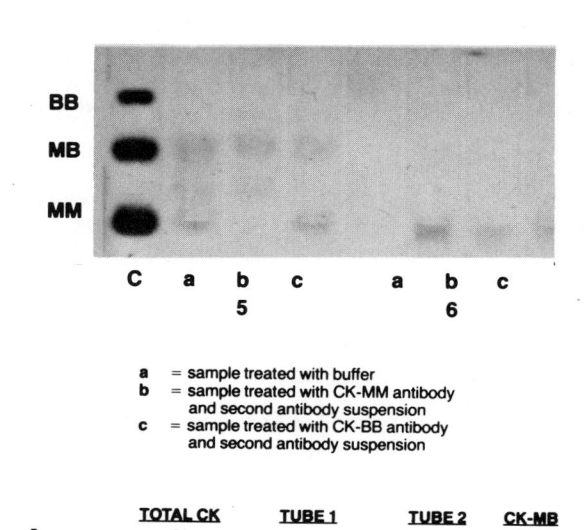

a = sample treated with buffer
b = sample treated with CK-MM antibody and second antibody suspension
c = sample treated with CK-BB antibody and second antibody suspension

	TOTAL CK	TUBE 1	TUBE 2	CK-MB
5	54	37.0	36.1	1.8
6	49	50.6	47.7	5.8

Fig. 13. Testing of atypical CK specimens in the immunochemical CK-MB assay.

elutes at a molecular weight corresponding to approximately 325,000 daltons. Binding studies have shown that this isoenzyme is precipitated by both CK-BB antiserum and protein A. In addition to this, normally migrating CK-BB can be recovered if the complex is exposed to high concentrations of urea to disrupt the IgG-BB bonds. Finally, if purified CK-BB is added to a patient's serum which has the type I atypical CK, the CK-BB is converted to the complexed form and will migrate between CK-MM and CK-MB. The type I atypical CK appears in the highest incidence in women, and usually is seen only in older patients. This isoenzyme commonly persists in the serum for months or years, and no significant clinical correlations have yet been made.

The second major type of atypical CK, commonly referred to as type II, migrates cathodic to CK-MM on electrophoresis. These bands have been shown to be predominantly CK of mitochondrial origin. A number of authors have shown that the mitochondrial form of CK does not react with antiserum to the MM or BB isoenzymes (Bohner, 1982 and Roberts, 1980). We have confirmed this with a limited number of type II atypical CK samples. There is also evidence to suggest the presence of two interconvertible forms of mitochondrial CK, (Hall, 1979 and Wevers, 1980), one which migrates cathodal to CK-MM and another which comigrates with CK-MM. The mitochondrial form of CK is occasionally present simultaneously with CK-BB, and a number of reports have indicated a correlation between this atypical CK and advanced stages of cancer (Kwong, 1981; James, 1979; Heinz, 1980; Stein, 1982; Bark, 1980; Pesce, 1982; Bayer, 1982).

In addition to the two classical types of atypical CK, we have observed two patients who had atypical isoenzymes comigrating with CK-MB on electrophoresis (Figures 12 and 13). These atypical CKs did not react with CK-MM or CK-BB antisera, and the nature of this particular atypical CK is unknown at present.

Clinical Evaluations of Isomune-CK^R

In addition to our original clinical studies with the specific immunochemical method (Wicks et al. 1982), a number of other authors have reported clinical evaluations of this methodology (Isomune-CK^R).

Ali, et al. (1982), studied 215 patients admitted consecutively to the coronary and intensive care units with suspicion of acute myocardial infarction. The diagnostic profile for these patients consisted of daily EKG's, total LD and LD-1 activity (Isomune-LD^R), total CK, and CK isoenzymes by both Isomune-CK and electrophoresis on cellulose acetate gels. The diagnosis of acute MI was established on the basis of clinical course, LD-1 activity, and EKG data. In some cases, technetium pyrophosphate scans or autopsy was also considered in the formulation of a diagnosis. Forty-five patients were diagnosed as acute MI by these criteria.

A cutoff value of 10 IU/L CK-MB was selected for the immunochemical CK-MB method on the basis of 100 assays on patients without acute MI (mean + 3 SD). For cellulose acetate electrophoresis, a sample was considered positive if a visible band was seen and confirmed by densitometry. A patient was classified positive by either assay if one or more of the serial specimens were positive. Using these criteria 43/45 MI patients (96%) and 162/170 non-MI patients (95%) were correctly classified by the Isomune procedure for a test efficiency of 95%. For the electrophoresis procedure, 42/45 MI patients (93%) and 161/170 non-MI patients were correctly classified, for an efficiency of 94%. This agrees very well with our original study which showed 100% sensitivity and 95% specificity in 83 patients.

Bruns, et al. (1983), also evaluated the Isomune-CK procedure in 99 patients admitted to a coronary care unit, and compared the results to electrophoresis on agarose gels. Samples were obtained on admission and at 4 to 8 hour intervals during the first 24 hours, and at 8 to 24 hour intervals until the fourth day after admission. The diagnosis in these patients was established by the clinical history, EKG changes, and serial changes in total CK, CK isoenzymes (electrophoresis) and LD isoenzymes. Total CK and CK-MB by the Isomune method were measured at 30°C.

In 27 patients with documented myocardial infarction, both the immunochemical and electrophoretic methods had at least one serial specimen positive, for sensitivities of 100%. In 72 patients without myocardial infarct, the specificity of the immunochemical method was 97%, and 94% by the electrophoretic method. Both assays provided laboratory evidence of infarction at about the same time. The first positive electrophoretic result occurred at 5.8 ± 4.4 hours after admission, and 5.3 ± 4.6 hours for the Isomune method.

The quantitative immunochemical assay for CK-MB also provided useful clinical information in three myocardial infarct patients who had re-infarctions during their hospital stay. An extension of the infarcts in these patients had been suspected based on EKG changes following a second episode of chest pain. Neither the total CK nor electrophoretic CK-MB results provided an indication of reinfarction, whereas the immunochemical method showed a secondary rise in CK-MB activity after the initial peak. This advantage of the quantitative immunochemical assay could be helpful in this frequent clinical problem (Marmor, 1981).

Wu and Bowers (1982) evaluated the Isomune-CK procedure and compared the results to immunoinhibition, using electrophoresis to confirm the presence of atypical forms of CK. Reference intervals of 8 IU/L CK-MB (immunoinhibition) and 5 IU/L CK-MB (Isomune-CK) were established at 29.8^{o}C based on analysis of 53 apparently healthy adults. In serum from 55 patients from an acute cardiology unit that were positive by the immunoinhibition method, all specimens had CK-MB activity greater than the cutoff value of 5 IU/L by the Isomune method ($y = 0.57x + 1.5$, y = Isomune CK, x = immunoinhibition). The correlation between the two methods was poor, however, in 34 patients that exhibited abnormal forms of CK by electrophoresis (atypical CK type I or II, or CK-BB). All specimens in this group were classified as positive by the immunoinhibition method, whereas no false positives were observed using the Isomune-CK assay, confirming the specificity of this method.

A clinical evaluation was done by Seckinger, et al. (1983), of the Isomune-CK procedure, and results were compared to immunoinhibition (Abbott A-gent CK-MB), radiometric CK-MB assay (Embria-CK, International Immunoassay Labs, Inc.), and electrophoresis on agarose gels (Corning Medical and Scientific). A reference interval was established for each method based on analysis of 40 apparently healthy subjects. Patients admitted to a coronary care unit to rule out myocardial infarction were investigated, with CK-MB determined on admission, and at 8 and 24 hours after admission. Fifteen samples that showed the presence of atypically migrating CK variants on electrophoresis were also included in the study to examine the specificity of the methods. All diagnoses were made retrospectively based on discharge diagnosis after reviewing the medical records. The following results were obtained:

	Immunoinhibition	Immunochemical	RIA	Electrophoresis
Sensitivity	18/18=100%	36/36=100%	5/5=100%	15/17=88%
Specificity	24/40=60%	138/148=93%	12/17=71%	32/36=89%
Efficiency	72%	95%	60%	89%

The sensitivities of the immunological methods were all 100% in this study. The electrophoretic method of CK-MB failed to detect CK-MB in two myocardial infarct patients, presumable due to poor sensitivity in detecting the presence of "trace" bands of CK-MB. The specificity of these four methods showed much greater variability. The immunoinhibition method showed frequent false positive results, predominantly due to the patients with atypical CK included in the study. The radiometric CK-MB assay theoretically should eliminate interference by the atypical CK's, however the poor precision of the assay in the normal range may account for these false positives. Overall, the immunochemical method demonstrated the highest efficiency of the four methods.

Summary

The well established clinical utility of the MB isoenzyme of creatine kinase has led to extensive studies on the immunochemistry of CK isoenzymes. These studies have led to many different assays for CK-MB including radioimmunoassay, immunoinhibition, "sandwich" type immunoassays, and a specific immunochemical assay (Isomune-CK^R). Due to the hybrid nature of CK-MB, only the "sandwich" assays utilizing both CK-BB and CK-MM antibodies, and the immunochemical method which combines immunoinhibition and immunoprecipitation, have been shown to measure CK-MB specifically. The greater awareness of a number of atypical CK isoenzymes has made the need for specificity in CK-MB assays more critical.

The simplicity, specificity, and excellent clinical correlation of these immunochemical methods for the analysis of the LD-1 and CK-MB isoenzymes are rapidly making them the system of choice for the analysis of cardiac isoenzymes.

References

Ali, M., Braun, E.V., Laraia, S., Olusegun, F., Nalebuff, D., and Palladino, P.H. (1981). Immunochemical assay for myocardial infarction. Amer. J. Clin. Path. 76:426-429.

Ali, M., Laraia, S., Angeli, R., et al. (1982). Immuno chemical CK-MB assay for myocardial infarction. Amer. J. Clin. Pathol. 77:573-579.

Bark, C.J. (1980). Mitochondrial creatine kinase. J. Amer. Med. Assoc. 243:2058-2060.

Bayer, P.M., Wider, G., Unger, W., et al. (1982). Atypische kreatinkinase isoenzyme. Klin. Wochenschr. 60:365-369.

Bohner, J., Stein, W., Steinhart, R., et al. (1982). Macro creatine kinases: Results of isoenzyme electrophoresis and differentiation of the immunoglobulin-bound type by radio assay. Clin. Chem. 28:618-623.

Boll, M., Backs, M., and Pfleiderer, G. (1974). Quantitative Immunologische bestimming von multipler formen der lactat-dehydrogenase in tierischem un humanen gewebe. Z. Physiol. Chem. 355:811-818.

Bruns, D.E., Emerson, J.C., Intemann, S., Bertholf, R., Hill, K.E., and Savory, J. (1981). Lactate dehydrogenase isoenzyme-1: Changes during the first day after acute myocardial infarction. Clin. Chem. 27:1821-1823.

Bruns, D., Chitwood, J., Koller, K., et al. (1983). Creatine kinase MB activity: Clinical and laboratory studies of specific immunochemical technique with optimized enzymatic assay. Ann. Lab. Clin. Sci. 13:59-66.

Burd, J.F., Usategui-Gomez, M., Fernandez De Castro, A., Mhatre, N.S., and Yeager, F.M. (1973a). Immunochemical studies on lactate dehydrogenase: Preparation and characterization of anti-M lactate dehyrogenase antiserum: Development of an immunochemical assay for lactate dehydro genase isoenzymes in human serum. Clin. Chim. Acta. 46:205-215.

Burd, J.F., and Usategui-Gomez, M. (1973b). Immunochemical studies on lactate dehydrogenase. Biochim. Biophys. Acta. 310:238-247.

Burgher, A., Richterich, H., and Aebi, H. (1964). Die Hetero genitat der Kreatin Kinase. Biochim. Z. 339:305-314.

Cahn, R.D., Kaplan, N.O., Levin, L., and Zwilling, E. (1962). Nature and development of lactic dehydrogenase. Science 136:962-969.

Fogh-Andersen, N., Sorensen, P., Muller-Petersen, J., and Ring., T. (1982). Lactate dehydrogenase isoenzyme I in diagnosis of myocardial infarction. J. Clin. Chem. Clin. Biochem. 20:291-294.

Foo, A.Y., Nemesanszky, E., and Rosalki, S.B. (1981). Comparison of procedures for the measurement of the anodal isoenzymes of lactate dehydrogenase in serum. Ann. Clin. Biochem. 18:232-236.

Galen, R.S., and Gambino, S.R. (1975). Isoenzymes of CPK and LDH in myocardial infarction and certain other diseases. Pathobiol. Annu. 5:283-315.

Galen, R.S. (1978). Isoenzymes and myocardial infarction. Diagn. Med. 1:40-52.

Gerhardt, W., Hofuendahl, S., Ljungdahl, L., Waldenstrom, J., Tryding, N., Pettersson, T., and Ohlsson, O. (1983). Serum LD-1 activity in suspected acute myocardial infarction. Clin. Chem. 29:1057-1060.

Hall, N., Addis, P., and DeLuca, M. (1979). Mitochondrial creatine kinase. Physical and kinetic properties of the purified enzyme from beef heart. Biochem. 18:1745-1751.

Heinz, J.W., O'Donnel, N.J., and Lott, J.A. (1980). Apparent mitochondrial creatine kinase in the serum of a patient with metastatic cancer to the liver. Clin. Chem. 26:1908-1911.

James, G.P., and Harrison, R.L. (1979). Creatine kinase isoenzymes of mitochondrial origin in human serum. Clin. Chem. 25:943-947.

Jockers-Wretou, E. and Pfleiderer, G. (1975). Quantitation of creatine kinase isoenzymes in human tissues and sera by an immunological method. Clin. Chim. Acta. 58:223-232.

Jockers-Wretou, E., and Plessing, E. (1979). Atypical serum creatine kinase isoenzyme pattern caused by complexing of creatine kinase-BB with immunoglobulins G and A. J. Clin. Clin. Biochem. 17:731-737.

Keutel, H.J., Okabe, K., Jacobs, H.K., et al. (1972). Studies on adenosine triphosphate transphosphorylases XI. Isolation of the crystalline adenosine triphosphate creatine transphosphorylases from the muscle and brain of man, calf and rabbit, and a preparation of their enzymatically active hybrids. Arch. Biochem. Biophys. 150:648-678.

Kwong, T.C., and Arvan, D.A. (1981). How many creatine kinase "isoenzymes" are there and what is their clinical significance? Clin. Chim. Acta. 115:3-8.

Liu, F., Belding, R., Usategui-Gomez, M., and Reynoso, G., (1981). Immunochemical determination of LDH-1. Amer. J. Clin. Path. 75:701-707.

Lott, J.A., and Stang, J.M. (1980). Serum enzymes and isoenzymes in the diagnosis and differential diagnosis of myocardial ischemia and necrosis. Clin. Chem. 26:1241-1250 Review.

Ljungdahl, L., and Gerhardt, W. (1978). Creatine kinase isoenzyme variants in human serum. Clin. Chem. 24:832-834.

Markert, C.L., and Appella, E. (1963). Immunochemical properties of lactate dehydrogenase isoenzymes. Ann. N.Y. Acad. Sci. 103:915-929.

Marmor, A., Sobel, B.E., and Roberts, R. (1981). Factors presaging early recurrent myocardial infarct ("extension") Amer. J. Cardiol. 48:603-610.

McCoy, M.T., Aguanno, J.J., and Ritzmann, S.E. (1983). Lactate dehydrogenase isoenzyme I; A rapid assay for ischemic myocardial disease. Arch. Intern. Med. 143:434-436.

Morin, L.B. (1979). Creatine kinase isoenzyme antibody reactions in immunoinhibition and immunonephelometry. Clin. Chem. 25:1415-1419.

Neumeier, D., Prellwitz, W., Wurzburg, U., et al. (1976). Determination of creatine kinase isoenzyme MB activity in serum using immunological inhibition of creatine kinase M subunit activity. Clin. Chim. Acta. 73:445-451.

Newman, E., Hager, H.J., and Hansen, H.J. (1980). Preparation of Kynar resin particles as a dispersed solid-phase antibody. Paper given at 72nd annual meeting of the Amer. Assoc. for Cancer Research, May 1980. Details of polyvinylidene fluoride coupling can be found in U.S. Patent No. 3,843,443.

Nisselbaum, J.S. and Bodansky, O. (1963). Immunochemical studies of human lactic dehydrogenase. Ann. N.Y. Acad. Sci. 103:930-937.

Pesce, M.A. (1982). The CK isoenzymes: Findings and their meaning. Lab. Mgmt. 20:25-37.

Pesce, M., and Bodourian, S. (1983). Effect of atypical creatine kinase isoenzymes and CK-BB on an immunochemical CK- MB assay. Clin. Chem. 29-584-585.

Roberts, R., Parker, C.W., and Sobel, B.E. (1977). Detection of acute myocardial infarction by radioimmunoassay for creatine kinase MB. Lancet ii, 319-322.

Roberts, R. and Grace, A.M. (1980). Purification of mitochondrial creatine kinase. J. Biol. Chem. 255:2280-2287.

Sax, S.M., Moore, J.J., Giegel, J.L., and Welsh, M. (1979). Further observations on the incidence and nature of atypical creatine kinase activity. Clin. Chem. 25:535-541.

Seckinger, D., Vasquez, A., Rosenthal, P., and Mendizabal, R. (1983). Cardiac isoenzyme methodology and the diagnosis of acute myocardial infarction. Amer. J. Clin. Path. 80:164-169.

Stein, W., Bohner, J., Steinhart, R., and Eggstein, M. (1982). Macro creatine kinase: Determination and differentiation of two types by their activation energies. Clin. Chem. 28:19-24.

Usategui-Gomez, M., Wicks, R.W., and Warshaw, M. (1979). Immunochemical determination of the heart isoenzyme of lactate dehydrogenase (LDH_1) in human serum. Clin. Chem. 25:729-734.

Usategui-Gomez, M., Wicks, R., Farrenkopf, B., et al. (1981). Immunochemical determination of CK-MB isoenzyme in human serum: A radiometric approach. Clin. Chem. 27:823-827.

Vessell, E.S., and Bearn, A.G. (1958). The heterogeneity of lactic and malic dehydrogenase. Ann. N.Y. Acad. Sci. 75:286-291.

Wang, T.Y., Godfrey, J.H., Graham, L.G., Haddad, M.N., and Hamilton, T.C. (1982). Clinical evaluation of immuno chemical assay of lactate dehydrogenase isoenzyme I in early detection of acute myocardial infarction. Clin. Chem. 28:2152-2154.

Weider, N. (1982). Laboratory diagnosis of acute myocardial infarction: Usefulness of determination of lactate dehydrogenase (LDH)-1 level and of ratio of LDH-1 to total LDH. Arch. Pathol. Lab. Med. 106:375-377.

Wevers, R., Mul-Steinbush, M., and Soons, J. (1980). Mitochondrial CK (EC 2.7.3.2) in the human heart. Clin. Chim. Acta. 101:103-111.

Wicks, R.W., Usategui-Gomez, M., Miller, M., and Warshaw, M. (1982). Immunochemical determination of CK-MB isoenzyme in human serum: II. An enzymic approach. Clin. Chem. 28:54-58.

Wu, A., and Bowers, G. (1982). Evaluation and comparison of immunoinhibition and immunoprecipitation methods for differentiating MB from BB and macro forms of creatine kinase isoenzymes in patients and healthy individuals. Clin. Chem. 28:2017-2021.

Yuu, H., Ishizawa, S., Takagii, Y., et al., (1980). Macro creatine kinase: A study on CK-linked immunoglobulin. Clin. Chem. 26:1816-1820.

Zweig, M.H., Van Steirteghem, A.C., and Schecter, A.N. (1978). Radioimmunoassay of creatine kinase isoenzymes in human serum: Isoenzyme BB. Clin. Chem. 24:422-428.

ENZYME-LINKED IMMUNOELECTROTRANSFER BLOT (EITB)

Victor C. W. Tsang,* George E. Bers,** and
Kathy Hancock*

*Division of Parasitic Diseases
Center for Infectious Diseases
Centers for Disease Control
Public Health Service
U.S. Department of Health and Human Services
Atlanta, GA 30333 USA

**Bio-Rad Laboratories, Research Product Group
32nd & Griffin Ave
Richmond, CA 94804 USA

INTRODUCTION

The enzyme-linked immunoelectrotransfer blot (EITB) (Tsang et al., 1983a) is a synergistic union of 3 independently powerful techniques. Complex mixtures of biological molecules are first separated by sodium dodecyl sulfate polyacrylamide gel electrophoresis (SDS-PAGE), then electrophoretically transferred to a solid matrix. Molecules thus "blotted" are visualized directly or indirectly with enzyme-labeled antibodies by enzyme-linked immunosorbent assays (ELISAs) or with isotope-labeled antibodies by radioimmunoassays (RIAs) and radiography. Variations within the 3 basic techniques are numerous; however, the end results are generally good in terms of resolution, sensitivity, fidelity, and specificity.

The usefulness of this synergistic method is evident from the explosive increase of published literature on the subject. From the first experiments in protein blotting by Towbin et al. (1979) to the almost uncountable variations of today, a MEDLARS II off-line bibliographic citation list generated in May 1984 retrieved over 250 involved with the blot technique and not dealing with nucleic acids. The actual number of articles on blot applications is most likely much higher.

Historically, electrophoretic transfer blots of separated biomolecules began with Southern (1975). The term "Western blot" was coined by Burnette (1981) to continue the tradition of "geographic" naming of transfer techniques. Detection of antigens separated by SDS-PAGE with peroxidase (EC 1.11.17)-labeled antibodies was suggested by Raamsdonk et al. (1977).

The scope of this article will be limited to the electrophoretic transfer of proteins and glycoproteins separated by SDS-PAGE and subsequently visualized with ELISA. Even with these constraints, it is not possible to cover comprehensively all published works in this area. For additional information on immunoblotting and dot immunobinding, the reader is referred to a recent review by Towbin and Gordon (1984). Reviews of blotting from 2-dimensional gels (Symington, 1984) and protein blotting in general (Gershoni and Palade, 1983) are also very helpful.

SDS-PAGE-separated proteins can be transferred onto solid matrices such as nitrocellulose and covalent linkage papers (Renart et al., 1979). Covalent linkage papers are generally believed to have higher capacities, however, at a sacrifice of higher backgrounds. By far the most frequently used blot matrix is nitrocellulose. Since the methods for protein separation by SDS-PAGE and/or isoelectric-focusing (IEF) are too numerous to review here, we will assume that the proteins have already been separated by one of these methods and that we have on hand the finished gel.

Since the publication of our detailed method (Tsang et al., 1983a) and the initial use of this procedure (Tsang et al., 1982), we have modified the procedure to include several labor-saving and resolution-improving steps. The method we will present herein is the one currently used in our laboratory and several others. Results obtained with this method have been most useful in elucidating complex antigen/antibody reactions in infections. The method has also been invaluable in resolving questions about specificities of monoclonal antibody productions.

METHODS

(A) Solutions

1. Blot buffer. 0.025 M Tris, 0.193 M glycine, 20% methanol (v/v), pH 8.35. The pH of this buffer may range from 7.8 to 8.4, depending on the source of reagents. Blot buffer pH should not be adjusted. Maintaining molarity, thus conductivity, is more important. We use ultra-pure Tris from Schwarz/Mann,* P.O. Box 965, Cambridge, MA 02142, and glycine from BDH,

Gallard-Schlesinger, Carle Place, NY 11514. Prepare 6 liters at a time and store at 4^oC. This solution can be used repeatedly for at least 20 blot operations.

2. PBS-Tween. 0.01 M Na_2HPO_4/NaH_2PO_4, 0.15 M NaCl, pH 7.20 phosphate-buffered saline (PBS), with 0.3% Tween 20 (polyoxyethylene sorbitan monolaurate from Sigma Chemical Co., P.O. Box 14508, St. Louis, MO 63178*). The PBS can be prepared in large quantities (20-40 liters), poured in liter bottles, sterilized, and stored at room temperature until use. Add the Tween (3 ml per liter) to the PBS slowly with stirring and prepare fresh daily.

3. Substrate solution. 50 mg of 3,3'-diaminobenzidine (3,3',4,4'-tetraaminobiphenyl tetrahydrochloride, 97%-99% pure from Sigma), 10 ul of H_2O_2 (30%), and PBS, pH 7.20 to 100 ml. Make fresh daily.

(B) Preparatory Actions at the SDS-PAGE Stage

It is essential to know the relative molecular weight positions of the resolved proteins bands on the transferred blot. The easiest way to achieve this is by incorporating prestained protein molecular weight standards in a separate lane in the original gel. These prestained standards are purchased from BRL, P.O. Box 6009, Gaithersburg, MD 20877. The use, treatment, and calibration of these standards have been described in detail by Tsang et al. (1985). Alternatively, unstained protein standards may be used and stained with India ink (Hancock and Tsang, 1983), after being blotted as previously described.

(C) EITB Procedure

This procedure requires a trans-blot cell (e.g., from Bio-Rad Laboratories).

1. Draw a line (with a fine-point ballpoint pen) 0.5 cm from the top edge of a 16- x 21-cm nitrocellulose sheet (BA 83, 0.2 um, in 3-m rolls from Schleicher & Schuell, Keene, NH 03431) and soak it in blot buffer for about 10 min before the SDS-PAGE ends. The line will be used to mark the top of the gel and blot contact surface of the nitrocellulose. Caution: Wet first one side of the nitrocellulose sheet, then the other. If both sides of the sheet are submerged simultaneously, air may be trapped in the matrix.

*Use of trade names and commercial sources is for identification only and does not imply endorsement by the Public Health Service or the U.S. Department of Health and Human Services.

2. Place about 300 ml of blot buffer in a tray with a piece of filter paper (15 x 20 cm, Cat # GB002 from Schleicher & Schuell).

3. Ease gel onto filter paper in blot buffer. Caution: initially the stacking gel portion must float down without touching the paper. Once it is touched, the gel tends to stick to the filter paper, making repositioning difficult. Place the filter paper to one side at the bottom of the tray containing buffer. Once the gel is settled at the bottom, pull it slowly, smoothly, and squarely onto the filter paper. Now with a single smooth motion, while gently holding gel and paper together, slide both out of the blot buffer.

4. Place the paper and gel onto one of the Scotch-Brite pads of the trans-blot cell, with the gel facing up.

5. Wet the gel surface with about 10 ml of blot buffer. Ease the wet nitrocellulose sheet onto the gel by bending it slightly and positioning the middle, then lowering one edge first towards the end of the running gel. Then lower the other edge of the nitrocellulose onto the stacking side. All air bubbles must be completely removed between gel and nitrocellulose. Caution: Once the nitrocellulose and gel touch, they cannot be separated easily. Thus, it is important that no air bubble exists between gel and nitrocellulose at this point. Rolling a large pipet gently over the nitrocellulose will express excess liquid and air bubbles, while ensuring good gel-to-nitrocellulose contact. If for some reason (e.g., particulates between gel and nitrocellulose sheet) it is necessary to separate the gel and nitrocellulose, attempt this at the running gel portion of the "sandwich."

6. Place another wet filter paper on top of the nitrocellulose sheet. Roll with the pipet again.

7. Now place the second Scotch-Brite pad on top. Cut away a small corner-piece from this last pad to show that it is at the nitrocellulose side of the "sandwich."

8. Complete the "sandwich" by placing all layers inside the plastic support frame. Secure the frame and its contents with a rubber band around the middle along the long axis. It is important to tightly secure the sandwich so that all layers are in good contact during blot. Newer models of blot apparatus (e.g., Bio Rad) have sliding latches, which will help to secure the whole sandwich.

9. Slowly push the "sandwich" into the blot buffer in the trans-blot cell with the nitrocellulose (pad with cut corner) side toward the positive electrode.

10. Electrophoretic transfer of SDS-protein complexes is accomplished in 3 hr, at 4°C with a constant 100 Vdc. Transfer may be allowed to continue overnight at 60 Vdc. We have not observed any significant differences in the long transfer blots.

Current for a typical run at 100 Vdc with fresh buffer

is usually 200 mA; this will increase to about 800 mA with old buffer. We recommend a normal schedule of replacement every 2 months or 20 blot sessions, provided that the blot buffer is stored covered at 4°C when not in use. Caution: Using old buffer for overnight transfers is sometimes hazardous, since conductivities of old buffers are generally higher (possibly due to SDS contamination); current is concomitantly higher. Temperature of buffer during overnight transfers at 60 Vdc can reach boiling point, and fires can start under these conditions. Thus, we do not recommend overnight transfer with old buffers without using high-capacity internal cooling. Using old buffer may also cause problems if large amounts of contaminating proteins are allowed to leak out (e.g., from overloaded gel).

11. After transfer, disassemble the "sandwich," leaving only nitrocellulose and gel still in contact with each other. Mark the position of the resolving area of the gel and then mark the locations of the molecular weight standards. When separating gel from nitrocellulose, start from the end of the resolving gel and work towards the stacking gel. Some of the stacking gel may stick, but may be removed by wiping first with dry gauze, then with buffer-wetted gauze.

12. Wash the blot 4 times (5 min each) with PBS/Tween, preferably at 37°C, but room temperature washes will also do. Wash volumes should be about 250 ml each.

13. When a single antigen sample is processed (resolved in a single-well chamber, as bands across the whole gel, and blotted), place the blot onto a plastic backing sheet (e.g., "Write-On" film, Cat. # 00-0015-1064-3, from 3M, St. Paul, MN 55144) and cut into 0.5-cm strips. If multiple samples are involved, do not cut the blot. Continue the processing with the sheet intact.

14. Dilute antibody solutions (either serum or monoclonal products) in PBS/Tween. The concentration varies with the antibody contents. For parasitic diseases, we usually use 1:200 to 1:1000 dilutions of patients' sera. For monoclonal culture supernatants, we use 1:2 to 1:100 dilutions, and for monoclonal ascites fluids, 1:200 to 1:50,000. A total volume of 5 ml of diluted antibodies is sufficient for each 0.5- x 200-mm strip and 100 ml for each intact 180- x 200-mm sheet.

15. After this point, process the intact sheets in plastic trays. Process the cut strips in slotted incubation trays (Tsang et al., 1983a) from Bio Rad Laboratories (Cat. # 170-4037).

16. Allow the strips or sheets to remain in the antibody solution for 1 hr at room temperature or overnight at 4°C, with gentle agitation on a rotary shaker. We prefer the overnight incubation for general-purpose runs in which the actual antibody contents and affinity are unknown.

17. Wash the blot 4 times for 5 min with PBS/Tween at room temperature.
18. Expose the washed blot for 1 to 2 hr with gentle agitation to an appropriate solution of horseradish peroxidase-labeled anti-antibody (POD-anti-Ig) in PBS/Tween. Since the blot assay is largely a qualitative test in which sensitivity is important, we recommend that a high molecular weight polymer conjugate be used (see Tsang et al., 1984a).
19. Wash the blot 4 more times (5 min each) with PBS/Tween and once with PBS only. The blot may be held in PBS alone for at least 4 hr without visible effects on enzyme activities.
20. Then expose the processed blot for 5-10 min to the substrate solution. Positive reaction bands usually appear within 10 min. Stop reaction by washing with water.
Caution: 3,3'-diaminobenzidine is a suspected carcinogen! Therefore wash the developed blots thoroughly with H_2O.

(D) Photographic Recording and Storage of Blots

Blot sheets or strips may be stored dry but must be rewet before being photographed. We have found that if dried blots are simply wiped with a damp sponge, the positive bands will increase in intensity. We routinely perform photography with a Polaroid MP3 photocopying system, with a Kodak 38A gelatin filter (blue) and Polaroid Type 55 positive/negative film. The negative of this film can be used to produce publication-quality prints with conventional printing processes. Direct lighting for photography may be provided with 4 150-watt photoflood lamps.

Before photographing, reassemble the blot strips with companion strips from the same blot, together with the molecular weight marker strip. Using the guide line that was drawn on the nitrocellulose sheet before blotting, align the cut strips to their original relative positions. Place the blots on typing paper or the "Write-On" film for support. Place another sheet of "Write-On" film on top of the blots for protection in storage.

(E) Densitometry

Densitometry of blots can be used to precisely correlate the blot pattern with the transmission trace of the corresponding protein-stained gel and to obtain quantitative information from different blots.

After fixation, staining, and destaining, the polyacrylamide gel stained for total protein is often larger or smaller than the corresponding blot, making match-up of the EITB band(s) with the protein-stained band in the gel difficult. A reflectance/transmittance densitometer can be used to optimize the presentation of the data so that the blot trace can be superimposed directly on the total protein trace.

When different samples are run on the same gel and subjected to the same EITB procedure, the relative amount of antigen in each sample can be assessed using reflectance densitometry and peak integration. Absolute quantitative data can be obtained if some of the samples subjected to EITB contain known amounts of antigen. After densitometry of the blot and integration, a standard curve can be established, and the integral of the sample antigen can be correlated with a specific quantity of antigen.

TYPICAL RESULTS

(A) Multiple Antigens/Single Antibody Sources

Figure 1 demonstrates the ability of the EITB to compare the antigenic activities of multiple antigen preparations. Fractions from a purification scheme (as in this case) or multiple antigens from different sources may be reacted against a single antibody source and the difference in antigenic activities examined.

Figures 1B and C compare the blots of several antigen fractions from adult *Schistosoma mansoni* (MAMA) (Tsang et al., 1983b,c), transferred in different buffers. Figure 1A is a silver-stained (Tsang et al., 1984b) SDS-PAGE of the antigens. Figure 1B was blotted for 2 hr in 0.025 M Tris, 0.193 M glycine, 20% methanol (v/v), pH 8.35 at 100 Vdc/200mA, with cooling (4°C). Figure 1C is an identical blot done in 0.424 M Tris/HCl, 20% methanol, pH 9.18. The conductivity of this buffer, which is our SDS-PAGE resolving buffer (Tsang et al., 1983a), is much higher, thus at the same 100Vdc the current was 1400 mA. The blot done in the SDS-PAGE buffer (Fig. 1C) appears to be just as clear as, if not better than, the one done in the usual blot buffer (Fig. 1B). Indeed, the resolution of Figure 1C at the lower molecular weight region appears to be better than that of Figure 1B. High molecular weight components (MAMA lane) also appeared to have been transferred better.

Figure 1D is an identical blot to Figure 1C but done in the absence of methanol. The blot here is again similar; however, the gel was somewhat swollen at the end of the 2-hr transfer period. Some distortion of bands may result if the swelling is allowed to continue (as in overnight transfers).

(B) Single Antigen/Multiple Antibody Sources

Figure 2 compares the reactivities of various antibody sources against a single antigen. The microsomal antigens were resolved in a single-well SDS-PAGE (Tsang et al., 1983b,c) and blotted. The resulting blotted sheet was cut into 0.5-cm strips and exposed to 7 patient sera. Sera A, G, H, J, and K were obtained from 5 acutely infected patients at different times during their infection and treatment. Sera with the same letter are from the same person. Numbers in the same letter series denote different times in that patient's infection course. Sera Z and "normal" are from a chronically infected person and an uninfected person. The blank strip was not exposed to serum. The antigen recognition spectra by the patients' antibodies changed with time for each patient. There are also individual differences among patients. The variations among the antigens recognized by different infection-phase sera and among sera taken at different times during one person's infection are even more pronounced in crude antigens (Tsang et al., 1983c).

Experiments such as Figure 2 are useful in defining the antibody specificities of monoclonal antibodies and in studying epitope mapping of any given antigen/antibody system.

(C) Sensitivity Limits

The sensitivity limits of the EITB have probably not yet been reached. With better conjugates and higher sensitivity substrates, the present limits will most likely be extended in the near future. With our procedure, using a bovine serum albumin (BSA) and anti-BSA system, we are able to detect antibodies at 10 ng/ml and antigen at 10 ng/0.5-mm strip. The antibodies used were polyclonal and produced in rabbits. With monoclonal antibodies and parasitic target antigens, these limits can be extended to approximately 50 pg. The detection limit of blotted antigens with enzyme-linked techniques is generally considered to be about 10 to 20 pg. With more sensitive substrates and enzymes, this limit will most likely improve in the near future.

DISCUSSION

Our own experience and the published data from many other laboratories indicate that the blotting techniques (and variations thereof) are extremely powerful and valuable tools in immunochemistry. Since its beginning (Southern, 1975), an explosive growth in the number of published reports citing the use of the blot has attested to its acceptance by and value to the scientific community. The enzyme-linked immunoelectrotransfer blot especially allows the combination of the extreme resolving power of electrophoresis with the high sensitivity and specificity of ELISA. From the number of variations to this technique, we are mistakenly led to believe that the blot is rather forgiving and thus easily deployable by even the most careless of laboratorians. However, special precautions must be observed in the use of blotting techniques and relying on the results obtained. We will review some of these below.

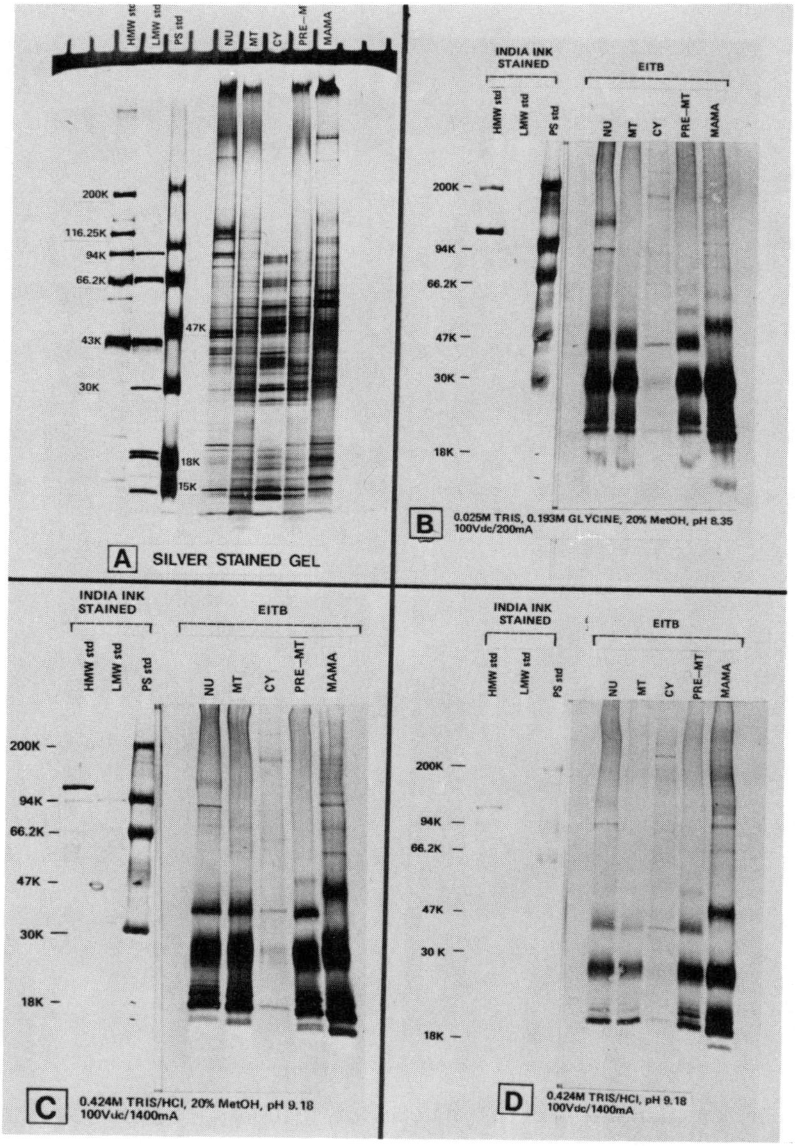

Fig. 1. (A) silver-stained gel of Schistosoma mansoni antigen fractions. (B), (C), (D) blots of identical gels as (A) transferred in different buffers as listed. Fraction NU=nuclear, MT=mitochondrial, CY=cytosolic, PRE-MT=pre-mitochondrial, MAMA=S. mansoni adult microsomal antigen (Tsang et al., 1983b). All lanes contain 1.5 ug of total protein. Numbers ending in k's indicate positions of molecular weight markers in 1000 daltons.

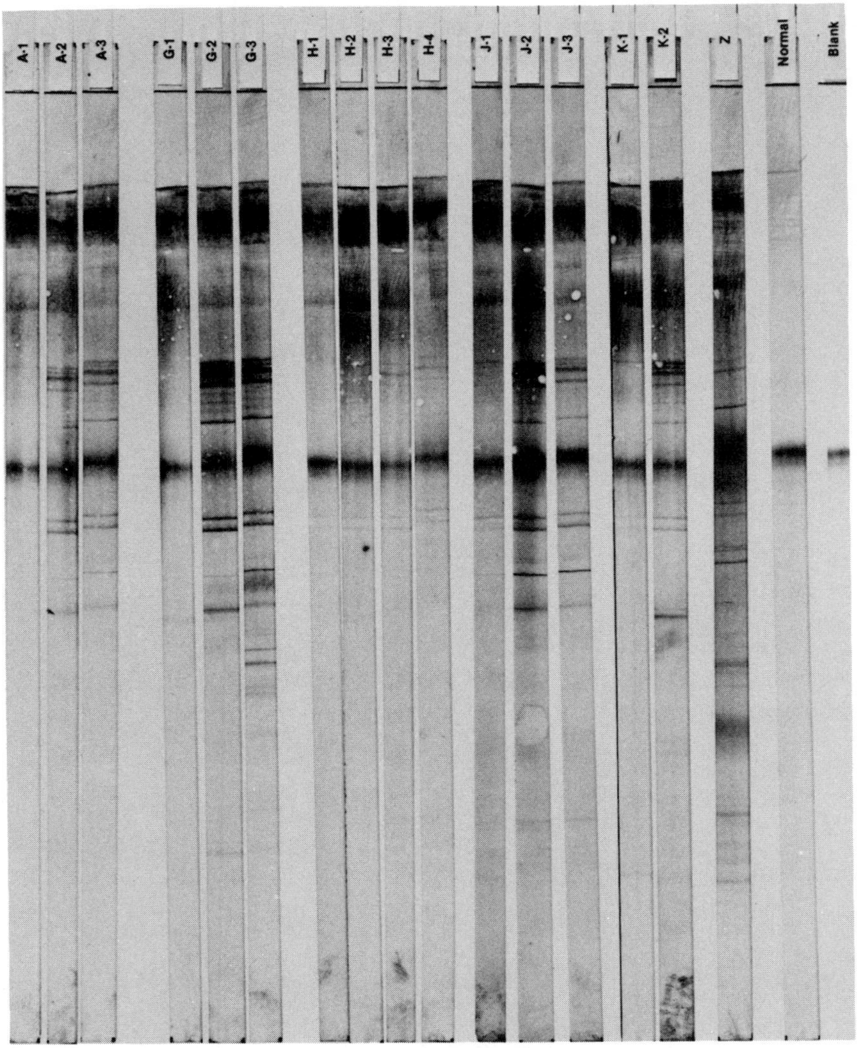

Fig. 2. A single blot of MAMA (100 ug) was cut into 0.5-mm strips and exposed to various patient infection sera. Series A, G, H, J, and K are acute-phase infection sera from 5 patients, obtained at various times in their infection course (range=6 weeks to 2 months). Serum Z is a chronic infection serum. Serum "normal" was obtained from a patient with no exposure to S. mansoni. Blank is a saline control. Numbers ending in K's indicate position of molecular weight markers in 1000 daltons.

Special Precautions in Interpretation of Results

SDS treatment is generally assumed to alter the conformation of most proteins; thus, antigenic and enzymatic active sites may be changed or lost. Yet without SDS, proteins will not all be negatively charged and migrate towards the positive electrode in PAGE or blot. The fact that SDS-treated antigens are transferable and remain active implies that either SDS is indeed removable by the post-blot washes and proteins are refolded sufficiently to allow recovery of antigenic activities or that all the antigenic activities demonstrable in blots are associated with epitopes of primary amino acid sequences. It is also possible that a combination of both these conditions allows the detection of SDS-treated and blotted antigens. Evidence for the first possibility that does not completely exclude the latter is the recognition of both native and SDS-treated/blot antigens in different assays using monoclonal antibodies. Furthermore, we have been able to treat peroxidase with SDS at 37°C, resolve it with PAGE, blot, and detect activity directly (Tsang et al., 1984a). This would imply that the SDS treatment is indeed reversible. Whether the protein-bound SDS-treated antigens are replaced or removed by the presence of methanol during transfer or by washing in the presence of non-ionic detergents, which are generally present in blot washes, is yet to be determined.

Special Problems Associated with Blot Detection Using Monoclonal Antibodies

Polyclonal or monospecific antibodies have been shown to bind with high specificity to a wide variety of protein antigens bound to blots following electrophoretic transfer from SDS-denaturing polyacrylamide gels. The similar use of monoclonal antibodies as blot-detection reagents, however, has posed some problems. Although denatured antigens can be successfully detected with monoclonal antibodies (i.e., single band staining) (O'Connor et al., 1982; Caceres et al., 1984; Colcher et al., 1983; Steineman et al., 1984; Braun et al., 1983; Turner, 1983; Ogata et al., 1983; Flaster et al., 1983), problems most often observed include: 1) no reactivity (Steineman et al., 1984; Braun et al., 1983; Turner, 1983; Mandrell et al., 1984; Irons et al., 1983); 2) variable staining intensity (Steineman et al., 1984; Braun et al., 1983; Turner, 1983; Mandrell et al., 1984; Irons et al., 1983); and 3) multiple band staining (Braun et al., 1983; Turner, 1983; Ogata et al., 1983; Flaster et al., 1983; Anderson et al., 1982). Such observations are not unexpected, since monoclonal antibodies are usually derived from native proteins, and most blots are carried out under denaturing conditions. Monoclonal antibodies derived from denatured proteins may circumvent some of these problems.

Monoclonal antibodies may fail to react with blotted proteins because of destruction of antigenic determinants. Such destruction may be a result of any combination of factors, such as detergents, urea, chaotropes, proteases, physical shearing, or heat, which may be encountered during sample preparation or electrophoresis. Similarly, detergents, pH and ionic changes, or methanol present during electrophoretic transfer may destroy or alter determinants. Partial destruction of antigenic determinants and low antibody titer or affinity would account for differences observed in staining intensity. Multiple banding suggests that monoclonal antibodies recognize short amino acid sequences common to other proteins or repeat determinants exposed upon antigen denaturation. An example of such a sequence might be a common posttranslational modification. In any event, reactivity of any monoclonal antibody with its specific antigen may be a hit-or-miss prospect.

Ideally, monoclonal antibodies should be pretested for suitability in detection of blotted proteins. This can be accomplished with the aid of a microtitration-formated vacuum filtration apparatus (i.e., BioDot from Bio-Rad). The immunogen from which the monoclonal antibody is derived is spotted and bound to nitrocellulose in both native and denatured conformations. Antibodies able to recognize the denatured antigen can be selected for electrophoretic transfer analysis; nonreactive antibodies can be rejected or retested using different binding or reaction conditions.

Types of Biological Molecules that are Blottable

Proteins, lipoproteins (Huang et al., 1983), glycoproteins (Glass et al., 1981; Schott et al., 1984; Hawkes, 1982) and even lipopolysaccharides (Peppler, 1984; Cousland and Poxton, 1983; Chart et al., 1984) are transferable and blottable. Nucleic acids, of course, were the first biological macromolecules to be blotted (Southern, 1975). In general, there is no reason to doubt that any charged molecules can be electrophoretically transferred and blotted onto some suitable matrix.

Matrixes for Blot

Several matrixes are currently in use for blots. Nitrocellulose sheets (BA83-0.2 um and BA85-0.45 um, Schleicher and Schuell), diazo paper (APT-paper, ABM-paper, Bio Rad Laboratories), and charged nylon sheets ("Zeta-Probe", Bio Rad Lab.) are the most commonly used. For a review of methods using these matrixes, the reader is referred to Symington (1984). In general, the diazo papers have high capacities and bind proteins covalently, thus allowing multiple probing and dissociations of the same blot. Backgrounds for these matrixes are generally high, and they are most complicated to use, requiring activation and/or extensive blocking. Diazo papers are generally unsuitable for EITB since they become orange or reddish brown upon activation, thus rendering subsequent substrate visualization difficult. The same problems of high backgrounds and blocking are also encountered with the charged nylon matrix. These membranes, however, may be useful when methanol is undesirable or when high binding capacities are required (480 ug/cm^2).

Nitrocellulose sheets are relatively easy to use and inexpensive. Blocking agents, aside from the incorporation of 0.3% Tween-20 in all buffers, are often unnecessary (Tsang et al., 1983a,b,c, 1984b). Multiple probing with a single nitrocellulose blot is, however, limited to 3 disassociation cycles (Legocki and Verma, 1981).

Buffers and Power Requirements for Electrophoretic Transfer Blot

The power requirements for transferring protein are usually considered in terms of field intensity or strength, which depends on the type of buffer used (the conductivity thereof) and on the distances between the electrodes and is expressed as Vdc/cm. Theoretically, higher field intensities will give higher efficiencies in transfer, especially for large molecules. Thus, either higher voltages or closer electrode distances will allow shorter blotting times and more complete transfers of large molecules. However, in buffers with high conductivities, raising field intensity will result in the concomitant increase of current, which invariably produces more heat. Cooling therefore is recommended. This principle has been demonstrated by transferring myosin (220,000 dalton) from a 10% SDS gel (Laemmli) using the transfer buffer of Towbin in a 3-hr experiment. Using a field strength of 55 V/cm (220 W) at 0°C, myosin transferred with nearly 60% efficiency. Recirculated coolant temperature rose to above 40°C by the end of 3 hours. This result was compared with only 10% efficiency observed using standard field strength 7.5 V/cm (9 W) at room temperature. In general, low field intensities are used for long (overnight) and/or uncooled transfers. We prefer high-intensity, cooled (4°C), short transfers whenever possible to reduce experimental times. Also, we have consistently observed that the backgrounds for short transfers appear to be lower.

The most frequently encountered buffer system is 0.025 M Tris, 0.193 M glycine, with 20% methanol (v/v), pH 8.35, as originated by Towbin et al. (1979). For transferring most proteins with this buffer, 60 Vdc (7.5 Vdc/cm) at 200mA to 800mA for 3 hr is sufficient. We prefer to transfer proteins of varying molecular weights (1,000,000 to 10,000 daltons) at 100Vdc, 200-800mA at 4°C for 3 hr. Erickson et al. (1982) incorporated 0.1% SDS into their Tris/glycine buffer and reported quantitative transfers of proteins from brain synaptosomes. Mandrell (1984) substituted 0.4% zwitterionic detergent into the Tris/glycine (Towbin) transfer buffer and 0.1% zwitterionic detergent in the denaturing gel to suitably renature membrane proteins following electrophoresis and transfer and allow detection with a monoclonal antibody. They could not detect activity when transfers were performed without Zwitterionic detergent, nor could they detect antigen with added SDS and a variety of non-ionic detergents. Peppler (1984) used 50 mM sodium phosphate buffer at pH 7.5 to transfer lipopolysaccharides onto nitrocellulose. Huang et al. (1983) transferred apolipoproteins in 25 mM phosphate, pH 6.5 at 0.5 Amp for 16 hr. For native gels in urea, resolved proteins in the absence of SDS were transferred in 7% acetic acid (Towbin et al., 1979).

In theory, there is no reason to doubt that SDS-treated proteins can be blotted with the same buffer used in the PAGE system from which the proteins were resolved. We have demonstrated this with the data in the results section. There we used our PAGE resolving buffer (0.424 M Tris/HCl, pH 9.18) with 20% methanol as our blot buffer. The results from this transfer are just as good as, if not better than, similar blots using the Tris/glycine/methanol buffer. The only disadvantage of using the PAGE buffer in transfer was the higher conductivity of this buffer. The current during blot was higher, so that more cooling was required. We dilute this buffer to 20% methanol, 0.212 M TRIS/HCl, pH 9.18, and are able to blot with equal efficiencies but at much less current (100 Vdc/1.2 Amp). Methanol was included in both buffers to prevent the gel from swelling during transfer. Methanol also may help extract SDS from proteins and thus participate in renaturation of transferred proteins. Unfortunately, methanol may cause some proteins to precipitate within the gel, thus contributing to the difficulty often encountered when quantitative elution of large molecular weight proteins is attempted.

General Discussion on the Choice of Buffers and Power Conditions

A general strategy in choosing the proper transfer buffer is to maximize field strength (V/cm) and minimize both current and heating. Buffer pH is usually not a concern when SDS-protein complexes are transferred to nitrocellulose membranes. It is important to note that methanol, which is a normal constituent of transfer buffers, presumably reverses the binding of SDS to proteins and thus plays a role in protein renaturation and binding to nitrocellulose. With this in mind, caution should be exercised in transferring basic SDS-proteins by the method of Towbin because they may become electrically neutral.

Elution of high molecular weight SDS proteins or partially soluble membrane proteins from polyacrylamide gels can be difficult. Two strategies are often used to aid in the transfer of such molecules. They include (1) eliminating methanol from the transfer buffer and (2) adding SDS or other detergents in the transfer buffer. While improving elution, elimination of methanol decreases the binding capacity of nitrocellulose for proteins. Detergents, while improving elution, may affect protein binding to the negatively charged nitrocellulose (at pH 8.3), alter or destroy antigenicity, and substantially increase buffer conductance and heating.

Tris-based buffers (i.e., Tris acetate, Tris borate, and Tris glycine) allow the highest voltages at the lowest currents and are therefore the most desirable when one is transferring to nitrocellulose or Zeta-Probe. These buffers are often diluted 2X to 4X to further lower current and allow generation of greater V/cm. Phosphate and acetate buffers are the least desirable because of their high conductivities at low field strengths, rapid degeneration with subsequent drastic ionic and pH decomposition, and excessive heating characteristics.

Some of the various buffers are summarized in Table I.

Blocking Agents and Nonspecific Reactions

Three distinct types of background may be encountered upon performing EITB: 1) overall membrane staining, 2) nonspecific staining of bands due to impurity or cross-reactivity of the first antibody, and 3) "nonsense" staining of bands due to reaction with the second antibody or detection reagent in the absence of first antibody.

Membrane staining may be a result of insufficient blocking but more often of using too high a concentration of second antibody, using impure second antibody (affinity absorbed as recommended), or lacking mild detergent in washing steps. Type 2 background is generally due to impure or cross-reactive first antibody or use of first antibody in excessively high concentration. "Nonsense" background is the most troublesome and not easily explained, since it may be present even with use of the most impeccably pure antibody preparations. It is presumed to be due to strong hydrophobic binding of large antibody-enzyme complexes to the bound protein. Overloads of antigen samples in SDS-PAGE may also contribute to this problem.

An immunoblotting strategy effective in eliminating or reducing this Type 3 background involves the sequential blocking of the protein blot, first with protein and second with Tween-20. In such cases where Type 3 background is evident, Tween-20 blocking alone may not be sufficient. It is also advisable to use Tween-20 (0.05% v/v) and carrier protein (i.e., 1% w/v BSA, gelatin, or normal serum) with both antibody reactions, and Tween-20 (0.05% v/v) in washing steps. Greater concentrations of Tween-20 or other detergents (NP-40, SDS, Triton X-100, etc.) may be deleterious to membrane bound protein and losses may result (Bjernum and Larsen 1983).

Table I
Examples of Buffers used in Transfers of SDS-Gels

Reference	Buffer
Towbin et al. (1979)	25 mM Tris, 192 mM Glycine pH 8.3, 20% Methanol (MetOH)
Erickson et al. (1982)	" + 0.1% SDS
Nielsen et al. (1982)	" + 0.01% SDS without MetOH
Mandrell et al. (1983)	" + 0.1-0.4% Empigen BB and other non-ionic or Zwitter-ionic detergents
Schott et al. (1984)	0.125 M Tris, 0.96 M Glycine, + 0.1% SDS + 20% MetOH
Glass et al. (1981)	0.125 M Tris, D.096 M Glycine + 20% MetOH
Anderson et al. (1982)	25 mM Tris, 192 mM Glycine without MetOH
Burnette (1981)	20 mM Tris, 150 mM Glycine pH 8.3 + 20% MetOH
Gershoni et al. (1982)	15.6 mM Tris, 120 mM Glycine pH 8.3 ± 20% MetOH
Davis et al. (1983)	20 mM Tris/acetate, 2 mM EDTA pH 7.5, + 0.01% SDS
Gershoni et al. (1982)	41 mM Tris, 40 mM borate pH 8.3
Reiser et al. (1981)	10 mM Na borate pH 9.2
Huang et al. (1983)	25 mM PO_4 pH 6.5
Nelson et al. (1984)	10 mM Tris/HCl, 20 mM Hepes pH 7.56, 0.1% SDS, 0.1 mM Dithiothreitol + 20% Ethanol
Granger et al. (1984)	20 mM Tris/acetate pH 8.3, 0.1% SDS, 0.1 mM Dithiothreitol + 20% Isopropanol

Stefansson et al. (1983)	20 mM Tris/HCl pH 7.5 + 20% MetOH
Sutton et al. (1982)	10 mM Tris, 77 mM Glycine pH 8.3, 0.1% SDS
Griesser et al. (1983)	10 mM Na Acetate pH 5.2
Carson et al. (1983)	10 mM Tris/HCl pH 8.5
Goldstein (1983)	50 mM NaH_2PO_4/Na_2HPO_4 pH 7.4
Fujita et al. (1983)	50 mM $NaBO_4$ pH 8.5
Tsang et al. (this article)	0.424 M Tris/HCl pH 9.18 + 20% MetOH, or 0.212 M Tris/HCl pH 9.18 + 20% MetOH

When covalent linkage blot matrixes such as DMB-paper are used (Renart et al., 1979), it is important to block or rather inactivate the remaining diazonium sites on the paper after blot. Renart et al. (1979) used Tris, ethanolamine, and gelatin for this operation. Blocking agents are also used in conjunction with other matrixes, ostensibly to block all "non-specific" reactions both on the matrixes themselves and on the blotted antigens. Some of the blocking agents include 3% bovine serum albumin (Towbin et al., 1979; Aubertin et al., 1983), 10% fetal bovine serum (Ramirez et al., 1983; De Blas et al., 1983), and 3% gelatin (Symington, 1984). Variations in concentration of these blocking agents are common. In addition, to prevent non-specific association of test ligands with hydrophobic groups on the blotting matrixes and on the blotted proteins, a variety of non-ionic detergents are used: 0.05% Tween-20 (Symington, 1984), and 0.05% NP-40 (nonidet P-40, Shell Chemical Co., West Orange, NJ; Burnette, 1981).

In our experience with nitrocellulose matrix without covalent linkages, even when extremely complexed and crude parasitic antigens are blotted (Tsang et al., 1982, 1983b,c, 1984b), protein blocking agents are unnecessary. We have observed that in some cases blocking with an esoteric protein such as BSA actually decreased specific activity, while not improving the noise-to-signal ratios of ELISA (Tsang et al., 1980). We believe that most of the so-called non-specific reactions are the results of nonspecific reagents or excessive concentrations. Thus, with reagents of good specificities, we recommend the use of 0.3% Tween-20 to prevent hydrophobic associations and dispense with all other protein blocking agents.

Another common cause of nonspecific reactions is the tendency to overload blots. In our experience, with most complex systems, a total protein load of 1 ug per 0.5-cm strip is quite sufficient for ELISA visualizations. Ochs (1983) has reported nonspecific reactions (artifacts) resulting from contamination by human skin proteins in SDS-PAGE. This is to be expected and avoided in high-sensitivity detection systems involving antihuman reagents.

Enzyme Conjugates and Substrates

The most commonly encountered enzyme label is horseradish peroxidase (POD; donor: H_2O_2 oxidoreductase, EC1.11.17). Ideally, the conjugate used should be high molecular weight polymer (Tsang et al., 1984a). Low molecular weight conjugates, such as those with a 1:1 antibody-to-enzyme molar ratio, will give extremely long ranges of quantitative linearity but low sensitivity. With the blot we are presumably more interested in sensitivity than quantitative range; therefore, a high-sensitivity, high molecular weight conjugate is more appropriate.

POD is visualized according to the following coupled reactions:

$$H_2O_2 \xrightarrow{POD} H_2O + 1/2\ O_2 \qquad (1)$$

$$2[1/2\ O_2] + DH_2 \longrightarrow D + H_2O \qquad (2)$$

Reaction (1), the actual enzyme reaction, is invisible. What we see is the oxidation of the reduced dye (DH_2) to its chromogenic form (D). The selection of the chromophasic dye is thus important. For the blot it is essential that a precipitating chromogen is selected, so that once oxidized, D will remain at the site of its oxidation. We prefer DAB (3,3'-diaminobenzidine) for this purpose. De Blas and Cherwinski (1983) reported the use of nickel and cobalt ions to intensify the DAB color reaction on blots. Other chromogens are also used: 4-chloro-1-napthol (Anderson et al., 1983; Aubertin et al., 1983); o-dianisidine (Towbin et al., 1979). Other common chromogens for the POD system, such as o-phenylenediamine, 5-amino-salicylic acid, 2,2'-azino-di-(3-ethyl-benzothiazoline-6-sulphonic acid), and 3,3',5,5'-tetramethylbenzidine, are not recommended for blot, since they are soluble after oxidation. Unfortunately, since all the useful chromogens listed above are implicated as carcinogens, due caution must be exercised when using these compounds.

Alkaline phosphatase (orthophosporic-monoester phosphohydrolase, EC 3.1.3.1) is another enzyme label used in conjunction with the blot (Turner, 1983; O'Conner, 1982). Detection of native and SDS-denatured rat liver phosphodiesterase I (EC 3.1.4.1) with a 2-site antibody test was also described (Muilerman et al., 1982). This article proposed an interesting potential for enzymatic detection with impure anti-enzyme antisera in SDS-denatured mixtures after blot.

Glucose oxidase has recently been observed to be an excellent alternative to horseradish peroxidase as a detecting label, its main advantage being lower backgrounds (Porter, 1984). Incidentally, direct staining of enzymes transferred to DEAE has been demonstrated for a wide variety of enzymes (McLellan, 1981).

SUMMARY

The EITB and other immunologic variations of the protein blotting techniques are invaluable in answering questions about such things as disease-related antigen/antibody reactions (Sharma et al., 1983; Partanen et al., 1984), molecular variances (Hersh et al., 1984), and detection of specific antigens in biochemical systems (Fish et al., 1984). Exciting advances are also being made with monoclonal antibodies and epitope mapping (Van Wyke et al., 1984; Dziadek et al., 1983), with domain mapping of cell surface and hormone receptors (Russell et al., 1984), with tumor-specific surface antigen identification (Colcher et al., 1983), and with oncogene product localization (Klempnauer et al., 1984). Potentially this technique, especially when coupled to the 2-dimensional gel systems (Symington, 1984), is virtually unparalleled in resolution and sensitivity. The versatility of the blot is also evident in its ability to couple with isoelectric focusing (Glynn et al., 1982) and agarose electrophoresis (Bittner et al., 1980), as well as the usual SDS-PAGE.

Future improvements of sensitivity through better chromogens, such as a possible insoluble version of the Ngo and Lenhoff (1980) chromogenic assay, will be useful. Systematic studies of conditions that will optimize blotting are also needed. Nearly all blotting conditions are now set somewhat empirically. Relationships between blotting buffer, electrical power, ionic strength, pH, and the type and sizes of molecules transferred are yet to be determined. Future improvements on an already powerful tool will depend on these considerations.

REFERENCES

Anderson, D.J., P.H. Adams, M.S. Hamilton, and N.J. Alexander (1983). Antisperm antibodies in mouse vasectomy sera react with embryonal teratocarcinoma. J. Immunol. 131: 2908-2912.

Anderson, N.L., S.L. Nance, T.W. Pearson, and N.G. Anderson (1982). Specific antiserum staining of two-dimensional electrophoretic patterns of human plasma proteins immobilized on nitrocellulose. Electrophoresis 3: 135-142.

Aubertin, A.M., L. Tondre, C. Lopez, G. Obert, and A. Kirn (1983). Sodium dodecyl sulfate-mediated transfer of electrophoretically separated DNA-binding proteins. Anal. Biochem. 131: 127-134.

Bittner, M., P. Kupferer, and C.F. Morris (1980). Electrophoretic transfer of proteins and nucleic acids from slab gels to diazobenzyloxymethyl cellulose or nitrocellulose sheets. Anal. Biochem. 102: 459-471.

Bjernum, O.J. and K.P. Larsen (1983). Some recent developments of the electroimmunochemical analysis of membrane proteins. Application of zwittergent, Triton X-114 and western blotting technique. In Modern Methods in Protein Chemistry (Tschesche, H., Ed.). Walter de Gruyter and Co., N.Y. 74-124.

Braun, D.K., L. Pereira, B. Norrild, and B. Roizman (1983). Application of denatured, electrophorectically separated, and immobilized lysates of Herpes Simplex virus-infected cells for detection of monoclonal antibodies and for studies of the properties of viral proteins. J. Virol. 46: 103-112.

Burnette, W.N. (1981). "Western Blotting:" Electrophoretic transfer of proteins from sodium dodecyl sulfate-polyacrylamide gels to unmodified nitrocellulose and radiographic detection with antibody and radioiodinated protein A. Anal. Biochem. 112: 195-203.

Caceres, A., L.I. Binder, M.R. Payne, P. Bender, L. Rebhun and O. Steward (1984). Differential subcellular localization of tubulin and the microtubule-associated protein MAP2 in grain tissue as revealed by immunocytochemistry with monoclonal hybridoma antibodies. J. Neurosci. 4: 394-410.

Carson, S.D., S.M. Carson., and W.H. Konigsberg (1983). Monoclonal antibody recognizing rabbit IgG (Fab). J. Biol. Chem. 258: 9510-9513.

Chart, H., D.H. Shaw, E.E. Ishiguro, and T.J. Trust (1984). Structural and immunochemical homogeneity of Aeromonas salmonicida lipopolysaccharide. J. Bactrtiol. 158: 16-22.

Colcher, D., P.H. Hand, D. Wunderlich, M. Nuti, Y.A. Teramoto, D. Kufe, and J. Schlom (1983). Monoclonal antibodies to human mammary carcinoma associated antigens and their potential uses for diagnosis, prognosis, and therapy. In Immunodiagnostics Alan R. Liss, Inc., N.Y., 215-258.

Cousland G. and I.R. Poxton (1983). Analysis of lipopolysaccharides of Bacteroides fragilis by sodium dodecyl sulphate-polyacrylamide gel electrophoresis and electroblot transfer. FEMS Microbiol. Lett. 20: 461-465.

Davis, J. and V. Bennett (1983). Brain Spectrin. J. Biol. Chem. 258: 7757-7766.

De Blas, A.L. and H.M. Cherwinski (1983). Detection of antigens on nitrocellulose paper immunoblots with monoclonal antibodies. Anal. Biochem. 133: 214-219.

Dziadek, M., H. Richter, M. Schachner, and R. Timpl (1983). Monoclonal antibodies used as probes for the structural organization of the central region of fibronectin. FEBS Lett. 155: 321-325.

Erickson, P.F., L.N. Minier, and R.S. Lasher (1982). Quantitative electrophoretic transfer of polypeptides from SDS polyacrylamide gels to nitrocellulose sheets: A method for their re-use in immunoautoradiographic detection of antigens. J. Immunol. Methods 51: 241-249.

Fish, A.J., M.C. Lockwood, M. Wong, and R.G. Price (1984). Detection of Goodpasture antigen in fractions prepared from collagenase digests of human glomerular basement membrane. Clin. Exp. Immunol. 55: 58-66.

Flaster, M.S., C. Schley, and B. Zipser (1983). Generating monoclonal antibodies against excised gel bands to correlate immunocytochemical and biochemical data. Brain Res. 277: 196-199.

Fujita, S.C., S.L. Zipursky, S. Benzer, A. Ferrus, and S.L. Shotwell (1982). Monoclonal antibodies against the Drosophila nervous system. Proc. Natl. Acad. Sci. USA 79: 7929-7933.

Gershoni, J.M. and G.E. Palade (1982). Electrophoretic transfer of proteins from sodium dodecyl-sulfate-polyacrylamide gels to a positively charged membrane filter. Anal. Biochem. 124: 396-405.

Gershoni, J.M. and G.E. Palade (1983). Protein blotting: Principles and applications. Anal. Biochem. 131: 1-15.

Glass, W.F. III, R.C. Briggs., and L.S. Hnilica (1981). Use of lectins for detection of electrophoretically separated glycoproteins transferred onto nitrocellulose sheets. Anal. Biochem. 115: 219-224.

Glynn, P., H. Gilbert, J. Newcombe, and M.L. Cuzner (1982). Rapid analysis of immunoglobulin isoelectric focusing patterns with cellulose nitrate sheets and immunoperoxidase staining. J. Immunol. Methods 51: 251-257.

Goldstein, M.E., L.A. Sternberger, and N.H. Sternberger (1983). Microheterogeneity ("neurotypy") of neurofilament proteins. Proc. Natl. Acad. Sci. USA 80: 3101-3105.

Granger, B.L., and E. Lazarides (1984). Membrane skeletal protein 4.1 of avian erythrocytes is composed of multiple variants that exhibit tissue-specific expression. Cell 37: 595-607.

Griesser, H.W., B. Muller-Hill, P. Overath (1983). Characterization of β-galactosidase-lactose-permease chimaeras of Escherichia coli. Eur. J. Biochem. 137: 567-572.

Hancock, K. and V.C.W. Tsang (1983). India ink staining of proteins on nitrocellulose paper. Anal. Biochem. 133: 157-162.

Hawkes, R. (1982). Identification of concanavalin-A binding proteins after sodium dodecyl sulfate gel electyrophoresis and protein blotting. Anal. Biochem. 123: 143-146.

Hersh, L.B., B.H. Wainer, and L.P. Andrews (1984). Multiple isoelectric and molecular weight variants of choline acetyltransferase. Artifact or real? J. Biol. Chem. 259: 1253-1258.

Huang, L.S., J.S. Jaeger, and D.C. Usher (1983). Allotypes associated with β- apolipoproteins in rabbits. J. Lipid Res. 24: 1485-1493.

Irons, L.E., L.A.E. Ashworth, and P. Wilton-Smith (1983). Heterogeneity of the filamentous haemagglutinin of Bordetella pertussis studies with monoclonal antibodies. J. Gen. Microbiol. 129: 1769-2778.

Klempnauer, K.H., G. Symonds, G.I. Evan, and J.M. Bishop (1984). Subcellular localization of proteins encoded by oncogenes of avian myeloblastosis virus and avian leukemia virus E26 and by the chicken c-myb gene. Cell 37: 537-547.

Legocki, R.P. and D.P.S. Verna (1981). Multiple immunoreplica technique: screening for specific proteins with a series of different antibodies using one polyacrylamide gel. Anal. Biochem. 111: 385-392.

Lin, W. and H. Kasamatsu (1983). On the electrotransfer of polypeptides from gels to nitrocellulose membranes. Anal. Biochem. 128: 302-311.

Mandrell, R.E. and W.D. Zollinger (1984). Use of a zwitterionic detergent for the restoration of the antibody-binding capacity of electroblotted meningocoecal outer membrane proteins. J. Immunol. Methods 67: 1-11.

McLellan, T. and J.A.M. Ramshaw (1981). Serial electrophoretic transfers: A technique for the identification of numerous enzymes from single polyacrylamide gels, Biochem. Genetics 19: 647-654.

Muilerman, H.G., H.G.J. Ter Hart, and W. Van Dijk (1982). Specific detection of inactive enzyme protein after polyacrylamide gel electrophoresis by a new enzyme-immunoassay method using unspecific antiserum and partially purified active enzyme: Application to rat liver phosphodiesterase I. Anal. Biochem. 120: 46-51.

Nelson, W.J. and E. Lazarides (1984). Goblin (ankyrin) in striated muscle: Identification of the potential membrane receptor for erythroid spectrin in muscle cells. Proc. Natl. Acad. Sci. USA 81: 3292-3296.

Ngo, T.T. and H.W. Lenhoff (1980). A sensitive and versatile chromogenic assay for peroxidase and peroxidase-coupled reactions. Anal. Biochem. 105: 389-397.

Nielsen, P.J., K.L. Manchester, H. Towbin, J. Gordon, and G. Thomas (1982). The Phosphorylation of ribosomal protein S6 in rat tissues following cycloheximide injection, in diabetes, and after denervation of diaphragm. J. Biol. Chem. 257: 12316-12321.

Ochs, D. (1983). Protein contaminants of sodium dodecyl sulfate polyacrylamide gels. Anal. Biochem. 135: 470-474.

O'Connor, C.G. and L.K. Ashman (1982). Applications of the nitrocellulose transfer technique and alkaline phosphatase conjugated anti-immunoglobulin for determination of the specificity of monoclonal antibodies to protein mixtures. J. Immunol. Methods 54: 267-271.

Ogata, K., M. Arakawa, T. Kasahara, K. Shiori-Nakano, and K. Hiraoka (1983). Detection of Toxoplasma membrane antigens transferred from SDS-polyacrylamide gel to nitrocellulose with monoclonal antibody and avidin-biotin, peroxidase anti-peroxidase and immunoperoxidase methods. J. Immunol. Methods 65: 75-82.

Partanen, P., H.J. Turunen, R.A. Paasivuo, and P.O. Leinikki (1984). Immunoblot analysis of Toxoplasma gondii antigens by human immunoglobulins G, M, and A antibodies at different stages of infection. J. Clin. Microbiol. 20: 133-135.

Peppler, M.S. (1984). Two physically and serologically distinct lipopolysaccharide profiles in strains of Bordetella pertussis and their phenotype variants. Infect. Immun. 43: 224-232.

Porter, D.D. and H.G. Porter, H.G. (1984). A glucose oxidase immunoenzymes stain for the detection of viral antigen or antibody on nitrocellulose transfer blots. J. Immunol. Methods 72: 1-9.

Raamsdonk, W.V., C.W. Pool, and C. Heyting (1977). Detection of antigens and antibodies by an immuno-peroxidase method applied on thin longitudinal sections of SDS-polyacrylamide gels. J. Immunol. Methods 17: 337-348.

Ramirez, P., J.A. Bonilla, E. Moreno, and P. Leon (1983). Electrophoretic transfer of viral proteins to nitrocellulose sheets and detection with peroxidase-bound lectins and protein A. J. Immunol. Methods 62: 15-22.

Reiser J. and J. Wardale (1981). Immunological detection of specific proteins on total cell extracts by fractionation on gels and transfer to diazophenylthioether paper. Eur. J. Biochem. 114: 569-575.

Renart, J., J. Reiser, and G. R. Stark (1979). Transfer of proteins from gels to diazobenzyloxymethyl-paper and detection with antisera: A method for studying antibody specificity and antigen structure. Proc. Natl. Acad. Sci. USA 76: 3116-3120.

Russell, D.W., W.J. Schneider, T. Yamamoto, K.L. Luskey, M.S. Brown and J.L. Goldstein (1984). Domain map of the LDL receptor: Sequence homology with the epidermal growth factor precursor. Cell 37: 577-585.

Schott, K.J., V. Neuhoff, B. Nessel, U. Potter, and J. Schroter (1984). Staining of concanavalin A. Reactive glycoproteins on polyacrylamide gels with horseradish peroxidase -- A critical evaluation. Electrophoresis 5: 77-83.

Sharma, S.D., J. Mullenax, F.G. Araujo, H.A. Erlich, and J.S. Remington (1983). Western blot analysis of the antigens of Toxoplasma gondii recognized by human IgM and IgG antibodies. J. Immunol. 131: 977-983.

Southern, E.M. (1975). Detection of specific sequences among DNA fragments separated by gel electrophoresis. J. Mol. Biol. 98: 503-517.

Stefansson, K., L. Marton, J.P. Antel, R.L. Wollmann, R.P. Roos, G. Chejfec and B.G.W. Arnason (1983). Neuropathy accompanying IgM monoclonal gamopathy. Acta Neuropathol. 59: 255-261.

Steinemann, C., M. Fenner, H. Binz, and R.W. Parish (1984). Invasive behavior of mouse sarcoma cells is inhibited by blocking a 37,000 - dalton plasma membrane glycoprotein with Fab fragments. Proc. Natl. Acad. Sci. USA 81: 3747-3750.

Sutton, R., C.W. Wrigley, and B.A. Baldo (1982). Detection of IgE- and IgG-binding proteins after electrophoretic transfer from polyacrylamide gels. J. Immunol. Methods 52: 183-194.

Symington, J. (1984). Electrophoretic transfer of proteins from two-dimensional gels to sheets and their detection. In Two-dimensional gel electrophoresis of proteins: Methods and applications (Celis, J.E., and R. Bravo, Eds.). Academic Press, New York.

Towbin, H., T. Staehelin, and J. Gordon (1979). Electrophoretic transfer of proteins from polyacrylamide gels to nitrocellulose sheets: Procedure and some applications. Proc. Natl. Acad. Sci. USA 76: 4350-4354.

Towbin, H. and J. Gordon (1984). Immunoblotting and dot immunobinding -- current status and outlook. J. Immunol. Methods 72: 313-340.

Tsang, V.C.W., B.C. Wilson, and S.E. Maddison (1980). Kinetic studies of a quantitative single-tube enzyme-linked immunosorbent assay. Clin. Chem. 26: 1255-1260.

Tsang, V.C.W., Y. Tao, L. Qui, and H. Xue (1982). Fractionation and quantitation of egg antigens from Schistosoma japonicum by the single-tube kinetic-dependent enzyme-linked immunosorbent assay (k-ELISA): Higher antigen activity in urea-soluble than in aqueous-soluble fractions. J. Parasitol. 68: 1034-1043.

Tsang, V.C.W., J.M. Peralta, and A.R. Simons (1983a). Enzyme-linked immunoelectrotransfer blot techniques (EITB) for studying the specificities of antigens and antibodies separated by gel electrophoresis. Methods, Enzymol. 92: 377-391.

Tsang, V.C.W., K.R. Tsang, K. Hancock, M.A. Kelly, B.C. Wilson, and S.E. Maddison (1983b). Schistosoma mansoni adult microsomal antigens, a serological reagent. I. Systematic fractionation, quantitation, and characterization of antigenic components. J. Immunol. 130: 1359-1365.

Tsang, V.C.W., K. Hancock, M.A. Kelly, B.C. Wilson, and S.E. Maddison (1983c). Schistosoma mansoni adult microsomal antigens, a serological reagent. II. Specificity of antibody responses to the S. mansoni microsomal antigen (MAMA). J. Immunol. 130: 1366-1370.

Tsang, V.C.W., K. Hancock, and S.E. Madison (1984a). Quantitative capacities of glutaraldehyde and sodium m-periodate coupled peroxidase-anti-human IgG conjugates in enzyme-linked immunoassays. J. Immunol. Methods 70: 91-100.

Tsang, V.C.W., K. Hancock, S.E. Maddison, A.L. Beatty, and D.M. Moss (1984b). Demonstration of species-specific and cross-reactive components of the adult microsomal antigens from Schistosoma mansoni and S. japonicum (MAMA and JAMA). J. Immunol. 132: 2607-2613.

Tsang, V.C.W., K. Hancock, and A.R. Simons (1985). Calibration of prestained protein molecular weight standards for use in the "Western" or enzyme-linked immunoelectrotransfer blot techniques. Anal. Biochem. 143: (in press).

Turner, B.M. (1983). The use of alkaline-phosphate-conjugated second antibody for the visualization of electrophoretically separated proteins recognized by monoclonal antibodies. J. Immunol. Methods 63: 1-6.

Van Wyke, K.L., J.W. Yewdell, L.J. Reck, and B.R. Murphy (1984). Antigenic characterization of influenza A virus matrix protein with monoclonal antibodies. J. Virol. 49: 248-252.

ENZYME IMMUNOASSAYS FOR TROPICAL PARASITIC DISEASES

D. de Savigny
Ontario Ministry of Health
Box 9000, Terminal A
Toronto, Canada

F. Speiser
Swiss Tropical Institute
Socinstrasse 57
4051 Basel, Switzerland

INTRODUCTION

The immunodiagnosis of tropical parasitic diseases has begun to benefit significantly from recent developments such as: *in vitro* cultivation of pathogenic parasites; new immunochemical methods for producing and purifying immunodiagnostic reagents; emergence of reagent monoclonal antibodies; and the development of sensitive immunoassay methodology for detecting low concentrations of circulating or excreted pathogens or their markers in body fluids.

However it is still true that many of these methods remain in the research laboratory and relatively few have found practical application where they are most needed - in the tropics. The arguments supporting the need for effective immunodiagnosis for tropical parasitic diseases and the specific problems of conducting such immunodiagnosis in a tropical environment have been reviewed by Voller and de Savigny (1981).

Currently there are two major options for *in vitro* immunodiagnosis of tropical parasitic diseases: 1) detection of circulating antibodies; and 2) detection of circulating or excreted antigens.

The first approach to detecting antibodies is the conventional one, however the quality and sensitivity of the new generation of immunoassays has outstripped the quality of the reagent antigens available to them. Hence these assays suffer increasingly from non-specifity due to non-availability of parasite-specific reagent antigens. This has given parasite serology a poor performance record in terms of diagnosis. Furthermore antibody detection, even when using pathogen specific antigens, has low predictive value since antibodies may not circulate in detectable amounts in early and/or light infections (false negatives) and antibodies tend to persist for years following effectively treated or self-cured infections (false positives). In many tropical populations the prevelance of antibodies may be so high due to repeated re-infections that the relevance of the magnitude of specific antibody response as a diagnostic tool may be slight. Unlike antigen quantitation, antibody quantitation is not strictly possible since heterogeneous immunoassays reflect both the quantity and the affinity constant of the antibody (Butler et al, 1978). At best the antibody result provides presumptive laboratory diagnosis.

The second approach of detection of specific circulating or excreted antigens in patient body fluids by immunoassay is attracting increasing interest because this minimizes the above problems of predictive value and can yield a definitive diagnosis without the need for highly trained parasitologists or microscopists.

In both antibody and antigen detection, immunoassay provides great potential for rapid, high volume through-put, high cost-effectivness and the possibility of vastly simplified methodology. In this chapter we describe an example of both approaches of using enzyme immunoassays for antibody <u>and</u> antigen detection in tropical parasitic diseases.

PART A. ENZYME IMMUNOASSAY OF ANTIBODY

Because highly specific parasite antigens are not yet economically produced or otherwise readily available for immunodiagnostic purposes, the value of applying a single test using non-specific antigen for a particular serum specimen is limited and questionable. However, if a sufficiently comprehensive battery of serologic tests using different parasite antigens is applied to each specimen, one can, with experience, interpret through the relative intensity of specific and non-specific reactions, the most probable diagnosis and hence increase the predictive value of the result. The high efficiency and economy of the indirect enzyme-linked immunosorbent assay (ELISA) has now made such an approach economically and practically feasible. For the past several years we have been offering a rapid multi-antigen ELISA at the Swiss Tropical Institute with satisfactory results even though for most of the diseases in the battery only, inexpensive and crudely prepared antigens are used.

METHODS

Rapid Multi-Antigen ELISA for Antibody Screening

Antigen preparation. Use crude antigen extracts. For protozoa use culture forms of: *Leishmania donovani* (1S Sudan strain); *Trypanosoma gambiense* (EATRO 210); and *Trypanosoma cruzi* (Y-strain) disrupted in 0.06M carbonate buffer solution (coating buffer) pH 9.6 by five cycles of freezing and thawing using dry ice-petroleumbenzene and 37C water bath. For amoebiasis use soluble antigens of *Entamoeba histolytica* (HK 9 strain). For helminths (*Dipetalonema viteae*, *Fasciola hepatica*, *Schistosoma mansoni* and *Schistosoma haematobium*) prepare total adult worm extract antigens by homogenizing the organisms with a glass homogenizer and extracting at 4°C for 24h in 0.13M phosphate buffered saline pH 7.2. For Toxocariasis use metabolic antigen from culture supernatants of second stage *Toxocara canis* (de Savigny, 1975). For echinococcosis use hydatid fluid from bovine fertile lung cysts of *E. granulosus* centrifuged at 200g for 15 min. and dialyzed overnight in coating buffer. Ultracentrifuge all antigens prior to use (100,000g, 4°C, 2h).

ELISA Procedure. Dilute all antigen preparations in 0.06M carbonate buffered saline pH 9.6 according to optimal antigen titrations (ca 1ug/ml) and dispense 200 ul per well into alternate horizontal rows in ELISA plates (Dynatech 29B or Petraplast 1.011). Cover and incubate plates for 2 days at 4°C, but use within 3 weeks. To commence the assay wash the plates with tap water containing 0.15% Tween 20 (Merck 822184). Dilute patients' sera to 1:160 in PBS containing 0.05% Tween 20 (PBS-T) and dispense 200 ul in each well of a vertical column. Thus, each serum is tested against 8 different antigens. Incubate at 37°C for 15 min. Empty the plates and fill with warm (37°C) PBS-T, incubate at 37°C for 5 min., and then wash in tap-water-Tween as above. Add conjugate to each well (goat anti-human IgG, H+L) (Miles 61-230) diluted 1:1500 in PBS-T and incubate at 37°C for 15 min. Wash again as above with a final rinse of distilled water before adding the substrate-chromogen solution of 0.03% hydrogen peroxide (Merck 8597) 0.1% ortho-phenylenediamine (Merck 7243) in 0.07M PBS pH 5.0). Stop the enzyme reaction with 8N sulphuric acid (50 ul/well) after the positive and negative control reactions reach the required absorbance values. In routine diagnostic work, the amoebiasis system serves as a reference system for the whole multi-antigen plate. Measure absorbance values at E_{492nm}. Sera with absorbance values over 2.0 should be diluted 1:5 with stopped blank substrate solution and measured again.

TYPICAL RESULTS

Rapid Multi-Antigen ELISA for Antibody Screening

Titrating different serum pools in the multi-antigen ELISA and incubating the sera for different times (15 min., 30 min., 120 min.) at 37°C reveals parallel curves for each antibody-antigen system (Fig. 1). Since generally higher background reactions are observed

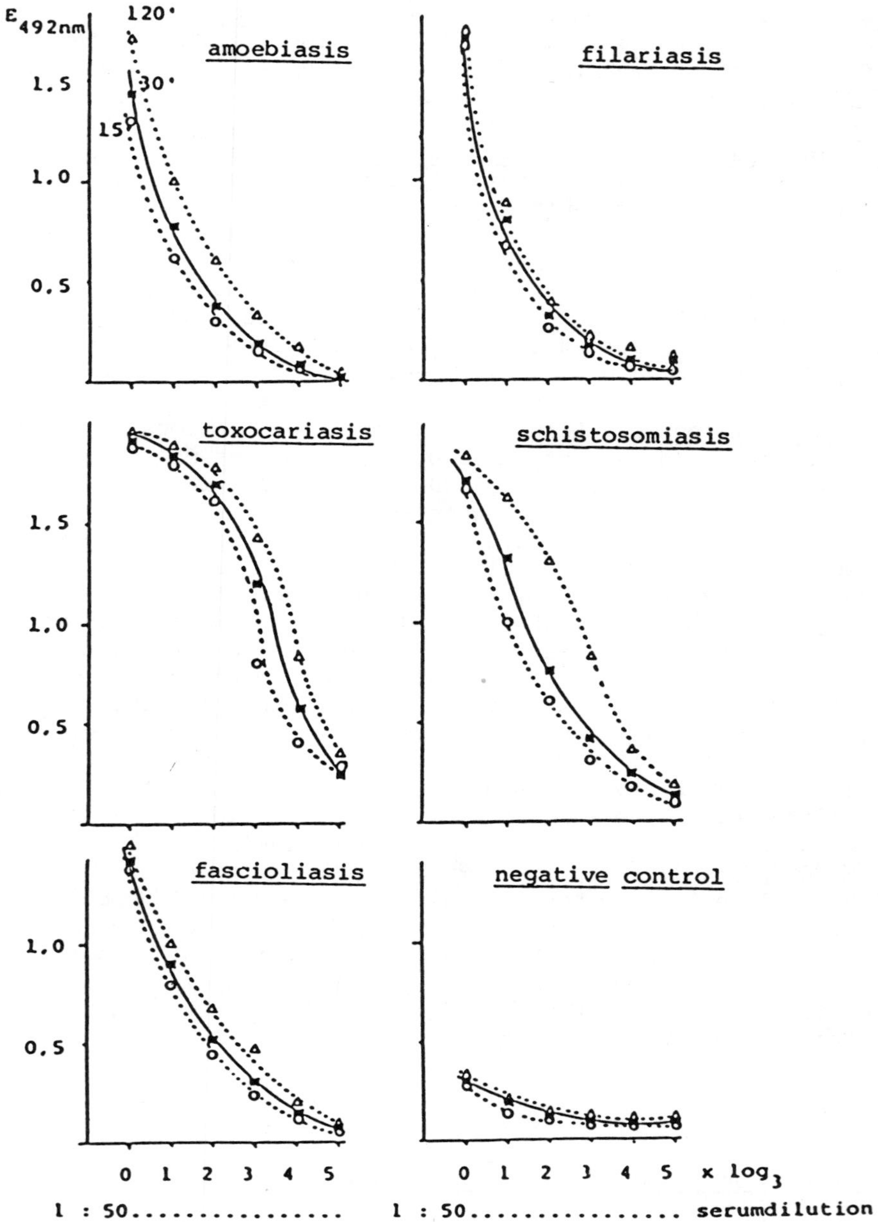

Fig. 1. Serum titration curves at different serum incubation times in a multi-antigen ELISA. Results are means of eight tests on serum pools.

with antigens from Echinococcus granulosus than with other helminthic antigens, the negative serum pool is shown against this antigen. Differences between absorbance values in regard to the serum incubation times were found to be no more than 50% depending on the antibody-antigen system. Only in the toxocariasis system is there evidence of an absorbance value plateau. The coefficients of variation in the 15 min. incubation system were typically less than 15% at all serum dilutions under 1:10,000. They were lowest at low serum dilutions and generally increased with increasing dilutions of sera. This tendency could be observed with all antibody-antigen systems and was independant of the serum incubation times.

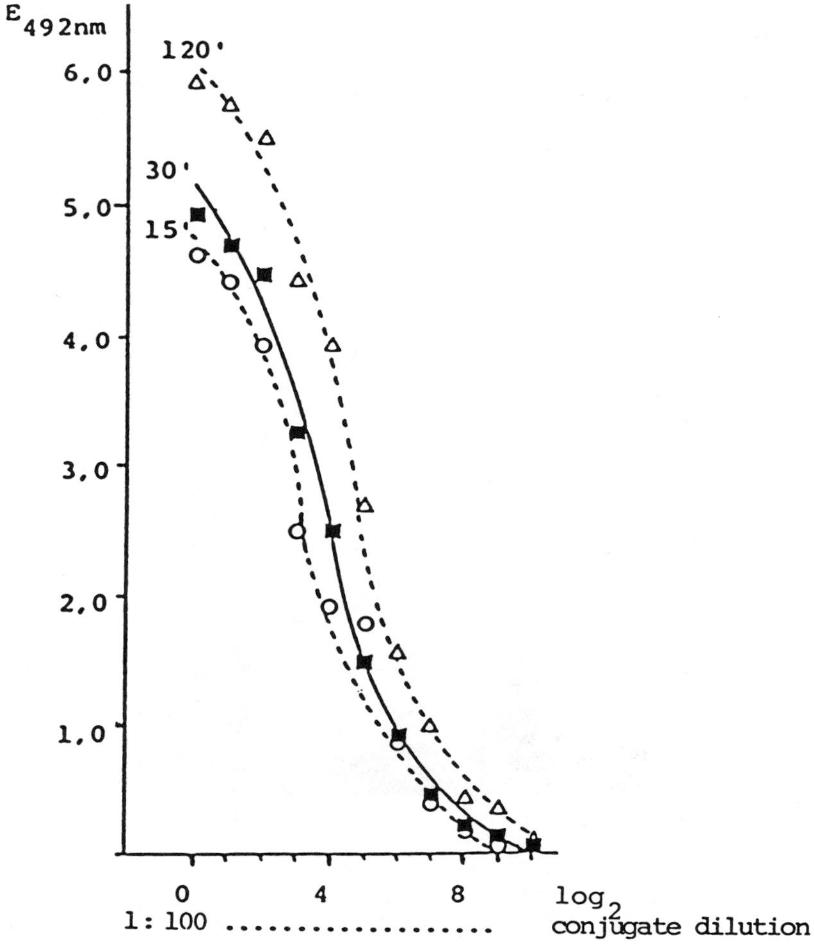

Fig. 2. Conjugate titration curves obtained in a direct ELISA using solid phase IgG.

Titrating and incubating conjugate for different times (15 min., 30 min., and 120 min.) at 37°C in a direct ELISA against solid phase IgG gives the pattern in Fig. 2. A plateau of the conjugate titration curves is not observed. This is also the case when the conjugate in titrated is the multi-antigen system. Differences between absorbance values in relation to the conjugate incubation time were no more than 50% in both the direct and in the indirect test systems. The differences between absorbance values as obtained by different incubation times could easily be minimized by monitoring the enzyme-substrate incubation step. The enzyme reaction is linear and very fast (0.25 absorbance units/minute). Only about 1 to 2 min. longer incubation in this final step was needed to obtain the same absorbance values after a 15 min serum/conjugate incubation as would have been obtained after 2 hours incubation time.

Fig. 3 shows the inter- and intra-assay variations of a multi-antigen helminth ELISA-plate, as obtained during a two month routine

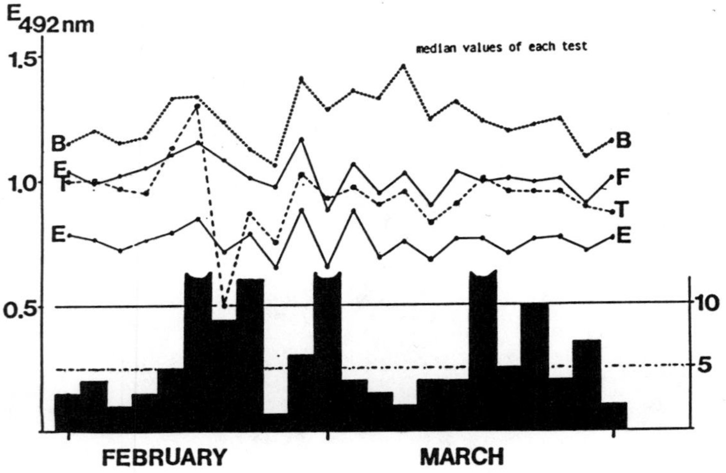

Fig. 3. Reproducibility of homologous control reactions in the Rapid Multi-Antigen ELISA over a two months' application in the routine serodiagnostic laboratory. The black histogram gives the percent maximum and minimum variation within one system (schistosomiasis) (scale in percent shown on right ordinate). (B) Schistosomiasis, (F) Filariasis (T) Toxocariasis, (E) Echinococcosis.

application. All curves show the same variations with the exception of the toxocariasis system where a coating error was identified at the end of February. The percent variation within one test (plate to plate variations) is given for the schistosomiasis system. It

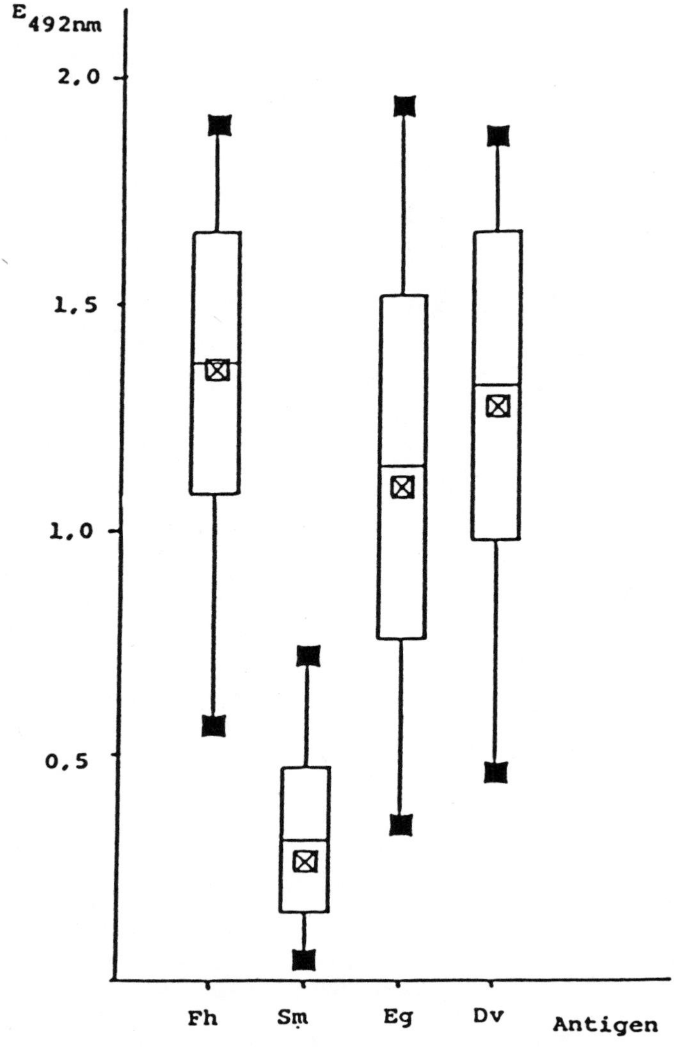

Fig. 4. Comparison of absorbance values in Multi-Antigen ELISA as obtained with 49 sera from parasitologically proven fascioliasis patients. The columns represent the mean and standard deviations, the closed squares give the extreme values and the open squares give the median values.
(Fh) F. hepatica, (Sm) S. mansoni, (Eg) E. granulosus, (Dv) D. viteae.

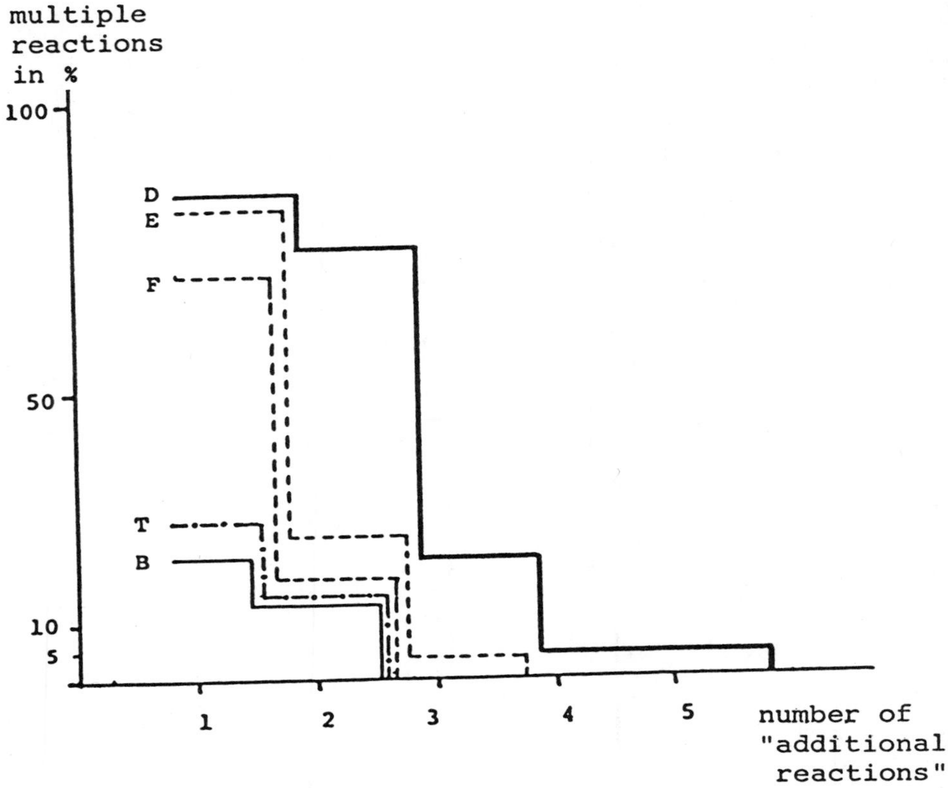

Fig. 5. Cumulative frequency of multiple positive reaction in various patients' sera tested against different helminthic antigens in the Rapid Multi-Antigen ELISA during one year of application (n = 3400 sera).

was over 10% in 4 out of 22 tests (16%, 12%, 14%, 14%). With other antibody-antigen systems of the multi-antigen plate the variations were in this same order. The inter-assay variations, expressed as coefficient of variation (CV%), were: 8.2% for the schistosomiasis; 6.9% for the filariasis; 8% for the echinococcosis; and 15.5% for the toxocariasis positive controls. The main source of test variation was the substrate-chromogen step . Since this solution is prepared in highly excess concentration, a rapid and linear enzyme reaction could be obtained. Thus, the monitoring of the enzyme reaction time was most critical to quality control.

When sera from different parasitologically proven parasitic infections were analyzed with the Multi-Antigen ELISA it was found that the homologous antibody-antigen system usually gave the highest absorbance values. Even if the differences of absorbance values between some homologous and the cross-reacting systems were small, these differences were reproducible. With sera from helminth infected patients "one way" cross-reactions can occur. Sera from fascioliais patients showed extensive cross-reactions with antigens from *Dipetalonema viteae* and *Echinococcus granulosus* (Fig. 4) while sera from filariasis or echinococcosis patients did not react with antigens from *Fasciola hepatica* in the Multi-Antigen ELISA (cf Fig. 5). Only one out of 44 sera from Swiss patients with parasitologically proven schistosomiasis, filariasis or echinococcosis, having high antibody activity against their corresponding antigens, was also reactive with antigens from *F. hepatica* (results not shown). Thus, the frequency of multiple reactions within a Multi-Antigen ELISA was dependant on the disease.

This is reflected by analyses of data from one year of routine application (3400 different sera tested) of the helminth Multi-Antigen ELISA (Fig. 5). Of all *F. hepatica* reactive sera (n = 24), 75% were also reactive with two additional antigens (mostly from *D.viteae* and *E. granulosus*). Sera reactive with *D. viteae* antigens (n = 334) showed additional reactions with another antigen in 60% (mostly with antigens from *E. granulosus*) and sera reactive with antigens from *E. granulosus* (n = 244) were also reactive with an additional antigen in 80% (mostly with antigens from *D. viteae*). Reactions with two additional antigens were much less common than with *F. hepatica* reactive sera. Multiple reactions were least pronounced in the toxocariasis and schistosomiasis systems.

DISCUSSION

Rapid Multi-Antigen ELISA for Antibody Screening

Technical considerations. Rapid ELISA systems for screening purposes have already been described (Saunders et al, 1978) These allow more efficient use of laboratory resources and serodiagnostic results can be based on comparative analyses of several tests without increase of laboratory personnel. Rapid ELISA systems may therefore be especially interesting for serodiagnosis of those diseases where cross-reactions occur frequently. However, rapid serological screening systems are often qualitative only. When antibodies are to be quantified, the antibody activity of a patient's serum should be the only limiting test factor (Tsang et al, 1980). In conjugate titration experiments no plateau of absorbance values could be obtained after a conjugate incubation of 15, 30, 120 minutes or even after 24 hours. This was also the case when conjugates from other suppliers were evaluated. Therefore a compromise (based on economical considerations) was made in chosing the optimal conjugate concentration to get a semi-quantitative test.

In this work we found that the concentration of immunoreactants has a greater impact on final absorbance values than does the incubation time. Initial antibody binding was very fast. Up to 15% of final absorbance values were obtained after a serum incubation of 1 minute at 20°C in comparison to an incubation of 15 minutes at 37°C with the same serum (results not shown). This fast binding can be understood from the Law of Mass Action showing that the rate of the forward reaction is higher than the rate constant of the reverse reaction at the beginning of the reaction and before reaching equilibrium (Maggio, 1981). Using short incubation times may favour test specificity since the most avid antibodies should bind first. The binding of high avidity antibodies may additionally be favoured by low serum dilutions as discussed by Lethonen and Eerola (1982). This can explain a higher discrimination capacity of absorbance values with a rapid ELISA system over an ELISA with long incubation times for antibody-antigen reactions where cross-reactions are very pronounced. Such cross-reactions are very common in parasitic diseases especially when crude antigen extracts are used.

Another factor influencing the final test results to a large extent, is the substrate-indicator solution. To get a quantitative test system, the enzyme-substrate reaction should be linear. Linearity of an enzyme reaction can be obtained with an excess concentration of the substrate which also leads to a very fast enzyme-substrate turnover. This has the disadvantage that the monitoring of enzyme reaction time becomes critical. All enzyme reaction curves were highly linear due to the short substrate incubation time and this results in more quantitative test values. The ELISA system, presented in this work, may be similar to the kinetic (K-ELISA) system described by Tsang et al (1980, 1981) where the slope of the initial 2 minutes of enzyme reaction is expressed in $\Delta A460$/min.

Intra-assay variations (plate to plate variations) were in the same order as the inter-assay variations. It is therefore important to test a serum within the same plate when absorbance values obtained with different antigens are to be compared. This is especially true when the differences between absorbance values of homologous and cross-reacting antibody-antigen system are small. When absorbance values from duplicate tests were analyzed from one year of routine ELISA application, the corresponding differences were found to be around 5% (results not shown). The reproducibility of the Multi-Antigen ELISA as expressed by coefficient of variation was between 5-19%. The tendency of decreased reproducibility with increased serum dilutions was independant of the incubation time. This is another feature favouring the use of low serum dilutions in ELISA. The higher backgrounds encountered in low serum dilutions can be controled by increasing the amount of Tween-20 in the diluting buffer solution or by additional incubations at 37°C with the diluting buffer solution.

Serodiagnostic considerations. The experience using the Rapid Multi-Antigen ELISA in the serodiagnostic routine laboratory shows that cross-reactions in parasitic diseases occur not only in infections with phylogenetically related organisms but also can be extremely pronounced among very different parasite etiologies (eg. between filariasis and echinococcosis, Speiser, 1980). However, sera of echinococcosis patients always react strongest with antigens from *Echinococcus granulosus* in the multi-antigen helminth ELISA (Speiser, 1982).

All *F. hepatica* reactive sera react strongly with antigens from *Dipetalonema viteae* and *E.granulosus*. Cross-reactions of fascioliasis sera with antigens from *S. mansoni* were less frequent and less pronounced. Thus, a kind of "one way" reaction pattern can occur with parasitic infections. Such "one way" reactions could also be experimentally confirmed in our laboratory with sera from mice, infected with *S. mansoni* or *F. hepatica* (results not shown).

Other antibody-antigen systems are found to be very specific. Specific metabolic antigens from second stage larvae of *Toxocara canis* have been described for ELISA by several authors (de Savigny et al, 1979, Carlier et al, 1982, Matsumura and Endo, 1982, Van Knapen et al, 1983). According to results from the routine laboratory, sera from presumed toxocariasis patients rarely cross-react with antigens from other helminths. Thus, the humoral immune response in toxocariasis patients seems to be mainly against very specific metabolic antigens released *in vivo* and *in vitro*. The Toxocariasis ELISA may therefore be a canditate for a standardized helminth serology (Speiser and Gottstein, 1984).

The consideration of combined results within a Multi-Antigen ELISA (degree of absorbance value, frequency of multiple reactions, reaction pattern) leads to a more conclusive serodiagnosis than could be achieved by using only a single antibody-antigen ELISA system. If a patient's serum reacts well with antigens from *T. canis* and from *D. viteae* a toxocariasis double infection with *filariasis* (or another helminthiasis eg. strongyloidiasis) is to be considered. If a high ELISA value with a patient's serum is observed with antigens from *F. hepatica*, *D. viteae* and *E. granulosus*, a strong indication for a distomatosis (fascioliasis, paragonimiasis) is given. If the reaction is only limited to the *F. hepatica* antigens, a non specific (not a cross-reaction) is to be considered. If a patient's serum reacts with all helminth antigens, non-specific causes must be considered.

The ELISA system presented in this work proved to be a very effective method. Up to 700 different sera can be tested against 4 different antigens within one day by two technicians and enables a routine serodiagnostic laboratory to provide a more complete and reliable laboratory report to the clinician. However, the goal of a multi-antigen test should remain the use of highly specific antigens. Such antigens have been described but they are not yet available economically for routine work. Therefore it may be more practical to look for semi-purified antigen preparations, leading to enhanced differences of ELISA values between homologous and cross-reacting antibody-antigen systems than obtained by using only crude antigen extracts. Such semi-purified antigens may increase the discriminative ability of combined ELISA results and thus lead to a more conclusive serodiagnosis of parasitic disease.

PART B. ENZYME IMMUNOASSAY OF PARASITE ANTIGEN

Some of the merits of the antigen detection approach to diagnosis of parasitic disease have been mentioned in the general introduction. Amebiasis is a common and potentially serious parasitic infection of man throughout the tropics and is often seen in travellers returning from visits to the developing world. Because of this it has attracted interest by diagnosticians in the developed world. Accurate diagnosis currently requires the preparation and microscopic examination of multiple faecal samples by highly trained and experienced workers. Cost-efficiency is low and this has been a further impetus for workers to develop a more simple and rapid indirect immunoassay to detect the presence of the pathogen, Entamoeba histolytica in faecal specimens. We describe here a modification of the recently developed method of Grundy, Voller and Warhurst (1985) using an indirect enzyme linked immunosorbent assay (ELISA) which is capable of detecting both trophozoites and cysts in faecal extracts.

METHODS

ELISA for E. histolytica Antigen Detection

Antigen. Prepare axenic cultures of E. Histolytica strain HK9 according to Diamond (1968). Harvest amoebae after 96 h by centrifugation at 1000 RCF for 5 minutes at 4°C. Wash pellet 3 times at 4°C and solubilize by sonication on ice (3 x 1 min).

Immunosorbent Columns. Activate 3 g of lyophilized CNBr-Sepharose 4B (Pharmacia) and couple with 20 ml of a 50% solution of heat inactivated normal sheep serum. Activate 10 g of lyophilized CNBr-Sepharose 4B and couple overnight with 60 ml of 2 mg/ml protein solution of E. histolytica homogenate as above. Activate 3 gm of lyophilized CNBr-Sepharose 4B and couple with 20ml of a 20% solution of rabbit anti-bovine serum Ig fraction (Dako).

Sheep Anti-E. histolytica Immunoglobulin. Suspend washed homogenized amoebae as prepared above to a concentration of 8 mg/ml wet-weight in PBS containing 0.1% triton X100. Pass the supension through an equilibrated rabbit anti-bovine serum immunosorbent column and collect the drop-through fraction at a flow rate of 10 ml/hr at 4°C. Immunize a sheep with 2 ml of drop-through fraction at 2.5 mg/ml in Freund's complete adjuvant administered subcutaneously on days 1, 6, 13, 20, and 34 and bleed for serum on day 41. Pass the serum through the _E. histolytica_ immunosorbent column at 10 ml/hr at 4°C and after washing, elute specific antibodies with 0.1 M glycine-HCl buffer pH 2.4, adjust eluate rapidly to pH 8.0 using Tris-HCl and dialyse overnight at 4°C against PBS containing 0.02% NaN_3. Store at 4°C.

Rabbit Anti-E. histolytica Immunoglobulin. Immunize two New Zealand White rabbits with 10^6 _E. histolytica_ trophozoites intraperitoneally on days 1, 14, and 21 bleeding and pooling sera on day 28. Pass the serum through the _E. hystolytica_ immunosorbent column at 10 ml/hr (4°C), wash, elute, and dialyze as above. Pass the affinity-purified immunoglobulin through the normal sheep serum immunoadsorbent column, collect the drop-through fraction and store at 4°C.

Sheep anti-rabbit Fc Immunoglobulin-Peroxidase. Conjugate sheep anti-rabbit Fc Ig fraction (DAKO) to peroxidase (Sigma, Type VI) using the method of Nilsson et al (1981) with an HRP:IgG molar ratio of 1.95 and store at -20°C in 50% glycerol.

Faeces Extracts for ELISA. Mix a portion of the faeces sample in 10 ml PBS (10-50% by wt) and shake extensively to suspend particles. Filter out large particles by passing the suspension through two layers of cotton gauze. Dilute stool filtrate solution into an equal volume of PBS-Triton X100 (0.1%) containing 20% normal sheep serum and test immediately or freeze at -20°C.

Enzyme-linked Immunosorbent Assay. Coat wells of polystyrene micro-ELISA plates (Greiner) with 200 ul of affinity purified sheep anti-_E. histolytica_ immunoglobulin at at 4-6 ug/ml in 0.05 M carbonate buffer, pH 9.6 for 18 h at 4°C (Test wells). Coat additional plates as above using the same concentration of normal sheep immunoglobulin (Control wells). Add 200 ul of stool filtrate each to both test and control wells and incubate 2.5 hours at 4°C. Wash plates 3 times for 3 minutes with PBS containing 0.05% Tween 20 (PBS-T). Wash as above, add 200 ul per well of rabbit anti-amoeba Ig at 24 ug/ml diluted in heat inactivated normal sheep serum and incubate 2.5 h at 4°C. Wash in PBS-Triton X100 (0.05%) (PBS-TX) as above, and add 200 ul per well of sheep anti-rabbit Fc-Peroxidase conjugate diluted in PBS-TX and incubate overnight at 4°C. Wash in PBS-TX as above and add 200 ul/well of 0.1% 5-aminosalicylic acid (Ellens and Gielkins, 1980) with 0.005% H_2O_2 in 0.1 M sodium phosphate buffer pH 6.0 containing 1 mM EDTA. Stop reaction after 30 minutes by addition of 50 ul/well of 1 M NaOH. Read absorbance at 450 nm and subtract control values from test values. Positive control consists of known concentration of antigen in PBS-TX and 20% normal sheep serum.

TYPICAL RESULTS

ELISA for E. histolytica Antigen Detection

Dose-response curves in this assay suggest quantitation of amoebae in the range of 20 to 5 x 10^3 per ml however results are reported qualitatively based on a minimal detectable dose calculation using at least six known negative samples run with each assay. Stool samples exert a non-uniform effect on solid phase antibody resulting in the subsequent release of antibody from the solid phase. Grundy et al (1985) examined the effect of different incubation buffers on positive to negative differential and found that a diluent containing 5 to 50% sheep serum was most effective in minimizing the deleterious action of the faecal extract on solid phase immunoglobulin. Sensitivity was also enhanced significantly by the substitution of PBS-TX 100 for PBS-T by decreasing non-specific interactions between the solid-phase antibody and the conjugate suggesting that this interaction may have a hydrophobic basis.

In work by Grundy et al (1985) on clinical samples defined parasitologically, 37 of 54 samples from asymptomatic carriers of *E. histolytica* cysts were positive while one of 24 parasitologically negative samples was positive. This latter patient was found to be parasitologically positive two days later and thus the Amoeba Antigen-ELISA result is probably not a false positive. Of 17 patients with other intestinal parasites three were positive. The parasites showing no significant cross-reactivity were *E. Nana* *Hymenolepis nana*, *Entamoeba coli*, *Giardia lamblia*, *Strongyloides stercoralis*, *Ascaris lumbricoides*, *Trichomonas*, and *E. hartmanii*. One sample that was strongly parasitologically positive for *Iodomoeba* but negative for *E. hystolytica* was strongly positive by ELISA. Whether this was a true cross-reaction is unclear however since other samples containing large numbers of *Iodamoeba* were negative by ELISA.

The detection of a large percentage of the relatively difficult to diagnose asymptomatic cyst carriers supports the use of this assay for both epidemiological and clinical applications. The assay is under evaluation regarding diagnostic performance in both symptomatic (invasive) and non-symptomatic amoebiasis.

DISCUSSION

ELISA for E. histolytica Antigen Detection

In this portion of the chapter we have described a recent development illustrating a new trend in immunodiagnosis for parasitic diseases, that of detecting antigen in body fluids or excretions. Such an approach has the potential to increase the cost-effectivness of parasite diagnosis through simplified methodology applied on a large scale which does not require professional parasitologists in the field.

Antigen detection frees the immunodiagnostician from being tied to the blood specimen. Many populations living in the tropics resent the taking of venous blood. In addition venepuncture under field conditions requires skilled technicians, expensive materials and poses a significant risk to the patient. Subsequent steps require a cold chain, centrifuge, and further costly containers and transfer devices for each specimen. The collection of faeces and urine is non-invasive and most parasitic-infections will be reflected by excretion of parasite antigens or metabolites in one or both of such specimens detectable by the new generation of immunoassays. However enzyme immunoassay as described here is still too sophisticated for most tropical applications except perhaps those in central laboratories. Still needed is the one step 'dip-stick' approach to immunoassay. As a necessary first step, experience is required to tell us what parasite antigens are excreted or circulating in which body fluids, in what concentration and, with what diagnostic or prognostic relevance. To this end EIA applications will play a major role.

To date parasite antigens have been detected by enzyme immunoassay in faeces for amoebiasis, in blood for malaria, in urine and blood for filariasis, in urine and blood for schistosomiasis, and in mosquitoes for malaria.

SUMMARY

Two emerging approaches to parasite immunodiagnosis have been presented: the Rapid Multi-Antigen ELISA using crude antigens for antibody detection; and the use of ELISA for the direct demonstration of the parasite or its products in clinical specimens.

For antibody detection we describe an ELISA system which tests sera against a comprehensive battery of antigens using relatively short (15 minutes) incubation steps. A more specific laboratory diagnosis is achieved by interpretation of the relative pattern and intensity of the reactions and cross-reactions observed leading to a more reliable result than obtained in tests applied singly against individual antigens. Up to 700 sera can be processed daily by two workers.

For antigen detection we describe an ELISA system which detects E. histolytica antigen in faecal specimens at levels as low as 20 organisms/ml in asymptomatic cyst passers and in which no significant cross-reactions with other intestinal parasites occur. A case is made for encouraging further efforts in the direction of antigen detection which is hoped will provide for more reliable large-scale epidemiology and immunodiagnosis.

ACKNOWLEDGEMENTS

We gratefully acknowledge Drs. M.S. Grundy, A. Voller and D. Warhurst for permission to quote unpublished work. This work was supported in part by the UNDP/World Bank/WHO Special Programme for Research and Training in Tropical Diseases.

REFERENCES

Carlier, Y., J. Yang, D. Bout, and A. Capron (1982). The use of excretory-secretory antigen for an ELISA specific serodiagnosis of visceral larva migrans. Biomedicine 36: 39-40.

de Savigny, D. (1975). *In vitro* maintenance of *Toxocara canis* larvae and a simple method for production of *Toxocara* ES antigen for use in serodiagnostic tests for visceral larva migrans. J. Parasitol. 61: 781-782.

de Savigny, D. A. Voller, and A.W. Woodruff (1979). Toxocariasis: serological diagnosis by enzyme-linked immunosorbent assay. J. Clin. Path. 32: 284-288.

Diamond, L.S. (1968). Techniques of axenic cultivation of *E. histolytica* (Schaudin 1903) and *E. histolytica*-like amoebae. J. Parasitol. 54: 1047-1056.

Ellens, D.J., and A.L.J. Gielkens (1980). A simple method for the purification of 5-aminosalicylic acid. Application of the product as substrate in ELISA. J. Immunol. Meth. 37: 325-332.

Grundy, M.S., A. Voller, and D. Warhurst (1985). An enzyme linked immunosorbent assay for the detection of *E. histolytica* antigens in faecal material. Trans. roy. Soc. trop. Med. Hyg. (in press).

Maggio, T. (1981). Enzyme Immunoassay. CRC Press, Inc. Boca Raton, Florida.

Matsumura, K., and R. Endo (1982). Enzyme-linked immunosorbent assay for toxocariasis: its application to the sera of children. Zbl. Bakt. Hyg. I Abt. Orig. A 253: 402-406.

Nilsson, P. N.R. Berquist, and M.S. Grundy (1981). A technique for the preparing of defined conjugates of horse-radish peroxidase and immunoglobulin. J. Immunol. Meth. 41: 81-83.

Saunders, G.C., E.H. Clinard, M.L. Bartlett, P.M. Peterson, Wm. Sanders, R.J. Pyne, and E. Martinez (1978). Serological test systems development. Report LA-7078-PR, Department of Energy, University of California.

Speiser, F. (1980). Application of the enzyme-linked immunosorbent assay (ELISA) for the diagnosis of filariasis and echinococcosis. Tropenmed. Parasit. 31: 459-466.

Speiser, F. (1982). Serodiagnosis of tissue dwelling parasites: application of a multi-antigen Enzyme-Linked Immunosorbent Assay (ELISA) for screening. Ann. Soc. belge Med. trop. 62: 103-120.

Speiser, F., and B. Gottstein (1984). A collaborative study on larval excretory-secretory antigens of *Toxocara canis* for the immunodiagnosis of human toxocariasis with ELISA. Acta Tropica. 41: (in press).

Tsang, V.C.W., Y. Tao, and S.E. Maddison (1981). Systematic fractionation of *Schistosoma mansoni* urea-soluble egg antigens and their evaluation by the single-tube, kinetic dependent enzyme-linked immunosorbent assay (K-ELISA). J. Parasitol. 67: 340-350.

Tsang, V.C.W., B.C. Wilson, and S.E. Maddison (1980). Kinetic studies of a quantitative single tube enzyme linked immunosorbent assay. Clin. Chem. 26: 1255-1260.

Tsang, V.C.M., B.C. Wilson, and J.M. Peralta (1983). The quantitative single tube, kinetic dependant Enzyme-Linked Immunosorbent Assay (K-ELISA). Methods. Enzymol. 92: 391-403.

Van Knapen, F., J. van Leusden, A.M. Polderman, and J.H. Franchimont (1983). Visceral larva migrans: Examination by means of enzyme-linked immunosorbent assay of human sera for antibodies to excretory-secretory antigens of the second stage larvae of *Toxocara canis*. Z. Parasitenkd. 69: 113-118.

Voller A., and D. de Savigny (1981). Diagnostic Serology of tropical parasitic diseases: A review. J. Immunol. Meth. 46: 1-29.

MONOCLONAL ANTIBODIES IN IMMUNOENZYMETRIC ASSAYS

Stanley Y. Shimizu, David S. Kabakoff and E. Dale Sevier

Hybritech Incorporated

11095 Torreyana Road, San Diego, CA. 92121

INTRODUCTION

A new world of immunodiagnostics emerged when Kohler and Milstein (1975) elevated antibodies to the status of "reagent grade chemicals" by producing hybrid cells capable of virtually unlimited synthesis of homogeneous antibody molecules. These monoclonal antibodies have rapidly found their niche in various versions of the immunometric -- or labeled antibody -- assay technique. Six years after the initial work of Kohler and Milstein (1975), routine monoclonal antibody-based immunoradiometric and immunoenzymetric assays were developed (Sevier et al., 1981). The favorable reaction conditions of these excess antibody systems (Ekins, 1981) facilitate rapid approach to equilibrium and determination of extremely low minimal detectable antigen concentrations.

ADVANTAGES OF MONOCLONAL ANTIBODIES

Conventional polyclonal antisera are rapidly being replaced by monoclonal antibodies because of their many advantages especially when used in labeled antibody methods (Sevier, 1982). It should be noted at this point that monoclonal antibodies confer very few advantages to labeled antigen (e.g. competitive binding immunoassay) systems (Wu et al., 1983). Those possible advantages include consistent supply and the virtual elimination of "crossreactivity" when using a monoclonal antibody with sufficient binding strength to produce an acceptable assay.

Before launching into a discussion of the assay systems, a

closer look at the nature of monoclonal antibodies is necessary for the understanding of their advantages. Monoclonal antibodies display homogeneous chemical characteristics. That is each antibody molecule is an exact replica of the next, and all share the identical binding site on the antigen of interest. By carefully screening and selecting an antibody that binds to a determinant that defines the antigenic molecule, unwanted reactivity is minimized. This ability to screen out defined "crossreactants" at the cellular level leads to the often quoted claim that monoclonal antibodies can "virtually eliminate crossreactions" (DiPetro, 1981).

The hybrid cells that produce monoclonal antibodies are essentially immortal. They are highly adapted to growing in culture or as tumors in host animals. Using standard culture techniques, a virtually unlimited supply of a monoclonal antibody can be produced. This is a major advantage of monoclonal antibodies over polyclonal antisera when using these antibody preparations in the immunometric technique.

The chemistry of labeling antibodies is no different than labeling any other protein. It is often empirical and can be difficult to control and reproduce. Since monoclonal antibodies are abundant reagent grade chemicals, the maximization of the labeling technique does not impinge on the supply of antibody as it would if the source were a finite batch of polyclonal antibody. Another major advantage of labeled monoclonal antibodies is that maximal concentrations of the reagent is permitted in the reaction mixtures because there is no limit to the amount that can be used.

ENZYMETRIC ASSAYS USING MONOCLONAL ANTIBODIES

Two categories of advantages exist when using labeled monoclonal antibodies in the immunoenzymetric technique, and they are interrelated and synergistic. First, we will discuss the advantages of antibody excess methods and then interrelate the advantages of using monoclonal antibodies as an excess reagent.

The immunoenzymetric technique utilizes high concentrations of a capturing antibody and an enzyme-antibody conjugate to detect limiting amounts of an analyte. The components of such a system are shown in Figure 1. The "capture" antibody is fixed to a solid phase and is directed against a unique site on the antigen. The enzyme-antibody conjugate is directed against a second site, one that does not interfere with the binding of the solid phase antibody. Since the binding of one antibody does not interfere with the binding of the second it is possible to simultaneously incubate all three components together. After the incubation period the solid phase is washed to remove any unbound material. The amount of enzyme-antibody conjugate bound is directly proportional to the amount of analyte bound to the solid phase.

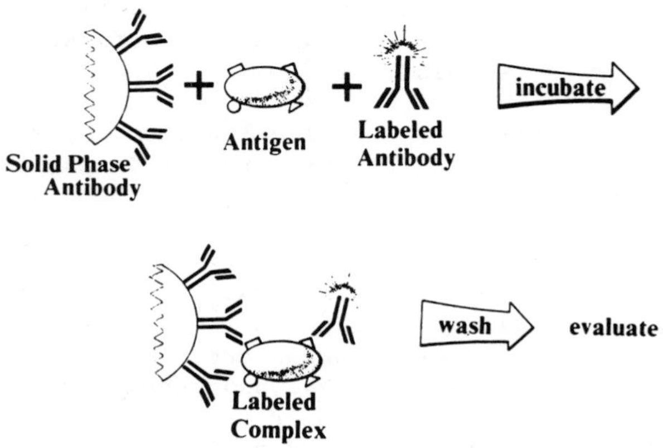

Figure 1. Typical protocol for an immunoenzymetric assay using monoclonal antibodies illustrating the simultaneous incubation of the solid phase "capture" antibody, antigen and enzyme labeled second antibody.

Most immunometric assays are diffusion limited because most use a solid phase separation procedure. An antigen must make its way to a solid phase antibody and then become labeled by an enzyme-antibody conjugate. Both of these molecules must travel through the reaction solution. Excess antibody on a large capacity solid phase will facilitate the adsorption of the antigen even at extremely low antigen concentration, again because the relatively high concentration of the solid phase antibody drives the reaction. The solution phase reaction is also driven by high concentrations of the antibody reagent. At the same time the problem of diffusion is minimized by the high concentration of reagents. Since the antibody concentration is high there is an adjacent antibody molecule that is ready to react without having to first travel any appreciable distance. The result is a fast, sensitive assay system.

Labeled antibody assays require excess binding reagents to yield systems that realize their theoretical advantages. The limited quantity of polyclonal antiserum makes it a less than ideal reagent for immunometric assays. Since polyclonal antiserum is a heterogeneous mixture of antibodies of differing affinities and specificities, isolation of an antibody for a specific site often requires exhaustive purification. The purification process may result in alteration of the antibody and substantial loss of material.

All of the advantages of monoclonal antibodies stem from their homogeneity. When reagents are present in excess, the advantage of the labeled antibody technique becomes apparent. First, high capacity solid phase "capture" antibodies minimize and in some cases virtually eliminate the saturation effect called the "high dose hook effect." (Ryall et al., 1982). This problem is common to all direct assay systems in that a saturation point exists. Designing an assay that overcomes the possibility of saturation is the key, and monoclonal antibodies easily allow very large capacity solid phase adsorbents and relatively high concentration labeled antibodies. The excess reagent system not only leads to rapid, sensitive assays, but also extremely wide useful ranges. The design of the labeled antibody assay system allows generally 3 to 10 times the range of labeled antigen competitive binding assays.

One final advantage of the labeled antibody technique is the virtual elimination of the unpredictable phenomenon of specimen-to-specimen variation, more commonly referred to as "matrix effects." (Ekins, 1981) When variations of protein, lipid or any other component occur, they can have disasterous effects on reliable antigen quantitation in serum. In urine, the effects come from differences in ionic strength, pH, protein concentration or specific gravity. Whenever a limiting reagent is used to quantitate an antigen, subtle matrix differences can alter the quantitation. Worse still, in the absence of the analyte, matrix effects can mimic its presence, leading to falsely elevated analyte estimation. The chances of affecting the reaction rate of antigen-antibody binding at low reagent concentrations are high because the molecules that make up the "matrix" are in the same concentration range. The effect is markedly reduced when excess reagents are used (Fritz and Bunker, 1982). The antigen-antibody binding will not be affected by small changes in the matrix because the reagents are at very high relative concentration.

Most of the characteristics of the labeled antibody techniques are enhanced by using monoclonal antibody reagents. Their homogeneous chemical reactivity ensures reliable, reproducible systems. By incorporating them in the immunometric technique and labeling them with enzymes another important characteristic is uncovered: approach towards first-order chemical kinetics (Barany, 1980). "Order" in chemical reactions refers to the exponents of the concentrations of reactants in the rate law. If the reaction rate is independent of a reaction concentration, the reaction rate is defined as the "zero order" rate constant. If one reactant is limiting, the rate of reaction will be proportional to the concentration of that limiting reactant. If two or three are limiting, the rate becomes a function of the concentration of each and at the same time quite complex.

Now let's examine the immunometric assay design. Excess reagents allow the possibility of first-order reactions. First-order reactions show a linear, directly proportional relationship between the measured response and analyte concentration, if the reactants are chemically homogeneous. The heterogeneity of polyclonal antibodies eliminate them as candidate reagents for linear immunometric assays because their chemical reactivities are not equivalent. A monoclonal antibody preparation, on the other hand, is the perfect reagent from a chemical view point: it can be used at excess and all the molecules react the same.

LINEAR IMMUNOENZYMETRIC ASSAYS

To achieve directly proportional calibration curves in immunodiagnostics, three criteria must be met. As seen above the first requirement is to use the excess reagent assay design. The second requirement is also listed above: chemical homogeneity of reagents. The third requirement is the linear translation of the chemical reactions into a measurable response. Enzyme labeled antibodies are used to accomplish this mainly because radiolabels cannot be incorporated at high enough specific activity to label the required excess amount of antibody.

Using immunoenzymetrics, monoclonal antibodies and alkaline phosphatase, linear immunodiagnostic assays are possible. Given linear calibration responses, the requirement of conventional immunoassays of multiple calibrator points to define the extrapolation curve is minized. Linear calibration curve for human chorionic gonadotropin is shown in Figure 2.

METHODS

Measurements of human chorionic gonadotropin (hCG) were done with the following commercial kit: TANDEM®-E HCG (Hybritech Incorporated, 11095 Torreyana Road, San Diego, CA 92121). All assays were done according to manufacturer's instructions.

TABLE 1. Criteria for Developing Immunoassays Having Linear Calibration Curves

1. Presence of large excess of the indicator antibody reagent. The analyte of interest should be the limiting factor.

2. Homogeneous reactivity of the indicator antibody for the analyte.

3. A direct transformation of the analyte-indicator antibody complex to a measurable response.

Briefly, a bead coated with a monoclonal antibody specific for the alpha subunit of HCG was placed in a 12 x 75 mm plastic tube, followed by 100 μl of sample, calibrator or control, followed by 100 μl of a second monoclonal antibody specific for the beta subunit of HCG conjugated to alkaline phosphatase. The reagents were incubated at 37°C for 2 hours. At the end of the incubation period unbound conjugate was removed using the wash reagent provided. Two hundred μl of substrate, p-nitrophenylphosphate, was added to each tube. After 15 minutes, 1.5 mls of a quench reagent was added to each tube and the absorbance measured at 405 nanometers.

CALIBRATION

Once linear reactions are obtained, a single positive calibrator system can be developed. The entire curve can be defined by a line through the zero calibrator and the single

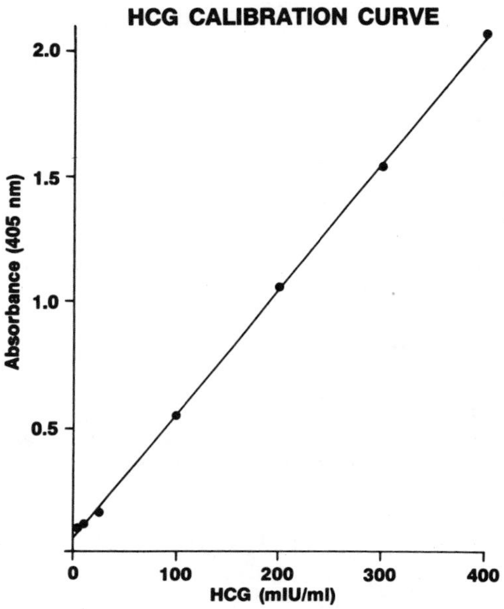

Figure 2. Linear calibration curve for human chorionic gonadotropin extending from 0 to 400 mIU/ml.

positive calibrator. Data reduction then becomes simple linear interpolation using the equation $y = mx + b$, where y = the unknown value, m = the slope of the calibrator curve, x = the optical density measured for the unknown and b = the intercept or the zero calibrator response. Figure 3 shows the correlation for the quantitation of HCG samples using either multiple calibration or single point calibration. No significant difference was observed between either method using single positive calibrator system.

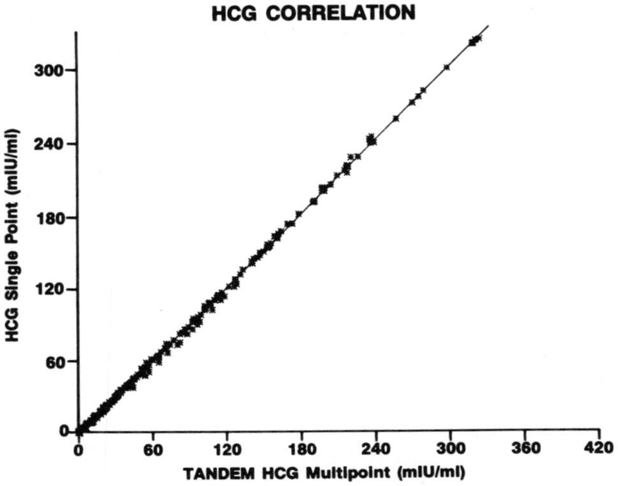

Figure 3. Quantitation of patient samples containing HCG using multiple point calibration compared to single positive calibrator system.

DUAL WAVE LENGTH ENHANCEMENT

Excess antibody-alkaline phosphatase conjugate will continue to generate colored product in proportion to the concentration of the enzyme regardless of the fact that absorbance cannot be measured accurately above two absorbance units. Since the colored product is produced in a linear fashion at lower extinction wavelengths, one can increase the effective range of the assay (Saunders, et.al., 1984). The absorbance maximum for para-nitrophenol phosphate is 405 nm. The extinction at 450 nm is approximately one fifth of that at 405 nm. Therefore, measuring both 405 and 450 nm will give proportional readings over five times the range of the readings at 405 nm alone. Using a spectrophotometer, this means switching the wavelengths from 405 to 450 nm for every specimen. Microprocessor controlled, multi-wavelength filter photometers can do this automatically. The microprocessor decides which wavelength to use to calculate the specimen by checking the relative readings at the two wavelengths. Usually, the assay system can be engineered to give excellent sensitivity at the higher extinction wavelength while using the secondary wavelength to give the desired range. Remember that the sensitivity of the immunoenzymetric assay is nearly totally a function of the specific activity of the enzyme-antibody conjugate. So any increase in the specific activity of the conjugate without concommitant increase in the non-specific binding of the conjugate will yield higher sensitivity with respect to minimal detectable concentration of the analyte.

TWO-SITE SELECTIVITY

When monoclonal antibodies are used in two-site immunometric assays an additional advantage over immunoassays using conventional polyclonal antibodies is realized. It relates to the concept of specificity of an antibody. A molecule may be measured based on unique antigenic determinants that define the molecule. Specificity generally describes how specific an antibody reacts with a given antigenic molecule. When the antibody preparation is polyclonal, this concept relates to the ability of the antibodies to react specifically to multiple antigenic determinants which may or may not uniquely define the molecule. When the antibody is a monoclonal preparation, then this relates to the ability of the antibody to bind a unique antigenic determinant associated with the molecule. The selective advantage occurs when the assay is the two-site monoclonal antibody immunometric system. Here, the design of the assay allows the effective measurement of mass because the system basically measures the continuity between two unique antigenic determinants. There are two main advantages of this type system over the conventional competitive binding immunoassay. First, since two antigenic determinants are required to complete the "sandwich" of antibodies to link the label to the solid phase, only those molecules that contain the two determinants will react. The advantage of this system is its increased specificity which is based on the use of two unique antigenic determinants, rather than one, to define a molecule. This problem is compounded in competitive immunoassays when calibrators and control pools are produced that contain antigenically active fragments of the antigen in addition to the antigen as it is found in serum. Therefore, it is easy to see that assay systems designed to measure antigenic determinants which are not unique to or associated with an analyte will deliver different quantitative estimates of the analyte than those assays designed to measure the whole molecule. It also follows that the two-site immunometric assay more closely relates to the whole molecule than the immunoassay where sometimes antigenic activity does not relate to the intact molecule. This problem in using competitive immunoassay is compounded when calibrators and control pools are produced that contain antigenically active fragments of the antigen in addition to the antigen as it is found in serum. Therefore, it is easy to see that assay systems designed to measure antigenic determinants which are not unique to or associated with an analyte will deliver different quantitative estimates of the analyte than those assays designed to measure the whole molecule. It also follows that the two-site immunometric assay more closely relates to the whole molecule than the immunoassay where sometimes antigenic activity does not relate to the intact molecule.

An example is the immunoassay for the MB isoenzyme of creatine kinase (CKMB) (Figure 4). Creatine kinase is an enzyme found in the highest concentrations in the intracellular fluid of the brain, skeletal muscle and cardiac muscle. It is a dimeric molecule composed of two subunits designated as M and B. As a result creatine kinase may exist as either the MM, MB or BB isoenzyme. CKMB is found almost exclusively in cardiac muscle and its presence in serum is indicative of cardiac damage as in a myocardial infarction. Using a "trapping" antibody specific for the M subunit and an enzyme-antibody conjugate specific for the B subunit it is possible to detect only the MB isoenzyme in the presence of the MM or BB isoenzyme.

VISUAL END-POINT IMMUNOENZYMETRIC ASSAYS

Once the enzyme-based system can be developed to produce visable color in the concentration range required to detect the presence of the antigen consistent with a positive state, a simple screening procedure can be developed. Using the two-site monoclonal immunoenzymetric assay design, pregnancy screens have been produced that rely on the ability to measure the pregnancy hormone human chorionic gonadotropin.

The mechanics of the two-site screening assay are no different than the quantitative versions: they both are basically diffusion limited because they require molecules to travel through solution to react on the solid phase. Attachment

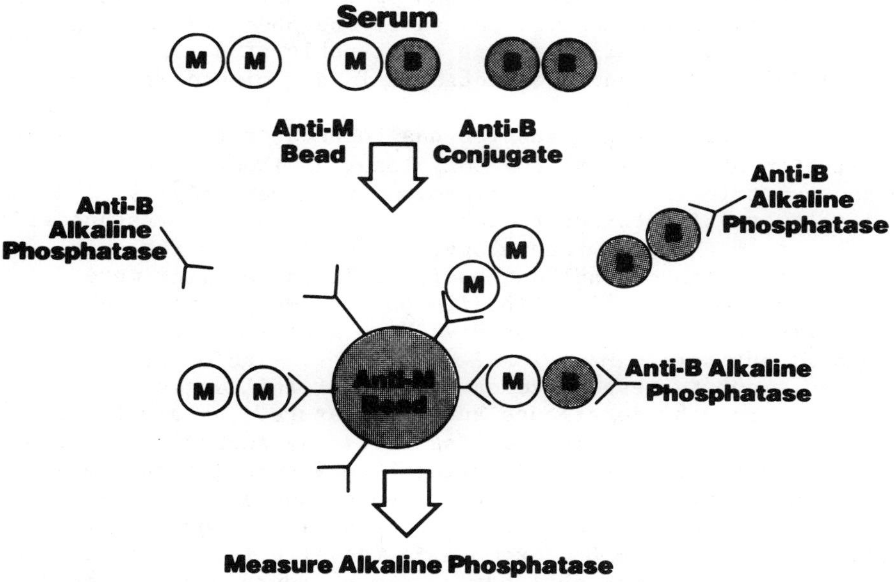

Figure 4. Components of the immunoassay specific for the MB isozyme of creatine kinase. Presence of MM or BB isozyme will not interfere in detection of CK-MB.

to the solid phase simplifies the removal of unbound labeled antibody. But the reaction rates in a heterogeneous system do not allow the rapid assays required for screening tests.

One solution is to increase the molecular collisions between antibody and antigen. This can be done by allowing the antigen to percolate through a membrane coated with capture antibody. The high concentrations of the capture antibody and the short diffusion distance results in short reaction times - generally the time it takes to flow through the membrane. The enzyme-labeled antibody is passed through the membrane in the same manner and reacts with the same speed. After rinsing the unbound labeled antibody from the filter, substrate is added. Since the system uses very high concentrations of the antibodies, very low concentrations of the analyte can easily be observed in a short time. Systems are available that are capable of detecting as low as 25 mIU/ml of chorionic gonadotropin in a 3 minute procedure.

To this point our discussion has focused on the benefits of substitution of monoclonal for polyclonal antibodies in immunoenzymetric assays. The improvement in sensitivity, specificity and lot-to-lot reliability represents a substantial improvement of existing technology, but will not be the only contribution of monoclonal antibodies. It is becoming more apparent that monoclonal antibodies possess many unique characteristics that could not have been identified with polyclonal antisera. Monoclonal antibodies, like other proteins, have been shown to respond to environmental influences such as pH, ionic strength, detergents and temperature. No doubt some of the antibodies in polyclonal antisera display such responses, but since that particular antibody molecule may comprise only a small part of the entire population its response would be unnoticed. Only when we have a homogeneous population is it possible to observe such effects. One example is the effect of pH on the binding of a monoclonal antibody to human Growth Hormone (hGH) (Figure 5). At pH 7 hGH is tightly bound to the antibody but exposure to pH 4 or lower, results in release of this hormone (Bartholomew, et al., 1983). Other monoclonal antibodies do not show this effect and are able to bind hGH regardless of pH. This sensitivity to pH may be due to a conformational change in a specific region of the Growth Hormone molecule or a change in the antibody combining site. It is possible to select monoclonal antibodies for sensitivity to specific conditions allowing release of analyte under defined conditions. Using this technique tests can be designed in which analyte would be released into the surrounding solution prior to quantitation. If the analyte were complexed with a second monoclonal antibody labeled with an enzyme, conditions could be adjusted such that the complex would be released into the solution during the enzyme turnover reaction allowing turnover to proceed in solution.

Another use of such an antibody would be for purification. An antibody sensitive to pH could be coupled to a solid phase and used to remove a particular molecule from a complex mixture. The solid phase would be washed with a buffer of a particular pH and the molecule released and collected (Figure 6).

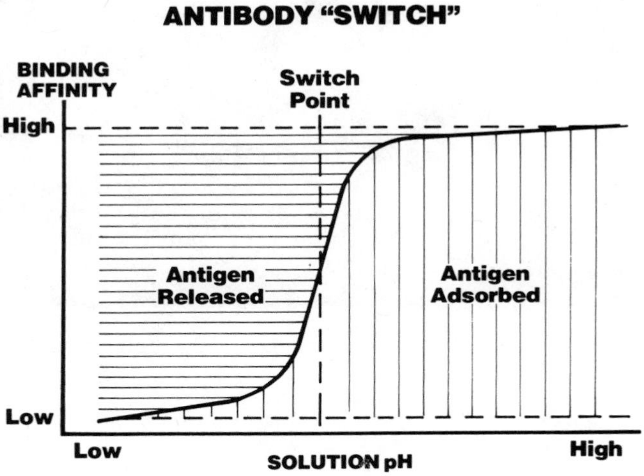

Figure 5. Dependence of antigen binding on solution pH.

PRODUCTION OF BIFUNCTIONAL MONOCLONAL ANTIBODIES

Normally antibodies contain two identical binding sites. Using monoclonal antibody technology it has been possible to construct a unique antibody molecule consisting of two different binding sites (Martinis, et al., 1982) (Figure 7). A hybridoma making mouse monoclonal antibody to prostatic acid phosphatase

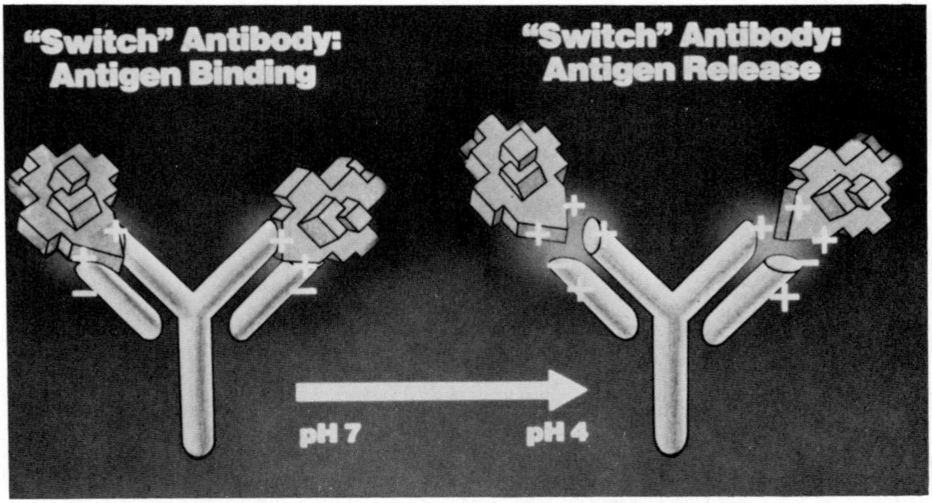

Figure 6. Binding of antigen to the "Switch" antibody binding at pH 7 and release at pH 4.

(PAP) was fused with spleen cells from a mouse immunized with hepatitis B surface antigen (HBsAg). Clones were isolated which secreted antibody capable of binding both antigens simultaneously. Antibody secreted by one such hybridoma could be separated into three well defined species by DEAE chromatography. Trace-labeled antigen binding experiments clearly demonstrated that one was the original antibody specific for only PAP, another a new antibody derived from a spleen cell specific for HBsAg, and the third a hybrid antibody with specificity for both PAP and HBsAg expressed in the same molecule. One possible use of bifunctional antibodies would be for the preparation of antibody enzyme conjugates without any chemical modification of either molecule. One binding site could be directed against an indicator enzyme such as horse radish peroxidase or alkaline phosphatase. The second to a site on the antigen of interest. The antibody enzyme complex could be used in any two-site immunoenzymetric assay as described above.

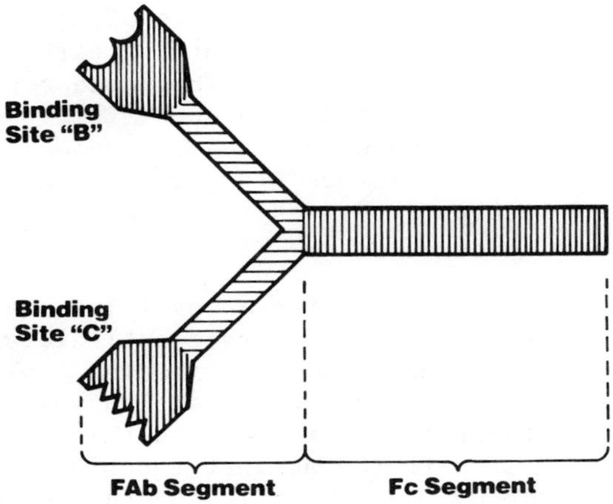

Figure 7. Bifunctional monoclonal antibody consisting of two unique antigen binding sites.

MONOCLONAL ANTIBODIES TO CANCER ANTIGENS

Another area in which monoclonal antibodies have been shown to be useful is in the area of cancer testing. It has been possible to isolate unique cancer markers using hybridoma technology. Such markers include CA 125 and CA 19-9 (Koprowski, et al., 1979). CA 125 is a new determinant defined by monoclonal antibodies produced from mice immunized with an ovarian carcinoma cell line. It has been found to be expressed in a majority of nonmucinous epithelial ovarian carcinomas and may be useful in monitoring patients with ovarian cancer. CA 19-9 is a determinant found to be present in gastrointestinal adenocarcinomas. It was obtained after immunization with a colon carcinoma cell line. It is currently under investigation as a marker for pancreatic and other GI carcinomas. Hybridoma technology and the ability to select for specific clones has made it possible to identify monoclonal antibodies capable of recognizing unique markers on these cell lines. Markers which may be useful in indicating the presence of tumors.

Recent work by Ehrlich and Moyle (1983) has described another unique feature of monoclonal antibodies. Using human chorionic gonadotropin (hCG) as a model, they have described a system in which mixtures of certain monoclonal antibodies bind antigen in a "cooperative" fashion by forming circular complexes. Some pairs of antibodies demonstrate greater affinity for hCG than seen in either antibody alone. Their work showed that at least two epitopes were required and monoclonal antibodies capable of simultaneous binding were required. Data also showed that intact antibody was required for the maximum effect.

Potocnjak (1982) have described a novel immunoassay named "4i", standing for inhibition of idiotype-antiidiotype interaction. It uses two monoclonal antibodies, only one of which recognizes the analyte, to quantitate an unknown sample. One monoclonal is directed to an epitope on the antigen of interest; the second is directed against the binding site of the first monoclonal antibody. As a result several competing binding reactions can occur. The first is the binding of the antigen by antibody A. The second is the binding of antibody B to antibody A at the antigen recognition site. The binding of antigen by the antibody A prevents the antibody A antibody B interaction. In this assay increasing concentration of the analyte results in lowered response of the response variable.

SUMMARY

Monoclonal antibodies have been shown to be a useful immunodiagnostic tool because of their chemical homogeneity, high affinities and availability. These characteristics coupled with the immunoenzymetric technique have enhanced the ability of the clinical chemist to quantify an analyte with greater accuracy, sensitivity and precision. The selection of the proper indicator enzyme extends the utility of the immunoenzyme technique by allowing the use of single positive calibrator system. In addition to improving existing technology, monoclonal antibodies have been instrumental in defining new antigenic determinants for cancer and other diseases. The homogeneity of monoclonal antibodies have unveiled many new properties, uses for which are limited by one's imagination.

ACKNOWLEDGEMENTS

The authors wish to express their thanks to Ms. Karyl LaPinska, without whose efforts and skill this manuscript would not have been possible.

REFERENCES

Barany, G., 1980, The explicit analysis of consecutive pseudo-first-order reactions: applications to kinetics of thiolysis of dithiosuccinoyl (Dts) amino acids, Anal. Biochem. 109:114-122.

Bartholomew, R., Beidler, D., David, G. and Hill, C., 1983, Switch immunoaffinity chromatography with monoclonal antibodies. Biotechniques, Volume 1.

DiPietro, D.L., 1981, Ectopic pregnancy: interpreting hCG levels. Laboratory Management 19:43-51.

Ekins, R., 1981, Towards immunoassays of greater sensitivity, specificity and speed: an overview, in: "Monoclonal Antibodies and Developments in Immunoassay," A. Albertini and R. Elkins, eds., Elsevier/North-Holland Biomedical Press, The Netherlands.

Erlich, P.H. and Moyle W.R., 1983, Cooperative immunoassays: ultrasensitive assays with mixed monoclonal antibodies, Science 221:279-281.

Fritz, T.J., Bunker, D.M., 1982, Hyperlipidemia interference in radioimmunoassays, Clin. Chem. 28:2325.

Kohler, G. and Milstein, C., 1975, Continuous cultures of fused cells secreting antibody of predefined specificity. Nature (London) 256:495-497.

Koprowski, H., Steplewski, Z., Mitchell, K., Herlyn, M., Herlyn, D., Fuhrer, P., 1979, Colorectal carcinoma antigens detected by hybridoma antibodies, Somatic Cell Genetics, 957:972.

Martinis, J., Kull, J.F., Franz, G., and Bartholomew, R.M., Monoclonal antibodies with dual antigen specificity, 1982, III Annual Colloquium of the Protides of Biological Fluids.

Potocnjak, P., Zavala,F., Nussenzweig, R., Nussenzweig, V., Inhibition of idiotype-anti-idiotype interaction for detection of a parasite antigen: a new immunoassay,

Science 215:1637-1639.

Ryall, R.G., Story, C.J., Turner, D.R., 1982, Reappraisal of the causes of the "hook effect" in two-site immunoradiometric assays, Anal. Biochem. 127:308-315.

Saunders, R.L., David, G.S., Sevier, E.D., 1984, Monoclonal antibodies: advances in in vitro and in vivo diagnostics. Pathologist 38:638-646.

Sevier, E.D., David, G.S., Martinis, J., Desmond, W.J., Bartholomew, R.M., Wang, R., 1981, Monoclonal antibodies in clinical immunology, Clin. Chem. 27:1797-1806.

Sevier, E.D., 1982, Monoclonal antibodies: expanded potential for labeled antibody ligand assays, J. Am. Med. Tech. 48:651-653.

Wu, T., Ballon, S.C., Teng, N.N.H., 1983, Monoclonal versus polyclonal antibody radioimmunoassay against the beta-subunit of human chorionic gonadotropin in patients with gestational trophoblastic neoplasia. Am. J. Obstet. Gynecol. 147:821-828.

THE NGO-LENHOFF (MBTH-DMAB) PEROXIDASE ASSAY

William D. Geoghegan

Department of Dermatology
The University of Texas Health Science Center-Houston

INTRODUCTION

The Ngo-Lenhoff peroxidase assay represents a new and sensitive procedure for the assay of horseradish peroxidase (HRP) and HRP-coupled reactions (Ngo and Lenhoff, 1980). The assay is sensitive enough to detect as little as 2-5 femtomoles of HRP per milliliter (Ngo and Lenhoff, 1980, Geoghegan et. al. 1983). Its sensitivity is based upon the oxidative coupling of 3-methyl-2-benzothiazolinone hydrozone (MBTH) and 3-(dimethylamino)benzoic acid (DMAB) to form an indamine dye with a molar extinction coefficient of 47,600 (Ngo and Lenhoff, 1980).

A brief history of the developments which led to the combination of these two reagents begins with the work of Hunig and Fritsch (1957). They described the oxidative coupling mechanism which results in the formation of an indamine dye. One of their examples was the combination of MBTH and N,N-dimethylaniline (DMA)(Figure 1). MBTH and DMA were oxidatively coupled by the addition of ferric chloride or one of several other oxidizing agents. Most significantly, they mentioned that the reaction could be driven by H_2O_2 plus $Fe^{2}+$. The product of the oxidative coupling of MBTH and DMA resulted in a dye with a molar absorptivity of 79,500 in methanol. The corresponding sensitivity of a dye with such a high molar absorptivity made the reaction very attractive.

MBTH was first used by Sawicki et. al. (1961) who needed a sensitive method for the detection of water-soluble aliphatic aldehydes in acid media. The reaction proceeded via two steps. In the first step (I), MBTH condenses with the aldehyde to form its azine (Figure 2).

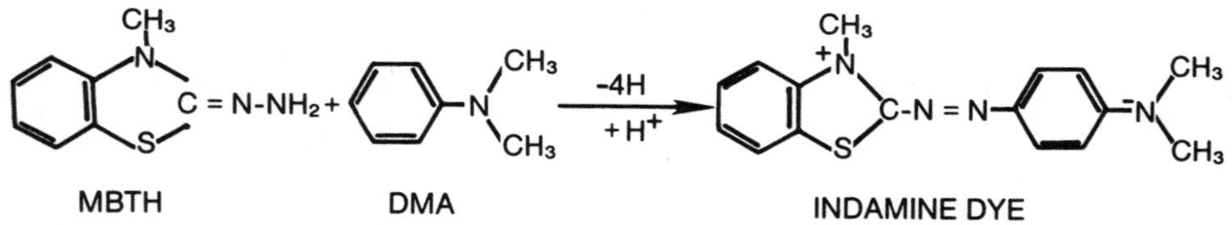

Fig. 1. Reaction between MBTH and DMA to form an indamine dye.

The azines have an active hydrogen which then reacts with the product of the second step. The second step (II) begins with the addition of $FeCl_3$ (Figure 2). This oxidizes the remaining MBTH to a reactive cation which then complexes to the azine to form a blue cationic tetraazapentamethine dye (Hunig, 1962). This dye has also been synthesized by Hunig and Fritsch (1957). They reported its molar absorptivity as 76,000 in acetone with a maximum absorbance at 670 nm. Sawicki et. al. (1961) reported that the dye produced in the analytical procedure had a molar absorptivity of 65,000 and maxima at 635 and 670 nm. Its production in an analytical procedure combined with its high extinction coefficient led others to experiment with it.

The reactivity of MBTH with other reagents usually required heat and an acidic pH. Under these conditions, MBTH was subsequently used to detect glycosaminoglycan hexosamine (Smith and Gilkerson, 1979) and a large number of carbonyl compounds (Paz et. al., 1965). In the detection of carbonyl compounds, Paz et. al. (1965) reported that the azine could be formed at 40°C at pH 4. Although colorless, the azine derivatives of each carbonyl compound had characteristic spectra at pH 1 which permitted their identification. Subsequently, a tetraazopentamethine dye was formed by heating the azine to 100°C followed by the addition of $FeCl_3$ after cooling. Sialic acid was quantitated in the absence of heating if

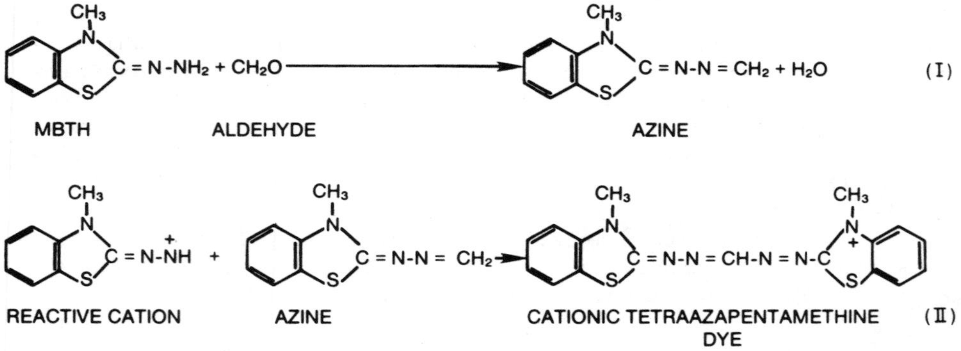

Fig. 2. Mechanism proposed by Sawicki et. al. (1961) for the reaction between MBTH and an aldehyde to form a cationic tetraazapentamethine dye.

first split between C8 and C9 by periodic acid. This released a molecule of formaldehyde which was then quantitated (Massamiri et. al., 1978). Most common sugars with pyranose hemiacetal structures do not react with MBTH. However, certain structures viz. erythrose with a furanose structure gave the color reaction at low pH values when heated.

Table 1: Substances Reactive with MBTH at Low pH Values

Aliphatic aldehydes	(Sawicki et. al. 1961)
Pyridine aldehydes	
Aniline	
N-alkylanilines	
N,N-dialkyanilines	
Polynuclear aromatic amines	
Diarylamines	
Indoles	
Carbazoles	
Phenothiazines	
Hydroxyaldehydes	(Paz et. al. 1965)
Dialdehydes	
Acid aldehydes	

An increase in the molecular weight of the aliphatic aldehyde results in a decrease in detectability (Sawicki et. al., 1961). It is reported to be sensitive enough for the rapid determination of 1-9 carbon aldehydes (Paz et. al., 1965).

Sawicki et. al. (1961) did not examine the effect of pH on the reaction between MBTH and the compounds listed in Table 1. Paz et. al. (1965) reported that pH was crucial in their modification of Sawicki's colorimetric determination. They reported that the pH of the solutions must be between 3 and 4. At these pH values, acetate, citrate and phosphate interfered at concentrations greater than 0.05 M, 0.05 M and 0.01 M respectively.

In 1971, Gochman and Schmitz (1971) described a procedure in which hydrogen peroxide was quantitated by the oxidative coupling of MBTH and N,N-dimethylaniline to form an indamine dye. Their contribution was to replace the chemical oxidizing agents used by Hunig and Fritsch (1957) with a system that generated hydrogen peroxide in the presence of horseradish peroxidase. The result of the peroxidase catalyzed reaction was the same water soluble blue indamine dye that Hunig and Fritsch (1957) had synthesized.

The production of the indamine dye was proportional to the quantity of hydrogen peroxide generated by a second enzyme viz. uricase, glucose oxidase together with the appropriate substrate uric acid and glucose respectively (Gochman and Schmitz, 1971,1972). In effect, the absorbance measured would be related to the quantity of substrate used. Finally, the method was resistent to many reducing substances viz. L-ascorbic acid, gentisic acid, which in other methods resulted in a more significant level of interference (Gochman and Schmitz, 1971, 1972).

In 1980, Ngo and Lenhoff described the use of 3-(dimethylamino) benzoic acid (DMAB) as an oxidizable substrate for coupling to the reactive cation generated from MBTH. While DMA was highly toxic, poorly soluble in water and had a stench, the DMAB was not reported to be toxic, did not have a stench and was soluble in water (Figure 3).

By this method, HRP was detectable at 2 femtomoles per milliliter. In addition to the detection of HRP, the reaction was coupled to other enzymes which generated H_2O_2 viz. glucose oxidase, α-glucosidase. The result was a very sensitive assay which may be coupled to other enzymes such as uricase (Gochman and Schmitz, 1971), oxalate oxidase (Obzansky and Richardson, 1983), putrescine oxidases (Wampler et. al. 1979), guanase and xanthine oxidase (Ando et. al. 1983), trehalase (Killick and Wang, 1980) and choline oxidase (Capaldi and Taylor, 1983).

In 1983, Geoghegan et. al. confirmed the sensitivity reported by Ngo and Lenhoff (1980) and made several modifications to: 1) lower the background color and the sensitivity to light; 2) block interference due to catalase; and 3) to stop the reaction at the desired time. These three improvements were made by 1) use of a citric acid phosphate buffer, 2) inclusion of sodium azide, and 3) by reducing the pH to 3.0 with a strong acid.

METHODS

3-methyl-2-benzothiazolinone hydrozone hydrochloride (MBTH) and 3-(dimethylamino)benzoic acid (DMAB) were obtained from Aldrich Chemical Co., Milwaukee, Wisconsin. It should be noted

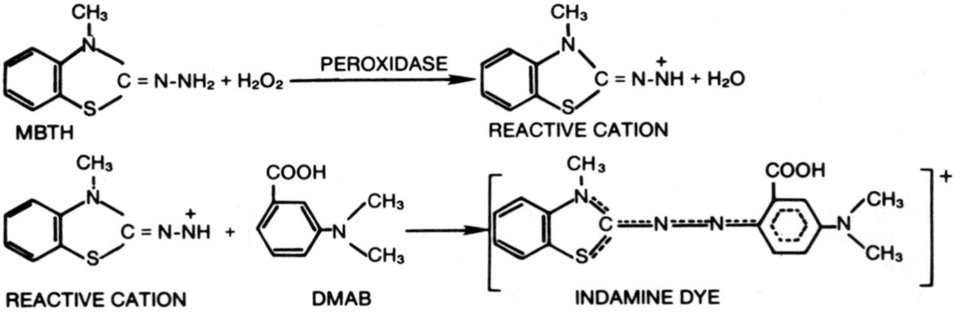

Fig. 3. The mechanism proposed by Ngo and Lenhoff (1980) for the generation of an indamine dye from MBTH and DMAB.

that several other compounds are often referred to as DMAB. Two of these are seen frequently, p-(dimethylamino)benzaldehyde (Ehrlich's reagent) and 4-(dimethylamino)benzoic acid. They should not be confused with the reagent 3-(dimethylamino)benzoic acid used here.

Reagent grade chemicals are used for all solutions. Except where noted, all stock solutions are made by dissolving reagents in citric acid phosphate buffer (CAP), pH 7.0. The buffer was prepared by diluting 120 ml of 0.1 M citric acid and 500 ml of 0.2 M dibasic sodium phosphate to 1 liter with distilled water (McIlvaine, 1921). This buffer was chosen because it resulted in a significant reduction in spontaneous or light induced dye production (Figure 4) (Geoghegan et. al., 1982, 1983).

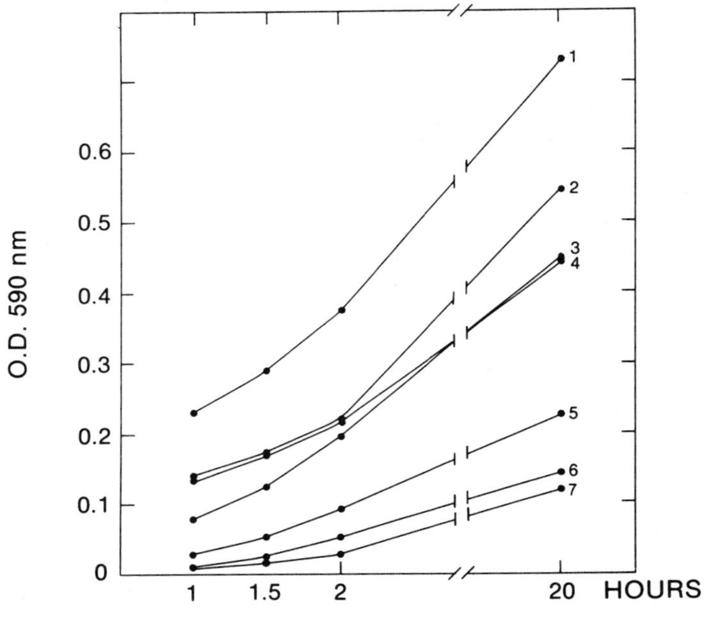

Fig. 4. Development of color in reagent blanks with different buffers. (1) 0.01 M Tris-HCl, pH 7.2; (2) 0.1 M Tris-citric acid, pH 8.0; (3) 0.1 M phosphate and citric acid, pH 7.0; (4) 0.1 M Tris-citric acid, pH 7.2; (5) 0.1 M phosphate-citric acid, pH 6.0; (6) 0.1 M phosphate and citric acid, pH 5.0; and (7) 0.1 M Hepes, pH 5.0

Ngo-Lenhoff reagent stock solutions DMAB, 3×10^{-2} M, MBTH, 6.0×10^{-4} M (Ngo and Lenhoff, 1980) and 1.0 M sodium azide are dissolved in CAP and stored in tightly sealed amber bottles at 4°C. Note that both DMAB and MBTH dissolve slowly.

The stock solutions should be discarded if they exhibit a yellow color. Immediately prior to use, the components of the working reagent are combined as follows: 156 ml of CAP, 22 ml of stock MBTH, 22 ml of stock DMAB, 0.24 ml of 30% H_2O_2 and 2 ml of 1.0 M azide are gently and briefly mixed.

The generation of the indamine dye from the Ngo-Lenhoff reagent by the action of HRP and H_2O_2 will continue until the substrates are exhausted. The reaction proceeds rapidly (Figure 5) and the rate of color development begins to slow after the first 10 minutes.

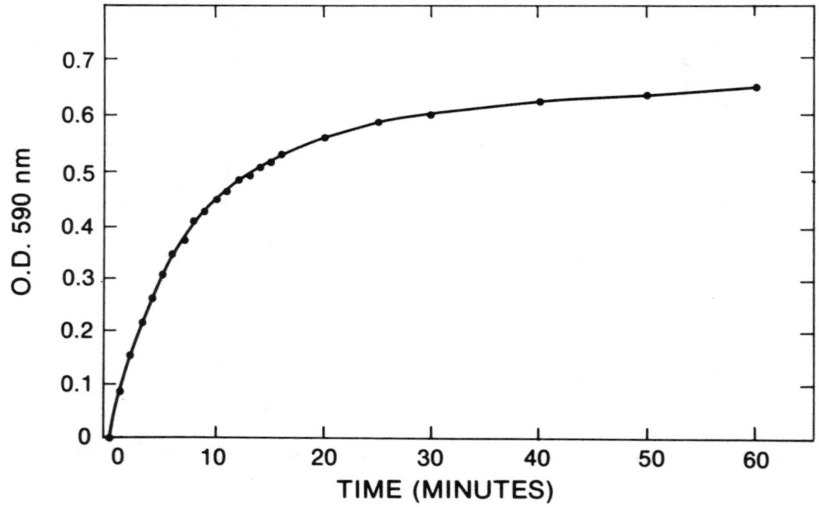

Fig. 5. Rate of color development: 7.05×10^{-9} g HRP in 0.1 ml of CAP was placed in a cuvette in a Cary double beam recording spectrophotometer. The reaction was started by the addition of 0.9 ml of the Ngo-Lenhoff reagent. The second cell contained the Ngo-Lenhoff reagent only.

At this time, 60% of the color has been generated (Geoghegan et. al., 1983).

Peroxidase at picomolar levels (femtomoles/ml) may be determined by either a fixed time or rate method. A linear relationship between the absorbance at 590 nm and the peroxidase concentration exists for a given time period (Ngo and Lenhoff, 1980). The reaction may be stopped by the addition of 1-3 N HCl or H_2SO_4 such that the final pH is 2.9-3.1 (more extreme pH changes, pH 0.06-1.7, result in as much as a 20% depression of absorbance). A typical example for stopping the reaction would be the addition of 0.1 ml of 1 N H_2SO_4 to 0.9 ml of the reagent. The addition of H_2SO_4 or HCl to the reagent stopped the reaction and resulted in a slight shift in the absorbance maximum from 590 nm at pH 7.0 to 595 nm at pH 1 to 3.0 (Figure 6). The shift in the maximum corresponded with a visual color change from purple-blue to a definite clear blue. Following the addition of the acid, a decrease in the absorbance

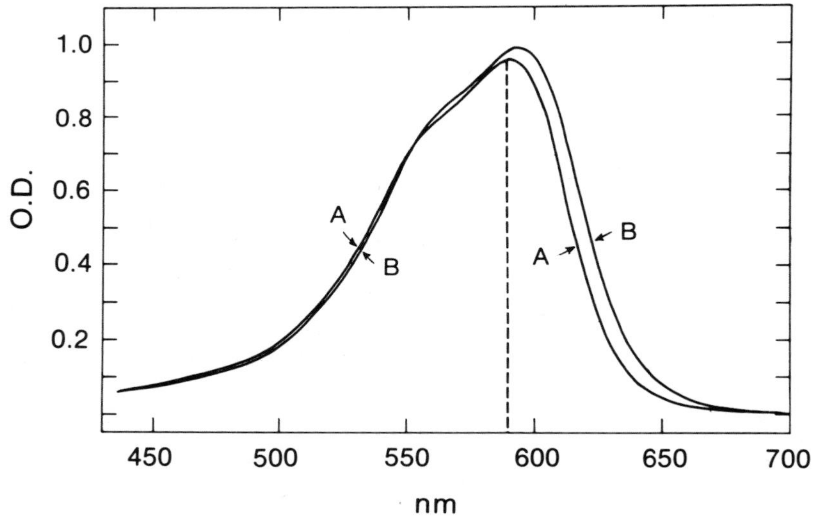

Fig. 6. Effect of pH reduction on the absorption spectrum of the enzyme product. (A) pH 7.0, (B) pH 2.9. Identical results are obtained with 1 N HCl or H_2SO_4. Broken line is 590 nm.

corresponding to the dilution occurred. No further decrease in absorbance was observed for at least 4 hours (Figure 7)(Geoghegan et. al. 1983).

Ideally, a 1 cm light path is used for the measurement of the absorbance. If a flat bottom 96 well microtiter plate is used, a maximum total volume of 330 µl may be used. This will leave the liquid level slightly below the top of the well. This is necessary to maintain a flat surface and to keep the Micro-ELISA reader dry. The reagent volume must be adjusted approximately viz. 267 µl of reagent followed after the desired time by 33 µl of 1 N HCl or H_2SO_4 to stop the reaction at pH 3.0.

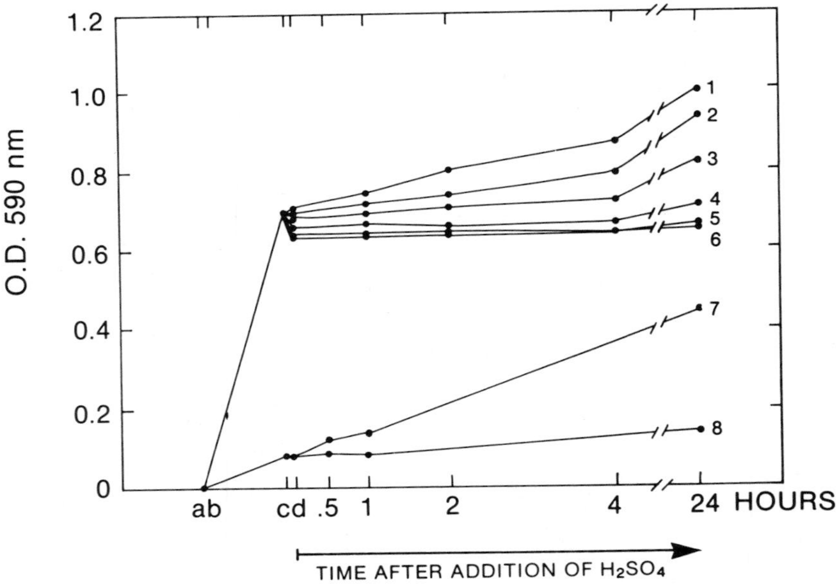

Fig. 7. The ability of increasing quantities of H_2SO_4 to stop color development. 0.1 ml of each of the following were added to 1 ml of Ngo-Lenhoff reagent with or without HRP as indicated. In the presence of HRP: 1) H_2O, 2) 0.2 N H_2SO_4, 3) 0.5 N H_2SO_4, 4) 1 N H_2SO_4, 5) 2 N H_2SO_4, 6) 3 N H_2SO_4. In the absence of HRP: 7) H_2O and 8) 3 N H_2SO_4. Letters a, b, c and d refer to the absorbance of buffer at time zero, after the addition of all reagents, after 1 h and after the addition of acid, respectively.

The solid phase procedure followed is basically that of Engvall (1980) and Hay et. al. (1976, 1979). Microtiter plates are coated with 267 µl of 1-10 µl/ml of antigen or antibody. The coating is performed for 3 hours at 37°C and overnight at 4°C. Unbound protein is aspirated and the wells rinsed three times and coated with BSA. A note of precaution is required on BSA. It should be free of competitive substances viz. hemoglobin, catalase, cytochromes or bilirubin (Perlstein et. al., 1977, 1978) which would consume H_2O_2 and/or give a positive result. Recently, Saravis (1984) recommended the use of a liquid gelatin, made from fish skin, as a blocking agent following blotting of protein onto nitrocellulose. It did not interfere with color development in a peroxidase coupled reaction. This is an inexpensive liquid gelatin with an average MW of 60,000. It contains methyl and propyl p-hydroxybenzoates as a preservative. These reportedly did not interfere. Following the coating or blocking step, the microtiter plates are rinsed with PBS plus 0.05% Tween 20 and dried.

Samples are generally incubated for 1 hour at 37°C and for 30 minutes at 4°C. This is followed by 3 rinses with PBS plus 0.05% Tween 20. An HRP labelled antibody is added and incubated as for the sample followed by washes. Following the washes, 267 µl of the Ngo-Lenhoff reagent is added and incubated. The time of incubation, usually 30 minutes or 1 hour, must be determined. The reaction is stopped by the addition of 33 µl of acid and the absorbance read at 595 nm. Values are determined by comparison to a standard curve (Figure 8).

Ngo and Lenhoff (1980) reported that the precision and accuracy of the MBTH-DMAB method in the glucose assay were satisfactory. The coefficient of variance averaged 4.4% and the comparison of the actual value with the determined value for glucose varied between 98.5% for the rate method and 101% for the fixed time method. The average standard deviation for absorbance in the peroxidase assay (comparison of triplicate values) was ± 0.008 S.D. (Geoghegan, unpublished).

Interfering Substances

In a solid phase ELISA, very few substances present in a serum, cerebrospinal or urine sample can interfere. This is because the sample is washed away from the surface before the dye is generated. In peroxidase coupled reactions, employing a second enzyme, viz. glucose oxidase (for the generation of H_2O_2 from glucose), the sample and the reagents must be in contact with each other. In this instance, interfering substances need to be considered.

Catalase is a ubiquitous enzyme in mammalian tissues. If present in the peroxidase assay, it will be a major competitor with HRP for H_2O_2. It is present in the serum and in red blood cells

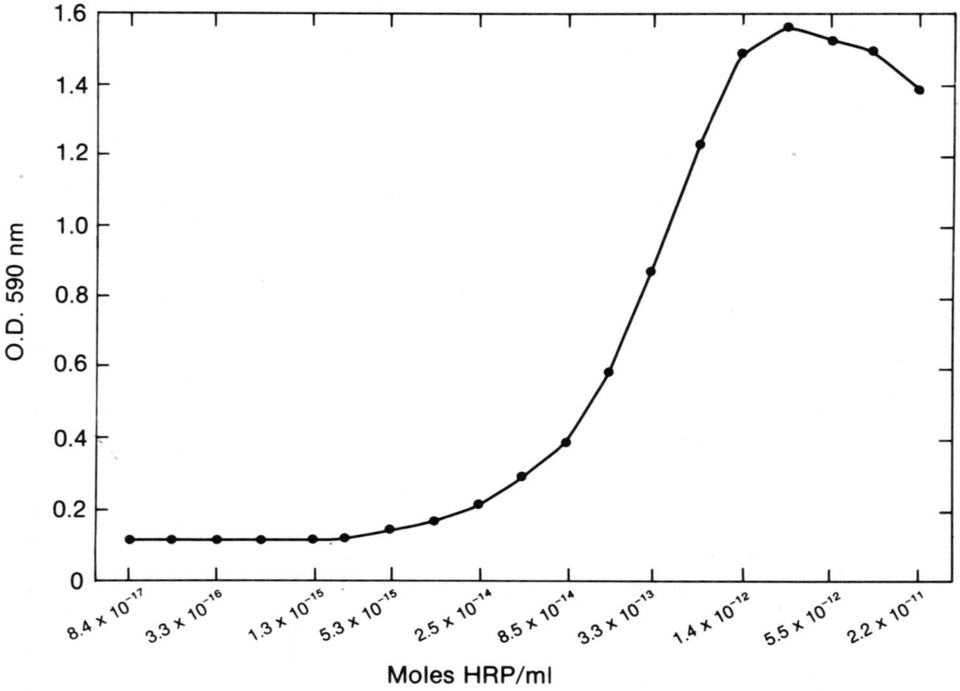

Fig. 8. Detection limits of HRP by Ngo-Lenhoff reaction.

(Deisseroth and Dounce, 1969). Its addition to the Ngo-Lenhoff reagent does not result in the production of the dye at pH 7.0 (Geoghegan et. al. 1983). Although no color is produced, its presence significantly reduces the generation of the dye by HRP. Sodium azide inhibits catalase with minimal effects on HRP at pH 7.5 (Lenhoff and Kaplan, 1955) and with apparently minimal effects at pH 7.0 (Geoghegan et. al. 1983). Its inclusion in the buffer is a reasonable precaution even when catalase contamination is not suspected.

Ascorbic acid has been reported to depress the glucose result in some test systems based on the use of peroxidase (White-Stevens and Stover, 1982). The depression, 3-5 mg/100 ml occurs at 10 mg ascorbic acid/100 ml, which is above the physiological concentration (Gochman and Schmitz, 1972). White-Stevens and Stover (1982) suggest that in systems employing MBTH, peroxidase with H_2O_2 catalyzes the oxidation and activation of MBTH to the oxidized species, $MBTH_{ox}$. This results in a lag time during which the color develops at a depressed rate. In some systems, viz. o-toluidine, the color already formed in bleached by the addition of ascorbic acid (White-Steven and Stover, 1982). Because MBTH is oxidatively coupled to DMAB, the resulting indamine dye is not reversible.

Ngo and Lenhoff (1980) reported that while ascorbate at high concentrations interferes with the rate method, it does not interfere with the fixed time method when MBTH and DMAB are used. This is because in glucose determinations, the sera must first be diluted to allow the intense color to be read. The dilution lowers the ascorbate concentration such that it no longer interferes. In addition, Gochman and Schmitz (1971) report that ascorbic acid disappears at a significant rate which results in a further reduction of its ability to interfere.

MBTH reacts with a number of substances at acidic pH values (Table 1). Capaldi and Taylor (1983) have taken advantage of this in their peroxidase color reaction which employs MBTH with its formaldehyde azine. By comparison, the Ngo-Lenhoff MBTH-DMAB assay has an advantage because the reaction occurs at pH 7.0. At this pH, reactions with other substances viz. aldehydes do not occur readily. Oxidizing agents such as potassium ferricyanide, lead tetraacetate and ferric chloride must be avoided as they will produce a positive interference (Sawicki et. al. 1961).

DISCUSSION

Artiss et. al. (1981) stated that MBTH coupled to either aromatic amines or alcohols appeared to have little obvious merit. Their conclusions were based on the coupling of only three compounds to MBTH: N,N-dimethylaniline HCl, N,N-diethylaniline HCl (DEA) and sodium 2-hydroxy-3,5-dichlorobenzene sulfonate (HDCBS) at pH 7.8 and 37°C. They reported that a large blank reaction (1.0 absorbance for DEA and 0.8 absorbance for HDCBS) resulted from the use of MBTH after only 30 minutes. In order to place their paper in its proper perspective, several comments should be made: 1) Gochman and Schmitz (1971,1972) reported their pH optimum for MBTH-DMA to be 4.2 in phosphate buffer; 2) Geoghegan et. al. (1982,1983) reported that MBTH-DMAB blanks decreased with decreasing pH values. An acceptable blank (less than 0.1 absorbance after 1 hour) was obtained by the use of a 0.1 M citric acid phosphate buffer pH 7.0 or by the use of 0.1 M Hepes pH 7.0.

The question of the effect of pH and buffer has not been thoroughly investigated. Capaldi and Taylor (1983) avoided some of the problem reported by Artiss et. al. (1981) by generating H_2O_2 at pH 7.8 with choline oxidase and then lowering the pH to 3.5 to effect coupling of MBTH with its formaldehyde azine. Geoghegan et. al. (1983) compared the effect of adding citrate to a number of buffers. Tris-Hcl pH 7.2 yielded a poor blank. Tris-citric acid yielded nearly a 50% improvement even at pH 8.0.

Bovaird et. al. (1982) state that the effect of different buffers on the stability of HRP has not been systematically investigated. They reported that HRP activity diminished when stored in phosphate buffer at pH 5.0 at 25°C. Of the buffers tested, citrate buffer was preferred. Citrate buffer (20 mM pH 7.0) yielded the highest activity of HRP. An increase to 100 mM reduced the HRP activity by nearly 20% at pH 7.0. As the pH increased above 8 or decreased below 7, HRP activity decreased for the citrate buffer. The optimum for the phosphate buffer was closer to pH 6.0 falling off at values acidic or alkaline to that value. The effects of the buffers were most pronounced at low HRP concentrations. As the HRP concentrations increased above 1.25 nmol/L, inactivation decreased.

The combined effects of pH and buffer on HRP argue for an assay utilizing a low molarity citrate buffer at pH 7 to 8. The Ngo-Lenhoff MBTH-DMAB assay is at its optimum at pH 7 in a phosphate or citric acid phosphate buffer (Ngo-Lenhoff, 1980, Geoghegan et. al. 1983). The phosphate in the citric acid phosphate buffer might be reduced and supplemented or replaced by an HRP compatible reagent.

The molar absorptivity of the dye may be enhanced by quenching the reaction with acetone. Acetone interferes with color development at pH values of 1-4 (Sawicki et. al. 1961, Paz et. al. 1965). The interference derives from the ease with which the MBTH reacts with acetone. A colorless product absorbing at 297 and 310 nm can be detected (Paz et. al. 1965). Undesired interference will occur unless all glassware used for making the reagents is kept free of acetone. Capaldi and Taylor (1983) reported that quenching the MBTH-MBTH azine at pH 3.5 with acetone increased the absorbance over that of an HCL quenched reaction. In order to take advantage of the acetone in the Ngo-Lenhoff assay, it may be necessary to use acid acetone. The optimal conditions required for the simultaneous quenching of the reaction and enhancement of the reaction are not known. A very preliminary test demonstrated that some 96 well microtiter plates will tolerate between 50% and 75% acetone. Pure acetone started to dissolve the plates.

In conclusion, the continued study and application of MBTH, contrary to the report by Artiss et. al. (1981), appears to have merit. Its application in the Ngo-Lenhoff (MBTH-DMAB) peroxidase assay (Ngo and Lenhoff, 1980) as applied to solid phase ELISA is evidence of this (Geoghegan et. al. 1983).

REFERENCES

Ando, T., T. Muraoka and H. Okuda (1983). A sensitive spectrophotometric assay for guanase activity. Anal. Biochem. 130: 295-301.

Artiss, J.D., R.J. Thibert, J.M. McIntosh, and B. Zak (1981). Study of various substrates for peroxidase-coupled oxidations. Microchem. J. 26:487-505.

Bovaird, J.H., T.T. Ngo and H.M. Lenhoff (1982). Optimizing the O-phenylenediamine assay for horseradish peroxidase: effects of phosphate and pH, substrate and enzyme concentrations and stopping reagents. Clin. Chem. 28:2423-2426.

Capaldi, D.J., and K.E. Taylor (1983). A new peroxidase color reaction: oxidative coupling of 3-methyl-2-benzothiazolinone hydrazone (MBTH) with its formaldehyde azine. Application to glucose and choline oxidases. Anal. Biochem. 129:329-336.

Cary, R.N., D. Feldbruegge and J.O. Westgard (1974). Evaluation of the adaptation of the glucose oxidase/peroxidase-3-methyl-2-benzothiazolinone hydrazone-N,N-dimethylaniline procedure to the technicon "SMA 12/60" and comparison with other automated methods for glucose. Clin. Chem. 20:595-602.

Deisseroth, A., and A.L. Dounce (1969). The purification and crystallization of beef erythrocyte catalase. Arch. Biochem. Biophys. 131:18-29.

Engvall, E. (1980). Enzyme immunoassay ELISA and EMIT. In Methods in Enzymology (Van Vunakis H. and Langone J.J., Eds.). Vol. 70, Academic Press, New York.

Geoghegan, W.D., M.F. Struve, and R.E. Jordon (1982). Adaptation of the Ngo-Lenhoff peroxidase assay for ELISA. Fed. Proc. 41:792 (abstract).

Geoghegan, W.D., M.F. Struve, and R.E. Jordon (1983). Adaptation of the Ngo-Lenhoff peroxidase assay for solid phase ELISA. J. Immunol. Methods 60:61-68.

Gochman, N., and J.M. Schmitz (1971). Automated determination of uric acid, with use of a uricase-peroxidase system. Clin. Chem. 17:1154-1159.

Gochman, N., and J.M. Schmitz (1972). Application of a new peroxidase indicator reaction to the specific, automated determination of a glucose with glucose oxidase. Clin. Chem. 18:943-950.

Hay, F.C., L.J. Nineham and I.M. Roitt (1976). Routine assay for the detection of immune complexes of known immunoglobulin class using solid phase Clq. Clin. Exp. Immunol. 24:396-400.

Hay, F.C., L.J. Nineham, R. Perumal and I.M. Roitt (1979). Intra-articular and circulating immune complexes and antiglobulins (IgG and IgM) in rheumatoid arthritis: correlation with clinical features. Ann. Rheum. Dis. 38:1-7.

Honda, S., Y. Nishimura, H. Chiba, and K. Kakehi (1981). Determination of carbohydrates by condensation with 3-methyl-2-benzothiazolinone hydrazone. Anal. Chim. Acta 131:293-296.

Hunig, S., and K.H. Fritsch (1957). Azofarbstoffe durch oxydative kupplung I farbstoffe aus N-methyl-benzthiazolon-hydrazon justus liebigs. Ann. Der. Chemie. 609:143-160.

Hunig, S. (1962). Heterocyclic azo dyes by oxidative coupling. Angew. Chem. Internat. Edit. 1:640-646.

Killick, K. and L.W. Wang (1980). The localization of trehalase in polyacrylamide gels with eugenol by a coupled enzyme assay. Anal. Biochem. 106:367-372.

Massamiri, Y., M. Beljean, G. Durand, J. Feger, M. Pays, and J. Agneray (1978). Colorimetric assay of sialic acid by a methyl-3-benzothiazolinone-2-hydrazone reactants. Anal. Biochem. 91:618-625.

Lenhoff, H.M., and N.O. Kaplan (1955). Cytochrome C and cytochrome C peroxidase from pseudomonas fluorescens. In Methods in Enzymology, Volume 2, (Colowick, S. and N. Kaplan, Eds). Academic Press, New York, p. 758-764.

McIlvaine, T.C. (1921). A buffer solution for colorimetric comparison. J. Biol. Chem. 49:183-186.

Ngo, T.T. and H.M. Lenhoff (1980). A sensitive and versatile chromogenic assay for peroxidase and peroxidase-coupled reactions. Anal. Biochem. 105:389-397.

Obzansky, D.M. and K.E. Richardson (1983). Quantification of urinary oxalate with oxalate oxidase from beet stems. Clin. Chem. 29:1815-1819.

Paz, M.A., O.O. Blumenfeld, M. Rojkind, E. Henson, C. Furfine, and P. M. Gallop (1965). Determination of carbonyl compounds with N-methyl benzothiazolone hydrazone. Arch. Biochem. Biophys. 109:548-559.

Perlstein, M.T., R.J. Thibert, and B. Zak (1977). Bilirubin and hemoglobin interferences in direct colorimetric cholesterol reactions using enzyme reagents. Microchem. J. 22:403-419.

Perlstein, M.T., R.J. Thibert, R. Watkins and B. Zak (1978). Spectrophotometric study of bilirubin and hemoglobin interactions in several hydrogen peroxide generating procedures. Microchem. J. 23:13-27.

Saravis, C.A. (1984). Improved blocking of nonspecific antibody binding sites on nitrocellulose membranes. Electrophoresis 5:54-55.

Sawicki, E., T.R. Hauser, T.W. Stanley and W. Elbert (1961). The 3-methyl-2-benzothiazolone hydrazone test. Anal. Chem. 33:93-96.

Sawicki, E., R. Schumacher and C.R. Engel (1967). Comparison of MBTH and other methods for the determination of sugars and other α-glycolic derivatives. Microchem. J. 12:377-395.

Smith, R.L. and E. Gilkerson (1979). Quantitation of glycosaminoglycan hexosamine using 3-methyl-2-benzothiazolone hydrozone hydrochloride. Anal. Biochem. 98:478-480.

Wampler, J.E., M.G. Mulkerrin and E.S. Rich (1979). Instrumentation and techniques for analysis of hydrogen peroxide and peroxide-producing reactions involving earthworm (diplocardia longa) bioluminescence. Clin. Chem. 25:1628-1634.

White-Stevens, R.H. and L.R. Stover (1982). Interference by ascorbic acid in test systems involving peroxidase. II. Redox-coupled indicator systems. Clin. Chem. 28:589-595.

Supported in part by Research Grant AM32974 of the National Institutes of Health.

Figures 4-8 are reproduced with the permission of the Publisher.

IMMUNOPEROXIDASE TECHNIQUES USING THE AVIDIN-BIOTIN SYSTEM

Su-Ming Hsu

Department of Pathology and Laboratory Medicine
University of Texas Health Science Center
Houston, Texas

INTRODUCTION

Although immunofluorescence techniques were first develpoed in the early 1960s, they did not gain widespread use and acceptance until the following decade, when they became invaluable tools for the resolution of a number of problems in experimental immunology. Since then, the method has been used clinically for the identification of surface and intracellular lymphocyte markers and immune complexes of glomerular diseases. Within the past 15 years, an alternative method using enzyme-labeled antibodies has been developed which is rapidly replacing immunofluorescence because of its simplicity and usefulness in formalin-fixed paraffin embedded tissue sections. Although many enzymes have been sucessfully conjugated to various antibodies, most recent studies have employed horseradish peroxidase. The amount of antigens in tissue sections is generally determined semi-quantitatively by examining the staining intensity using serial diluted antibodies (Hsu et al.,1981a,b). More recently, the immunoperoxidase techniques have been adapted to other solid phase systems, such as enzyme-linked immunoassys or western blots of antigens (Azen and Yu, 1984; Madri and Barwick, 1983; Rosenkoetter et al., 1984). Quantitation of the antigens is made possible by measurement of the soluble color endproducts in a spectrophotometer (Madri and Barwick, 1983).

In this chapter, we describe a different approach to localize antigen in the tissue sections or other solid phase systems based on an interaction between avidin and biotin (review see Bayer and Wilcheck, 1980; Hsu and Raine, 1983). Avidin is an egg white glycoprotein (M.W. 68,000) that has a high affinity for the small molecule vitamin biotin. In the early 1960s, avidin has been used for the characterization of biotin-requring enzymes. During the past 15 years, biotin-labeled substances have had numerous applications in cellular biochemistry. Biotin-labeled insulin or adrenocorticotropic hormone (ACTH), for example, have been used for the isolation and purification of specific hormone receptors (Hofmann and Kiso, 1976; Hofmann et al., 1977). Moreover, Gusedon et al., (1979) have recently suggested an application of the avidin-biotin interaction in immunohistochemistry, however, the avidin-biotin techniques were not widely accepted as sensitive and specific immunohistochemical method until the methology was fully refined (Hsu et al., 1981a and b).

APPLICATIONS OF AVIDIN-BIOTIN TECHNIQUES

One of the major advantages of the avidin-biotin technique is that any substance which is conjugated with biotin can be detected by avidin. Unlike other immunological methods, the technique may not require an antigen-antibody reaction for the detection. Several examples are shown here;

(1). Biotin-labeled hormone can be used to detect hormone receptor. (Hofmann et al., 1977).
(2). Biotin-labeled toxin can be used to detect toxin receptor.
(3). Biotin-labeled lectin can be used to detect lectin receptor or specific glycoconjugate.(Hsu et al.,1982,a,c, d and e).
(4). Biotin-labeled lectin when inject into axon can be used to study axonal retrograde transport.
(5). Biotin-labeled viral DNA can be used for the in-situ hybridization to localize viral genomes (Myerson et al.,1984).

All biotin-labeled substances, including immunoglobulins can be detected by avidin-conjugated or complexed substances . Depending on the detecting systems, the substances can be either Texas red, fluorescin, rhodamin, ferritin, colloid gold, peroxidase, alkaline hosphatase or other enzymes. Therefore, the avidin-biotin techniques can be applied in all test systems, including immunohistochemistry, electronmicroscopy, western blots, enzyme linked immunoassay, flow cytometry, hybridoma screening and DNA is-situ hybridization. All techniques involve similar procedures. We will discuss in details only the applications of avidin-biotin techniques in immunohistochemistry. The same principle can be applied easily in other systems.

PRINCIPLES OF AVIDIN-BIOTIN TECHNIQUES

Basically, three avidin-biotin techniques can be used, including (1) labeled avidin-biotin(LAB) technique. (2). bridged avidin-biotin (BRAB) technique. (3). Avidin-biotin complex (ABC) technique. In the LAB method, the biotin-labeled substance is directly visulized by an avidin conjugate, such as avidin-fluorescin or avidin-peroxidase conjugates. Although, the use of an avidin-peroxidase conjugate produces a reasonable sensitivity, one of the major drawbacks is that the avidin-conjugates usually yield a high back-ground as well as non-specific staining of some tissue elements such as knidney.

In the BRAB method, avidin is used as a link between two or more biotin-conjugated molecules; such as biotin-labeled antibody and biotin-labeled peroxidase. This technique is based on the biochemical characteristics of avidin, a substance which has four binding sites for biotin. Theoretically, the BRAB technique may produce a higher sensitivity than the LAB methd because of the amplification effected through the biotin-avidin-biotin interaction. The avidin molecules that bind to biotin-labeled antibody potentially can react with an additional 2-3 biotin-peroxidase molecules. However, in our experience, the BRAB technique does not produce as maximal a sensitivity as would be expected theoretically (Hsu, Raine and Fanger, 1981b). The steric hindrance between molecules may play an important role in decreasing the sensitivity of the BRAB technique.

METHOD OF THE LAB TECHNIQUE

The followings are the procedure using the LAB technique to detect the peanut agglutinin receptor in formalin-fixed paraffin-embedded tissue sections;

1. Deparaffinize and rehydrate sections with xylene and graded alcohol, and wash in distilled water briefly.
2. Wash slides in Tris-buffered saline (TBS), 0.05M, pH 7.6.
3. Transfer slides into TBS containing 0.1% (w/w) mouse liver powder for 5 min.
4. Wash briefly in TBS.
5. Incubate with biotin-labeled peanut agglutinin, 50 ug/ml for 1 hr.
6. Wash in TBS for 5 min.
7. Incubate with avidin-peroxidase conjugate(Vector Laboratory, Burlingame, CA), 25 ug/ml for 30 min.
8. Wash in TBS for 5 min.
9. Develop in a substrate, such as diaminobenzidine-hydrogen peroxide (see below) for approximately 5 min.
10. Wash in TBS for 5 min.
11. Dehydrate through graded alcohol and xylene.
12. Mount.

METHOD OF THE BRAB TECHNIQUE

The followings are the procedure using the BRAB technique to detect insulin-secreting cells in the pancreas.

1. Deparaffinize and rehydrate the sections routinely.

2. Wash in TBS for 5 min.
3. Incubate with 1% normal rabbit serum in TBS for 5 min.
4. Incubate with guinea pig anti-insulin antiserum, 1:200 for 30 min.
5. Wash in TBS for 5 min.
6. Incubate with biotin-labeled rabbit anti-guinea pig antibody, 1:200 for 30 min.
7. Wash in TBS for 30 min.
8. Incubate with avidin (Vector Lab.) 10 ug/ml for 30 min.
9. Wash in TBS for 5 min.
10. Incubate with biotin-labeled peroxidase (Vector Lab.), 10 ug/ml for 30 min.
11. Wash in TBS for 5 min.
12. Develop in diaminobenzidine-hydrogen peroxide solution for approximately 5 min.
13. Wash in TBS for 5 min.
14. Dehydrate through graded alcohol and xylene.
15. Mount.

PRINCIPLES OF THE ABC TECHNIQUE

To enhance the reaction sensitivity further, we developed a technique using an avidin-biotin-peroxidase complex (ABC). The same idea is also recently applied to the formation of avidin-biotin-alkaline phosphatase complex and avidin-biotin-glucose oxidase complex. The formation of an ABC is based on the following biochemical characteristics of avidin and biotin-conjugates.

1). Avidin has four binding sites for biotin-conjugates. Thus, avidin can form a link between possibly at least two or more biotin-conjugates.

2). Multiple biotin moleculs can be coupled to one peroxidase molecule. Thus, biotin-peroxidase conjugates are able to react with several avidin molecules.

3). With a proper ratio of avidin and biotin-peroxidase conjugate, the formation of a complex containing multiple peroxidase molecules is possible.

4). The sensitivity of the detecting system is increased because of an increased number of peroxidase molecules are present.

A proper ratio of avidin and biotin-peroxidase should be determined empirically. The degree of biotinylation of peroxidase can affect the sensitivity greatly. A reliable ABC can be obtained from a commercial source (Vector Laboratories). Alternatively, the ABC can be prepared according to a previously published method. (Hsu et al.,1981a,b). In general, the amounts of avidin and biotin-peroxidase is about 10 ug and 2.5 ug per 1 ml, respectively. One should note that the availability of free biotin-binding sites in the complex is created by incubating relative excess amounts of avidin with biotin-peroxidase molecules. An excess of free avidin molecules will decrease the sensitivity because of a competition between free avidin molecules and ABC for the biotin-molecules associated with antibody. When there is an excess of biotin-peroxidase, the biotin-binding sites in avidin molecules are likely to become saturated and therefore, the complex, deficient in free biotin-binding sited, is rendered unsuitable for the ABC technique.

The following is an example of the effect of various ratios of avidin and biotin-peroxidase on the staining results; the staining intensities were evaluated by an anti-IgG reaction in paraffin sections of thyroid (Hashimoto's thyroiditis) tissue sections. A diaminobenzidine-hydrogen peroxide mixture was used as the substrate for peroxidase which produces a brown colored precipitate. The results were read semi-quantitatively as brownish black, 3+; brown, 2+; yellowish brown, 1+ and -.

TABLE 1

concentrations of biotin-peroxidase(ug/ml)	concentrations of avidin (ug/ml)					
	40	20	10	5	2.5	1.25
1.25	-	-	2+	2+	1+	-
2.5	1+	2+	3+	1+	1+	-
5	2+	3+	3+	1+	-	-
10	3+	2+	1+	-	-	-
20	3+	1+	-	-	-	-

Since the ratio of 10 to 2.5 (avidin to biotin-peroxidase) produces a maximal sensitivitiy and minimal background staining, we have used this combination routinely for immunohistochemistry. The ABC obtained from commerical source (Vector Laboratory, Burlingame, CA) contains two separate reagents which have been prediluted. An optimal concentration of ABC can be prepared by mixing equal volumes of A (avidin)(10 ul) and B (biotin-conjugated peroxidase) (10 ul) in 1 ml buffer.

ABC IMMUNOPEROXIDASE TECHNIQUE

The following procedure is an example to detect immunoglobulin G in the tissue sections.

1. Deparaffinize and rehydrate section routinely, if paraffin sections are used.
   Fix the sections in cold acetone or other mild fixatives (eg., methanol or ethanol) for 5 min., if frozen sections are used.
2. Wash slides in Tris-buffered saline (TBS), or phosphate buffered saline for 5 min.
3. Incubate with 1% nonimmune normal rabbit serum in TBS.

Prefer serum from the same animal species as that of the secondary antibody.

4. Incubate with rabbit anti-IgG at a proper dilution (usually at 1:400 in paraffin sections, and 1:2,000 or 3,000 in frozen sections) for 30 min.
5. Wash in TBS for 5 min.
6. Incubate with biotin-labeled goat anti-rabbit antiserum at a dilution of 1 to 200 for 30 min.
7. Wash in TBS for 5 min.
8. Incubate with the ABC reagent for 30 min.
9. Wash in TBS for 5 min.
10. Develop in a substrate, such as diaminobenzidine-hydrogen peroxide.

The substrate is prepared as the following (Hsu and Soban, 1983).

Dissolve 100 mg diaminobezidine-HCl (sigma) in 200 ml TBS, stir and filter.

add 1 ml of 8% Nickel chloride (aqueous solution).

add 25 ul of 3% hydrogen peroxide.

11. Terminate the reation by washing the sections in TBS.

When biotin-labeled primary antibody is available, the above procedures can be shortened by omitting the secondary antibody. Any biotin-labeled substance can be adapted in the system. For example, biotin-labeled lectin can be applied and the localization can be visualized by the following ABC and substrate reactions(Hsu et al., 1982a,c,d and e).

ADVANTAGES OF AVIDIN-BIOTIN SYSTEMS

Biotin is a small molecule and the conjugation of biotin to protein or other substances involves a relatively mild chemical reaction, thus, the biotinylation usually does not affect the biological or physiological affinity of the substance. Since biotin is a small molecule, the biotin-labeled substance maintains a size comparable with that of original. In the event(eg. in immunoelectron microscopy), the molecular size of antibody is very important in order to having a good penetration into tissues, the best reagent to be used is perhaps the biotin-labeled Fab fragments of antibody. (Childs and Unabia, 1980 and 1982).

Biotin-labeled substances can be detected by any avidin-conjugated or complexed substance, such as avidin-fluorescin, avidin-ferritin or ABC. Therefore, once the substance labeled by biotin can be easily adapted into all system. In other systems, such fluorescin-labeled antibody or peroxidase labeled antibody, the application is restricted. The avidin-biotin system offers a greater flexibility than other enzyme-conjugated detecting systems.

The biotin-labeled immunoglobulins and lectins are very stable.

They maintaine the same reactivities up to 3 years when kept at 4C. The same is true for the avidin and biotin-conjugated peroxidase. We used to prepare the ABC before its application onto slides. The ABC can be stored at 4C for more than 2 days. However, we did not test the stability of this complex.

ADAPTION OF ABC INTO SYSTEMS OTHER THAN IMMUNOHISTOCHEMISTRY

Depending on the detecting systems, the most important part of the reaction is to select a suitable substrate. In tissue sections that will be dehydrated through alcohol and xylene, the selection of a substrate that forms a alcohol/xylene insoluble endproduct is necessary. Diaminobenzidine, despite a potential carcinogen, is still the substrate of choice. In some testing systems, the end-product should be soluble with distinct color so that it can be read in a spectrometer. In this case, orthophenylenediamine (20 mg in 50 ml of phosphate buffer with 50 ul of 30% hydrogen peroxide) is probably the substrate of choice.

Details of the application of ABC technique in enzyme-linked immuno-assay (Madri and Barwick, 1983), immunelectronmicroscopy (Childs and Unabia, 1982), nucleic acid hybridization (Myerson et al.,1984) and immunological screening of c-DNA clones (Hsu et al., 1985) has been described in details.

BACKGROUND AND NONSPECIFIC STAINING

Biotin is a coenzyme for decarboxylase, and exist in many tissues, such as liver, pancreas, kidney and intestine. In fixed tissues, the biotin-containing proteins are likely destroyed by the formalin-fixation or paraffin embedding procedures, thus, the use of ABC does not produce unwanted bindings to tissues or cells. In frozen sections of liver, kidney, pancreas, etc., the sections have to be pre-treated with avidin (25 ug/ml) for 15 min and followed by biotin at saturated concentration for 15 min with an interval washing in buffer. The pre-treatment of tissue sections with avidin can sucessfully block the endogenous biotin molecules. Since avidin is a tetramer which has four binding sites, the sections require an additional biotin-treatment to saturate all free biotin-binding sites associated with avidin so that the added avidin will not react with biotin-labeled antibody or ABC. Biotin when bound to avidin will not react with second avidin molecule (Hsu and Raine, 1982b).

ABC can react with mast cells in both fixed or frozen sections. The reaction can not be prevented by the avidin-biotin pretreatment. One can prepare the ABC in high pH solution (pH 8.5) or in high inoic strenghth buffer such as in 0.5M NaCl buffer. The high pH or ionic strenghth solutions will not affect the ABC-biotin reaction but can prevent or minimize the unwanted binding to mast cells.

In test systems, such as enzyme-linked immunoassays or western blots, the background staining of ABC can be effectively minimized by the preincubations of the cellulose paper or plastic wells with 0.01% Triton X.

BIOTINYLATION OF PROTEINS

Biotinylation of immunoglobulins, lectins or other glyco-proteins can be prepared by the following procedures:

1. Dissolve N-hydroxysuccinimide biotin (sigma) in dimethylsulfoxide (DMSO) at a concentration of 1 mg/ml
2. Adjust immunoglobulin or other glycoprotein at a concentration of 1 mg/ml in 0.1M $NaHCO_3$, pH 9.0
3. Mix above two reagents at a volume ratio of 1:8 (activated biotin:protein) at room temperature for 4 hours.
4. Dialyze against phosphate buffer.
5. Concentrate the solution by vaccum dialysis.

Biotinylation of nucleic acid has been described in details by Leary et al., (1983).

SUMMARY

The avidin-biotin complex method offers an extraordiary sensitivity which is due to the strong affinity between avidin and biotin. The most important advantage of this technique, however, is the flexibility that can be adapted in all systems. In this review, we did not address the quantitative aspects of this technique, however, the principle and technical aspects of quantitation using avidin-biotin-complex would be the same as that employing antigen-antibody reactions described in details in this book.

REFERENCE

Azen, E.A., Yu, P.L., 1984, Genetics polymorphism of CON 1 and CON 2 salivary proteins detected by immunologic and Concanavalin A reactions on nitrocellulose with linkage of CON 1 and CON 2 genes to the SPC (salivary protein gene complex), Biochem Genetics 22:19.

Bayer, E.A., Wilcheck, M.,1980, The use of the avidin-biotin complex as a tool in molecular biology, Methods Biochem Anal 26:1.

Childs, G., Unabia, G., 1982, The application of the avidin-biotin-peroxidase complex (ABC) method to the light microscopic localization of pituitary hormones, J Histochem Cytochem 30:713.

Childs, G., Unabia, G., 1982, Application of a rapid avidin-biotin peroxidase complex (ABC) technique to the localization of pituitary hormones at the electron microscopic level, J Histochem Cytochem, 30:1320

Guesdon, J.L., Ternynck, T., Avrameas, S., 1979, The use of avidin-biotin interaction in immunoenzymatic techniques. J Histochem Cytochem 27:1131.

Hofmann, K., Kiso, Y., 1976, An approach to the targeted attachment of peptides and proteins to solid supports. Proc Natl Acad Sci USA 73:3516.

Hofmann, K., Finn, F.M., Friesen, H.J., Diaconescu, C., Zahn, H.,1977, Biotinyl insulins as potential tools for receptor studies. Proc Natl Acad Sci USA 74:2697

Holland, P.C., Pena, S.D.J., Guerin, C.W., 1984, Developmental regulation of neuramidase-sensitive lectin binding glycoproteins during myogenesis of rat L6 myoblasts. Biochem J 218:465.

Hsu, P.L., Hsu, S.M., and Appela, E., 1985, Immunological screening of a cDNA library using avidin-biotin-peroxidase complex method. Gene Anal Tech, in press.

Hsu, S.M., Raine, L., Fanger, H., 1981a, A comparative study of the peroxidase-anti-peroxidase method and an avidin-biotin complex method for studying polypeptide hormones with radioimmunoassay antibodies. Am J Clin Pathol 75:734.

Hsu, S.M., Raine, L., Fanger, H., 1981b, Use of avidin-biotin-peroxidase complex (ABC) in immunoperoxidase techniques: A comparison between ABC & unlabeled antibody (PAP) procedures. J Histochem Cytochem 29:577.

Hsu, S.M., Raine, L., 1981c, Protein A, avidin and biotin in immunohistochemistry. J Histochem Cytochem 29:1349.

Hsu, S.M., Raine, L.,1982a, Discovery of high affinity of Phaseolus vulgaris agglutinin (PHA) with gastrin-secreting cells. Am J Clin Pathol 77:396.

Hsu, S.M., Raine, L.,1982b, Unwanted binding associated with avidin-biotin-peroxidase complex in frozen tissues. Am J Clin Pathol 77:775.

Hsu, S.M., Raine, L., 1982c, Versatility of biotin-labeled lectins and avidin-biotin-peroxidase complex for localization of carbohydrates in tissue sections. J Histochem Cytochem 30:157.

Hsu, S.M., Raine, L., 1982d, Warthin's tumor-epithelial cell differences. Am J Clin Pathol 77:78.

Hsu, S.M., Raine, L., 1982e, Concanavalin A (Con A) reaction with plasma membrane. Am J Clin Pathol 77:508.

Hsu, S.M., Soban, E., 1982, Color modification of diaminobenzidine (DAB) precipitation by metallic ions and its applications for double immunohistochemistry. J Histochem Cytochem 30:1079.

Hsu, S.M., Ree, H., 1983, Histochemical studies on lectin binding in reactive lymphoid tissues. J Histochem Cytochem 31:538

Hsu, S.M, Raine, L., 1983, The use of avidin-biotin-peroxidase-complex (ABC) in diagnostic and research pathology. In Advances in Immunohistochemistry. Ed: R.A. DeLellis, Masson Publishing USA, Inc., New York, pp 31-42.

Leary, J.J., Brigati, D.J., and Ward, D.C., 1983, Rapid and sensitive colorimetric method for visualizing biotin-labeled DNA probes hybridized to DNA or RNA immobilized on nitrocellulose: Bio-blots. Proc Natl Acad Sci USA 80:4045.

Leipold, B., Remy, W., 1984, Use of avidin-biotin peroxidase complex for measurement of UV lesions in human DNA by micro-ELISA. J Immuno Methods 66:227.

Madri, J.A., Barwick, K.W., 1983, Use of avidin-biotin-complex in ELISA system. A quantitative comparison with two other immunoperoxidase detection systems using keratin antisera. Lab Invest 48:98

Myerson, D., Hackman, R.C., Meyers, J.D., 1984, Diagnosis of cytomegaloviral pneumonia by in situ hybridization. J Infect Diseases. 150:272.

Rosenkoetter, M., Reder, A.T., Oger, J.J.F., Antel, J.P.,. 1984, T cell regualtion of polyclonally induced immunoglobulin secretion in humans. J Immunol 132:1779.

CONTRIBUTORS

BERS, GEORGE E.
Bio-Rad Laboratories, Research Product Group, 32nd & Griffin Ave., Richmond, CA 94804

BUTLER, J. E.
Department of Microbiology, The University of Iowa Medical School, Iowa City, IA 52242

CASTRO, ALBERT
Department of Pathology, University of Miami School of Medicine, Miami, FL 33101

FREYTAG, WILLIAM J.
E. I. du Pont de Nemours & Co., Inc., Biomedical Products Department, Glasgow Research Laboratory, Wilmington, DE 19898

GEOGHEGAN, WILLIAM D.
Department of Dermatology, The University of Texas Health Science Center-Houston, 6431 Fannin, MSMB 1-204, Houston, TX 77030

GIBBONS, IAN
Syva Research Institute, 900 Arastradero Road, Palo Alto, CA 94304

GIEGEL, JOSEPH L.
American Dade Company, American Hospital Supply Corp., 1851 Delaware Parkway, Miami, FL 33152

HALBERT, SEYMOUR P.
Cordis Laboratories, 2140 North Miami Avenue, Miami, FL 33127

HANCOCK, KATHY
Division of Parasitic Diseases, Center for Infectious Diseases, Centers for Disease Control, Public Health Service U. S. Department of Health and Human Services, Atlanta, GA 30333

HSU, SU-MING
Department of Pathology and Laboratory Medicine, University of Texas Health Science Center, Houston, TX 77030

JAKLITSCH, ANNA
Syva Company, 900 Arastradero Road, Palo Alto, CA 94304

KABAKOFF, DAVID S.
Hybritech Inc., 11095 Torreyana Road, San Diego, CA 92121
KELLY, TIM A.
Bartels Immunodiagnostic Supplies, Inc., P. O. Box 3093, Bellevue, WA 98009
KOERTGE, T.E.
Department of Periodontics, Medical College of Virginia, Virginia Commonwealth University, Richmond, VA 23298
LENHOFF, H. M.
Department of Developmental and Cell Biology, University of California, Irvine, CA 92717
LIN, TSUE-MING
Cordis Laboratories, 2140 North Miami Avenue, Miami, FL 33127
LITCHFIELD, WILLIAM J.
Diagnostic & Bioresearch Systems Division, E. I. Du Pont de Nemours & Co., Inc., Glasgow Research Laboratory, Wilmington, DE 19898
LITMAN, DAVID J.
Syntex Medical Diagnostics Division, 3221 Porter Drive, Palo Alto, CA 94304
MILBY, KRISTIN H.
Monsanto Company, 800 Lindbergh Blvd., St. Louis, MO 63167
MONJI, NOBUO
Genetic Systems Corporation, 3005 First Avenue, Seattle, WA 98121
NGO, T. T.
Department of Developmental and Cell Biology, University of California, Irvine, CA 92717
PETERMAN, J. H.
Department of Microbiology, The University of Iowa Medical School, Iowa City, IA 52242
de SAVIGNY, D.
Ontario Ministry of Health, Box 9000, Terminal A, Toronto, Canada
SEVIER, E. DALE
Hybritech Inc., 11095 Torreyana Road, San Diego, CA 92121
SHIMIZU, STANLEY
Hybritech Inc., 11095 Torreyana Road, San Diego, CA 92121
SPEISER, F.
Swiss Tropical Institute, Socinstrasse 57, 4051 Basel, Switzerland
STANDEFER, JIM C.
Department of Pathology, University of New Mexico School of Medicine, Albuquerque, NM 87131
TSANG, VICTOR C. W.
Division of Parasitic Disease, Center for Infectious Diseases, Centers for Disease Control, Public Health Service, U. S. Department of Health and Human Services, Atlanta, GA 30333

USATEGUI-GOMEZ, MAGDALENA
Hoffmann-LaRoche, Inc., Department of Diagnostic Research, 340 Kingsland Street, Nutley, NJ 07110

WICKS, RICHARD
Hoffmann-LaRoche, Inc., Department of Diagnostic Research, 340 Kingsland Street, Nutley, NJ 07110

WONG, R. C.
Department of Developmental and Cell Biology, University of California, Irvine, CA 92717

INDEX